Engineering Principles and Concepts for Active Solar Systems

Engineering Principles and Concepts for Active Solar Systems

SERI

Solar Energy Research Institute

HEMISPHERE PUBLISHING CORPORATION

A member of the Taylor & Francis Group

New York Washington Philadelphia London

This permanent edition includes the complete text of the Solar Energy Research institute publication, *Engineering Principles and Concepts for Active Solar Systems,* prepared as part of the U.S. Department of Energy Solar Building Research and Development Program. Much earlier research on this project was done in collaboration with the Los Alamos National Laboratory. The final result involved over ten authors and illustrators.

1 2 3 4 5 6 7 8 9 0 BCBC 898

Library of Congress Cataloging-in-Publication Data

Engineering principles and concepts for active solar systems / Solar Energy Research Institute. —Authorized hardbound ed.
p. cm.
Previous ed. published as: Solar design workbook. 1978.
Bibliography: p.
Includes index.
1. Solar heating. I. Solar Energy Research Institute.
II. Science Information Resource Center (Philadelphia, Pa.)
III. Solar design workbook.
TH7413.E54 1988
697'.78—dc19 87-35483
ISBN 0-89116-855-9 CIP

Contents

Foreword

In keeping with the national energy policy goal of fostering an adequate supply of energy at a reasonable cost, the U.S. Department of Energy (DOE) supports a variety of programs to promote a balanced and mixed energy resource system. The mission of the DOE Solar Building Research and Development Program is to support this goal by providing for the development of solar technology alternatives for the buildings sector. It is the goal of the program to establish a proven technology base to allow industry to develop solar building products and designs that are economically competitive and can contribute significantly to building energy supplies nationally. Toward this end, the program sponsors research activities related to increasing the efficiency, reducing the cost, and improving the long-term durability of passive and active solar systems for building water and space heating, cooling, and daylighting applications. These activities are conducted in four major areas: (1) advanced passive solar materials research, (2) collector technology research, (3) cooling systems research, and (4) systems analysis and applications research.

Advanced Passive Solar Materials Research — This activity area includes work on new aperture materials for controlling solar heat gains and for enhancing the use of daylight for building interior lighting purposes. It also encompasses work on low-cost thermal storage materials that have high thermal storage capacity and can be integrated with conventional building elements, and work on materials and methods to transport thermal energy efficiently between any building exterior surface and the building's interior by non-mechanical means.

Collector Technology Research — This activity area encompasses work on advanced low-to-medium temperature (up to 180°F useful operating temperature) flat-plate collectors for water- and space-heating applications, and medium-to-high temperature (up to 400°F useful operating temperature) evacuated tube/concentrating collectors for space-heating and space-cooling applications. The focus is on design innovations using new materials and fabrication techniques.

Cooling Systems Research — This activity area involves research on high-performance dehumidifiers and chillers that can operate efficiently with the variable thermal outputs and delivery temperatures associated with solar collectors. It also includes work on advanced passive cooling techniques.

Systems Analysis and Applications Research — This activity area encompasses experimental testing, analysis, and evaluation of solar heating, cooling, and daylighting systems for residential and nonresidential buildings. This involves system integration studies; the development of design and analysis tools; and the establishment of overall cost, performance, and durability targets for various technology or system options.

This publication is a much refined and updated version of a solar design handbook originally prepared in 1978 to accompany a series of week-long courses conducted in support of the Solar Federal Buildings Program. The 1978 material was published in 1981 as the *Solar Design Workbook* (SERI/SP-62-308). This current document represents the culmination of an eight-year effort to compile a comprehensive state-of-the-art reference and instructional tool for practicing design professionals, architects, and engineers. It is intended to cover all phases of the design and installation of active solar energy systems for buildings. Although it contains many design guidelines, the emphasis is on providing sufficient knowledge of how these systems work to allow an engineer or architect to make well-informed decisions. It is aimed primarily at commercial building applications, but most of the material is also applicable to residential buildings.

This publication originally was a cooperative effort between the Solar Energy Research Institute (SERI) and the Los Alamos National Laboratory (LANL). The project has had many leaders during its lifetime, including Bruce D. Hunn (originally at LANL, now at the University of Texas at Austin), who has served as chief editor; Greg Franta (originally at SERI, now with the ENSAR Group, Lakewood, Colo.); Nancy Carlisle of SERI; and William Kolar (originally at SERI, now with McDonnell Douglas Architectural Engineering and Construction Co.).

The authors of this most current version include Peter Armstrong, Nancy Carlisle, William Davis, Bruce D. Hunn, Thomas A. King, Charles F. Kutscher, Jefferson G. Shingleton, Theodore Swanson, Stephen Weinstein, and C. Byron Winn. Most of the architectural-style drawings were done by Steve Hogg.

It was the authors' intent to distill in these pages the wealth of solar energy design information accumulated over the years by the U.S. Department of Energy. It is our hope that future design efforts will thereby benefit from the hard work of the many solar pioneers whose contributions made this document possible.

We particularly want to acknowledge the contribution made by Dr. John Yellott. Dr. Yellott did not live to see this document in print, but his wise counsel, especially in the project's formative stages, was invaluable. Those who worked with him are richer for the experience.

Part I
Fundamentals

Chapter 1

Energy and Buildings: A Perspective

The United States accounts for 32% of the world's annual energy consumption even though it has only 6% of the world's population and 20% of its fossil fuel reserves. Since 1950, U.S. population has increased by about 50% while energy consumption has more than doubled.

The U.S. energy use, classified by economic sector, is illustrated in Fig. 1-1. Residential and commercial buildings are significant energy consumers in the United States, using 10 quads* per year of oil and gas and 15 additional quads per year of primary energy to generate electricity in 1983 (DOE 1986). As shown in Fig. 1-1, 36% of the total U.S. energy is for operating residential and commercial buildings. This is in addition to approximately 15% of energy used in the industry classification for building construction and an unknown amount used for transporting construction materials and equipment.

Table 1-1 illustrates the breakdown of primary energy use by fuel type and end use in residential and commercial buildings. In both building types, the dominant energy use is for space heating. In residential buildings, water heating is the second largest use. In commercial buildings, lighting and air-conditioning follow space heating as the major energy uses. In large commercial buildings, typically office and retail buildings, where the major energy load is often determined by the uses and activities within the space and not by the building envelope, cooling and lighting rather than space heating are the dominant energy uses. Note that a substantial portion of the heating and cooling functions can use thermal energy supplied at less than 200°F (93°C). Such low-temperature applications are well suited to active solar systems. Nationally, in both the residential and commercial sectors, electricity is the dominant fuel type, accounting for more than 50% of total primary energy supplied. Energy used in buildings costs consumers $165 billion annually (ACEEE and ECC 1986).

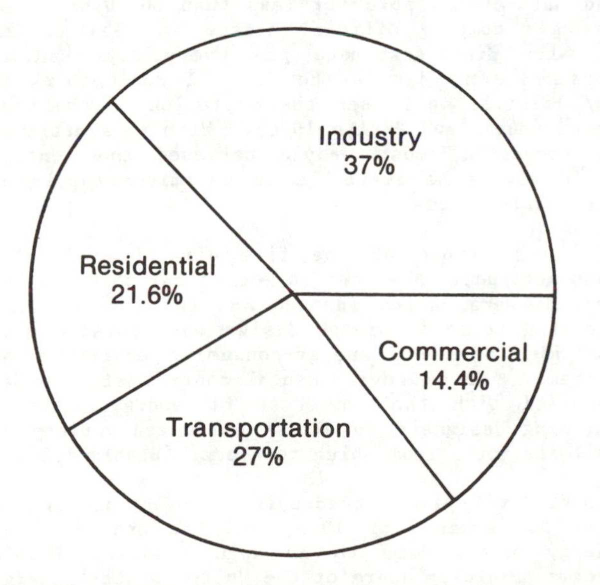

Fig. 1-1. Division of U.S. total energy use by economic sector (DOE 1986)

Since the 1973 energy crisis, numerous buildings have been built and monitored to demonstrate that it is possible to design, build, and operate buildings that incorporate conservation and solar technologies and use significantly less energy than conventional buildings. The DOE Passive Solar Commercial Buildings Program successfully demonstrated that buildings can be built that use an average of 45% less energy at little or no added first cost (Gordon et al. 1985). The SERI Class B residential building program, in which 56 residences were monitored over a 2-year period, demonstrated a 39% passive solar contribution to the total heating load (SERI 1984). Hirst et al. (1986) in their book, Energy Efficiency in Buildings: Progress and Promise, document numerous case studies and demonstration projects where energy savings in new buildings represent up to a 65% improvement in energy efficiency over typical U.S. buildings. These savings result from various combinations of conservation and passive and active solar strategies. In addition, numerous demonstration projects have also shown that it is possible to make existing buildings more energy efficient. Busch et al. (1984) examined the performance of low-energy homes in the Building Energy Use Compilation and Analysis (BECA) data base at the Lawrence Berkeley Laboratory, the largest data base on the topic, and concluded that almost all such homes are cost-effective.

*1 quad = 10^{15} Btu

Table 1-1. Primary Energy Consumption (in Quads) by Fuel Type; 1983 Data for Commercial and Residential Buildings (DOE 1985a)

	Electricity	Gas	Oil	Other	Total	
					Absolute	Percentage
Residential Sector						
Space Heating	2.06	3.36	1.13	0.54	7.09	45.0
Water Heating	1.27	0.80	0.08	0.08	2.23	14.2
Refrigerators	1.49				1.49	9.5
Lighting	1.18				1.18	7.5
Air Conditioners	1.10				1.10	7.0
Ranges/Oven	0.56	0.29			0.85	5.4
Freezers	0.65				0.65	4.1
Other	0.72	0.44			1.16	7.3
TOTAL	9.03	4.89	1.21	0.62	15.75	100.0
Commercial Sector						
Space Heating	1.01	1.82	1.03	0.35	4.21	39.8
Lighting	2.89				2.89	27.4
Air Conditioning	2.09	0.12			2.21	20.9
Water Heaters	0.04	0.08	0.09		0.21	2.0
Other	0.87	0.18			1.05	9.9
TOTAL	6.90	2.20	1.12	0.35	10.57	100.0
Total Residential and Commercial Consumption 1983	15.93	7.09	2.33	0.97	26.32	100.0

Because of the great expense and time required to install new power plant capacity, reduction in peak electricity demand has become a major need of most utilities. Energy-efficient buildings nearly always reduce this peak demand. Furthermore, active solar water heating systems reduce peak demand by providing heat during summer peak hours (Vliet and Askey 1984).

History has shown that it is possible to have climate-responsive architecture that uses energy frugally. The southwestern adobe dwellings, designed around central courtyards to provide a cool, shaded, outdoor environment; the New England saltbox, a very compact building form generally protected on the north side and open on the south side; and the midwestern prairie houses designed by Frank Lloyd Wright as elongated single-story building forms with broad overhangs to block the hot summer sun all evolved in response to local climatic conditions. Recent publications (AIA 1981; Ternoey et al. 1984) highlight several examples of large historic buildings with numerous energy-conscious design features. One example cited is Thomas Jefferson's design of Monticello near Charlottesville, Virginia, which uses a series of natural cooling techniques including cross ventilation, partially underground passageways, operable skylights, and thermal mass. Another is the Wainwright Building designed in 1890 by Louis Sullivan, which was a U-shaped, multistory, commercial building designed to provide all work stations with access to windows for natural ventilation and light.

More recently designed buildings have reached a point where occupants almost totally control atmospheric variables of temperature, humidity, and air purity through the use of mechanical and electrical equipment, powered by cheap and centralized energy forms. During the late 1950s and throughout the 1960s, the price of oil was less than $3/barrel, and natural gas sold for less than $0.10/ccf. As one gas company official stated in 1954: "The industry discovers more gas every day than is consumed every day in the U.S. I don't think in our lifetime we'll see the depletion of the product" (Butti and Perlin 1980). With this attitude so prevalent, most people believed that energy would always be available in unlimited supply at affordable prices.

The architecture of the time directly reflected that attitude. Advances in mechanical conditioning systems were easily adopted and the idea of controlling comfort through design was replaced with the idea of using energy-consuming environmental systems. These environmental-control strategies, combined with the low cost of energy, allowed building designers an almost unlimited variety of building forms from which to choose (Banham 1984).

Imported oil prices quadrupled as a result of the Arab oil embargo of 1973, and the era of cheap energy prices came to an abrupt halt. People became painfully aware of the United States' energy dependence. The increase in energy prices during the 1970s caused total energy use in buildings to reach a plateau or in some cases to decline even though the total number of homes and offices increased every year.

Between 1973 and 1984 energy became more important in building design, and consumers became more energy conscious. The American public achieved significant savings in energy due primarily to improved energy efficiency. On a per building basis, residential energy use declined 17% between 1973 and 1983. This savings can be attributed to improved envelope efficiency, changes in household size, increased wood use, appliance efficiency, migration, and other factors, including the use of solar energy. In the commercial sector, individual building energy use declined 10% between 1973 and 1983. Some of the savings can be attributed to the application of energy conservation measures (Adams et al. 1985).

With the availability of the federal solar tax credit, approximately 950,000 active solar systems, primarily for water heating, were installed; and 200,000 passive solar homes and 15,000 passive nonresidential buildings were built by 1984 (DOE 1984a). During this time, building designers and engineers also became more interested in methods of energy conservation. HVAC and lighting systems or components were modified to work at much lower energy levels, for example, by replacing electric chillers with two-stage evaporative coolers in semi-arid regions of the Southwest, and by replacing incandescent lighting with more efficient fluorescent lighting. Other approaches to energy conservation were also tried, such as cascading conventional energy sources through a series of processes (e.g., using waste heat from computer rooms to warm other parts of a building or to preheat ventilation air).

This increased activity in the use of solar and conservation was not all positive. The early cycles of the Federal Solar Heating and Cooling Demonstration Program showed designers that even when active systems were installed and functioning properly, the buildings on which the systems were installed were often energy inefficient. Also, many solar systems were unreliable. In many early passive designs, the architecture suffered because the passive elements were poorly integrated into or awkwardly attached to the building. Many of these designs were unsuccessful because the building's occupants were too hot. These early demonstration projects taught designers that the achievement of an energy-conserving building requires the integration of energy conservation technologies, passive solar design, and active solar systems to address a building's specific energy needs. The integration of energy and building design is not an easy task to accomplish. It requires the designer to understand many interrelated factors including energy end use requirements for the building under design, energy conservation and passive and active solar technologies and advances in research and development, solar system design and sizing techniques and construction details, climatic factors, economic incentives, regulations, and occupant behavior and attitudes.

Energy-efficient building design is generally the result of a team of architects and engineers working together toward a common set of goals. For a successful result, energy considerations must be addressed by the design team at the earliest possible design stage. Figure 1-2 illustrates the impact of decisions in terms of energy use as a function of stages in the design process. Energy considerations must be integrated with other considerations that affect design and need to be meshed with programmatic, spatial, and cost requirements. Frequently, these energy considerations then can broaden the range of possible solutions to the design problem.

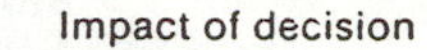

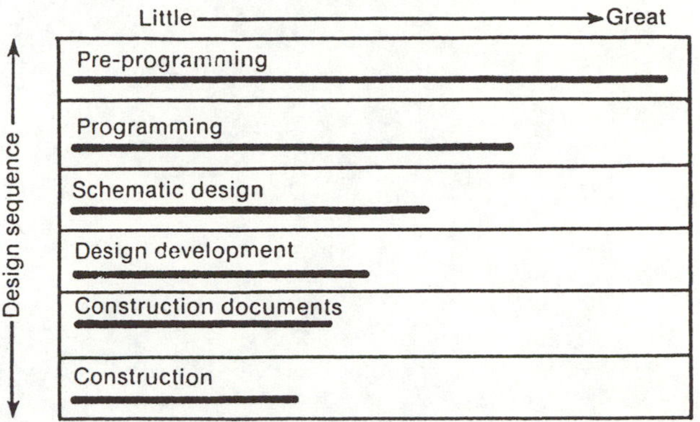

Fig. 1-2. Impact of energy-related decisions at various stages of the design process (AIA 1981)

Since the federal solar tax credits have expired and energy prices have stabilized, interest in solar energy and energy conservation is not as high as it was during the early 1980s. Nevertheless, energy-conscious design is still a vital issue. Continued advances in energy-saving technology can play a key role in assuring the competitiveness of the U.S. industry, and it is important that the American public not ignore the future consequences of a finite supply of oil reserves.

Unlike the early 1970s, a tremendous wealth of information on energy-efficient building and solar design exists today. In order to reassess the methods of environmental control in buildings, design professionals need sufficient technical knowledge derived from experience and experimentation. This book presents architects and engineers with the technical information necessary to design active solar heating and cooling systems that reduce fuel use and peak demand.

The book is divided into three parts. Part I, "Fundamentals," provides a perspective on energy-efficient design, gives details on building energy consumption and energy costs, and introduces the basic principles of human comfort and solar energy availability. Part II, "Active Solar System Design and Sizing," explains the various solar heating and cooling systems, describes the major components and subcomponents including sizing information, and presents both detailed and simplified analysis methods. Part III, "Active Solar System Installation, Construction, and Operation," provides detailed information needed to avoid the typical installation problems of the past, gives system and subsystem cost data, and covers topics aimed at ensuring long-term success, such as acceptance tests, and operation and maintenance manuals.

Chapter 2

Energy Use, Cost, and Load Calculations in Commercial Buildings

INTRODUCTION

Many conservation or renewable resource strategies considered in the building design process or in retrofit analysis emphasize energy savings but do not result in an equivalent or proportional energy dollar savings. However, using measures that save energy dollars for new or existing buildings can result in the reduction of both energy consumption and energy costs in excess of those associated with the energy consumed. The key often lies in the utility's rate structures for commercial buildings, which reflect both energy use and demand charges.

A solar design or analysis team should understand demand-intensive, energy-intensive, and time-differentiated rates, as well as other special rate structures unique to the utility service territory in which the building is located. By applying these structures to the analysis of a building design, a design team can set priorities for conservation and use of renewable resource strategies to reduce energy costs.

To determine how utility rate structures affect decisions about renewable resource strategies, the design/analysis team needs to understand the building's energy use characteristics. Furthermore, to determine the correct size of an active solar heating and/or cooling system, and its performance once properly sized, monthly or annual building heating and cooling loads (those imposed on the building plant) are needed.

This chapter generally characterizes the energy end-use patterns in commercial buildings and then presents methods for estimating energy use, and the resulting energy costs, in these buildings. Finally, several levels of building load and energy-use calculation methods and computer programs are presented.

NATIONAL PERSPECTIVE - END USE AND ENERGY COSTS

Energy Use Patterns

Recent data indicate that commercial buildings use 14% of the total annual national energy consumption, or about 10.6 quads of resource energy (DOE 1985a). A current estimate for commercial building floor area is 52.3 billion ft^2 (DOE 1985b); therefore, the total existing stock of commercial buildings uses an average of 202,000 Btu/ft^2-yr of resource energy.

Figure 2-1 illustrates commercial energy use disaggregated by building type for several categories of commercial buildings. Four building types listed (retail/services, office, health care, and education) account for more than 70% of total estimated commercial energy use. Table 2-1 presents commercial energy use by fuel type and end use, and shows that the largest end use is space heating (40%) followed by lighting (27%) and cooling (21%) (DOE 1985a). If the data were presented in terms of cost, the percentage attributed to lighting and cooling would be even greater: Electricity is used for lighting and most cooling applications while natural gas or oil is used predominantly for space heating. Building water heating energy is only about 2% of the total energy use in the commercial sector.

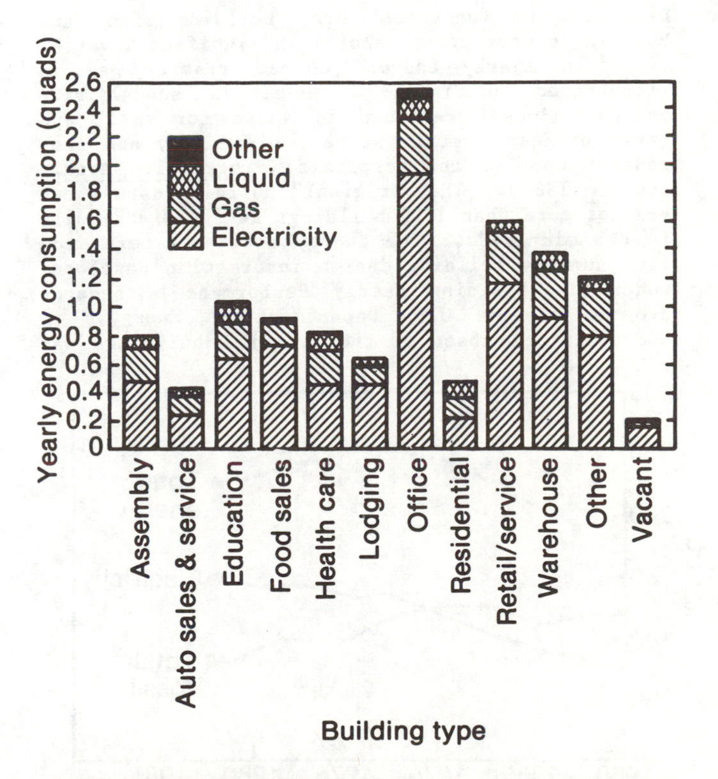

Fig. 2-1. Breakdown of primary energy consumption in commercial buildings (MacDonald et al. 1986)

Table 2-1. Commercial Energy Use (in Quads) by Fuel and End Use, 1983 (DOE 1985a)

	Space Heating	Cooling	Water Heating	Lighting	Other[a]	Total
Electricity	1.01	2.09	0.04	2.89	0.87	6.90
Gas	1.82	0.12	0.08		0.18	2.20
Oil	1.03		0.09			1.12
Other[b]	0.35					0.35
Total	4.21	2.21	0.21	2.89	1.05	10.57

[a]Other end uses include cooking and electromechanical uses.
[b]Other fuels include coal and liquid natural gases.

Between 1960 and 1973 total commercial building energy use per square foot rose at an annual rate of 1.8%. As Fig. 2-2 illustrates, most of this growth was attributed to increased use of electricity. From 1974 to 1977 electricity use per square foot rose nearly as fast as it had before the 1973 oil embargo, but since 1978 the growth rate has averaged less than 1% per year. Primary commercial energy use per square foot reached its peak in 1973 at 251 kBtu/ft^2 (average) and dropped to 227 kBtu/ft^2 in 1982. This represents an approximate 10% decline in energy use during this period. As Fig. 2-2 illustrates, this decline occurred primarily in fossil fuel consumption, which in 1982 was about 25% lower than its 1973 level. Some of this savings can be attributed to the application of energy conservation.

Diversity in functional use, building size, and building construction results in significant variations in energy end-use characteristics, as is illustrated in Fig. 2-3. Here the annual site energy end-use breakdown is shown for small and large offices, retail stores, elementary and secondary schools, and hospitals (Progressive Architecture 1982). The "original" building data shown are for more than 1600 buildings designed and built in the mid-1970s. The energy estimates were made from summary building design information developed under the Building Energy Performance Standards program of the U.S. Department of Energy. A 168-building subset of the original buildings was

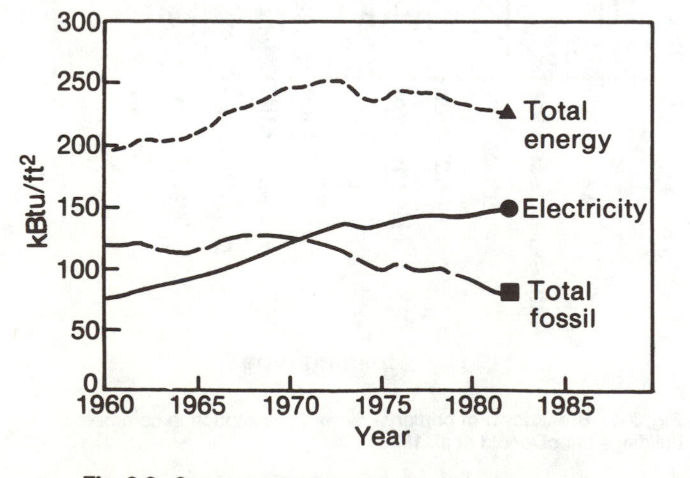

Fig. 2-2. Commercial building energy use (Adams et al. 1985)

statistically selected for redesign by their original architect/engineer design teams. The redesigns aimed at maximum feasible energy reductions, while maintaining the original program, site, and construction budget range. Note the wide range in heating, cooling, lighting, and fan energy use per unit of floor area and the significant energy reductions achievable by the redesigns.

Although these calculated energy savings are substantial, little measured data are available on a submetered basis to substantiate these end-use patterns and conservation potentials. However, measured data are now systematically being collected (Gardiner and Piette 1985) and are summarized nationally and regionally (DOE 1983a).

Energy Cost Patterns

End-use energy costs have risen dramatically since the early 1970s. The national average retail price of natural gas to the commercial sector was $0.73 per thousand cubic feet (10^3 ft^3) ($0.02/10^3 m^3) in 1967, $0.94 per 10^3 ft^3 ($0.03/10^3 m^3) in 1973, $2.23 per 10^3 ft^3 ($0.06/10^3 m^3) in 1978, (DOE 1979b) and $5.17 per 10^3 ft^3 ($0.15/10^3 m^3) in 1982 (DOE 1983b). From 1973 to 1983, the price of natural gas escalated nationwide at an average compound rate of over 21% per year. From 1978 to 1983, the annual compound escalation rate for natural gas prices averaged over 23%.

During 1983, the combination of reduced industrial and commercial business and increased energy conservation resulted in an "oversupply" of natural gas in many regions of the country. This situation has forced many natural gas suppliers to reduce the price of their commodity to distributing utilities.

Both the DOE Energy Information Administration and the Gas Research Institute projected that between 1983 and 1985 natural gas prices would decline slightly in all sectors. Between 1985 and 1990 prices are expected to rise by almost 20% (Ashby, Holtberg, and Woods 1985).

With respect to the price of electricity for commercial buildings, the national average cost per kilowatt-hour was $0.027 in 1967, $0.023 in 1973, $0.046 in 1978 (DOE 1979b), and $0.068 in 1982 (DOE 1983b). The average compound escalation rate for electricity prices from 1973 to 1982 was slightly more than 13% per year with the largest portion

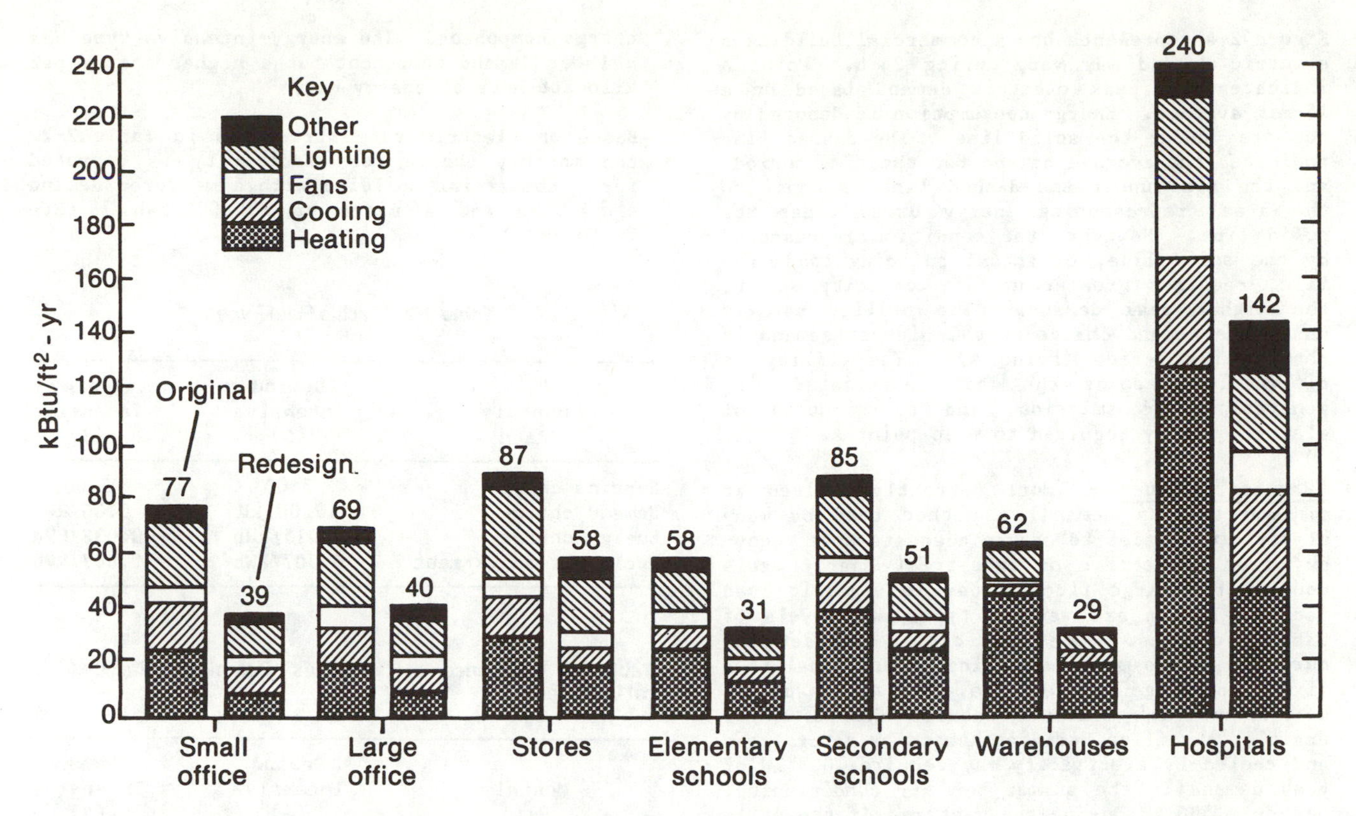

Fig. 2-3. Energy end use characteristics for selected commercial buildings (adapted from *Progressive Architecture* 1982)

resulting from 1973 to 1978 due to dramatic increases in the cost of oil for generation.

Electricity escalation rates for commercial customers billed on demand-energy rates may vary significantly, however, as a function of the building electric load factor, or the relationship between average and peak demand. (This concept is discussed in detail in later sections of this chapter.)

The Energy Information Administration and the Gas Research Institute also project the price of electricity. For the commercial sector, the 1982 to 1985 escalation rate was projected to be fairly stable. From mid-1985 to mid-1990 and beyond, electricity prices are projected to remain fairly stable as well (Ashby, Holtberg, and Woods 1985).

A similar historic perspective can be developed for domestically produced oil and imported oil. Commercial building facilities managers may recall the prices of the late 1960s for #2 distillate and #6 residual of $0.25/gal and as low as $0.15/gal, respectively. Prices in 1982 were $1.08/gal for commercial sector distillate and $0.78/gal for residual.

This information on fuel prices and price escalation may be useful for long-term planning or for producing feasibility studies or financial analyses. Energy escalation rates are a necessary input to any life-cycle cost economic model that discounts the value of future energy dollars saved to current or present value. Regional fuel prices and escalation factors can be obtained from the local utility or energy supplier.

ELECTRIC ENERGY COST AND USE CHARACTERISTICS

Utilities regulated by state public utility commissions must obtain approval of both the rate of return they may make on "rate base" and the rate structure used to recover their costs. Although rate structures applicable to commercial buildings are somewhat similar among utilities, how the actual rate recovers costs can differ significantly, affecting the designer's choice of systems, systems control, and solar application. By understanding the method (or rate structure) that an energy supplier uses to recover costs, a team can develop appropriate energy conservation strategies.

Electric Energy Costs

The following terms are used to describe electric energy use and costs:

- **Electric Consumption** - the quantity of energy used, measured in kilowatt-hours (kWh).

- **Electric Demand** - the rate of energy use, or kilowatt-hours per hour, usually expressed in kilowatts (kW). The electric meter records the highest consumption for a given interval during the monthly billing period, typically 15 or 30 min in the United States. This is called the peak demand.

Figure 2-4 represents how a commercial building's electric demand may vary during 24 h. Point A indicates the peak electric demand based on a 15-min average. Energy consumption is denoted by the area under the solid line. The dashed line indicates the average demand for the time period, and the area under the dashed line is equal to the area (representing energy used) under the solid line. However, the condition represented by the solid line, or actual building load profile, requires greater utility capacity due to the higher peak demand. The utility usually bases the demand charge on the highest demand in the billing period (Point A). The utility is allowed to recover the costs associated with generation, transmission, and distribution of electric energy required to meet point A.

- "Ratchet" Clause - (more correctly defined as minimum billing demand) a method used by many electric utilities to ensure adequate cost recovery from commercial or industrial energy users who exhibit large fluctuations in electric load factor. There are various types and levels of ratchet clauses. One such clause may set the current month's minimum billing demand equal to a fixed percentage (for example, 75%) of the previous 12 months' maximum measured demand. For example, an office building heated by natural gas and cooled by electricity may require 1000 kW of peak demand in the summer for air conditioning, but only 500 kW during the winter. If the utility's rate tariff has a 75% demand ratchet, even though the winter's measured demands are near 500 kW, the utility bills for a demand of 0.75 × 1000 kW, or 750 kW. In this case, the office building is paying for the utility's investment to serve the 1000-kW demand required during the summer, of which 50% sits idly (financially and functionally) during off-peak months.

- Rates - provide for charges according to power demand, energy use, and other factors, including time and accessibility. The demand-intensive type is characterized by a predominant demand charge component. The energy-intensive type has a lower demand component but a higher charge per kilowatt-hour of energy use.

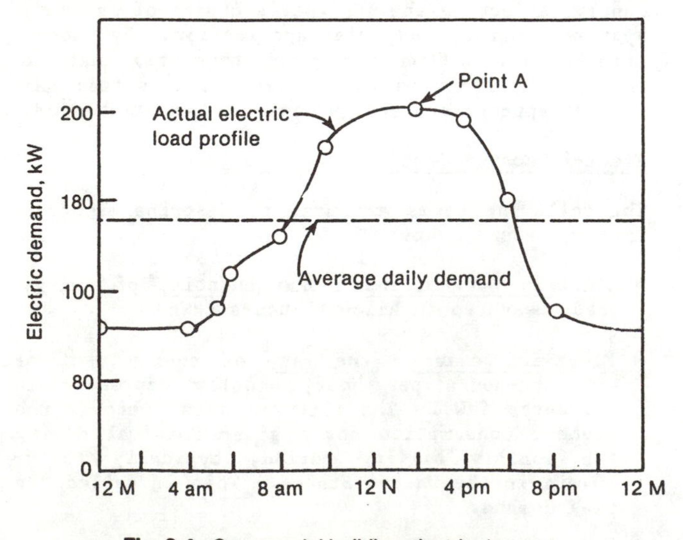

Fig. 2-4. Commercial building electric demand

Based on electric rate structures in Table 2-2, the monthly charge for electricity is computed for a commercial building with a measured demand of 400 kW and a usage of 130,000 kWh. (See Table 2-3.)

Table 2-2. Electric Rate Types

Monthly Rate	Demand Intensive ($)	Energy Intensive ($)
Service charge	3.00	3.00
Demand charge	12.00/kW	5.00/kW
Energy charge	0.015/kWh	0.032/kWh
Fuel cost adjustment	0.007/kWh	0.007/kWh

Table 2-3. Comparison of Rate Types - Commercial Facility (Typical Load Factor)

Monthly Rate	Demand Intensive ($)	Energy Intensive ($)
Service charge	3.00	3.00
Demand charge	4,800.00	2,000.00
Energy charge	1,950.00	4,160.00
Fuel cost adjustment	910.00	910.00
Total	7,663.00	7,073.00
Average cost/kWh	0.059	0.054

Although the total electric charges for both rates are very close, and the average cost per kilowatt-hour is similar, the example shows the significance of the demand charges in the demand-intensive rate. They represent 63% of the total bill.

Table 2-4 shows energy costs based on the same two rate types with the same component values for another building with a measured demand of 800 kW and a consumption of 100,000 kWh.

For this case, the average cost per kilowatt-hour is higher for the demand-intensive rate by 49%. In the previous example, the average cost per kilowatt-hour is higher by 9.3% when using the demand-intensive rate.

Time-Differentiated Rates are offered by some electric utilities. Building design teams should be aware of these rates. Time-differentiated rates may include off-peak and time-of-use rates, and rates for interruptible service.

A utility may offer an off-peak rate. Generally, an off-peak rate provides incentive for the

Table 2-4. Comparison of Rate Types - Commercial Facility (Low Load Factor)

Monthly Rate	Demand Intensive ($)	Energy Intensive ($)
Service charge	3.00	3.00
Demand charge	9,600.00	4,000.00
Energy charge	1,500.00	3,200.00
Fuel cost adjustment	700.00	700.00
Total	11,803.00	7,903.00
Average cost/kWh	0.118	0.079

customer to use thermal storage to shift electric heating or cooling to utility off-peak hours during the night. The off-peak portion of the electric rate may have a demand charge in addition to the energy charge. The off-peak charges are substantially less. Although thermal storage can be used for heating, a more appropriate storage application for commercial buildings may be chilled water or ice for air conditioning.

With time-differentiated rates, the utility generally requires two meters or meter registers. One meter records base-load electric use in the building, and the second meter measures electric energy for the thermal storage device only. A time clock integral to the meters separates the off-peak and on-peak use.

Time-of-Use electric rates vary as a function of the time of day. The utility computes its cost to deliver energy based on the historical daily and seasonal load curves that the utility generation system experiences. Electric rates may be categorized as off-peak, shoulder-peak, and on-peak. Generally, the cost per unit of energy is the same for all three periods, but the demand or capacity charge varies substantially for each period.

Interruptible electric rates are another form of time-differentiated rates. Under this rate, the utility may offer a 50% reduction in demand charges if the commercial building allows the utility to interrupt electric service during peak load conditions. Some interruptible rates do not specify how long the interruption will last during a period of peak load conditions, but generally the utility will suggest that it be no longer than 10 h. The customer usually must incur the cost of a service interruption device able to communicate with the utility's load dispatcher. In addition, a severe penalty is levied if the customer does not disconnect load when requested.

As with the demand-energy electric rates, special utility rate structures may provide designers of solar buildings with opportunities to apply unique designs in ways that maximize energy cost avoidance. Modeling building thermal and electrical energy profiles and applying all appropriate energy rate structures are suggested for ensuring that the energy cost is properly analyzed.

- **Electric Load Factor** - the ratio of the average electric demand to the peak electric demand in a given time period. Because 30 days is the usual billing period, most evaluations are done on a monthly basis. Electric load factor (LF) can be calculated with adequate accuracy for utility energy analysis by using the following formula:

$$LF = \frac{\text{kWh/mo}}{\text{kW peak demand} \times 730 \text{ h/mo}} \quad (2\text{-}1)$$

This calculated value should be a number less than 1.0. Hospitals with a large number of operating hours typically exhibit a more even or flat electric load profile than do office buildings. A hospital's monthly electric load factor may be between 0.77 and 0.85, while an office building may produce a value of between 0.40 and 0.50.

The commercial facilities used to produce Tables 2-3 and 2-4 have electric load factors of

$$\frac{130{,}000 \text{ kWh/mo}}{400 \text{ kW} \times 730 \text{ h/mo}} = 0.445, \text{ and}$$

$$\frac{100{,}000 \text{ kWh/mo}}{800 \text{ kW} \times 730 \text{ h/mo}} = 0.171, \text{ respectively.}$$

These examples show that the monthly electric load factor can be used as an indicator for electric utility costs in the following way:

- The higher the load factor, the lower the unit cost of electricity (compare the average costs of electricity in Tables 2-3 and 2-4).
- The average cost of electricity is more sensitive to load factor for demand-intensive rates than for energy-intensive rates.

An evaluation of energy management options, therefore, should consider the electric demand cost savings (in kilowatts) as well as the energy cost savings. Many electricity cost avoidance calculations are based only on the average cost per kilowatt-hour, leading to possible miscalculation of expected savings. Table 2-5 illustrates this point.

In this example, using the average cost (Method 2) results in an error of $1,980, or 43%. The discrepancy would be greater for higher demand-intensive rates.

So far the examples have emphasized the importance of understanding key parameters related to electric energy costs: peak electric demand (kW), electric energy use (kWh), and electric load factor (average demand/peak demand). Also important is the understanding of how electric energy use and electric demand might vary from one month to another for the building being designed, and what loads are the major contributors to energy use and peak demand. Preferably, a designer can use an hourly building energy analysis computer program in the conceptual design phase to quantify the load components. This

Table 2-5. Comparison of Methodologies to Calculate Energy Cost Savings

	Before Energy Management	After Energy Management
Measured demand, kW	900	450
Use, kWh	200,000	155,000
Load factor, %	30	47
Demand charge (@ $8.00/kW), $	7,200	3,600
Energy charge (@ $0.017/kWh), $	3,400	2,635
Fuel cost adjustment (@ $0.005/kWh), $	1,000	775
Total Energy Cost, $	11,600	7,010
Average cost/kWh (Total/Use)	$0.058	$0.045

Dollar cost avoidance = $11,600 - $7,010 = $4,590 (Method 1)

Dollar cost avoidance based on average cost/kWh = (200,000 - 155,000 kWh) × $0.058/kWh = $2,610 (Method 2)

analysis enables sensitivity analyses to be performed systematically. Electric demand/energy rates can be applied to establish the utility cost avoidance potential of various design strategies and the level of emphasis that should be placed on electric demand reduction versus electric energy conservation.

Figure 2-5 shows how monthly electric charges might vary for a large office building. The demand charges are significant because of a demand-intensive rate structure. Figure 2-5 also illustrates the effect of one possible ratchet clause.

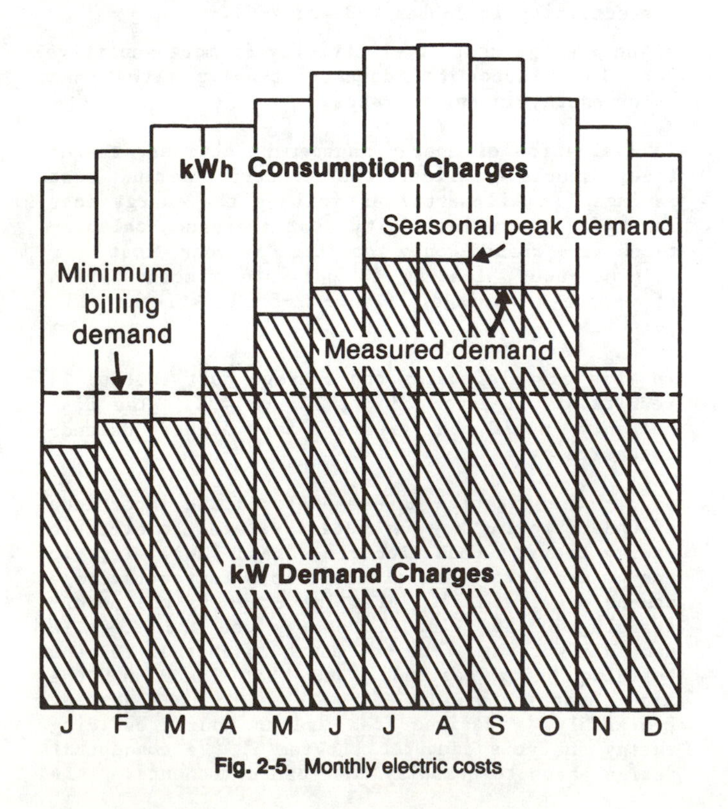

Fig. 2-5. Monthly electric costs

Note that the measured demand falls below the ratchet billing demand during the winter. The probability of this happening is greater if the building is heated by a fossil fuel and cooled by electricity-driven chillers, thereby maximizing the summer peak demand and minimizing the winter demand. In such a case, the building occupant pays demand charges during off-peak months that are greater than what the actual measured demand would require. Also, any measures designed to reduce demand do not reduce the electric demand cost in months where ratchet demand is greater than the measured demand.

Estimating Electric Energy Use

To estimate the major contributors to electric demand and electric consumption without a computer simulation tool, a designer can use a simple form (see Table 2-6) and some characteristic building design parameters in the predesign phase. The design parameters can be taken from similar recent designs or obtained from the Commercial Industrial Divisions of major utilities or from an experienced consulting engineer. Table 2-6 lists major electric components that make up a commercial electric load profile.

The use of the form is illustrated by considering a proposed new 100,000 ft^2 office building with base case building parameters, established before any significant building or system changes that might be developed in the predesign phase. The approach to this calculation is

1. Define the loads and their power use (see section on "Connected Electric Load").
2. Work out reasonable values for hourly energy use in winter and summer (see section on "Coincident Electric Load") using the information from Item 1.
3. Estimate the yearly energy consumed (see section on "Estimating Procedures for Annual

Table 2-6. Electric Energy Utilization Form

Component	Connected kW	Winter Peak Coincident kW	Winter Peak % of Total	Summer Peak Coincident kW	Summer Peak % of Total	Estimated Annual kWh	% of Total
Lighting	220	198	53.8	198	40	626,000	48
Electric Space Heating	N/A	--	--	--	--	--	--
Distribution Equipment, Heating	40	40	10.9	0	0	173,000	13
Electric Cooling	190	20	5.4	190	38	228,000	17
Distribution Equipment, Cooling	80	80	21.7	80	16	192,000	15
Processing	N/A	--	--	--	--	--	--
Water Heating	N/A	--	--	--	--		
Other	75	15	4.1	15	3	47,000	4
	30	15	4.1	15	3	47,000	4
Total	635	368		498		1,313,000	

Electric Energy Use"), using the information developed for Item 2.

A similar approach will be used later in this chapter for calculating fossil fuel costs. The base case building parameters in the example will vary for different climatic regions.

Connected Electric Load (kW)

Lighting - 2.2 W/ft^2 (= 220 kW).

Electric Space Heating - Building is heated with natural gas, and there is no electric use for heat. Therefore N/A.

Distribution Equipment, Heating - Assume hot water circulation pumps for possible perimeter system. Use 40 kW if no other information exists.

Electric Cooling - Assume 2 tons/1000 ft^2 (latent plus sensible) of cooling if loads have not been calculated. At 0.7 kW/ton of electric centrifugal chiller capacity, this converts to 140 kW. Also, include chilled water pumps, cooling tower pumps at 0.5 W/ft^2 (= 50 kW).

Distribution Equipment, Cooling - Air handling, or air terminal, units for ventilation. Assume 1.1 cfm/ft^2 and 0.75 kW/1000 cfm. This computes to 80 kW.

Processing - This category is meant for special loads such as heat treating, spray booths, electric furnaces, etc., typical of industrial applications or a computer facility. Assume N/A.

Water Heating - Domestic hot water is supplied by natural gas boilers. Therefore N/A.

Other - Assume 75 kW for elevators. Assume 0.3 W/ft^2 (= 30 kW) for office equipment.

Coincident Electric Load (kW)

An estimate is made of the coincident demand (or simultaneous use) among all connected electric loads. This is the best estimate of what level of demand will actually be registered by the demand meter, and considers that not all connected electric load will be on at full load simultaneously. Because few buildings have been submetered and because the types and functions of commercial buildings vary considerably, this estimate is very uncertain. Utility data on similar existing buildings can provide a total peak watts per square foot demand and monthly or annual energy use (kWh) per square foot, but generally these are not known by electric component. So judgment must be applied. Columns for winter and summer coincident peak loads are provided on the form to facilitate ratchet demand analysis.

Lighting - For both summer and winter peaks, assume 90% of the connected lighting load will be on. This allows for conference rooms, storerooms, etc.

Distribution Equipment, Heating - Assume all pumps will be on during the winter peak but not during the summer peak.

Electric Cooling - Assume full cooling capacity requirements during the summer peak. Assume outside air and cooling tower cycle will meet cooling requirements during the winter.

Distribution Equipment, Cooling - This represents air handling unit energy. Assume full load all year for ventilation purposes.

Other - Elevators operate on a stop-and-start basis, and although they operate year round, the demand meter only registers a 15- or 30-min integrated peak--not an instantaneous peak. Elevator operation has a diversity factor with respect to other elevators and with respect to the demand meter interval. Assume only a 20% coincidence. Office equipment, including typewriters, copy machines, word processors, and other small equipment, have coincident factors of about 50%.

If data on similar buildings are available, it is a good practice to check the computed coincident total demand per square foot with these metered demands. In this case, the total coincident summer load calculates as 498 kW, an average of 4.98 W/ft^2, which is typical of large, new office buildings with electric drive cooling.

Percentages can be calculated and entered on the form to help determine which loads should be addressed. As can be seen, lighting is a major contributor to peak demand. Also note that the estimated winter demand is 74% of the summer peak. If a ratchet of more than 75% were in effect, during the winter the building's occupants would pay for demand not imposed on the utility grid.

Estimating Procedures for Annual Electric Energy Use (kWh)

The estimated annual energy use in kilowatt-hours is the product of coincident kilowatts times hours of use. Bin method calculations or computer simulation offers a more accurate procedure than hours of use. However, to be sure it is reasonable, the estimate of energy use should be compared to utility data for similar buildings. This level of accuracy is probably satisfactory for the predesign phase.

Lighting

$$198 \text{ kW} \times \frac{12 \text{ h}}{\text{day}} \times \frac{5 \text{ days}}{\text{week}} \times \frac{52 \text{ weeks}}{\text{year}} + 20 \text{ kW} \times \frac{8 \text{ h}}{\text{day}} \times \frac{52 \text{ Saturdays}}{\text{year}} = 626{,}000 \text{ kWh}$$

(In the above calculation, occupancy on Saturday is assumed to be 10% of occupancy on a weekday.)

Distribution Equipment, Heating

$$40 \text{ kW in circulating pumps} \times \frac{24 \text{ h}}{\text{day}} \times \frac{180 \text{ days}}{\text{year}} = 172{,}800 \text{ kWh}$$

Electric Cooling - Equipment full load hours is assumed as 1200. Since full cooling load is 190 kW, seasonal energy use is

$$190 \text{ kW full load} \times 1200 \text{ equivalent full load hours} = 228{,}000 \text{ kWh}$$

Distribution Equipment, Cooling - Air handling fans (1 hp/1000 cfm) are usually cycled during unoccupied hours on a night set-back or set-up cycle, but usually are all operated during occupied hours.

Energy use during occupied hours:

$$80 \text{ kW} \times 0.5 \text{ average load} \times \frac{12 \text{ h}}{\text{day}} \times \frac{5 \text{ days}}{\text{week}} \times \frac{52 \text{ weeks}}{\text{year}} = 125{,}000 \text{ kWh}$$

Energy use during unoccupied hours:

$$80 \text{ kW} \times 0.5 \times 0.3 \text{ duty cycle } \left[\left(\frac{12 \text{ h}}{\text{day}} \times \frac{260 \text{ days}}{\text{year}}\right) + \left(\frac{24 \text{ h}}{\text{day}} \times \frac{104 \text{ days}}{\text{year}}\right)\right] = 67{,}000 \text{ kWh}$$

Total Distribution Equipment for Cooling Energy Use: 192,000 kWh

Other Loads

$$\text{Elevators: } 15 \text{ kW} \times \frac{12 \text{ h}}{\text{day}} \times \frac{260 \text{ days}}{\text{year}} = 47{,}000 \text{ kWh}$$

Office Equipment: 47,000 kWh

The estimated total annual kilowatt-hours of energy for the 100,000 ft^2 office building computes to 13.1 kWh/ft^2-yr, which is reasonable for office buildings without a data processing facility. As can be seen from the percentage, lighting is also the major electric energy user.

Interestingly, the estimated electric use within the building is 44,500 Btu/ft^2-yr, of which lighting is 21,000 Btu/ft^2-yr. The building energy use index goal of many modern passive solar office buildings including auxiliary space heating energy (the previous example did not include space heating energy) has been widely published to be between 25,000 and 40,000 Btu/ft^2-yr. Obviously, because lighting energy is proportional to hours of occupancy and also directly affects cooling requirements in the form of internal heat gain, increased use of daylighting and high-efficiency light sources can be key to reaching these goals.

Figure 2-6 illustrates the major contributors to electric demand and consumption for the above example. Identifying the contributors is the conceptual starting point for selecting general energy conservation and electric cost avoidance strategies. Thereafter, the problem is how to modify the proposed new building to reduce either electric demand charges or energy consumption or both, depending on whether utility rates are demand or energy intensive.

FOSSIL FUEL COST AND USE CHARACTERISTICS

Fossil Fuel Costs

The principal source of space heating energy in many commercial buildings is a fossil fuel such as natural gas, fuel oil, or utility-supplied low-pressure steam. A close relationship exists between the space heating energy required and the internal electric energy loads in the building: a substantial reduction in electric lighting energy will reduce cooling loads but can increase heating energy use. Similarly, increased solar aperture may allow daylighting and result in reduced electric lighting requirements but may also increase peak cooling and heating loads. To understand the interaction among energy conservation strategies in a building that uses multiple fuels at different conversion efficiencies (all of which are priced

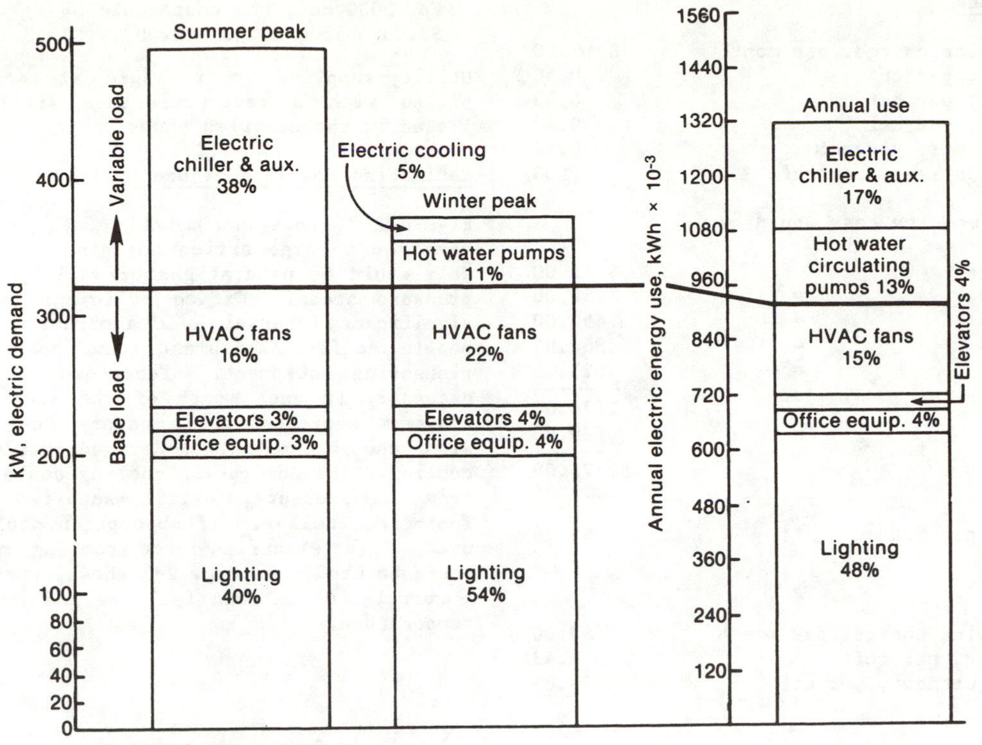

Fig. 2-6. Electric demand and consumption breakdown for office building example

differently per end-use energy), one needs to know about natural gas utility rates, as well as the cost of fuel oil, district steam, or other applicable fossil fuels. Such knowledge is also needed to ensure that energy utilization tradeoffs result in energy cost savings.

Some terms relative to fossil fuel utilization and utility rates are listed below:

- ccf - 100 cubic feet; a billing unit used by natural gas utilities to measure gas volume usage, also approximately equal to 1 therm.
- Therm - 100,000 Btu.
- Mcf - 1000 cubic feet.
- Heating value - The amount of heat produced by the complete combustion of a unit quantity of fuel. Representative heating values for common fuels are shown in Table 2-7.
- Firm gas - A utility rate classification for natural gas supplied on the basis of no service interruption.
- Interruptible gas - A utility rate classification for natural gas supplied with provision for service interruption during peak utility loads.
- M lb - 1000 pounds (of steam).

Table 2-7. Representative Heating Values for Common Fuels

Fuel	Btu/Std. Meas.	Btu/lb
Natural Gas	1030/ft^3	24,000
#2 Fuel Oil	138,000/gal	20,000
#6 Fuel Oil	148,000/gal	19,000
Propane	21,500/lb	21,500
Coal	9000-14,000/lb	9000-14,000

Unlike electric rates, natural gas and other fuels for nonindustrial customers are typically priced to the end use based only on the quantity of fuel consumed; peak hourly demand charges are not included. However, the natural gas industry uses demand metering for some industrial customers. Demand charges based on peak daily use of gas may be included in the customer's rates. For commercial buildings on firm gas rates, however, the commodity charge is the major portion of the cost; therefore, there is no demand charge.

Two brief examples of natural gas utility rates are shown below. The first type is a declining block rate, which has received criticism in recent years for not encouraging energy conservation. The second type is a flat rate.

Declining Block

Customer service charge, per month	$ 3.00
First 500 ccf, per ccf	0.50
Next 1000 ccf, per ccf	0.45
Next 2000 ccf, per ccf	0.43
All over 2000 ccf, per ccf	0.41
Fuel cost adjustment, per ccf	0.08

For 5000 ccf/mo, the cost would be

Service charge		$ 3.00
500 × 0.50	=	250.00
1000 × 0.45	=	450.00
2000 × 0.43	=	860.00
1500 × 0.41	=	615.00
		2178.00
5000 × 0.08	=	400.00
		$2578.00
Average cost per Mcf	=	$5.16

Flat Rate

Customer service charge, per month	$3.00
All cubic feet, per ccf	0.43
Fuel cost adjustment, per ccf	0.08

For 5000 ccf, the cost would be
$3.00 + 5000 (0.43 + 0.08) = $2553.

Utility-supplied low-pressure steam is generally billed with a rate structure similar to those listed in the examples above.

Estimating Fossil Fuel Use

Figure 2-7 shows how monthly fuel consumption may vary for a large office building. The fuel probably would be natural gas or utility-supplied low-pressure steam. Office buildings typically have significant internal heat generation from lights, people, office equipment, and possibly computer processing equipment. Therefore, cooling may be necessary in each month of the year. Figure 2-7 shows a representative monthly fuel use profile with and without fuel-derived steam absorption cooling. The absorption cooling could be generated from low-pressure, utility-supplied steam or a fuel-fired boiler. If absorption cooling were not used, then electric drive cooling most probably would be used. As Fig. 2-7 shows, there is usually a correlation between fuel use and average ambient temperature.

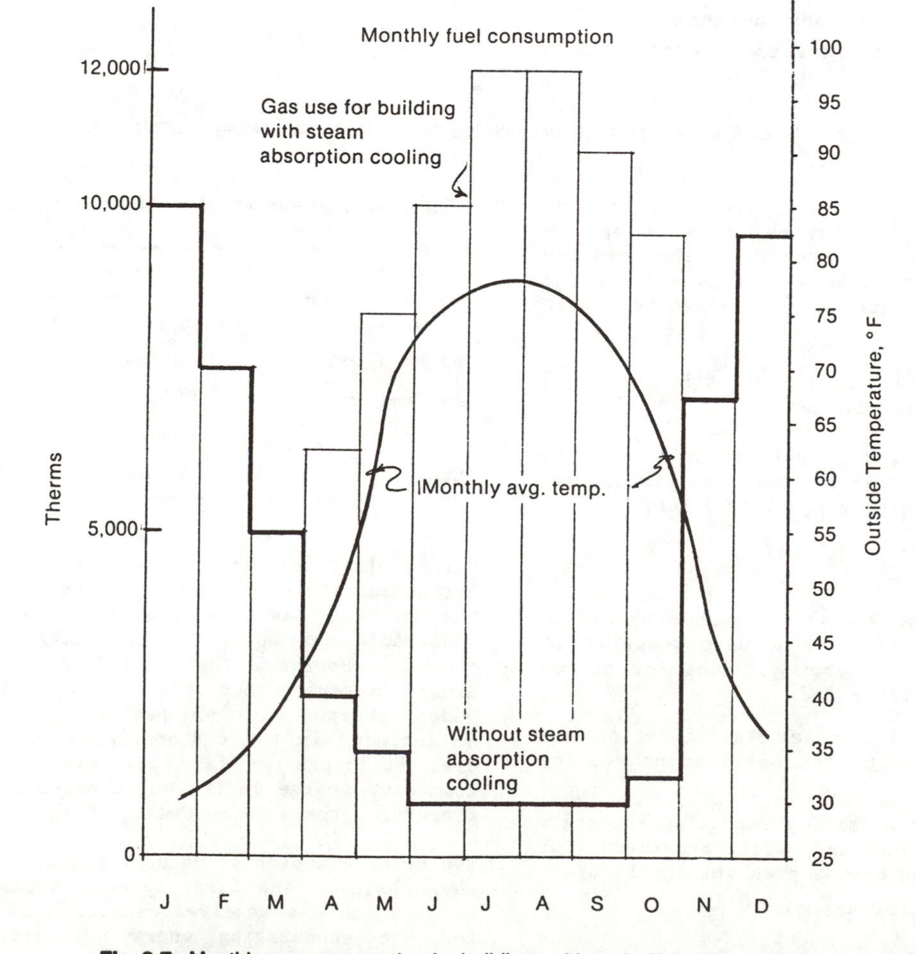

Fig. 2-7. Monthly gas consumption for buildings with and without steam absorption cooling

The most typical contributors to annual fuel consumption in commercial buildings are space heating, service hot water, and possibly absorption cooling. If a computer simulation program or other energy calculation method is not readily available, a simple estimating form and ASHRAE handbook data, coupled with building design parameters, may be sufficient in the predesign phase.

The following example of the major components of natural gas consumption in a 100,000 ft^2 office building illustrates use of an estimating form. [See Table 2-8 to determine the peak connected (design) loads.]

Table 2-8. Gas Energy Utilization Form

End Use	Connected Load (ft^3/h)	Estimated Annual Use* (10^3 ft^3)	% of Total Annual Use
Space heating	3000	4500	42
Absorption cooling	4900	5880	55
Service hot water	250	250	3
Process loads	N/A		
Kitchen	N/A		
Laundry	N/A		
Other	N/A		
Total	8150	8350	
Coincident Total	5150		

*Use appropriate method: hourly simulation, bin, equipment full-load hours, etc.

Connected Load (Input Rating):

Space Heating - determined by heat loss calculation at design conditions, including transmission load and ventilation load. If the heating load is not calculated, begin base case with estimate of 30 $Btu/h\text{-}ft^2$ total floor area. (This value is for a 5000 degree-day climate; use other values practical for the local climatic region.)

$$\frac{30\ Btu}{h\text{-}ft^2} \times 100{,}000\ ft^2 \div \frac{1000\ Btu}{ft^3}$$
$$= 3000\ ft^3/h \text{ of natural gas}$$

Space Cooling (gas-fired steam absorption) - determined by heat gain calculation at design conditions. If the load has not been calculated, estimate 1 ton/500 ft^2 as base case input.

$$\frac{1\ ton}{500\ ft^2} \times 100{,}000\ ft^2 \times \frac{20\ lb\ steam^*}{ton\text{-}h} \times \frac{980\ Btu}{lb\ steam}$$
$$\times \frac{1\ ft^{3**}}{1000\ Btu} \times \frac{1}{0.8\ eff.} = 4900\ ft^3/h$$

*Energy conversion efficiency for absorption units from manufacturer's literature, low-pressure steam.

**1000 Btu/ft^3 for natural gas is used here as an example. Check with local utility for local heating value.

Service Hot Water - use data from Ch. 34 of ASHRAE Handbook, Systems (ASHRAE 1984), or appropriate load calculation.

Estimating Procedures for Natural Gas Use

An hourly computer simulation tool or another automated method is appropriate for estimating seasonal or annual natural gas use (see below). A bin method could also be used, but for equipment efficiencies that vary significantly as a function of part-load, a profile of load versus outside temperature should be included in the calculation. To simplify the initial estimate, data derived from a billing record data base can be obtained from the local utility regarding full load heating and cooling hours. Using any of the above methods will still require experience and judgment.

Space Heating: $3000\ ft^3/h \times 1500$ full load h/yr $= 4.5$ million ft^3/yr

Space Cooling: $4900\ ft^3/h \times 1200$ full load h/yr $= 5.88$ million ft^3/yr

Service Hot Water: $250\ ft^3/h \times 1000$ h/yr $= 0.25$ million ft^3/yr

In this example, absorption cooling is the major contributor to annual natural gas use. No peak demand analysis is necessary unless the applicable natural gas rate contains a demand charge component. If space cooling had been provided by electric chillers, then space heating energy would have been the predominant natural gas use.

Multiple Fuel Cost Considerations

The energy cost application discussion and examples included in the preceding sections have been used to illustrate that:

- Electric demand charges can be very significant, depending on rate structure.
- Natural gas and other fossil fuels are generally billed on rates without demand charges.
- Manual techniques can be used to develop ballpark estimates for contributions to peak electric demand, annual electric use, and fuel energy use for each major building end-use component.
- End-use energy (and demand) contributions can be related to utility rate structure to help determine cost-effective energy conservation strategies.

These factors should be reviewed frequently when dealing with multiple fuels. For many building sites, there are at least two conventional energy sources available: utility-supplied natural gas and electricity.

One way to select a primary energy source or a backup source for a solar system is to use the average cost per unit of delivered energy and compare costs for achieving equivalent results.

The comparisons then are as straightforward as the following:

Electricity Cost:

$$\frac{\text{Avg. Cost (E), \$/kWh}}{\text{3413 Btu/kWh} \times \text{Eff.}} = \frac{293 \times \text{E, \$}}{10^6 \text{ Btu} \times \text{Eff.}}$$

Natural Gas Cost:

$$\frac{\text{Avg. Cost (NG), \$/}10^3 \text{ ft}^3}{1{,}000{,}000 \text{ Btu/}10^3 \text{ ft}^3 \times \text{Eff.}} = \frac{\text{NG, \$}}{10^6 \text{ Btu} \times \text{Eff.}}$$

Utility Steam Cost:

$$\frac{\text{Avg. Cost (US), \$/}10^3 \text{ lb}}{980{,}000 \text{ Btu/}10^3 \text{ lb} \times \text{Eff.}} = \frac{1.12 \times \text{US, \$}}{10^6 \text{ Btu} \times \text{Eff.}}$$

#2 Fuel Oil Cost:

$$\frac{\text{Avg. Cost (\#2), \$/gal}}{\text{138,000 Btu/gal} \times \text{Eff.}} = \frac{7.23 \times \text{\#2, \$}}{10^6 \text{ Btu} \times \text{Eff.}}$$

#6 Fuel Oil Cost:

$$\frac{\text{Avg. Cost (\#6), \$/gal}}{\text{146,000 Btu/gal} \times \text{Eff.}} = \frac{6.85 \times \text{\#6, \$}}{10^6 \text{ Btu} \times \text{Eff.}}$$

If the following average costs

E: \$0.05/kWh
NG: \$5.00/10^3 ft^3
US: \$6.50/10^3 ft^3
#2: \$1.05/gal
#6: \$0.75/gal

are inserted into the foregoing equations, and all efficiencies (except for that for electricity = 1) are set at 0.75, the following results are obtained per million Btu:

Electricity Cost	=	\$14.65
Natural Gas Cost	=	6.67
Utility Steam Cost	=	8.84
#2 Fuel Cost	=	10.14
#6 Fuel Cost	=	6.85

As emphasized in previous sections, the average cost per kilowatt-hour of electricity must be used with caution, with consideration given to demand charges and the electric load factor.

For some applications, the fuel cost must be combined with equipment energy conversion performance characteristics and consumption factors to develop a fuel choice decision. One example is the selection of primary cooling equipment. Although more accurately evaluated by simulation techniques, the following example illustrates this issue by comparing electric centrifugal chillers to a natural-gas boiler supplying steam for absorption chillers. For a 200-ton electric centrifugal chiller,

200 tons × 0.68 kW*/ton
× 1800 equivalent full load hours/yr
= 244,800 kWh/yr

200 tons × 0.68 kW/ton = 136 kW

*Manufacturer's data for high-pressure steam, high efficiency

Electric Cost:

$$244{,}800 \text{ kWh/yr} \times \$0.02\text{/kWh*} = \$\ 4{,}896\text{/yr}$$

$$136 \text{ kW*} \times \frac{\$8.00\text{/kW}}{\text{mo}} \times 9 \text{ mo/yr} = \$\ 9{,}792\text{/yr}$$

$$\text{Total} \quad \$14{,}688\text{/yr}$$

For a 200-ton steam absorption cold generator, the gas consumed is

$$\frac{200 \text{ ton} \times \frac{12 \text{ lb steam}}{\text{ton-h}} \times \frac{980 \text{ Btu}}{\text{lb}} \times 1800 \text{ h/yr}}{1{,}000{,}000 \text{ Btu/}10^3 \text{ ft}^3 \times 0.75 \text{ eff.}}$$

$$= 5645 \times 10^3 \text{ ft}^3\text{/yr}$$

Natural Gas Cost:

$$\text{At } \$5/10^3 \text{ ft}^3, = 5645 \frac{10^3 \text{ ft}^3}{\text{yr}} \times \frac{\$5}{10^3 \text{ ft}^3}$$

$$= \$28{,}255\text{/yr}$$

For a space heating application, electricity may be more expensive per unit of energy delivered. However, for the assumptions made in this cooling comparison, electricity is less expensive than natural gas.

These examples are quite simplified, but they illustrate that energy costs must be evaluated to address energy conservation strategies in commercial buildings. Such an evaluation requires an understanding of utility rate structures, building and use characteristics, and energy conversion parameters.

CALCULATION OF BUILDING LOADS AND ENERGY USE

In contrast to design heating or cooling load calculations for sizing conventional mechanical equipment, which are based on steady-state peak load conditions, solar system design and component selection should be based on annual system performance. This annual performance can be predicted by using procedures based on either the monthly or annual building thermal load. In the collector sizing and performance determination procedures that follow (Ch. 7), either monthly or annual building heating and cooling loads (those imposed on the building plant) are needed. These space conditioning loads, besides being influenced by the weather, are also influenced by the building envelope, internal loads, and the operation of the HVAC system.

When hourly solar system simulation computer programs are used, the hourly space and service hot water (SHW) loads are usually calculated in the program itself. These internally calculated loads are then used as input to the solar system simulation. If the solar simulation program does not generate its own loads, any of the computer dynamic energy analysis programs (see below) can be used to generate the hourly loads required as input.

*Typical commercial rate

This section discusses procedures and tools available for the calculation of heating, cooling, and SHW loads. The required or appropriate application and level of detail of these tools depend on the design phase involved and the complexity of the analysis. Several levels of tools, and their appropriate applications, are discussed below.

Effect of Building Load Characteristics on Solar System Performance

The building thermal load, in conjunction with the storage characteristics, influences the temperature of the fluid in the collectors, so the seasonal load variations strongly influence annual system performance. Furthermore, the performance of particular types of solar systems (SHW, space heating, space cooling, heat pump, or combinations thereof) is strongly influenced by the building thermal load.

Monthly SHW loads are relatively uniform, so properly sized SHW systems can operate effectively despite monthly weather and solar irradiation fluctuations. Furthermore, because SHW collector systems can preheat the supply water, there is no month when they are rendered inoperative by low collector temperatures. Because of their fairly uniform monthly loads, such systems should be designed to operate effectively year-round within a fairly narrow range of collector temperatures.

Likewise, combined space heating and cooling systems of the active type must meet a fairly uniform monthly load and can be used effectively year-round. During the winter heating season, collector temperatures of approximately 170°F (77°C) are needed for conventional duct heating coils, although heating coils that use from 120° to 140°F (50°-60°C) inlet water could be selected to improve collector efficiency. During the summer cooling season, 160° to 210°F (71°-99°C) collector temperatures are needed to drive absorption chillers.

Because of their fairly uniform load characteristics throughout the year, the performance of SHW systems and combined active heating and cooling systems can be predicted using annual load parameters.

On the other hand, the size of space heating or combined space and water heating systems is based on winter heating loads, so the collectors are oversized for the summer months. This requires shading or dumping of excess heat during the summer and the mild spring and fall seasons when heating loads are minimal. For such systems, weather and solar irradiation fluctuations are important in winter but not in summer. This fact, coupled with the varying length of the heating season with location, requires that the performance of these systems be predicted using monthly load parameters. However, annual correlations of monthly results have been developed that allow the use of annual load parameters (see Ch. 7).

Transient Response Methods

Many methods for estimating the energy transfers to and within buildings are based on steady-state or steady-periodic heat transfer. Where thermal mass is important, or where fluctuating hourly climatic conditions and energy use schedules need to be accounted for, or both, techniques for approximating the transient effects are used.

A technique often used to evaluate the heat conducted through multilayer building elements under transient (nonsteady and nonperiodic) exposure conditions is the response factor (or transfer function) method (ASHRAE 1985, Ch. 23). In this method, the heat flux through a wall at the current hour is expressed as a summation of outdoor and indoor temperatures at previous hours, multiplied by sets of response factors for the wall under consideration. Response factors for a variety of walls are tabulated in Ch. 23 of the ASHRAE Handbook of Fundamentals (ASHRAE 1985) and elsewhere. Many of the computer dynamic loads calculative methods discussed below use wall response factors.

To account for thermal mass effects within the space under transient conditions, the weighting factor (or room transfer function) method has been developed (ASHRAE 1985, Ch. 26). In the weighting-factor method, a two-step process is used to determine the air temperature and heat-extraction rate of a room or building zone for a given set of conditions. In the first step, the air temperature of the room is assumed to be fixed at some reference value. Instantaneous heat gains are calculated on the basis of this constant air temperature; several types of heat gains are considered. Various levels of sophistication can be used to calculate these instantaneous heat gains. Most automated procedures use response factor or conduction transfer functions to describe heat transfer in massive walls. Information about solar intensity is usually input as part of the weather data, but solar position, window transmission and absorption coefficients, the presence or absence of direct solar radiation on exterior walls, and shading by other objects are generally calculated on an hourly basis. Heat gains from lighting, people, or equipment in the room are usually specified in terms of a schedule.

In the second step, the total cooling load is combined with information on the heating, ventilating, and air-conditioning system attached to the room and a set of air-temperature weighting factors in order to calculate the actual heat-extraction rate and air temperature.

Tabulated values of weighting factors for typical building rooms are presented in the ASHRAE Handbook of Fundamentals (ASHRAE 1985, Ch. 26). One of the three groups of weighting factors, for light, medium, and heavy construction rooms, can be used to approximate the behavior of any specific room. Some automated simulation procedures allow weighting factors to be calculated specifically for the building under consideration. This option improves the accuracy of the calculated results, particularly if the building is not of conventional design. The weighting-factor method is used in some of the computer dynamic load methods discussed below. It is also used indirectly in the modified bin method.

<u>Determination of Monthly and Annual Building Loads</u>

Load calculation methods at three basic levels of sophistication and accuracy are recommended, depending on the design phase involved and the analytical tools available. (See ASHRAE 1985, Ch. 28 for a complete discussion.)

The three methods listed below appear in increasing order of accuracy and complexity.

- Single-measure methods (manual) use only a single variable, such as annual degree days, as the measure of the load. These methods are based on steady-state models that use long-term average weather data and consequently are only appropriate for simple buildings and applications.
- Simplified multiple-measure methods (manual, handheld calculator, or microcomputer) result in improved accuracy by including several of the variables on which the load depends, such as the number of hours anticipated under particular operating conditions. These methods generally are quasi-steady-state but may include some transient effects in an approximate manner.
- Detailed simulation (computer dynamic) methods include energy balance calculations at each hour over a period of analysis, usually one year. These transient methods, which account for hourly data variations, are more accurate, but they require more detailed input in their use. Hourly calculations account for the variation in solar loading, varying occupant and equipment schedules, and the thermal capacity of the building and its contents, the effects of which are averaged in daily or monthly calculations. Because capacitance can introduce time lags in the heating or cooling loads imposed on the HVAC system, it must be accounted for if the energy consumption calculation is to be accurate. This is especially true if thermal storage components are available to redistribute the loads. Transient effects often are important, resulting in reduced loads and possible changes in operating techniques.

The three methods are described below. For active systems the load imposed on the heating and cooling coils (secondary equipment load) is always the end computation goal.

Single-Measure Methods

During the conceptual design phase, when only estimates of solar system performance, size, and fuel savings are needed, a single-measure load method can be used. Unfortunately, the simplest method, the modified degree-day procedure (see ASHRAE 1985, Ch. 28), is restricted to small, single-story, envelope-dominated structures. For larger commercial or industrial buildings, where internal cooling-only zones are prevalent, the bin method (ASHRAE 1985, Ch. 28) should be used. These steady-state calculative procedures cannot account for energy requirements of HVAC systems using reheat, mixing boxes, and other central HVAC equipment providing for individual room control.

The Cooling Degree-Days and Equivalent Full-Load Hours (ASHRAE 1985, Ch. 28) procedures are not recommended because there have been insufficient verification data to enable use of either method with any confidence for solar system design (Stamper 1979).

Space heating loads in envelope-dominated structures, without internal-heating-only zones, can be estimated using the degree-day procedure. Residences and warehouses are in this category. This procedure assumes that the steady-state heat loss or building heat load, Q_L, is proportional to the equivalent heat loss coefficient (UA factor, including infiltration) of the building envelope; that is

$$Q_L = (UA)(t_B - t_A) \ , \qquad (2\text{-}2)$$

where t_B is the building temperature (reduced to account for internal heat generation and called the balance-point or base temperature) and t_A is the ambient temperature.

Traditionally, for residential-type buildings, t_B = 65°F (18°C) has been used. Recent research, however, indicates that monthly average internal gains can offset residential heat loss at a mean daily temperature below 65°F (18°C). The balance-point temperature is determined by the thermal integrity of the building envelope as well as the magnitude of internal gains. Furthermore, for commercial buildings characterized by relatively large internal loads, t_B may be closer to 50°F (10°C). Therefore, use of the modified degree-day procedure described in Ch. 28 of the <u>ASHRAE Handbook of Fundamentals</u> (1985) is recommended. Degree-days at 50°, 55°, 60° and 65°F (10°, 13°, 16°, and 18°C) base temperatures are tabulated in <u>Passive Solar Heating Analysis: A Design Manual</u> (Balcomb et al. 1984) and <u>Degree Days to Selected Bases for First-Order Type Stations</u> (DOC 1973).

The building heat loss coefficient, UA in Eq. (2-2), is the space heating loss (in Btu/h or kJ/h) at design conditions, estimated by ASHRAE procedures (ASHRAE 1985, Ch. 23) divided by the design temperature difference. Where ventilation is significant, UA should include the ventilation load based on the design temperature difference, calculated by traditional methods.

The building heat loss coefficient may also be expressed in Btu/DD-ft^2 (kJ/°C-day-m^2) of building. Multiplying UA (Btu/h-°F) by 24 and dividing by the building area gives the heat loss parameter in Btu/DD-ft^2 (kJ/°C-day-m^2). The building thermal load is then given by the heat loss parameter, times the building area, times the heating degree-days either per month or per year. Heating degree-days, referenced to a balance-point temperature of 65°F (18°C), are given by ASHRAE (1985, Ch. 28) or by the Climatic Atlas of the United States (DOC 1968). If 65°F (18°C) base degree-day data are used for a building with large internal heat sources (occupants, lights, or equipment), the internal heat gained from these sources should be subtracted from the heating load. Otherwise, degree-days corresponding to a lower balance-point temperature should be used.

Recent research, particularly by ASHRAE, has resulted in a refinement of the modified degree-day procedure that is termed the Variable-Base Degree-Day Method (ASHRAE 1985, Ch. 28). This method is available on hand-held calculators and microcomputers (Costello, Kusada, and Aso 1982). Another example is the interactive computer program (Beckman et al. 1982) that calculates monthly heating loads.

Simplified Multiple-Measure Methods

Bin Method. Both heating and cooling loads can be estimated using the bin (temperature frequency) method. The bin method consists of taking outdoor dry-bulb temperatures and multiplying the results by the number of hours of occurrence of each temperature bin. The bins are usually 5°F (2.8°C) in size, and they are collected into three daily 8-h shifts or time groups, labeled 02-09, 10-17, and 18-01. Because this method is based on hourly weather data rather than on daily averages, it is considerably more accurate than the degree-day method. In addition, the bin method takes into account both occupied and unoccupied building conditions and gives credit for internal loads by adjusting the balance point. The average coincident wet-bulb temperatures in the bins should be used to calculate the enthalpy of outdoor air for ventilation and the infiltration loads for cooling. Weather data for the bin method are given in Air Force Manual AFM 88-29 (1978). The following example shows the use of the bin method to calculate heating loads only (adapted from DOE 1978a).

EXAMPLE

Consider a three-story office and laboratory building of 16,800 ft^2 (1562 m^2) (see Fig. 2-8) located in Albuquerque, New Mexico, at 35.0°N latitude. The building is a well-insulated, brick-faced, concrete block structure with a flat roof (U = 0.040 Btu/h-ft^2-°F [0.23 W/m^2-°C] for the opaque walls, U = 0.078 [0.44] for roof). Since one floor is a basement, the total window area is only 157 ft^2. Because of several exhaust hoods, minimum outside air requirements are 2500 cfm/ft^2 (12.7 m^3/s-m^2). Lighting levels are about 2.5 W/ft^2, and equipment loads are minimal. Full occupancy is 118 people.

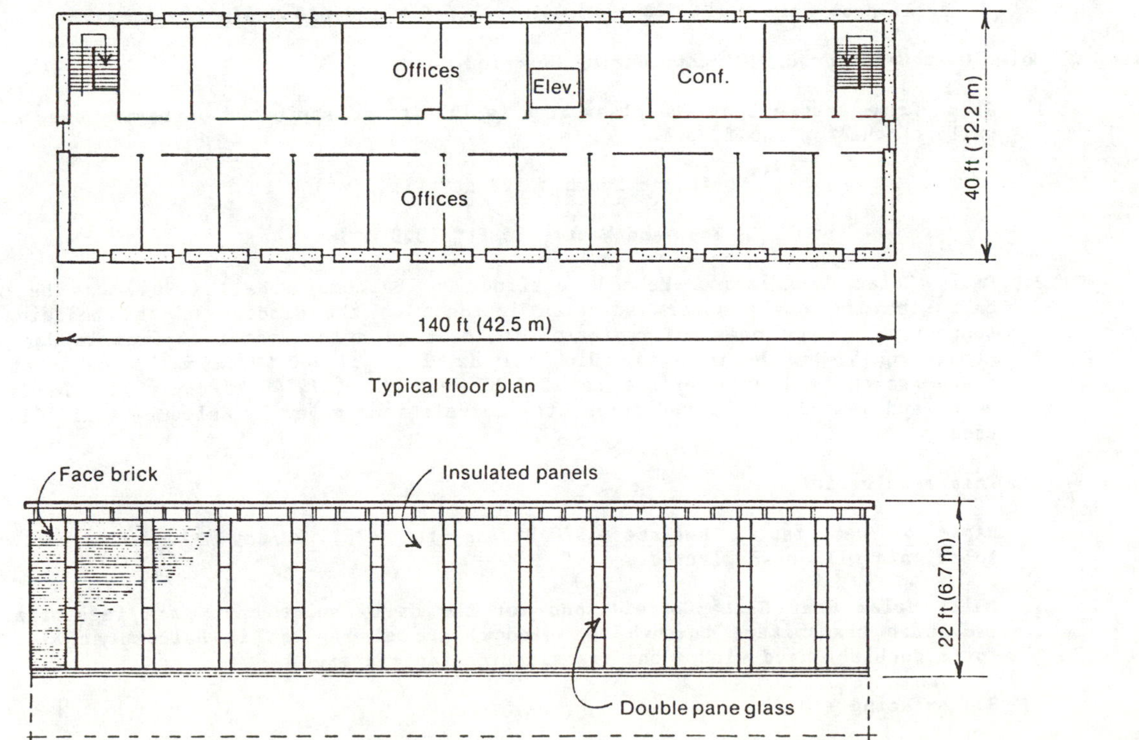

Example (continued on next page)

Example (continued)

We begin by dividing the building into perimeter and interior zones and obtaining overall heat loss coefficients and balance point temperatures of each zone.

Step A. Perimeter and Interior Zone Partition

Assume that the perimeter zone includes areas 15 ft (4.6 m) from the exterior walls and the entire top floor areas, which are affected by the outside environment. The remaining central part of the building, not affected by the outside environment and requiring cooling only, is the interior zone. Once the extent of the interior zone is determined, it no longer enters into the heating load calculation.

$$\text{Interior zone area} = 2 \text{ floors} \times (110 \text{ ft} \times 10 \text{ ft}) = 2200 \text{ ft}^2 \ (204 \text{ m}^2)$$

$$\text{Perimeter zone area} = 3 \text{ floors} \times (140 \text{ ft} \times 40 \text{ ft}) - 2200 \text{ ft}^2 = 16{,}800 - 2200 = 14{,}600 \text{ ft}^2 \ (1357 \text{ m}^2)$$

Step B. Perimeter Zone Sensible Internal Heat Gain During Occupied Hours

The sensible internal heat gains in the perimeter zone from lights, equipment, and people during occupied hours (7 a.m. to 5 p.m.) are

$$\text{Lights} = 2.5 \text{ W/ft}^2 \times 3.412 \text{ Btu/W-h} \times 14{,}600 \text{ ft}^2 = 124{,}540 \text{ Btu/h} \ (36.5 \text{ kW})$$

$$\text{Equipment} = 21{,}217 \text{ Btu/h} \times (14{,}600 \text{ ft}^2/16{,}800 \text{ ft}^2) = 18{,}440 \text{ Btu/h} \ (5.4 \text{ kW})$$

$$\text{People} = 250 \text{ Btu/h-person} \times 118 \text{ persons} \times (14{,}600 \text{ ft}^2/16{,}800 \text{ ft}^2) = 25{,}640 \text{ Btu/h} \ (7.5 \text{ kW}).$$

It is assumed that these loads are distributed uniformly over the entire floor area.

The total internal heat gain is the sum of these three or

$$\text{Internal heat gain} = 124{,}540 + 18{,}440 + 25{,}640 = 168{,}620 \text{ Btu/h} \ (49.4 \text{ kW}).$$

Step C. Solar Heat Gain Through Windows During Occupied Hours

1. **Glass Area**--Assume that the glass area is 157 ft^2, distributed uniformly over all four walls of the top two floors.

   North and South: 122 ft^2 (11.3 m^2)

   East and West: 35 ft^2 (3.3 m^2)

2. **Daily Solar Irradiation**--Methods outlined by Balcomb et al. (1984) can be used to calculate the daily solar radiation incident on the windows of the building. The long-term average measured radiation incident on a horizontal surface in January for Albuquerque, New Mexico, is 1016 Btu/ day-ft^2 (11,540 kJ/day-m^2), and that for a south-facing vertical surface is 1562 Btu/day-ft^2 (17,740 kJ/day-m^2). To determine east- and west-facing irradiation, the correlations given by Balcomb et al. (1984) are used.

   This results in:

   East- or west-facing surface = 670 Btu/day-ft^2 (7615 kJ/day-m^2). The north-facing solar gain will be neglected.

3. **Daily Solar Heat Gain**--Correlations for the daily solar heat gain (essentially the radiation transmitted through the window) are also given in Balcomb et al. (1984). For a double-glazed window the transmission factors are

   South-facing = 0.71

Example (continued on next page)

Example (continued)

East- or West-facing = 0.64. (Alternatively, ASHRAE values (ASHRAE 1985, Ch. 27) for transmission and shading coefficients could be used.)

$$\text{Daily Solar Gain} = [\sum (\text{Irradiation} \times \text{Area} \times \text{Transmission Factor})]$$
$$= [(1562\ \text{Btu/day-ft}^2) \times 122\ \text{ft}^2 \times 0.71 + (670\ \text{Btu/day-ft}^2) \times 35\ \text{ft}^2 \times 2 \times 0.64]$$
$$= 164{,}500\ \text{Btu/day}\ (173{,}500\ \text{kJ/day})$$

4. **Average Solar Heat Gain on Hourly Basis**

$$\text{Hourly Solar Gain} = \left(\frac{164{,}500\ \text{Btu/day}}{10\ \text{h/day}}\right) = 16{,}450\ \text{Btu/h}\ (4.8\ \text{kW})$$

Step D. Total Heat Gain During Occupied Hours

This is the sum of the internal heat gains (Step B) and the solar heat gains (Step C),

$$\text{Total Heat Gain} = 168{,}620 + 16{,}450 = 185{,}100\ \text{Btu/h}\ (54.2\ \text{kW})$$

Step E. Equivalent Heat Loss Coefficient (UA) of Perimeter Zone

1. **During Occupied Hours**

This is the sum of the building envelope loss and ventilation loss coefficients. The envelope loss coefficient is

$$UA_t = \frac{\text{design heating load attributed to transmission losses}}{\text{indoor design temperature} - \text{outdoor design temperature}}\ .$$

The design transmission loss is determined by standard procedures to be 94,000 Btu/h (27.6 kW) for an indoor design temperature of 70°F (21.1°C) and outdoor design temperature of 8°F (-13.3°C). Thus,

$$UA_t = (94{,}000\ \text{Btu/h})/(70 - 8)°\text{F} = 1516\ \text{Btu/h-°F}\ (0.80\ \text{kW/°C}).$$

The ventilation load during occupied hours is

$$1.08 \times \text{cfm} \times (\text{density correction for altitude}).$$

The density correction for altitude at Albuquerque, New Mexico, is 0.82. If the ventilation load is distributed over the entire building,

$$UA_v = 1.08 \times 2500\ \text{cfm} \times (14{,}600\ \text{ft}^2/16{,}800\ \text{ft}^2) \times 0.82$$
$$= 1924\ \text{Btu/h-°F}\ (1.02\ \text{kW/°C}).$$

The equivalent UA for the perimeter zone during occupied hours is

$$UA = UA_t + UA_v = 1516 + 1924 = 3440\ \text{Btu/h-°F}\ (1.82\ \text{kW/°C}).$$

2. **During Unoccupied Hours**

The envelope loss coefficient of the unoccupied building is the same as that during occupied hours. However, the ventilation requirement is set back to 50% for unoccupied hours

$$UA_v = 1924 \times 0.5 = 962\ \text{Btu/h-°F}\ (0.51\ \text{kW/°C})$$

and the equivalent UA for the perimeter zone during unoccupied hours is

$$UA = 1516 + 962 = 2478\ \text{Btu/h-°F}\ (1.31\ \text{kW/°C}).$$

Example (continued on next page)

Example (continued)

Step F. Balance Point for Perimeter Zone

1. **During Occupied Hours**

$$t_{BP} = \text{Indoor Temperature} - (\text{Total Heat Gain/UA}) = 70 - (185{,}100/3440)$$
$$= 16°F\ (-9°C).$$

2. **During Unoccupied Hours**

Because there is no internal heat gain during unoccupied hours, the balance point is the night setback temperature of 60°F (15.6°C).

Step G. Heating Degree-Hours During Occupied Hours

The values in Table 2-9 are determined as follows. The building is occupied from 7 a.m. to 5 p.m. for 5.5 days/week.

Table 2-9. Heating Degree-Hours for Occupied Hours in January

(1) Bin Temp. Range (°F)	(2) Observations/ Hour Group 02-09	(3) Observations/ Hour Group 10-17	(4) Average Bin Temp. (°F)	(5) = 16°F - (4) Balance Point Minus Average Bin Temp. (°F)	(6)=(2)×5.5/28 Observation in Occupied Hours 02-09 Hour Group	(7)=(3)x5.5/7 Observation in Occupied Hours 10-17 Hour Group	(8)=(6)+(7) Total Observations in Occupied Hours	(9)=(5)×(8) Heating Degree °F Hours
10/14	8	1	12	4	1.6	0.8	2.4	9.6
5/9	3	0	7	9	0.6	0	0.6	5.4
0/4	0	0	2	14	0	0	0	0
-5/-1	1	0	-3	19	0.2	0	0.2	3.8
							Total	18.8

Heating requirement for January occupied hours

= UA × heating degree-hours, = 3440 Btu/h-°F × 18.8°F-h (1.81 kW/°C × 10.4°C-h) = 64,700 Btu (18.9 kWh).

1. **Temperature Bin Observations Below Balance Point**

Because the balance point temperature is 16°F (-9°C), heating is required for ambient temperatures below this point.

From Air Force Manual AFM-88-29 (1978) for Albuquerque, enter the bin-hours for the temperature bins (°F) 10/14, 5/9, 0/4, and -5/-1 as in Table 2-9. Only those hour groups that are part of the occupied hours, 02-09 and 10-17, are entered.

2. **Temperature Differences (Δt) Below Balance Point**

The average temperature of each bin is determined. Thus the temperature difference is

$$\Delta t = 16°F - (\text{average bin temperature below } 16°F)$$
$$\times [-9°C - (\text{average bin temperature below } -9°C)].$$

3. **Fraction of Observations in Hour Groups When Building is Occupied**

There are two occupied hours in the 02-09 hour group each working day and 5.5 working days/week. Thus,

Occupied Fraction in 02-09 Group = 2/8 × 5.5/7 = 5.5/28.

All of the 10-17 hour groups are occupied; thus,

Example (continued on next page)

Example (continued)

Occupied Fraction in 10-17 Group = 5.5/7.

Multiply the observations by the appropriate fractions for each hour group and enter them as in Table 2-9.

4. **Degree-Hours Determination**

The degree-hours for each bin are obtained by multiplying the total observations for the bin by the Δt between the balance point and the average bin temperature. The total degree-hours for January during occupied hours is the sum of all degree-hours below the balance point temperature, 16°F (-9°C).

Step H. Monthly Heating Load, Occupied Hours

The monthly heating load is the equivalent UA of the perimeter zone during occupied hours multiplied by the total heating degree-hours for the occupied hours.

Heating Load = 3440 Btu/°F-h × 18.8 °F-h (1.81 kW/°C × 10.4°C-h)
= 64,700 Btu (18.9 kWh)

Step I. Heating Degree-Hours for Unoccupied Hours

Determine the values to list in Table 2-10 as follows. The building temperature is set back to 60°F (15.6°C) from 5 p.m. to 7 a.m. on working days and during all hours of the rest of the weekend.

Table 2-10. Heating Degree-Hours for Unoccupied Hours in January

(1) Bin Temp. Range (°F)	(2) Observations/h Group 02-09	(3) 10-17	(4) 18-01	(5) Average Bin Temp. (°F)	(6) = 60°F - (5) Set-Back Temp. Minus Average Bin Temp. (°F)	(7)=(2)×22.5/28 Observation in Occupied Hours 02-09	(8)=(3)×1.5/7 10-17	(9)=(4) 18-01	(10)=(7)+(8)+(9) Total Observations in Unoccupied Hours	(11)=(6)×(10) Heating Degree Hours
55/59	0	17	1	57	3	0	3.6	1	4.6	13.9
50/54	0	35	8	52	8	0	7.5	8	15.5	124.0
45/49	4	52	21	47	13	3.2	11.1	21	35.3	459.6
40/44	16	52	46	42	18	12.9	11.1	46	70.0	1260.0
35/39	42	40	65	37	23	33.8	8.6	65	107.4	2468.4
30/34	62	29	57	32	28	49.8	6.2	57	113.0	3165.0
25/29	61	12	29	27	33	49.0	2.6	29	80.6	2659.4
20/24	39	5	13	22	38	31.3	1.1	13	45.4	1725.6
15/19	12	2	5	17	43	9.6	0.4	5	15.0	648.1
10/14	8	1	2	12	48	6.4	0.1	2	8.5	409.0
5/9	3	0	1	7	53	2.4	0	1	3.4	180.8
0/4	0	0	0	2	58	0	0	0	0	0
-5/-1	1	0	0	-3	63	0.8	0	0	0.8	50.7
									Total	13,164.5

Heating requirement for January unoccupied hours

= 2478 Btu/h-°F × 13,164.5°F-h (1.31 kW/°C × 7314°C-h) = 32.621 × 10^6 Btu (9560 kWh).

1. **Temperature Bin Observations Below Balance Point**

The balance point temperature is 60°F (15.6°C), and heating is required for ambient temperatures below this point. From Air Force Manual AFM-88-29 (1978) for Albuquerque, enter the bin-hours for the temperature bins (°F) from 55/59 down to -5/-1 as in Table 2-10. Note that all hour groups 02-09, 10-17, and 18-01 are part of the unoccupied hours.

2. **Temperature Differences (Δt) Below Balance Point**

The temperature difference is

Δt = 60°F - (average bin temperature below 60°F)
× [15.6°C - (average bin temperature below 15.6°C)].

Example (continued on next page)

Example (continued)

3. **Fraction of Observations in Hours Groups When Building is Unoccupied**

The unoccupied fraction of observations in the 02-09 hour group is

$$= 1 - \text{fraction in hour group that is occupied (Step G.3)}$$
$$= 1 - (5.5/28) = 22.5/28.$$

Likewise, for the 10-17 hour group

$$= 1 - (5.5/7) = 1.5/7.$$

All the observations in the 18-01 hour group are unoccupied. Multiply the observations by the appropriate fractions for each hour group and enter them as in Table 2-10.

4. **Degree-Hour Determination**

The degree-hours for each bin are obtained by multiplying the total observations for the bin by the Δt between the balance point and the average bin temperature. The total degree-hours for January during unoccupied hours is the sum of all degree hours below the balance point temperature, 60°F (15.6°C).

Step J. Monthly Heating Load During Unoccupied Hours

The monthly heating load is the equivalent UA of the perimeter zone during unoccupied hours multiplied by the total heating degree-hours for the unoccupied hours.

$$\text{Heating Load} = 2478 \text{ Btu/°F-h} \times 13{,}164.5 \text{ °F-h } (1.31 \text{ kW/°C} \times 7314 \text{ °C-h})$$
$$= 32.621 \times 10^6 \text{ Btu } (9560 \text{ kWh}).$$

Step K. Monthly and Annual Heating Requirements

Determine the heating loads for the other months by repeating Steps C through J for February through December. Enter the results as in Table 2-11. Sum all monthly loads to obtain the annual load. The entire procedure and all the intermediate functions calculated for Steps A-K are summarized in Table 2-12.

Table 2-11. Monthly Heating Requirement for Bin-Hour Method Example Design Problem

Month	Heat Requirement [10^6 Btu (kW-h)]					
	Occupied Hours		Unoccupied Hours		Monthly Total	
January	0.06	(19)	32.61	(9560)	32.67	(9579)
February	0.02	(6)	25.18	(7380)	25.20	(7386)
March	0		20.38	(5973)	20.38	(5973)
April	0		9.99	(2928)	9.99	(2928)
May	0		3.77	(1105)	3.77	(1105)
June	0		0.33	(97)	0.33	(97)
July	0		0.01	(3)	0.01	(3)
August	0		0.02	(6)	0.02	(6)
September	0		1.02	(299)	1.02	(299)
October	0		8.14	(2386)	8.14	(2386)
November	0		21.91	(6421)	21.91	(6421)
December	0.01	(3)	32.44	(9508)	32.45	(9511)
Annual	0.09	(28)	155.80	(45,666)	155.89	(45,694)

Example (continued on next page)

Example (concluded)

Table 2-12. Bin-Hour Heating Load Calculation for January

Item	Function	Units	Source
(1)	Zone areas	ft^2 (m^2)	Step A
(2)	Internal heat gain, perimeter zone	Btu/h (kW)	Step B
(3)	Solar heat gain through windows	Btu/h (kW)	Step C
(4)	Total heat gain, perimeter zone	Btu/h (kW)	Step D
(5)	Equivalent UA, perimeter zone	Btu/h-°F (kW/°C)	Step E
(6)	Balance point	°F (°C)	Step F
(7)	Heating degree hours, occupied hours	°F-h (°C-h)	Step G, Table 2-9
(8)	Monthly heating load, occupied hours	Btu (kWh)	Step H, Table 2-9
(9)	Heating degree hours, unoccupied hours	°F-h (°C-h)	Step I, Table 2-10
(10)	Monthly heating load, unoccupied hours	Btu (kWh)	Step J, Table 2-10
(11)	Total monthly heating load	Btu (kWh)	(8) + (10), Table 2-11

Modified Bin Method. ASHRAE Technical Committee 4.7 (Energy Calculations) has developed a simplified procedure for estimating building loads and energy consumption (Knebel 1983).

The modified bin method has the advantage of allowing off-design calculations by use of diversified, rather than peak, load values to establish the load as a function of outdoor dry-bulb temperature. Furthermore, the modified bin method allows the incorporation of HVAC secondary system and plant equipment effects into the energy calculation. This approach permits the user to predict more accurately effects such as reheat and heat recovery that can only be assumed with the degree-day or conventional bin methods.

In the modified bin method, average solar gain profiles, average equipment and lighting-use profiles, and Cooling Load Temperature Difference (CLTD) values (ASHRAE 1985, Ch. 26) are used to characterize the time-dependent diversified loads. The CLTDs approximate the transient effects of building mass. Loads resulting from solar gains through glazings are calculated by determining a weighted-average solar load for a summer day and a winter day, each of average cloudiness and with average solar conditions, and then establishing a linear relationship of this solar load as a function of outdoor ambient temperature.

Several load and energy use microcomputer programs are available in the private sector; for the most part, these programs are based on variations of the modified bin method. A number of these procedures are reviewed by Kusuda (1980).

Graphical Methods. A simplified energy calculation procedure called Energy Graphics (Booz-Allen and Hamilton, n.d.) has recently been developed. It is a steady-state load analysis procedure to which empirical HVAC system and plant performance factors are applied. An advantage of this method is that it uses a graphical display format that allows the designer to examine the simultaneous interaction of hourly load components for typical days for each season.

Another graphical method is the DUBEAM method (Sud et al. 1979). This method uses graphical procedures to simplify the estimation of annual energy requirements for intermediate- and large-sized commercial and institutional buildings. The information required to use these procedures can provide relatively reliable results without laborious manual calculations.

DUBEAM is based on three sets of graphs that correspond to the three elements of any energy estimating method: (1) space load, (2) secondary equipment load, and (3) primary equipment energy requirements. A separate set of graphs is required for each weather zone, building type and size range, and HVAC system type. Parameters used in the graphs include high or low minimum ventilation rates for three types of economizer cycles, high or low interior loads with combination of high or low thermostat settings, and continuous or setback HVAC system operation.

Detailed Simulation Methods

Massive buildings, commercial buildings with complex HVAC systems and controls, and those in the design development or construction document phase require more sophisticated load determination procedures. In such cases, the hourly differences in building thermal capacity (transients); solar loading on the building's envelope; and heat generated internally by lights, equipment, and occupants significantly affect the building energy consumption. The diversity and sophistication of modern energy distribution and control systems, particularly solar energy systems, further contribute to energy consumption differences that can be found only by hourly calculation of the loads on the system and its response to them.

However, some approaches to transient space heating and cooling load determinations approximately include the transient effects by treating select hourly periods (usually one period per season). Some computer programs calculate loads for two conditions on one day in each month and extrapolate for annual loads. Most computer dynamic methods include calculations for the full 8760 h/yr; such methods are recommended for accurate results.

Building heating and cooling loads can be determined using computer dynamic analysis techniques. Several computer programs that predict hourly, monthly, and annual loads on the basis of hourly weather data are available in both the public and private domain. These programs also predict peak monthly and annual loads.

The earliest public domain programs are NBS's NBSLD (Kusuda 1974) and NASA's NECAP (Henniger 1975). NBSLD is strictly a load program and primarily a research tool, whereas NECAP is a full energy analysis program containing load, system, and equipment simulations. Both give good results; but the input is rather involved, and computer run times are long.

More recently, the Lawrence Berkeley Laboratory and the Los Alamos National Laboratory, under DOE sponsorship, developed a fast, easy-to-use, flexible building energy analysis computer program called DOE-2 (1981a,b). (Earlier versions were called Cal-ERDA and DOE-1.) DOE-2 typically operates on large, main-frame computers, but a microcomputer version is now available.

A similar public domain building energy analysis computer program that can be used to calculate hourly, monthly, or annual loads is the BLAST program developed by the U.S. Army Construction Engineering Research Laboratory (Hittle 1979, 1981).

Other programs, some of which are load and some both load and energy analysis programs, are privately available. Some examples are AXCESS, TRACE, ECUBE, and ESP-1. Because some are designed to compute annual heating and cooling energy requirements, they may have to be modified to give requirements on a monthly basis. Bibliographies of building energy analysis and loads programs have been published by ASHRAE (Crall 1981), SERI (1980), and the Electric Power Research Institute (Feldman and Merriam 1979). Assessments and evaluations of loads and energy use procedures are discussed by Kusuda (1980) and specifically for residential building applications by Merriam and Rancatore (1982).

All of the computer dynamic analysis programs require the input of some type of weather data. The more complex programs accept an input of NOAA weather data either in card form or on magnetic tape. These raw data are generally for hourly observation periods before 1965 and for 3-h observation periods after 1976 (except for a few stations that have continued with hourly observations). Although hourly data are available for specific weather years, it is preferable to use edited or composite weather years that are representative of long-term weather conditions for the geographic location in question. SOLMET and Typical Meteorological Year (TMY) data are examples of these edited or composite weather years (Hall et al. 1979); Test Reference Year (TRY) data, which are specific weather years, are also widely used (Arens, Nall, and Carroll 1979). They are available from the National Oceanic and Atmospheric Administration (NOAA). Some programs have a subroutine for processing these data and reducing them to a series of typical days ranging from one per month to one per week, while others use hourly readings directly for one representative year.

The less complex programs use weather data from ASHRAE tables (1985) or bin data from AFM 88-29 (1978). These sources deal with weather conditions averaged over the year, thus reducing the precision of the simulation.

Service Water Heating Load Estimation

Hot water loads vary according to building type, use, and occupancy, but an average hot water demand in residential buildings has been established (Mutch 1974). Furthermore, ASHRAE has published average hot water demand schedules for several types of commercial and/or institutional buildings (see ASHRAE 1984, Ch. 34).

The required temperature rise for the heated water is ($T_{HW} - T_M$), where T_{HW} is the minimum hot water supply temperature and T_M is the temperature of available main supply water. The available supply water temperature can be approximated on a monthly schedule by ground temperatures at the site. Such temperatures are given for selected cities in the *Handbook of Air Conditioning, Heating and Ventilating* (Strock 1959).

Building Energy Requirements

The above discussion has dealt with the thermal *load* imposed upon the building plant for space conditioning or service hot water. The actual *energy requirements* of the building (at the building boundary) will be the energy input to the plant equipment, as it reacts to meet the load, plus the energy input for lights, equipment, and other energy uses. All building energy analysis methods simulate the total operation of the building and its HVAC system, including their reaction to weather, occupancy, and equipment use, by first calculating the loads and then simulating the operation of the plant equipment (including part-load efficiencies) to meet those loads. Finally, the hourly or monthly plant energy consumption is integrated over the year to obtain the annual energy consumption.

One end result in computation is a prediction of energy end use (at the building boundary) expended per building area per year. As a general guideline for a new office building, the Federal Energy Administration has suggested an energy budget target of less than 55,000 Btu/gross ft^2-yr (625,000 kJ/m^2-yr). A realistic energy budget goal for retrofitting existing office buildings is 75,000 Btu/gross ft^2-yr (850,000 kJ/m^2-yr). Buildings designed for solar energy heating and/or cooling should at least meet these guidelines. However, note that buildings are being designed that meet energy budgets of 25,000-30,000 Btu/gross ft^2-yr (285,000-340,000 kJ/m^2-yr).

Chapter 3

Fundamentals: Solar Energy, Climate, and Human Comfort*

SOLAR RADIATION AT THE EARTH'S SURFACE

The Solar Constant

The solar constant is defined as the intensity of solar radiation, between 0.3 and 5 μm wavelength beyond the earth's atmosphere, on a surface normal to the sun's rays at the average earth-sun distance (93 million miles). Until recently, the accepted value was 429.2 Btu/h-ft^2 (of irradiated area) or 1353 W/m^2. Data on which this value is based, however, have recently been reexamined in light of new information obtained with advanced instrumentation. A more probable value is 435.5 Btu/h-ft^2 (1373 W/m^2) (Duffie and Beckman 1980).

The emission of energy by the sun is not precisely constant. Because of this inconstancy and because the earth's orbit is slightly elliptical, the normal incidence intensity on an extraterrestrial surface can vary by ±3% from the solar constant. Also, the variable transmittance of the earth's atmosphere causes variations in the amount of radiation reaching the earth.

Attenuation of Extraterrestrial Solar Radiation

The intensity of solar radiation at the earth's surface varies due to atmospheric scattering by air molecules, water droplets, and dust, and absorption by O_2, ozone (O_3), H_2O, and CO_2. This is illustrated by the solar spectral distribution curve shown in Fig. 3-1. The atmospheric path length is generally expressed in terms of the air mass (m), the ratio of the mass of atmosphere in the actual earth-sun path on a clear day to the mass that would exist at sea level if the sun were directly overhead at sea level (m = 1.0). (See Fig. 3-2.) For all practical purposes, on a clear day $m \cong 1.0/\sin\beta$ at sea level for β values between 70° and 90°. Beyond the earth's atmosphere, m = 0. The annual variation of clear-day direct normal irradiation with solar altitude, β, is given in the ASHRAE Handbook, Applications (ASHRAE 1982, Ch. 57, Fig. 5).

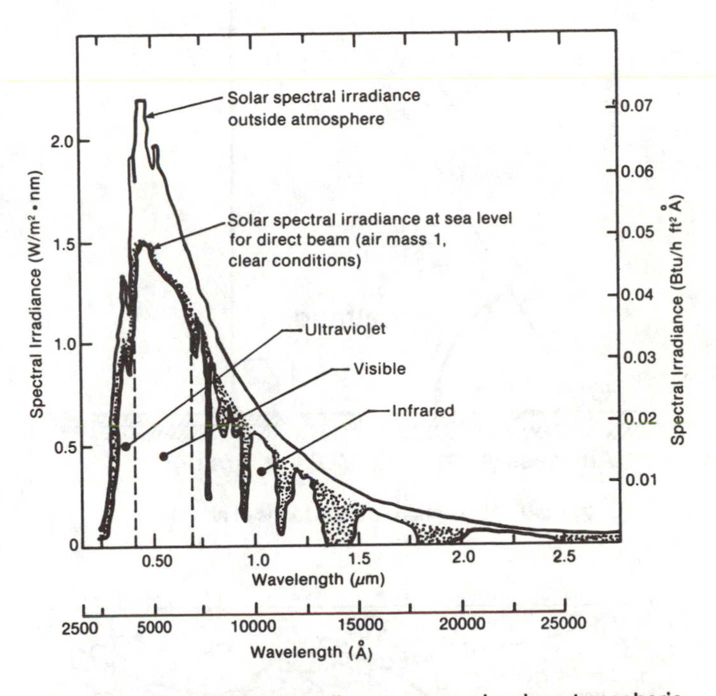

Fig. 3-1. Solar spectral irradiance curves showing atmospheric attenuation of direct radiation (adapted from Kreith and Kreider 1978)

Because local atmospheric water content and elevation vary markedly from the sea level average, the concept of "clearness number" was introduced to express the ratio between the actual clear-day total radiation intensity at a specific location and the intensity calculated given the standard atmosphere at the same location and date.

Figure 3-3 is a map of winter and summer clearness numbers for the continental United States. Total irradiation values given in the ASHRAE Handbook, Applications (ASHRAE 1982, Ch. 57, Tables 2 and 3), and in greater detail in Jordan and Liu (1977, Ch. IV) should be adjusted by the clearness numbers applicable to each location for more reliable clear-day values.

*For an in-depth discussion of the topics introduced here, see Duffie and Beckman (1980), Kreith and Kreider (1978), DOE Facilities Solar Design Handbook (DOE 1978a), Dickinson and Cheremisinoff (1980a), Mazria (1979), ASHRAE Handbook, Applications (ASHRAE 1982) and Jordan and Liu (1977).

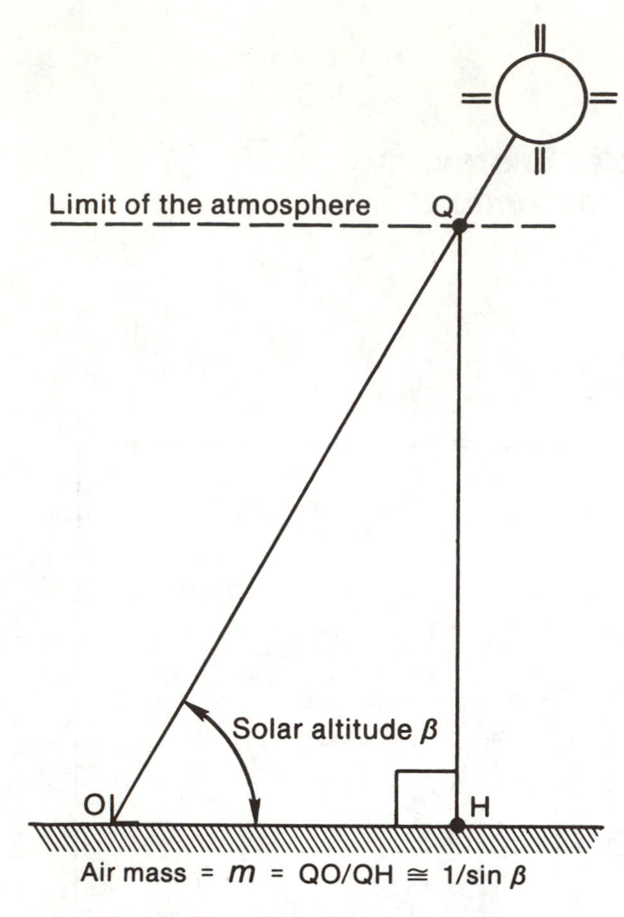

Fig. 3-2. Atmospheric path in terms of air mass

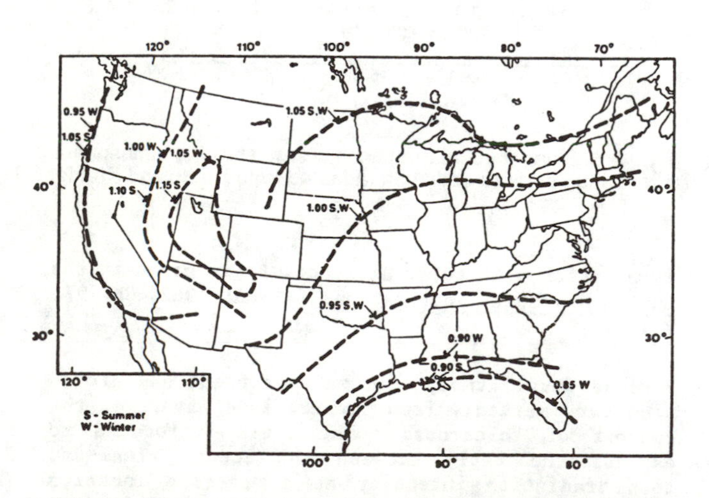

Fig. 3-3. Summer and winter clearness numbers (ASHRAE 1982)

Two types of solar radiation, direct and diffuse, reach the earth's surface. Direct, or beam, radiation comes from the sun without changing direction. Diffuse radiation is received from all parts of the sky (see Fig. 3-4) and is caused by both molecular scattering and diffusion by dust and aerosols. Solar radiation can also reach collecting surfaces by reflection from the foreground or from adjacent structures.

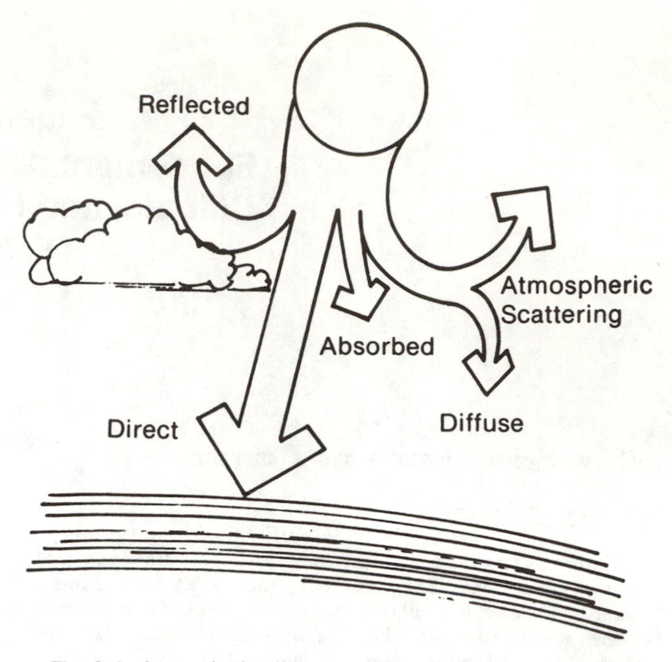

Fig. 3-4. Atmospheric effects on solar radiation (Mazria 1979)

Diffuse Radiation at the Ground

Because atmospheric components scatter part of the solar radiation toward the earth, there is always some diffuse radiation. Even when the sky is clear, the diffuse component may be between 5% and 15% of the total radiation. All radiation reaching the ground is diffuse when the sun is completely obscured by heavy clouds. The distribution of diffuse radiation is highly variable, depending upon the nature, extent, and location of cloud cover. As a first approximation, one can assume that diffuse radiation is isotropic (uniformly distributed) over the sky dome when the sky is either completely clear or completely overcast.

On clear days, the diffuse radiation can be estimated with good accuracy (ASHRAE 1982, Ch. 57; Jordan and Liu 1977, Ch. VI). For more detailed hourly, daily, or monthly values of diffuse radiation on average days, the methods described by Jordan and Liu (1977, Ch. V) should be used. At noon on a typical clear day, a characteristic value of diffuse radiation on a horizontal surface is approximately 30 Btu/h-ft^2 (95 W/m^2) under average annual humidity conditions for the United States.

Hourly, daily, and monthly values of total solar radiation incident on tilted surfaces will be discussed later in this chapter.

A detailed and thorough treatment of terrestrial solar radiation availability is found in Dickinson and Cheremisinoff (1980a, Ch. 5).

SOLAR GEOMETRY

The angle at which a beam of direct solar radiation strikes the earth's surface is influenced by several factors--principally declination δ, latitude L, and hour angle H.

Solar declination is defined as the angle between the earth-sun line at solar noon and the earth's equatorial plane (see Fig. 3-5). Because the earth's rotational axis is tilted at an angle of 23 1/2° from the axis of the plane of the annual orbit around the sun, the solar declination varies from 23 1/2° north of the equator on June 21 to 23 1/2° south of the equator on December 21. The solar declination is 0 at the spring and fall equinoxes, approximately March 21 and September 21.

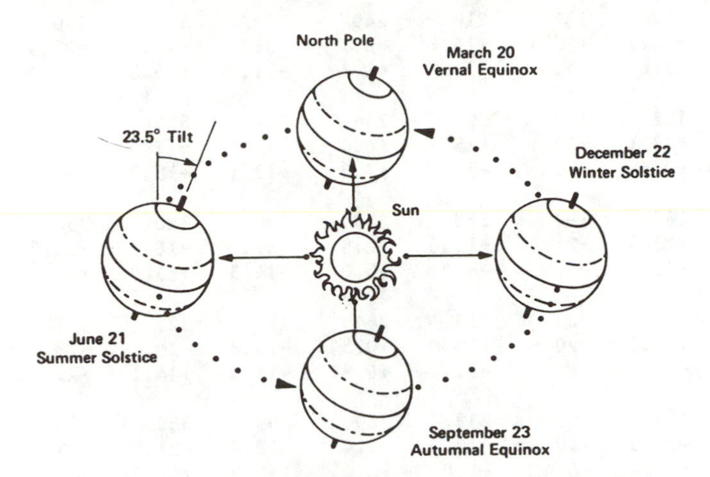

Fig. 3-5. The earth's annual orbit around the sun, showing the seasonal variation in solar declination (Dickinson and Cheremisinoff 1980a)

The influence of latitude on solar altitude, which is measured from the horizontal plane, is illustrated in Fig. 3-6 for the spring and fall equinoxes. At noon (solar time) on March 21 and September 21, the sun appears directly overhead to an observer standing on the equator (β_{noon} = 90°). To someone standing at 60°N latitude, the sun is only 30° above the horizon at noon on those days (β_{noon} = 30°).

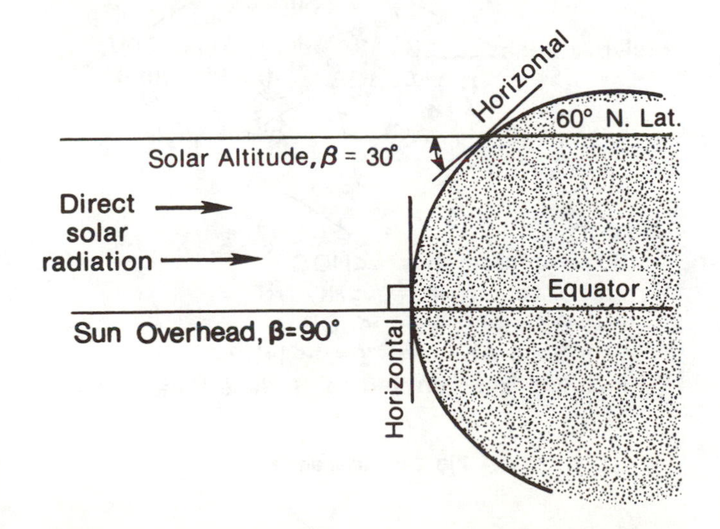

Fig. 3-6. Influence of latitude on noon solar altitude, shown for March 21 and September 21

The hour angle is the product of the earth's angular rate of movement, 15° per hour, and the number of hours from solar noon. At solar noon, H is zero, with mornings positive and afternoons negative. As an example, H = 15° at 11:00 a.m., apparent solar time (AST), and -37.5° at 2:30 p.m. (AST).

AST usually varies from local standard times; the difference can be significant, particularly when Daylight Savings Time is in effect. The procedure for finding AST is

$$\text{AST} = \text{local standard time} + \text{equation of time}^* + 4 \text{ minutes} \times (\text{local standard time meridian} - \text{local longitude}). \quad (3\text{-}1)$$

As an example, find AST for Madison, Wis., given:

10:30 a.m. central standard time
Date: February 1
Longitude: 89.38°.

From Table 3-1, equation of time = -13.7 min.

From Eq. (3-1): $\text{AST} = 10{:}30 - 13.7 + 4(90 - 89.38) = 10{:}19$ a.m.

Another useful quantity is the length of the solar collection day. The usual estimate for the effective winter solar collection day is from 9:00 a.m. to 3:00 p.m. (AST) because the vast majority of the daily solar radiation is received on a tilted collection surface during this period. However, the precise solar day is determined from the sunrise hour angle (H_{sr}), which is defined by the following (Dickinson and Cheremisinoff 1980a, Ch. 3):

$$\cos H_{sr} = -\tan L \tan \delta \quad (3\text{-}2)$$

where L is latitude and δ is declination.

For example, at 36°N latitude on December 21, Table 3-1 shows that δ = -23.4°. Thus

$$\cos H_{sr} = -\tan 36° \tan(-23.4°) = -0.727(-0.433) = 0.315$$
$$H_{sr} = 71.6°.$$

Therefore, the sun rises at this latitude on December 21 at (71.6°/15°) = 4.78 h before solar noon, or at 7:13 a.m. The solar day is thus 9.55 h long. General relationships for the length of the solar day are given in Dickinson and Cheremisinoff (1980a, Ch. 3).

A detailed discussion of solar geometry is given in Duffie and Beckman (1980).

SOLAR ANGLES

Solar angles must be known for determining both the intensity of direct solar radiation incident on a collector at a given time and the shading of an

---

*Equation of time converts standard time into solar time by correcting for longitude and latitude.

Table 3-1. Solar Position Data for 1977 (ASHRAE 1982)

Month Date		Jan	Feb	Mar	Apr	May	Jun	Jul	Aug	Sep	Oct	Nov	Dec
	Year Day	1	32	60	91	121	152	182	213	244	274	305	335
1	Declination	-23.0	-17.0	-7.4	+4.7	+15.2	+22.1	+23.1	+17.9	+8.2	-3.3	-14.6	-21.9
	Eq. of Time	-3.6	-13.7	-12.5	-4.0	+2.9	+2.4	-3.6	-6.2	+0.0	+10.2	+16.3	+11.0
	Year Day	6	37	65	96	126	157	187	218	249	279	310	340
6	Declination	-22.4	-15.5	-5.5	+6.6	+16.6	+22.7	+22.7	+16.6	+6.7	-5.3	-16.1	-22.5
	Eq. of Time	-5.9	-14.2	-11.4	-2.5	+3.5	+1.6	-4.5	-5.8	+1.6	+11.8	+16.3	+9.0
	Year Day	11	42	70	101	131	162	192	223	254	284	315	345
11	Declination	-21.7	-13.9	-3.5	+8.5	+17.9	+23.1	+22.1	+15.2	+4.4	-7.2	-17.5	-23.0
	Eq. of Time	-8.0	-14.4	-10.2	-1.1	+3.7	+0.6	-5.3	-5.1	+3.3	+13.1	+15.9	+6.8
	Year Day	16	47	75	106	136	167	197	228	259	280	320	350
16	Declination	-20.8	-12.2	-1.6	+10.3	+19.2	+23.3	+21.3	+13.6	+2.5	-8.7	-18.8	-23.3
	Eq. of Time	-9.8	-14.2	-8.8	+0.1	+3.8	-0.4	-5.9	-4.3	+5.0	+14.3	+15.2	+4.4
	Year Day	21	52	80	111	141	172	202	233	264	294	325	355
21	Declination	-19.6	-10.4	+0.4	+12.0	+20.3	+23.4	+20.6	+12.0	+0.5	-10.8	-20.0	-23.4
	Eq. of Time	-11.4	-13.8	-7.4	+1.2	+3.6	-1.5	-6.2	-3.1	+6.87	+15.3	+14.1	+2.0
	Year Day	26	57	85	116	146	177	207	238	269	299	330	360
26	Declination	-18.6	-8.6	+2.4	+13.0	+21.2	+23.3	+19.3	+10.3	-1.4	-12.6	-21.0	-23.4
	Eq. of Time	-12.6	-13.1	-5.8	+2.2	+3.2	-2.6	-6.4	-1.8	+8.6	+15.9	+12.7	-0.5

(Units for declination are angular degrees; units for Equation of Time are minutes of time.)

Standard Time Meridians:

Eastern Standard Time	75°W	Pacific Standard Time	120°W
Central Standard Time	90°W	Yukon Standard Time	135°W
Mountain Standard Time	105°W	Alaska-Hawaii Standard Time	150°W

object, surface, or collector. The angular relationships between the incoming beam of solar radiation and a plane, such as a wall surface or solar collector, oriented arbitrarily relative to the earth can be described in terms of several angles. These angles and their interrelationships, shown in Figs. 3-7 and 3-8, can be calculated when the latitude, the date (and hence the declination), and the AST are known.

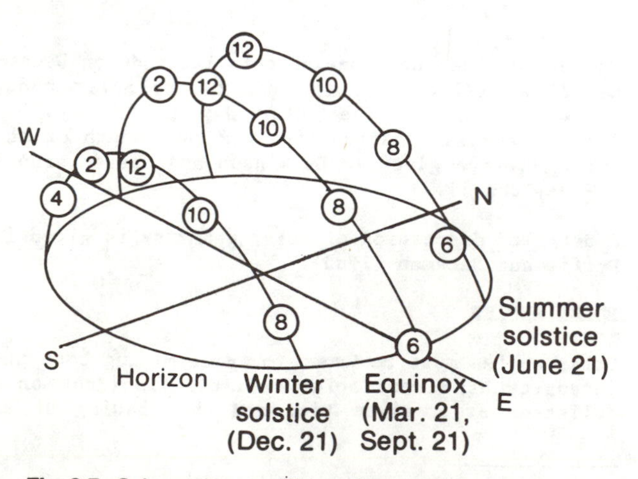

Fig. 3-7. Solar position at different times of day and year

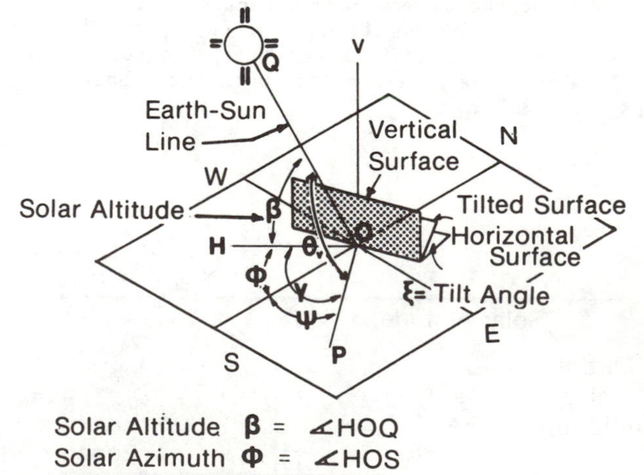

Fig. 3-8. Solar angles

solar altitude (β) = the angle between the horizontal plane and the direct solar beam. At noon,

$\beta = 90° -$ latitude L + declination δ.

At other times of the day,

$\sin \beta = \cos L \cos \delta \cos H + \sin L \sin \delta$.

surface tilt angle (Σ) = the angle between the horizontal and the surface in question.

solar azimuth (ϕ) = the sun's position measured east or west of true (solar) south. At solar noon, $\phi = 0$. At other times of the day,

$\cos \phi = (\sin \beta \sin L - \sin \delta)/(\cos \beta \cos L)$.

wall or surface azimuth (Ψ) = the angle between true south and the normal to the surface, or the angle between true south and the projection of the surface normal in the horizontal plane.

surface-solar azimuth (γ) = the angle between projections onto a horizontal plane of a line to the sun and the normal to the irradiated surface.

incident angle (θ) = the angle between the direct solar beam and a line normal to the irradiated surface. To calculate the angle of incidence for surfaces of any orientation, the following equation is used:

$$\cos \theta = \cos \beta \cos \gamma \sin \Sigma + \sin \beta \cos \Sigma .$$

The incident angle between the direct solar beam and a line normal to the irradiated surface determines the intensity of the direct component striking the surface and the ability of the surface to reflect, transmit, or absorb solar energy. Knowledge of this component is necessary for determining design values of total solar irradiation.

Sun path diagrams show solar altitude and azimuth angles for different latitudes (Mazria 1979; ERDA 1976; Anderson 1977). Figure 3-9 shows a typical sunpath diagram for 36°N latitude. Tables of solar altitude, azimuth, and incident angle are given in Appendix C of the <u>DOE Facilities Solar Design Handbook</u> (DOE 1978a). A detailed treatment of solar angles is presented in Duffie and Beckman

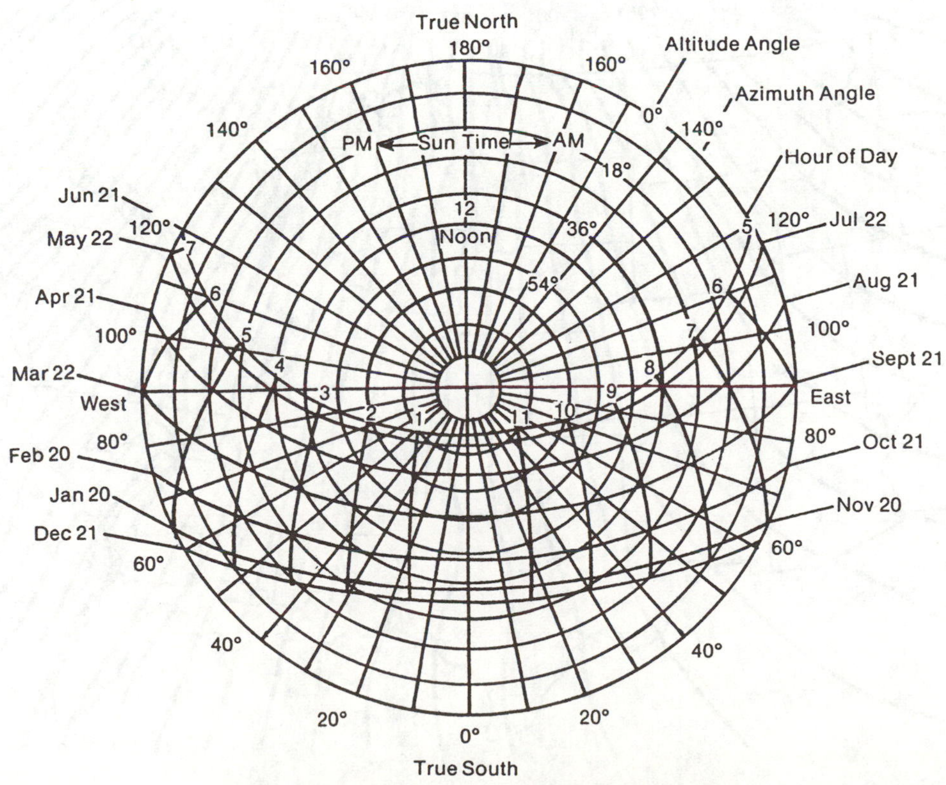

Fig. 3-9. Sun path diagram for 36°N latitude (ERDA 1976)

(1980), and values of all of the important solar angles for surfaces with various tilt angles and orientations are given in Jordan and Liu (1977, Ch. IV).

MAGNETIC DECLINATION

Solar and surface azimuth angles are measured with respect to true south. The most common method of locating south is with a magnetic compass. However, in most parts of the world, magnetic south and true south differ by several degrees. This difference is called magnetic declination and can be read from the isogonic chart (Fig. 3-10). The letters E, for east, and W, for west, indicate the direction that the compass reads off true north. The numbers indicate the angle distance in degrees that the compass reads off true north. For example, Los Angeles has a magnetic declination of about 15°E, indicating that a compass needle points approximately 15° east of true north. Thus, true geographic would be 15° east of magnetic south.

RADIATION INCIDENT ON TILTED SURFACES

Hourly Values of Total Solar Irradiation

Hourly solar irradiation values are required for collector efficiency calculations. In such cases, the total irradiation on a tilted surface is given by

$$I_t = \text{direct} + \text{diffuse} + \text{reflected} \quad (3\text{-}3)$$
$$= I_{DN} \cos\theta + I_{dH} \frac{(1 + \cos\Sigma)}{2} + I_{tH}\rho \frac{(1 - \cos\Sigma)}{2},$$

where

I_{DN} = direct normal solar irradiation, Btu/h-ft^2 (W/m^2)

I_{dH} = diffuse irradiation on a horizontal surface

I_{tH} = total horizontal irradiation

ρ = ground reflectance

This equation is based on the usual assumption of isotropic diffuse radiation from the sky dome; an anisotropic diffuse radiation model is proposed by Klucher (1979). To account for shading, the first two terms should be multiplied by appropriate view factors. The direct component view factor is calculated by shadow-mapping techniques, as discussed in the section on shading by detached objects. The diffuse component view factor is calculated as discussed in the section on adjustments for shading from attached surfaces.

The intensity of reflected radiation depends on the nature of the reflecting surface and the incident angle. Year round, a reasonable, conservative ground reflectance value is 0.20. Values for various characteristic landscapes are given in Hunn

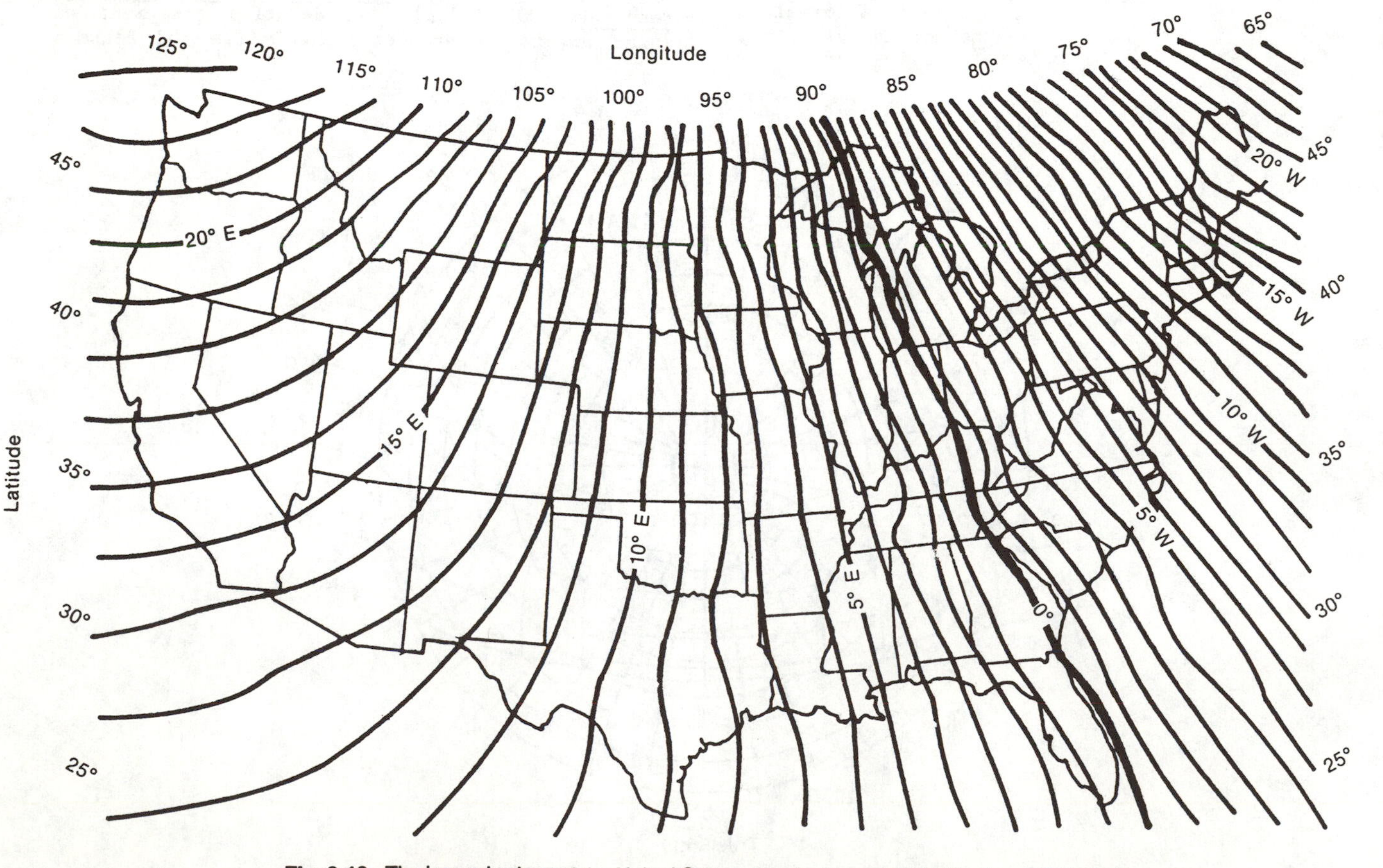

Fig. 3-10. The isogonic chart of the United States showing magnetic declination (DOC 1965)

and Calafell (1977) and the ASHRAE Handbook of Fundamentals (ASHRAE 1985, Ch. 27, Table 12).

Hourly values of direct normal and total horizontal radiation are available for selected cities in SOLMET magnetic tape format from the National Oceanic and Atmospheric Administration (NOAA), National Climatic Center, Asheville, N.C. Although less accessible, these values are more accurate than the correlational and tabulated values described below.

Lacking readily available measured direct and diffuse irradiation data on an hourly basis, one can use empirical methods to estimate the direct and diffuse components from total horizontal data. The most widely used technique is that developed by Jordan and Liu (1977, Ch. V) (correlations for horizontal and tilted surfaces are included). The basis of this technique is an empirical relationship between the daily diffuse fraction and the atmospheric attenuation coefficient (cloudiness index) K_T calculated on a daily (monthly average) basis. This relationship is usually assumed to apply on an hourly basis; however, a study by Orgill and Hollands (1977) was used to develop a refined correlation based on measured hourly data. Erbs, Klein, and Duffie (1982) validated this refined correlation with recent measurements from four U.S. stations (see Fig. 3-11). Dickinson and Cheremisinoff (1980a, Ch. 5) summarize and present an application of this technique and other related ones.

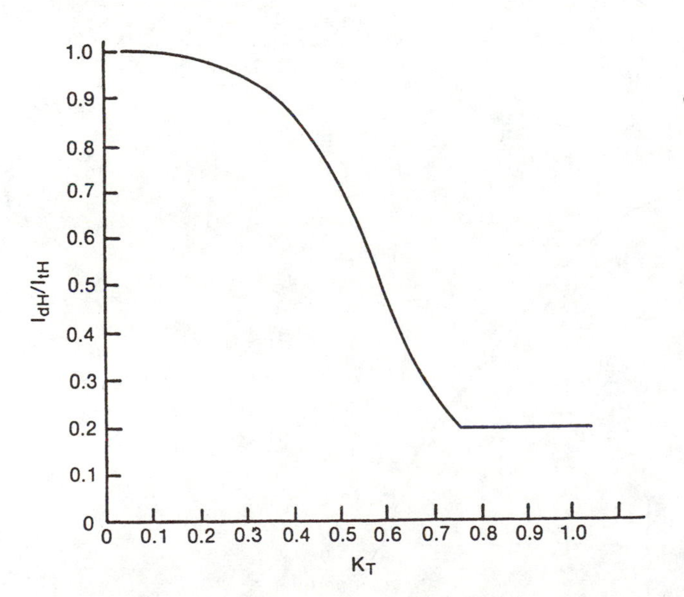

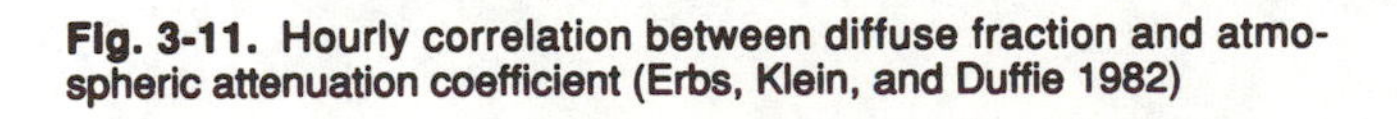

Fig. 3-11. Hourly correlation between diffuse fraction and atmospheric attenuation coefficient (Erbs, Klein, and Duffie 1982)

Hourly solar irradiation can also be estimated by using Table 2 in the ASHRAE Handbook, Applications (1982, Ch. 57) or Ch. IV of Jordan and Liu (1977), which show total hourly and daily clear-day irradiation on the 21st of each month at selected latitudes and tilt angles. These values do not include any reflected component. The solar irradiation values listed, when corrected by the clearness number, can be used as conservative estimates of clear-day irradiation. To account for cloud cover, the hourly clear-day values in Appendix C must be reduced by the hourly "cloud cover factor." The cloud indexes to be used with the cloud cover factor are the "total sky cover" (or mean sky cover) and the amount of cover by three different types of clouds (see Kimura and Stephenson 1969).

Mean Daily and Monthly Average Solar Irradiation

Most simplified solar heating and cooling system design methods require monthly or annual total radiation incident on the collector. Measured solar irradiation data are preferred but usually are not available. However, other meteorological measurements and solar data can be used to estimate solar irradiation. These data produce calculated values of radiation incident on surfaces oriented at various tilt angles. Most often, monthly average values are obtained from mean daily values for a characteristic day of the month.

Characteristics

Characteristics of clear-day solar irradiation incident on surfaces tilted at various angles are shown in Fig. 3-12. Note that on a clear day a

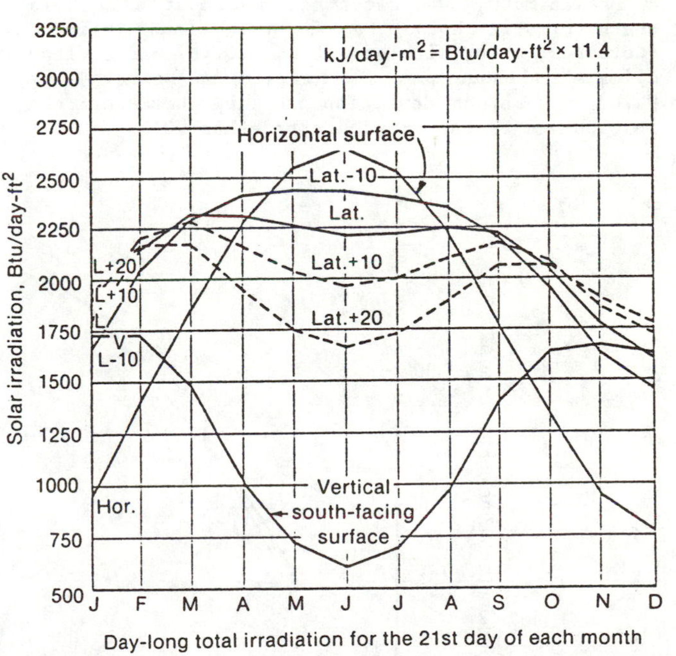

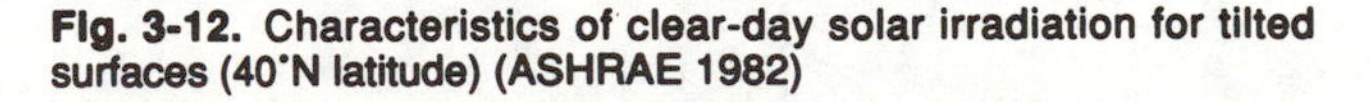

Fig. 3-12. Characteristics of clear-day solar irradiation for tilted surfaces (40°N latitude) (ASHRAE 1982)

vertical south-facing surface (such as a wall) receives the most energy in the winter and about half that amount in the summer, hence its effectiveness for heating season applications. In contrast, a horizontal surface (such as a roof) receives about twice the radiation during the summer as it receives during the winter. A south-facing surface tilted at the local latitude receives fairly uniform radiation throughout the year. In fact, from a solar radiation standpoint, a tilt angle of latitude plus 5° is optimal for service hot water applications where the monthly load is constant throughout the year (ERDA 1976).

The low winter solar altitude angles favor vertical surface irradiation, whereas the high altitude angles in summer result in little radiation on a vertical surface, substantial radiation on east- and west-facing surfaces, and a large amount of radiation on a horizontal surface. Therefore, horizontal, near-horizontal, west-facing, or east-facing windows are sources of usually unwanted solar heat gain in the summer. (See Fig. 3-7.)

Figure 3-13 shows the characteristic variation in incident solar radiation with azimuth for a vertical surface. These measured data were normalized by the incident solar radiation for a south-facing surface (Balcomb et al. 1982). Two cities are shown, each characterized by the annual atmospheric attenuation coefficient, K_T. K_T, which is defined as the ratio of the total radiation on a horizontal surface to the extraterrestrial radiation on a horizontal surface at the same time and latitude, is a measure of annual cloudiness at each location. The figure shows that for a 30° azimuth from solar south, the incident radiation decrement is 5%; for a 60° azimuth, the decrement is about 17%. This characteristic holds for both a cloudy climate (Columbus, Ohio, with annual $K_T = 0.46$) and a clear climate (Albuquerque, N. Mex., with annual $K_T = 0.70$). A slight deviation in this characteristic is observed as the azimuth approaches 90°.

SOLMET Measured Data

The most reliable and widely used mean daily solar irradiation data are those developed by NOAA from the SOLMET data (Cinquemani, Owenby, and Baldwin 1978). Measured and derived data for total solar irradiation on a horizonal surface for 248 U.S. stations are included. They are average values based on 25-year records. In addition, measured or calculated long-term average total horizontal irradiation and attenuation coefficients, K_T, are given for these same stations in the Insolation Data Manual (Knapp, Stoffel, and Whitaker 1980).

Where data for the stations mentioned above do not cover a particular location, monthly (mean daily) and annual radiation incident on a horizontal surface H_{MH} can be estimated from maps in the Climatic Atlas of the United States (DOC 1968); more recent data are shown on the maps published in the Solar Radiation Atlas (Hulstrom 1981). Note that the irradiation variations within a region, especially near bodies of water, can be up to 30% within a distance of 100 miles (161 km) (Atwater, Ball, and Brown 1976). For a given month, the site in question is located on the map, and the radiation in langleys per day is interpolated. Multiplying this number by 3.69 Btu/ft^2 (42.0 kJ/m^2)-langley gives the horizontal surface total solar irradiation, $\overline{H}_H$, in Btu/ft^2-day (kJ/m^2-day). If $\overline{H}_H$ is multiplied by N, the number of days in the month, the monthly solar irradiation is determined in Btu per square foot per month (kJ/m^2-mo).

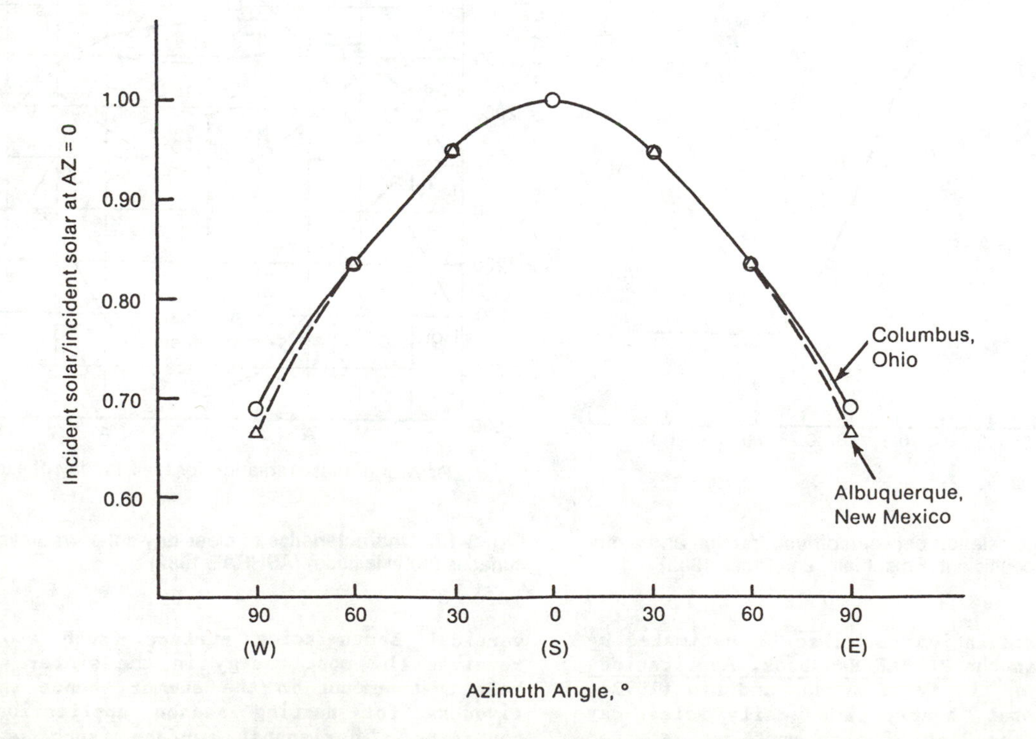

Fig. 3-13. Variation in incident solar irradiation with azimuth angle for vertical surface during heating season (September - May)

Correction for Tilt Angle

Mean daily, monthly, or annual values can be adjusted for the tilt of the collector by using any of several procedures. The Los Alamos National Laboratory has developed a correlation for horizontal solar irradiation $H_{M\Sigma}$ data available on a monthly basis (ERDA 1976). The total monthly radiation $H_{M\Sigma}$ on a surface tilted at an angle of latitude plus 10° is given by

$$H_{M\Sigma} = 1.025d - 8200 \quad (3\text{-}4a)$$

where

$$d = \frac{H_{MH}}{\cos\,(L - \text{declination at midmonth})} \quad (3\text{-}4b)$$

and H_{MH} is the horizontal solar irradiation in Btu/ft^2-mo.

Declination at midmonth is approximately 23.45°cos(30 M - 187), where M = month number (January = 1, December = 12). Table 3-2 gives expressions for $H_{M\Sigma}$ as a function of latitude for each month. Note that this correlation loses accuracy at more northern latitudes and for tilt angles other than latitude plus 10°. For latitudes below 34° and above 46°, use Eqs. (3-4a) and (3-4b).

As an example of this calculation, let us assume that the March monthly irradiation on a horizontal surface is 50,000 Btu/ft^2-mo (5.67 × 10^5 kJ/m^2-mo) and the latitude at the site is 40°N. For this case, d in Eq. (3-4b) is

$$d = \frac{50{,}000 \text{ Btu/ft}^2\text{-mo}}{\cos\,(40^0 - \text{declination at midmonth})}$$

where declination at midmonth = 23.45° cos (30 × 3 - 187) = -2.86°. Thus

$$d = 50{,}000/\cos\,(40° + 2.86°) = 68{,}211$$

and

$$H_{M\Sigma} = 1.025\,(68{,}211) - 8200$$
$$= 61{,}716 \text{ Btu/ft}^2\text{-mo } (7.00 \times 10^5 \text{ kJ/m}^2\text{-mo})$$

on a surface tilted at an angle equal to latitude +10°, or 50°, from the horizontal. Using Table 3-2 we get $H_{M\Sigma}$ = 1.3983 (50,000) - 8200 = 61,715 Btu/ft^2-mo, which is essentially the same result.

The U.S. Army Construction Engineering Research Laboratory (CERL) has developed a simple empirical correlation for converting from horizontal surface to tilted surface solar irradiation on an annual basis (Hittle, Holshouser, and Walton 1976). The radiation incident on a south-facing surface tilted at an angle of approximately latitude minus 10° from the horizontal (optimal for year-round active systems combined heating and cooling performance) is given by

$$H_{A\Sigma} = \frac{H_{AH}, \text{Btu/ft}^2\text{-yr}}{\cos\,(L - 8)} \quad (3\text{-}5)$$

where L is the site latitude in degrees.

Correlation Methods

Several methods have been developed for calculating the solar irradiation on a tilted surface using correlations of measured data. One common method described in Passive Solar Heating Analysis (Balcomb et al. 1984) uses SOLMET data from 26 U.S. cities recorded on Typical Meteorological Year (TMY) tapes. From these data, a least-squares correlation of monthly radiation quantities has been developed as a function of a solar-position parameter (latitude minus midmonth declination) and the atmospheric attenuation coefficient K_T. Correlations of surface incident-to-horizontal ratios are presented for tilt angles of 90°, 60°, and 30° and for azimuth angles of 0°, 30°, 60°, and 90°. Sky diffuse radiation (assuming an isotropic sky) and ground-reflected radiation (assuming a reflectance of 0.3) are included. The figures in the

Table 3-2. Tilted Surface Solar Irradiation Correction Factors (DOE 1978a)

$$H_{M\Sigma} = (C)(H_{MH}) - 8200,$$

where

C = values shown below.

H_{MH} = horizontal solar irradiation in Btu/ft^2-mo

Month	Latitude 34	36	38	40	42	44	46
1	1.8135	1.9122	2.0247	2.1541	2.3041	2.4799	2.6882
2	1.5349	1.5984	1.6691	1.7486	1.8384	1.9404	2.0570
3	1.2811	1.3163	1.3552	1.3983	1.4460	1.4990	1.5580
4	1.1295	1.1487	1.1701	1.1938	1.2199	1.2488	1.7945
5	1.0625	1.0734	1.0858	1.0999	1.1158	1.1335	1.1532
6	1.0433	1.0508	1.0598	1.0703	1.0823	1.0959	1.1112
7	1.0496	1.0583	1.0685	1.0803	1.0937	1.1088	1.1256
8	1.0900	1.1046	1.1210	1.1393	1.1597	1.1822	1.2072
9	1.1976	1.2241	1.2535	1.2858	1.3216	1.3611	1.4047
10	1.4053	1.4537	1.5075	1.5675	1.6345	1.7096	1.7945
11	1.6925	1.7750	1.8683	1.9746	2.0963	2.2369	2.4010
12	1.8960	2.0062	2.1327	2.2793	2.4506	2.6533	2.8963

handbook are easily used with tables of measured SOLMET values of total horizontal irradiation for 248 U.S. stations also published in the Insolation Data Manual (Knapp, Stoffel, and Whitaker 1980).

Another widely used method for computing solar irradiation on tilted surfaces is that of Liu and Jordan (1963) as modified by Klein, Beckman, and Duffie (1977). This method uses an empirical correlation of the ratio of monthly average daily diffuse radiation to total horizontal radiation ($\bar{H}_d/\bar{H}_H$) as a function of the monthly average daily atmospheric attenuation coefficient, $\bar{K}_T$:

$$\frac{\bar{H}_d}{\bar{H}_H} = 1.39 - 4.03\,\bar{K}_T + 5.53\,\bar{K}_T^2 - 3.11\,\bar{K}_T^3 \quad (3\text{-}6)$$

To obtain the monthly average daily radiation on a tilted surface, $\bar{H}_\Sigma$, the following equation is used:

$$\frac{\bar{H}_\Sigma}{\bar{H}_H} = \left(1 - \frac{\bar{H}_d}{\bar{H}_H}\right) R_D\,\bar{f}_i + \frac{\bar{H}_d}{\bar{H}_H}\left(\frac{1 + \cos\Sigma}{2}\right) + \rho\left(\frac{1 - \cos\Sigma}{2}\right) \quad (3\text{-}7)$$

where

$\bar{H}_H$ = monthly average daily total horizontal radiation

R_D = ratio of monthly average daily beam radiation on the tilted surface to that on a horizontal surface

$\bar{f}_i$ = monthly average direct radiation on a shaded receiver to that on an unshaded receiver (calculated as discussed in the section on adjustments for shading from attached surfaces)

Σ = surface tilt angle

ρ = ground reflectance

Tables of $\bar{H}_\Sigma$ for a range of values of $\bar{K}_T$ for collectors at various tilt angles are given in Klein, Beckman, and Duffie (1977). To correct for shading by an overhang or side fins, the diffuse surface sky view factor, $(1 + \cos\Sigma)/2$ can be modified by the methods discussed in the section on adjustments for shading from attached surfaces. Ground-reflected radiation can be treated as in the hourly solar irradiation discussion above. Typical values of monthly average daily irradiation on a tilted surface produced by this method are presented in Table 3-3. Note that these daily values must be multiplied by the number of days in a given month to obtain monthly values.

Approximate Methods Using Calculated Irradiation

A more approximate method for calculating irradiation on a tilted surface is to start with the calculated clear-day mean daily values for several tilt angles plus horizontal. This method is given in the ASHRAE Handbook, Applications (1982, Ch. 57). These tilted-surface values should be multiplied by the ratio of measured horizontal solar irradiation (Cinquemani, Owenby, and Baldwin 1978; Knapp, Stoffel, and Whitaker 1980) to calculated clear-day horizontal solar irradiation in order to account for cloud cover. This estimate then should be further adjusted by applying the clearness number for the site and adding a factor for the ground-reflected irradiation.

Alternatively, mean daily calculated clear-day values from ASHRAE sources (ASHRAE 1982, Ch. 57; Jordan and Liu 1977, Ch. IV) can be adjusted for cloud cover using cloud cover factors, such as the monthly "percent possible sunshine" from the Climatic Atlas (DOC 1968; ITT 1977). A word of caution: when percent possible sunshine is low, the calculated average daily irradiation values are far below measured values; however, when "percent possible sunshine" is above 50%, the correlation is considerably better (Bennett 1969).

Adjustments for Shading from Attached Surfaces

To obtain the total radiation incident on a south-facing vertical surface that is shaded by an opaque overhang, the direct and diffuse radiation components must be multiplied by appropriate shading and view factors, respectively. Consider first the sky view factor for the diffuse component.

If one assumes that the diffuse radiation is isotropic, an unshaded, vertical surface has a view factor to the sky of $F_{rs} = (1 + \cos\Sigma)/2 = (1 + \cos 90)/2 = 1/2$. Modification of F_{rs} for shading by a fixed overhang is determined with reference to Fig. 3-14. If the overhang extends beyond the surface for a distance E on each side,

Table 3-3. Monthly Average Daily Radiation on a Tilted Surface (Klein, Beckman, and Duffie 1977)

$\bar{H}_\Sigma$(Btu/ft^2-day) for K_T = 0.4, Σ = Latitude + 15°

Lat	Jan	Feb	Mar	Apr	May	Jun	Jul	Aug	Sep	Oct	Nov	Dec
20	1173	1219	1245	1215	1178	1144	1151	1200	1219	1217	1182	1160
25	1136	1178	1215	1215	1178	1143	1158	1186	1206	1181	1141	1038
30	1008	1130	1173	1193	1171	1147	1157	1177	1171	1141	1015	979
35	948	1089	1142	1175	1156	1144	1149	1159	1137	1093	965	923
40	890	1059	1095	1148	1147	1135	1135	1134	1103	1050	909	863
45	837	928	1060	1111	1132	1121	1129	1113	1066	1012	855	846
50	709	878	1013	1088	1117	1102	1103	1085	1033	900	816	686
55	666	821	971	1044	1086	1094	1088	1061	990	854	680	661
60	516	780	928	1009	1058	1072	1058	1030	950	802	652	515

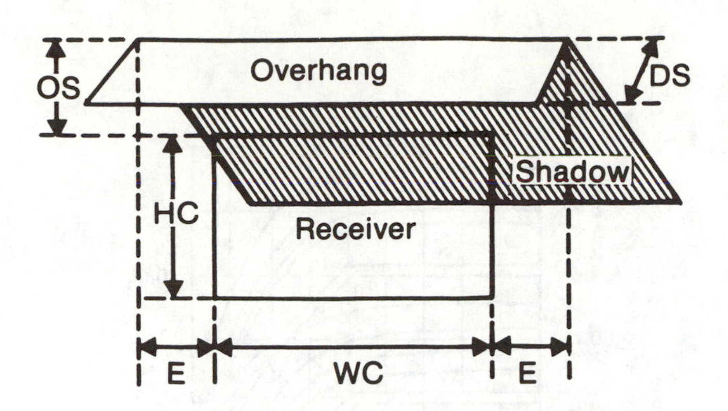

Fig. 3-14. Shading of a receiver surface by an overhang

and all dimensions are normalized by the height of the receiving surface, then the following relationships are defined: relative width, w = WC/HC; relative extension, e = E/HC; relative gap, g = OS/HC; and relative projection, p = DS/HC. Values of F_{rs} for various values of these parameters have been calculated by Utzinger and Klein (1979). Table 3-4 presents the results.

Because of the varying position of the sun, determining shading factors for the direct radiation component is more difficult. Fortunately, values have been calculated for $\overline{f_i}$, the ratio of direct radiation received by a shaded surface in a month to the direct radiation received by an unshaded surface in the same month (Utzinger and Klein 1979). This shading factor is multiplied by the unshaded surface direct component for the month to obtain the direct component for the shaded surface. Figures of monthly values of f_i for 35°, 45°, and 55°N latitude, relative widths of 1, 4, and 25, and relative gap widths of 0, 0.2, and 0.4 are given by Utzinger and Klein (1979). In addition, the effect of surface azimuths up to 45° from south is presented for two U.S. cities.

The effects of side fins are also important, particularly for east- or west-facing surfaces. Shading by overhangs will reduce the solar gains somewhat, although often only on the order of one-half. Side fins can reject much of the direct radiation in the summer while still allowing much of the beam radiation to reach the surface during the winter. Jones and Worley (1982) have developed a method for calculating monthly average radiation on surfaces shaded by relatively tall vertical fins. The method works for arbitrary azimuth angles and fin orientations. In general, side fins are preferable to overhangs for surfaces having azimuth angles between ± 45° and ± 90° (ESE to E and WSW to W).

Glazing Reflectance and Absorptance Effects

For many design calculations, one needs to know the amount of solar radiation transmitted through a glazing system, thus ultimately absorbed by the collection aperture (for a detailed discussion see Duffie and Beckman 1980). The amount of radiation transmitted through a transparent medium depends on the incident angle θ and the optical properties of the medium. The fraction of the incident radiation

Table 3-4. Receiver Radiation View Factor of the Sky, F_{rs} (Utzinger and Klein 1979)

			p								
e	g	w	0.10	0.20	0.30	0.40	0.50	0.75	1.00	1.50	2.00
0.00	0.00	1.0	0.46	0.42	0.40	0.37	0.35	0.32	0.30	0.28	0.27
		4.0	0.46	0.41	0.38	0.35	0.32	0.27	0.23	0.19	0.16
		25.0	0.45	0.41	0.37	0.34	0.31	0.25	0.21	0.15	0.12
	0.25	1.0	0.49	0.48	0.46	0.45	0.43	0.40	0.38	0.35	0.34
		4.0	0.49	0.48	0.45	0.43	0.40	0.35	0.31	0.26	0.23
		25.0	0.49	0.47	0.45	0.42	0.39	0.34	0.29	0.22	0.18
	0.50	1.0	0.50	0.49	0.49	0.48	0.47	0.44	0.42	0.40	0.38
		4.0	0.50	0.49	0.48	0.46	0.45	0.41	0.37	0.31	0.28
		25.0	0.50	0.49	0.47	0.46	0.44	0.39	0.35	0.27	0.23
	1.00	1.0	0.50	0.50	0.50	0.49	0.49	0.48	0.47	0.45	0.43
		4.0	0.50	0.50	0.49	0.49	0.48	0.46	0.43	0.39	0.35
		25.0	0.50	0.50	0.49	0.48	0.47	0.44	0.41	0.35	0.30
0.30	0.00	1.0	0.46	0.41	0.38	0.35	0.33	0.28	0.25	0.22	0.20
		4.0	0.46	0.41	0.37	0.34	0.31	0.26	0.22	0.17	0.15
		25.0	0.45	0.41	0.37	0.34	0.31	0.25	0.21	0.15	0.12
	0.25	1.0	0.49	0.48	0.46	0.43	0.41	0.37	0.34	0.30	0.28
		4.0	0.49	0.47	0.45	0.42	0.40	0.34	0.30	0.24	0.21
		25.0	0.49	0.47	0.45	0.42	0.39	0.33	0.29	0.22	0.18
	0.50	1.0	0.50	0.49	0.48	0.47	0.45	0.42	0.39	0.35	0.33
		4.0	0.50	0.49	0.48	0.46	0.44	0.40	0.36	0.30	0.26
		25.0	0.50	0.49	0.47	0.46	0.44	0.39	0.34	0.27	0.22
	1.00	1.0	0.50	0.50	0.49	0.49	0.48	0.47	0.45	0.42	0.40
		4.0	0.50	0.50	0.49	0.48	0.48	0.45	0.43	0.38	0.34
		25.0	0.50	0.50	0.49	0.48	0.47	0.44	0.41	0.35	0.30

that is reflected is given by the Fresnel Equation (Duffie and Beckman 1980) as a function of incident angle. For normal incidence the reflectance is given by

$$\rho(0) = \left[\frac{n_1 - n_2}{n_1 + n_2}\right]^2 \qquad (3\text{-}8)$$

where n_1 and n_2 are the indices of refraction for the two media. Normally, one medium is air ($n_1 = 1$), and the other is glass ($n_2 \approx 1.5$). For this case $\rho(0) \approx 0.04$; thus 4% of the incident direct radiation is reflected at normal incidence.

For multiple glazing systems and for light incident on the glazing at an angle θ, each glazing reflects radiation and also interacts with the other glazings. For single or multiple glazing, all of the same material, the transmittance, taking reflectance into account, is

$$\tau_r = \frac{1 - \rho}{1 + (2n - 1)\rho} \qquad (3\text{-}9)$$

where n is the number of glazings, and ρ is

$$\rho = \frac{\sin^2(\theta_1 - \theta_2)\ \tan^2(\theta_1 - \theta_2)}{2\ \tan^2(\theta_1 + \theta_2)} \qquad (3\text{-}9a)$$

Transmitted radiation is also diminished by absorption in the glazing material. Transmittance, taking absorptance into account, is

$$\tau_a = e^{-nKt/\cos\theta_2} \qquad (3\text{-}10)$$

where

n = number of glazings

K = extinction coefficient [varies from 0.1/in. (0.04/cm) for good, low-iron glass to 0.8 (0.32) for poor, high-iron glass]

t = thickness of one glazing

θ_2 = refracted angle, from Snell's Law

The overall transmittance is simply the product of the reflective and absorptive transmittances, or $\tau = \tau_r\ \tau_a$. A plot of the transmittance of one, two, three, and four glazed systems as a function of incident angle is given in Fig. 3-15 for three glass extinction factor (Kt) values.

For heating purposes the net transmitted radiation is usually absorbed and then reradiated or stored. For absorber surfaces positioned directly behind the glazing the effective transmittance-absorptance product is given by

$$(\tau\alpha)_e = \frac{\tau\alpha}{1 - (1 - \alpha)\ \rho_d} \qquad (3\text{-}11)$$

where

α = absorptance of the absorber surface, assumed to be independent of direction (for flat black paint, $\alpha \approx 0.95$)

ρ_d = diffuse reflectance of the glazing system (for one, two, three, and four glazings, ρ_d is approximately 0.16, 0.24, 0.29, and 0.32, respectively.)

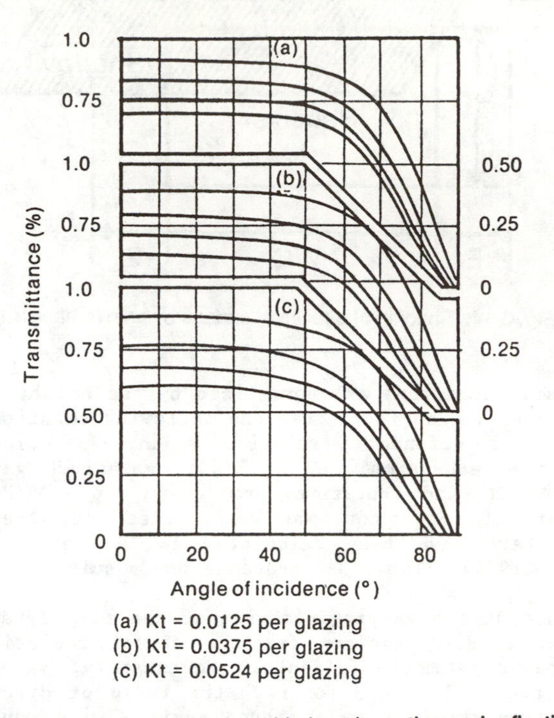

Fig. 3-15. Transmittance (considering absorption and reflection) of one, two, three, and four glazings for three glass extinction factor (Kt) values (Duffie and Beckman 1980)

The effective absorptance for certain passive solar systems, where the absorber is located some distance from and not parallel to the glazing, is often assumed to be unity. This is not always the case, as discussed below.

In addition to the correlation curves for incident-to-horizontal radiation for various surface tilts and orientations presented in the Passive Solar Design Handbook (Balcomb et al. 1982) correlations are also given for radiation transmitted through one, two, and three glazings. Corrections are given for a range of extinction coefficients and diffuse transmittances. Also presented are monthly correlations for the effective absorptance of several types of passive solar systems as a function of the number of glazings, the extinction coefficient, and the absorptance of the thermal storage mass (usually a wall or a floor). These absorption factors give the fraction of the monthly transmitted radiation absorbed by the building.

CLIMATOLOGY

The various climates in which we build are characterized by four major elements: temperature, wind, humidity, and thermal radiation (solar and infrared). Human comfort is directly influenced by these elements; therefore, they should be considered throughout the design process and eventually reflected in the built environment. Air movement (wind) for convective cooling, moisture addition for evaporative cooling, and solar radiation for radiant heating all affect human comfort in varying degrees.

The overall objective of identifying a climatological data base is to provide the designer with both general and specific weather data to promote climate-responsive design. Data should be presented in a form that communicates the essence of a specific climate to the designer. An excellent example of such a graphic technique is the Climatic Data Base series published by the Tennessee Valley Authority (TVA 1980). The phasing of climatic data needs for building design is shown on the design phase chart in Fig. 3-16. Various data types and formats are needed, depending on the design phase.

Design phase	Predesign	Schematic	Design development	Construction documents
Types of climatic data needed	Trends, general climatic patterns, seasonal typical day profiles.	Monthly averages and summaries, daily profile and peak analysis (means and ranges), microclimatic characteristics, solar access characteristics.	Design peak analysis, specific conditions, long-term summaries.	

Fig. 3-16. Climatic data needs for design

Dry-Bulb Temperature

Data on dry-bulb temperature, a primary determinant of building heat loss and heat gain, are summarized by month in terms of degree-days, average ambient temperature, or maximum and minimum design temperature. Design data for major cities throughout the United States and other nations are available in the ASHRAE Handbook of Fundamentals (ASHRAE 1985, Ch. 24). Maximum and minimum temperatures, average monthly temperatures, and long-term averages of degree days are given by Cinquemani, Owenby, and Baldwin (1978). Heating degree-days at 50°, 55°, 60°, and 65°F (10°, 13°, 16°, and 18°C) base temperatures are tabulated in the DOE Passive Solar Design Handbook (Balcomb et al. 1982) and national climatic data from the U.S. Department of Commerce (DOC 1973).

In addition to average weather conditions, diurnal and seasonal characteristics are the basis of the data base. Because of the varying pattern of heating and cooling loads, internal gains, and occupancy, it is important to know the fluctuational characteristics and persistence of weather patterns. Furthermore, patterns of coincidental occurrence are important. For example, the coincidental occurrence of high dry- and wet-bulb temperatures is a more appropriate determinant of discomfort than is dry-bulb temperature alone. Likewise, the simultaneous occurrence of low dry-bulb temperatures and high wind speeds produces high heating loads.

Extensive data on design values of dry-bulb temperatures and coincident wet- and dry-bulb temperatures for summer conditions are given in the ASHRAE Handbook of Fundamentals (ASHRAE 1985, Ch. 24). Similar data in 5°F (2.8°C) temperature intervals (bins), as well as heating and cooling degree days, are available in Air Force Manual 88-29 (1978) for many stations. The National Weather Service provides local monthly and annual climatic data summaries for many cities throughout the United States. These contain detailed data on cloud cover, wind regimes, percent possible sunshine, wet- and dry-bulb temperatures, dew point temperatures, and percent relative humidities. These are available from the National Climatic Center in Asheville, N.C.

Sol-Air Temperature

To account for the combined effects of radiation and convection on a sunlit opaque surface, the concept of the sol-air temperature has been devised (ASHRAE 1985, Ch. 26). This concept is most useful for calculating the load resulting from heat gain through exterior roofs and walls.

The sol-air temperature is an equivalent temperature of outdoor air that includes the effects of incident solar radiation, emitted infrared radiation, and convection to the air. An energy balance on a sunlit opaque surface results in the following expression for the sol-air temperature (t_{sa}):

$$t_{sa} = t_o + (\alpha I_t/h_o) - (\varepsilon \Delta R/h_o), \qquad (3\text{-}12)$$

where

t_o = outdoor air temperature, °F

α = surface absorptance for solar radiation

I_t = total solar radiation incident on the surface, (Btu/h-ft^2)

h_o = convective heat transfer coefficient at the surface, (Btu/h-ft^2-°F)

ε = hemispherical emittance of the surface

ΔR = difference between emitted and incident longwave radiation at the surface, (Btu/h-ft^2)

It is common practice to assume that for horizontal surfaces that exchange radiation with the sky but not the ground, $\Delta R \approx 20$ Btu/h-ft^2 (63 W/m^2). If $\varepsilon = 1$ and $h_o \approx 3.0$ Btu/h-ft^2-°F (17 W/m^2-°C), then for horizontal surfaces $\Delta R/h_o$ = 7°F (-14°C). For vertical surfaces that see both the cold sky and warm ground, $\Delta R \approx 0$. Therefore, for vertical surfaces $\varepsilon\ \Delta R/h_o = 0$.

Tables of sol-air temperature for July for a 40°N latitude are given in the ASHRAE Handbook of Fundamentals (ASHRAE 1985, Ch. 26, Table 2) for several values of α/h_o (see Table 3-5). Thus at noon on a sunny July day at 40°N latitude when the air temperature is 90°F (32°C), a light-colored, south-facing surface may be considered to be at an effective temperature of 112°F (44°C), whereas a dark-colored surface may be considered to be at an effective temperature of 134°F (57°C).

Table 3-5. Air-Conditioning Cooling Load (ASHRAE 1985)

Sol-Air Temperatures, t_e for July 21, 40°N Latitude (F)

Time	Air Temp, F	N	NE	E	SE	S	SW	W	NW	HOR
		$\alpha/h_o = 0.15$								
1	76	76	76	76	76	76	76	76	76	69
2	76	76	76	76	76	76	76	76	76	69
3	75	75	75	75	75	75	75	75	75	68
4	74	74	74	74	74	74	74	74	74	67
5	74	74	74	74	74	74	74	74	74	67
6	74	82	95	97	86	75	75	75	75	74
7	75	82	103	109	97	78	78	78	78	85
8	77	82	103	114	105	83	81	81	81	96
9	80	85	101	114	110	92	85	85	85	106
10	83	89	96	110	112	100	89	89	89	115
11	87	93	94	104	111	108	96	93	93	123
12	90	96	96	97	107	112	107	97	96	127
13	93	99	99	99	102	114	117	110	100	129
14	94	100	100	100	100	111	123	121	107	126
15	95	100	100	100	100	107	125	129	116	121
16	94	99	98	98	98	100	122	131	120	113
17	93	100	96	96	96	96	115	127	121	103
18	91	99	92	92	92	92	103	114	112	91
19	87	87	87	87	87	87	87	87	87	80
20	85	85	85	85	85	85	85	85	85	78
21	83	83	83	83	83	83	83	83	83	76
22	81	81	81	81	81	81	81	81	81	74
23	79	79	79	79	79	79	79	79	79	72
24	77	77	77	77	77	77	77	77	77	70
Avg.	**83**	**86**	**89**	**91**	**90**	**89**	**90**	**91**	**89**	**91**
		$\alpha/h_o = 0.30$								
1	76	76	76	76	76	76	76	76	76	69
2	76	76	76	76	76	76	76	76	76	69
3	75	75	75	75	75	75	75	75	75	68
4	74	74	74	74	74	74	74	74	74	67
5	74	74	74	74	74	74	74	74	74	67
6	74	90	117	121	99	77	77	77	77	81
7	75	90	131	144	120	82	82	82	82	102
8	77	87	130	151	134	89	86	86	86	122
9	80	91	122	148	141	105	91	91	91	140
10	83	95	109	137	141	118	96	95	95	155
11	87	100	101	122	136	129	105	100	100	166
12	90	103	103	104	125	134	125	104	103	172
13	93	106	106	106	111	135	142	128	107	172
14	94	106	106	106	107	129	152	148	120	166
15	95	106	106	106	106	120	156	163	137	155
16	94	104	103	103	103	106	151	168	147	139
17	93	108	100	100	100	100	138	162	149	120
18	91	107	94	94	94	94	116	138	134	98
19	87	87	87	87	87	87	87	87	87	80
20	85	85	85	85	85	85	85	85	85	78
21	83	83	83	83	83	83	83	83	83	76
22	81	81	81	81	81	81	81	81	81	74
23	79	79	79	79	79	79	79	79	79	72
24	77	77	77	77	77	77	77	77	77	70
Avg.	**83**	**89**	**95**	**100**	**99**	**95**	**99**	**100**	**95**	**107**

Prevailing Wind Conditions

Topography affects wind patterns by restricting the wind, thus increasing its velocity in certain areas and sheltering others (Leckie et al. 1975). Of all climatic variables, wind is the most affected by local conditions. General climatic data are probably insufficient for building design; site-specific conditions must be observed. Wind velocities appearing in climatic data are normally recorded at a height of 30 ft (9 m) above the ground. Because of the boundary layer near the ground and the effect of surface objects, these wind velocities may not be representative of actual site conditions. Thus, local wind data must be modified to account for site conditions. Standard wind engineering formulae can be used to translate the wind velocity in one terrain and at one height to the equivalent wind in another terrain and at another height (Sherman and Grimrud 1980). Geiger (1957) details the relationship among climate variables as they interact with site characteristics to produce the microclimate. Local wind speed and direction are needed to calculate building or solar collector heat loss, especially infiltration, as well as to calculate air movement potential for natural ventilative cooling.

Data on wind velocity and direction are presented in monthly average form in the Climatic Atlas of the United States (DOC 1968). Maximum velocities, as well as daily and yearly variations, are important. The local climatological data annual summary from the National Climatic Center in Asheville, N.C., is also a good source of local wind data.

Humidity

Relative humidity is an important index of human comfort. Vapor pressure is exerted by the moisture contained in the atmosphere and is a measure of the evaporative cooling potential for passive, hybrid, and mechanical cooling systems. It also strongly influences the radiative transmittance of the sky and is therefore an important parameter for radiative cooling. Humidity can also be a factor in the deterioration of solar collectors, piping, and insulation.

Design values for outdoor relative humidity for selected cities can be determined from the design dry-bulb and wet-bulb temperatures listed in the ASHRAE Handbook of Fundamentals (ASHRAE 1985, Ch. 24) and Air Force Manual AFM 88-29 (1978); coincident wet-bulb temperatures for dry-bulb temperature bins are tabulated for many cities in the Air Force manual.

Milne and Givoni (1979) present a series of bioclimatic charts, which relate human comfort to humidity and dry-bulb temperature conditions on a psychrometric chart. These charts, which define climate design strategies for heating and cooling as functions of outdoor air conditions, allow the designer to determine to what extent natural energy systems can be used to provide for human comfort. The charts are based on the expanded comfort zone concept developed by Olgyay (1963).

SHADING BY DETACHED OBJECTS

Solar Access

A careful study of solar access at the building site is critical since (1) solar radiation must not be obstructed from collecting surfaces during the winter, and (2) proper control of solar heat gain is essential to minimizing overheating during the summer. In addition, solar SHW and solar cooling systems may require maximum solar exposure during the summer. Building sites should be chosen to avoid winter shading of collecting surfaces by nearby structures or vegetation. Partial shading of solar collectors may significantly reduce the day-long efficiency of the system. Locating collecting surfaces in the northern part of a site (in the northern hemisphere) will decrease the probability of shading by future off-site development. If solar access is at all in question, winter shadow patterns should be studied (see Fig. 3-17).

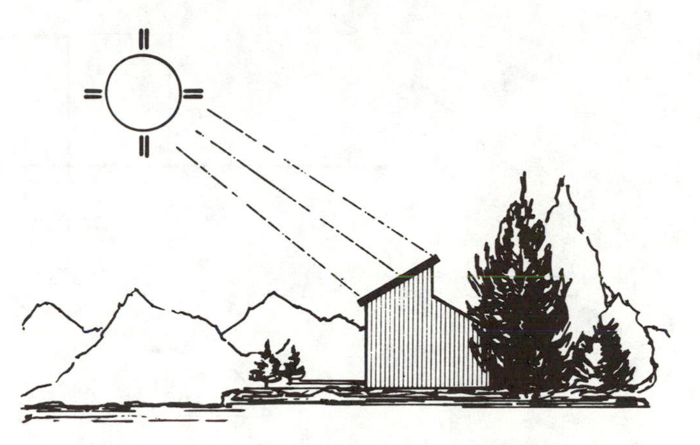

Fig. 3-17. Solar access

Shadow Mapping

As a rule of thumb, collecting surfaces should be exposed to direct sunlight between 9 a.m. and 3 p.m. Using this time span, the designer should trace the shadow patterns cast on the site and determine the optimal location for solar collection. The following example illustrates the simple process of shadow mapping (see Fig. 3-18). Note that solar time must be used.

PROBLEM

Trace the shadows cast by an object on December 21 on a site at 40°N latitude.

SOLUTION

1. Determine the solar altitude and azimuth values for this date and latitude (ASHRAE 1985, Ch. 27, Table 7). (The necessary values are shown here as Table 3-6.)

2. Project the solar altitude angles on the elevation to determine the shadow length (measured normal to the surface that is casting the shadow).

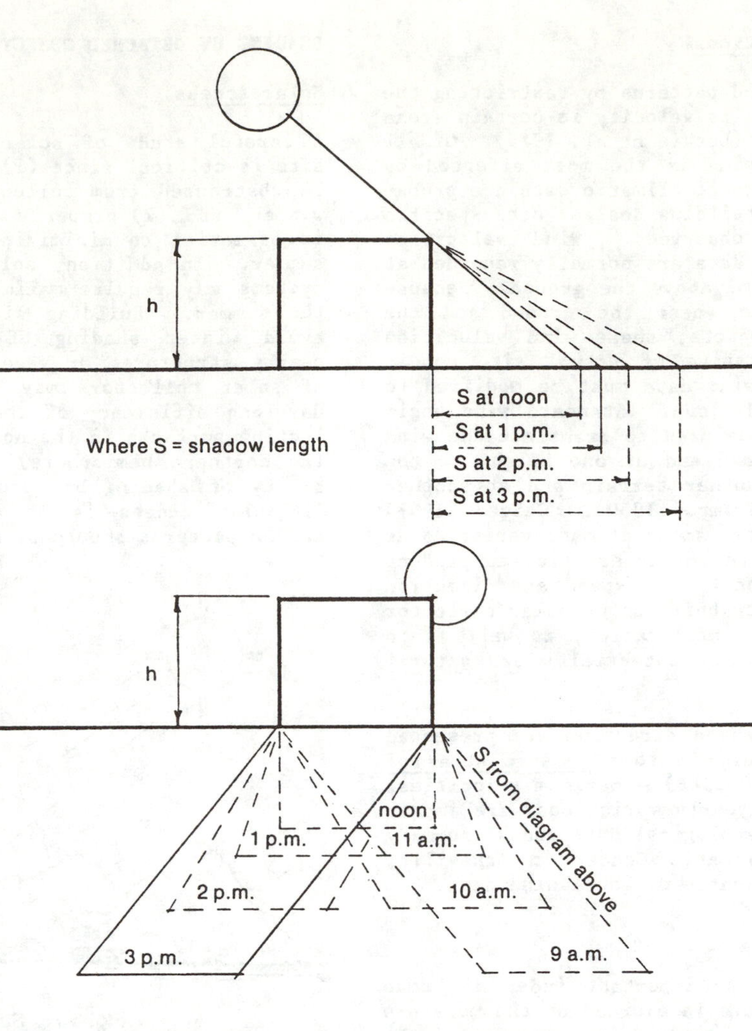

Fig. 3-18. Plotting shadows in plan and elevation

3. Draw the shadow position by plotting the azimuth angles on the plan drawing. (Note: Shadows will be symmetrical about solar noon.)
4. Scale shadow lengths (determined in Step 2) on azimuth lines to determine the shadow pattern.

Shadow lengths can be determined in two ways: (1) if the shadow mapping has been drawn accurately to scale, shadow lengths can be measured directly from the drawing, or (2) the following formula can be used:

$$S = \frac{h}{\tan \beta} \quad (3\text{-}13)$$

where

S = shadow length
h = height of object
β = solar altitude

Special protractor devices allow the easy determination of altitude and azimuth angles. Overlays allow determination of the incident angle and the profile angle for a vertical surface of arbitrary orientation.

Table 3-6. December 21 Altitude and Azimuth Values for Example Problem

Time AST	Altitude (β) deg	Azimuth (ϕ) deg
8:00	5.5	53.0 E
9:00	14.0	41.9 E
10:00	20.7	29.4 E
11:00	25.0	15.2 E
12:00	26.6	0
1:00	25.0	15.2 W
2:00	20.7	29.4 W
3:00	14.0	41.9 W
4:00	5.5	53.0 W

Site Survey

Several solar site survey devices are available to aid in shadow mapping. These devices are designed to be used at the site to determine the seasonal effect of potential solar obstructions (Clarke 1980).

ILLUMINATION DATA FOR DAYLIGHTING STUDIES

A logical design approach is to regard windows and other solar apertures as a source of natural illumination (daylighting). In commercial buildings, light fixtures create a major portion of the cooling load. Daylighting lessens this load. Also, daylighting lowers the electricity requirement for lighting. The careful use of daylighting can result in considerable cost and energy savings (Griffith 1978).

To use daylight to advantage, the building designer should consider the following general factors:

- Variations in the amount and direction of the incident daylight
- Luminance (photometric brightness) and luminance distribution of clear, partly cloudy, and overcast skies
- Effect of local terrain, landscaping, and nearby buildings on the availability and quality of light
- Use of controls for artificial illumination to supplement daylighting
- Period of building occupancy; for example, a school that is unoccupied during the summer will be a less economic application.

Illumination from daylighting is normally measured in footcandles, usually at the solar aperture and on the work surface. Illumination data are available from correlations of calculated or measured solar data (position and intensity) with solar and sky condition (overcast, clear, and partly cloudy) at the site (Evans 1981). A set of empirical charts and graphs showing illumination levels on surfaces of varying orientation, for several sky conditions, is also available (IES 1979; Libbey-Owens-Ford 1976).

Until recently, the most commonly used daylight availability data, found in the Illuminating Engineering Society Handbook (1981), were based on research conducted in the early 1900s. The most recent data are those developed by the Solar Energy Research Institute (Robbins and Hunter 1982); these data are based on TMY weather data and are applied to all TMY stations. The SERI model includes the effect of atmospheric turbidity and has been checked extensively with measured illuminance data.

Solar position and intensity, as well as average cloud cover conditions, are determined as discussed earlier in this chapter. Percent of cloud cover and percent possible sunshine are important determinants of available daylight. For sites with high percent sunshine, the design will focus on bright, sunny conditions with high exterior intensity levels. Analysis is usually based on extrapolations of typical day behavior for each season. Some attempts have been made to correlate diffuse and direct solar intensity data with illumination data, but with limited success. Not enough is yet known about these correlations to extend their application beyond a few specific sites.

HUMAN COMFORT

Human comfort is a complex interaction of conductive, convective, radiative, and evaporative energy transfers between a human and the environment. These energy transfers interact with human metabolism and a person's physical, physiological, and psychological states to produce the effect of perceived comfort. Although comfort is typically defined as a steady-state energy balance condition, a variety of transient environmental conditions can provide comfort. Detailed discussions of the principles of thermal comfort are given by ASHRAE (1985, Ch. 8), Fanger (1972), and Arens et al. (1980). An extensive annotated bibliography of thermal comfort in passive solar buildings is given in Rubin (1982).

The factors that control human comfort are dry-bulb temperature, vapor pressure and velocity of the surrounding air, and mean radiant temperature of the enclosing surfaces. These factors determine the rate at which the body's metabolic heat can be removed by convection, evaporation, and radiation. Each individual's metabolic rate depends in turn on the activity in which he or she is engaged. The ability to dissipate heat from the body is determined by the heat-transfer characteristics of the clothing being worn. Quantitatively, the metabolic rate is expressed in thermal units called mets, where 1 met = 58.2 W/m^2 = 18.4 Btu/h-ft^2. The metabolic rate in met units for a wide range of activities is given in the ASHRAE Handbook of Fundamentals (ASHRAE 1985, Ch. 8, Table 4A). Selected values of energy metabolism in units of Btu per hour are given in Table 3-7.

Environmental Indices

Several indices are in use that relate to thermal comfort. Direct indices, such as wet- and dry-bulb temperature, dew point, relative humidity, and air movement, are commonly understood. The derived index of mean radiant temperature (MRT) is an additional measure particularly important in passive solar applications. MRT is defined as the uniform surface temperature of an imaginary black enclosure with which a person (also assumed to be a blackbody) exchanges the same heat by radiation as in the actual environment. Thus MRT expresses the condition of net radiation interchange of a person with the environment and is the effective radiant temperature of the surroundings. Because the human body is very sensitive to the thermal radiation it receives or emits, MRT is more indicative of thermal comfort than is dry-bulb temperature. This concept is illustrated in Table 3-8; note how MRT can dramatically compensate for uncomfortable air temperatures.

An empirical index is the **effective temperature.** It combines the effect of dry-bulb, wet-bulb, and air movement to yield equivalent sensations of warmth or cold for lightly clothed, sedentary individuals. Another empirical index is the **black-globe temperature**, the equilibrium temperature of a black globe placed in a space. This index measures the combined effect of dry-bulb temperature, air movement, and the radiant heat of surrounding

Table 3-7. Estimates of Energy Metabolism (M) for Various Types of Activity (ASHRAE 1963)

Kind of Work	Activity	M Btu/h
None	Sleeping	250
	Sitting quietly	400
Light Work	Sitting, moderate arm and trunk movement (typing, desk work)	450-550
	Sitting, moderate arm and leg movement (operating word processor)	550-650
	Standing, light work at machine or bench	550-650
Moderate Work	Sitting, heavy arm and leg movements	650-800
	Standing, moderate work at machine or bench, some walking	750-1000
	Walking with moderate lifting or pushing	1000-1400
Heavy Work	Intermittent heavy lifting, pushing or pulling (pick and shovel work)	1500-2000
	Hardest sustained work	2000-2400

Table 3-8. The Balance Between Mean Radiant Temperature (MRT) and Dry-Bulb Ambient Air Temperature that Provides Comfort Conditions on the Psychrometric Chart

	Temperature (°F) for Winter Comfort				Summer Comfort			
Ambient Air	58	61	64	67	81	84	87	90
Mean Radiant	82	80	78	76	72	70	68	66
Equivalent Experience	70-72				76-78			

surfaces. It is often used as a device to measure MRT. A corrected effective temperature uses a black-globe temperature in place of the dry-bulb temperature to correct for radiant effects.

Prediction of Thermal Comfort

ASHRAE has studied thermal comfort for many years and has developed the ASHRAE Comfort Standard 55-74, which has been used widely in the HVAC industry. This standard defines thermal comfort as "that state of mind that expresses satisfaction with the thermal environment." Thus comfort is viewed as a perception rather than a strictly objective condition.

ASHRAE studies have correlated a vote of thermal comfort with temperature, humidity, sex, and length of exposure for normally clothed, sedentary individuals. The comfort problem may be simplified by assuming that the mean radiant temperature of the surrounding surfaces is the same as the dry-bulb temperature of the indoor air. It is further assumed that the air velocity is below 40 ft/min and the individuals are sedentary. Most of those individuals will be comfortable if the ambient air conditions lie within the ASHRAE comfort zone shown in Fig. 3-19.

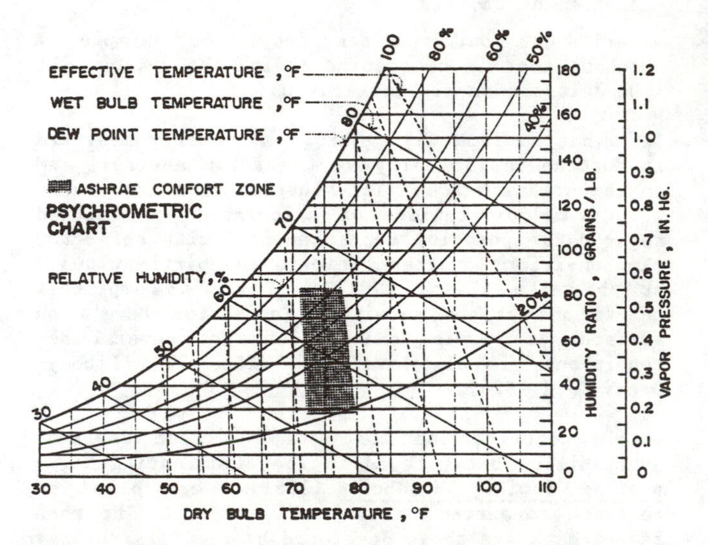

Fig. 3-19. ASHRAE-type psychrometric chart, 1981 edition, showing lines of effective temperature and the comfort zone (adapted with permission from ASHRAE 1981a, Ch. 8, Fig. 16)

A more recent version of the ASHRAE comfort chart is recommended in ASHRAE Standard 55-81. This chart, shown in Fig. 3-20, takes account of the changing clo-values typical of winter and summer. A clo-value is defined as the insulation value of various garments worn by men and women. When different types of clothing are taken into account, the upper and lower vapor pressure and dew-point temperatures remain unchanged; but the minimum winter temperature has been reduced to about 68°F (20°C), and the maximum summer temperature has been raised to 81°F (27°C). The slanting side boundaries in Fig. 3-20 correspond to effective temperatures of 68° and 74.5°F (20° and 24°C) in winter and 73° and 79°F (23° and 26°C) in summer.

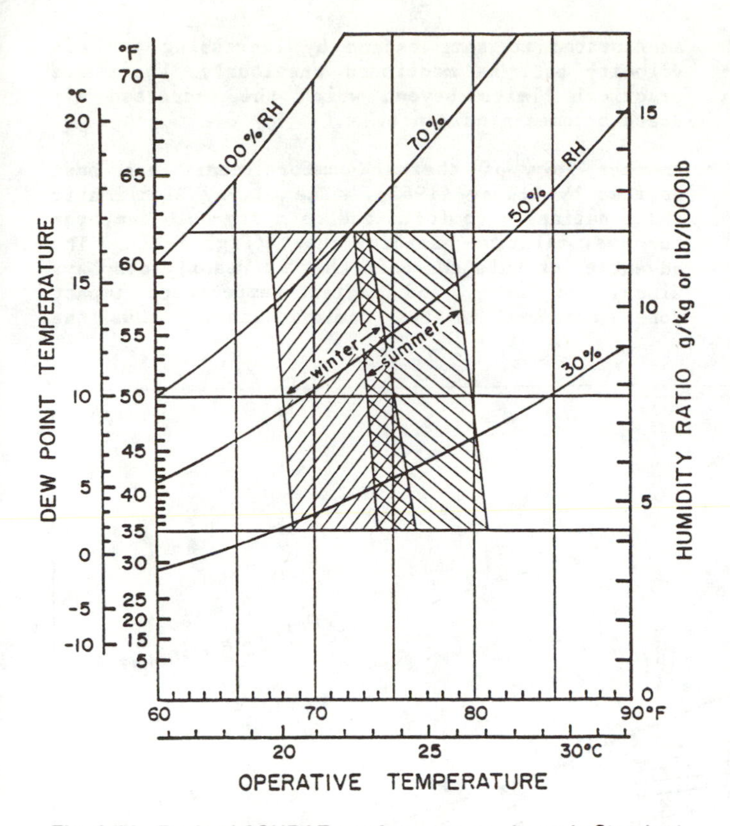

Fig. 3-20. Revised ASHRAE comfort zone as shown in Standard 55-81 (ASHRAE 1981b)

Fanger (1972) has recently developed a set of General Comfort Charts that apply to other clothing and activities where MRT is not equal to the dry-bulb temperature. Figure 3-21 summarizes the new summer conditions and shows how increasing the linear velocity of the air to 160 ft/min will raise the upper temperature limit to 82.4°F (28°C).

The upper and lower limits of the comfort zone should have dew-point temperatures of 62°F (16.7°C) and 35°F (1.7°C). It should be noted that many institutions such as hospitals and libraries and many industrial operations such as color printing require much higher and more constant absolute humidity levels. Humans are much more tolerant of variations at the low end of the humidity scale than are materials such as paper. Static electricity can also present serious problems when the dew-point temperature is very low.

It is at the high end of the humidity scale that most people begin to experience discomfort, and the upper corner of the 1981 ASHRAE comfort zone is defined by the 60% relative humidity line and 77°F dry-bulb temperature. At 72°F, the relative humidity may be as high as 60% before most people begin to be uncomfortable. The upper limit of the comfort zone can be raised significantly by increasing the air velocity, but air flowing continuously at velocities much above 50 fpm (0.25 m/s) can cause annoyances such as blowing papers. ASHRAE Standard 55-81, Thermal Environmental Conditions for Human Occupancy (ASHRAE 1981b) in commenting on air velocity, uses the expression "without regard to direction," but this needs to be qualified in at least two particulars.

The first is the almost universal feeling that air blowing toward one's face may be pleasant, but the same air blowing against the back of the neck may be quite unpleasant. Recent tests (Rohles, Jones, and Konz 1983) have shown conclusively that the varying directionality of air produced by a ceiling fan can lift the upper limit of the comfort zone to as high as 85°F (30°C) at 50% relative humidity.

The heavy dashed lines in Fig. 3-19 represent "effective temperatures," and each line takes its value from the dry-bulb temperature at which it

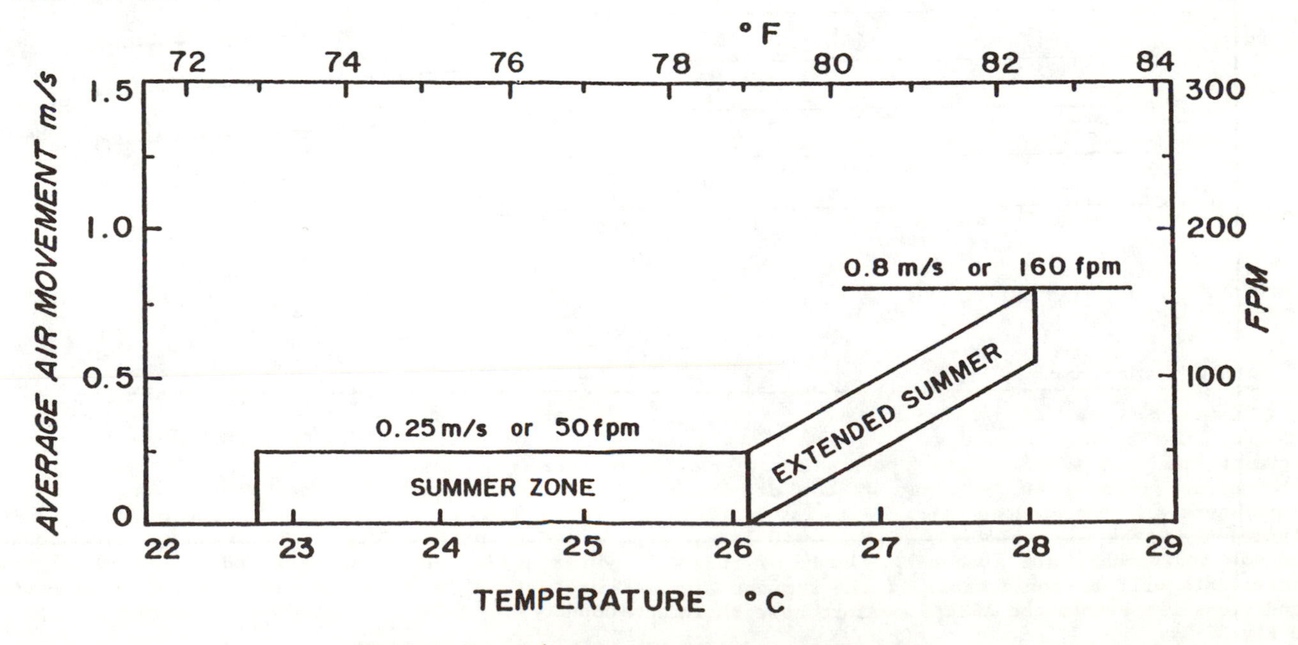

Fig. 3-21. Extended summer comfort zone (ASHRAE 1981b)

crosses the 50% relative humidity line. Most people will be comfortable at 75°F dry-bulb temperature and 50% relative humidity, and this point is located at about the center of the comfort zone. Approximately the same degree of "physiological strain" will be experienced at 76°F, 20% relative humidity and at 74°F, 80% relative humidity. At the upper condition, discomfort will be experienced by individuals engaged in vigorous activity. This higher humidity condition suppresses evaporation of moisture from the skin to such an extent that the individual will be uncomfortable. This can be ameliorated to some extent by increasing the air velocity but, as mentioned previously, there are practical limits beyond which the increased air speed becomes a nuisance.

Another form of thermal comfort chart has been devised by Olgyay (1963). The Olgyay Bioclimatic Chart defines a comfort zone on a dry-bulb temperature vs. relative humidity chart (Fig. 3-22). The advantage of this chart is that it readily displays wind-, moisture-, and radiant-temperature impact zones adjacent to the comfort zone. Thus the

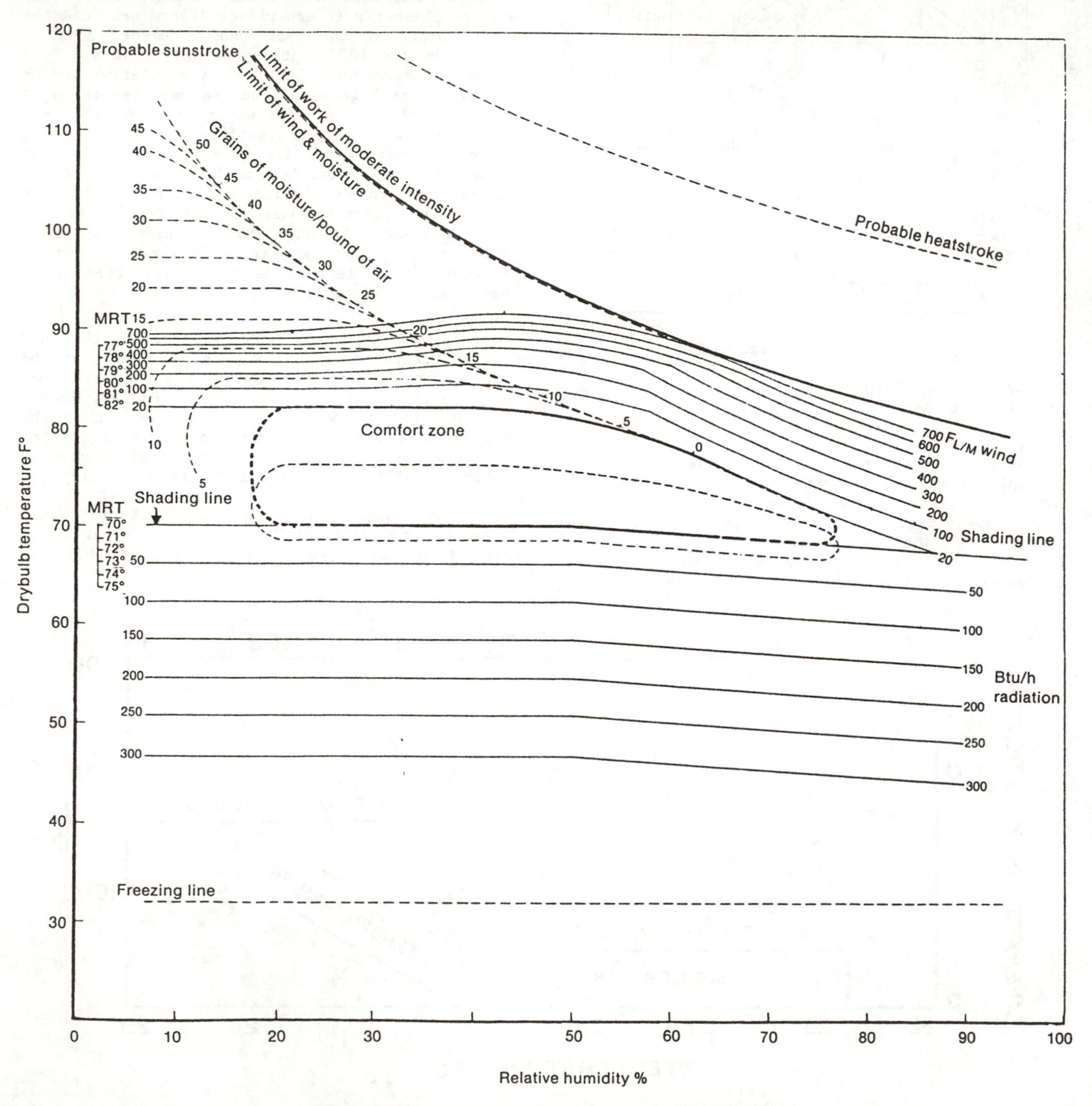

Fig. 3-22. Bioclimatic registration of climatic data (Olgyay 1963)

effects of air movement, evaporative cooling at the skin, and MRT on comfort can be estimated. MRT values offset temperatures that are too low, and the air velocity and evaporative cooling values offset temperatures that are too high; combinations of these variables are not considered. This chart can be used to define an expanded thermal comfort zone and to assess the potential for some solar heating and cooling strategies.

Because Olgyay's bioclimatic chart is not particularly appropriate in settings where buildings are not coupled closely to climate, as in large commercial buildings that create their own internal climate, a modified version of this chart has been developed by Arens et al. (1980). This new bioclimatic chart updates the original chart to reflect current physiological data and has been placed in a psychrometric chart format to increase its usefulness to engineers (see Fig. 3-23). Although the still-air comfort zone on this new chart closely resembles the original one, the air motion and radiation lines are markedly different, reflecting recent comfort studies.

Environmental Control Strategies and Human Comfort in Buildings

The concept of a thermal comfort zone superimposed on a psychrometric chart has been proposed by Milne and Givoni (1979) as a means of determining the potential for passive solar heating and cooling in a given climate. One variation of the Milne and Givoni comfort zone over those discussed above is the identification of distinct zones for winter and summer conditions.

One of these charts (similar to Fig. 3-24) has been divided into zones in which the outside ambient conditions can be brought within the comfort zone by passive heating and humidification strategies.

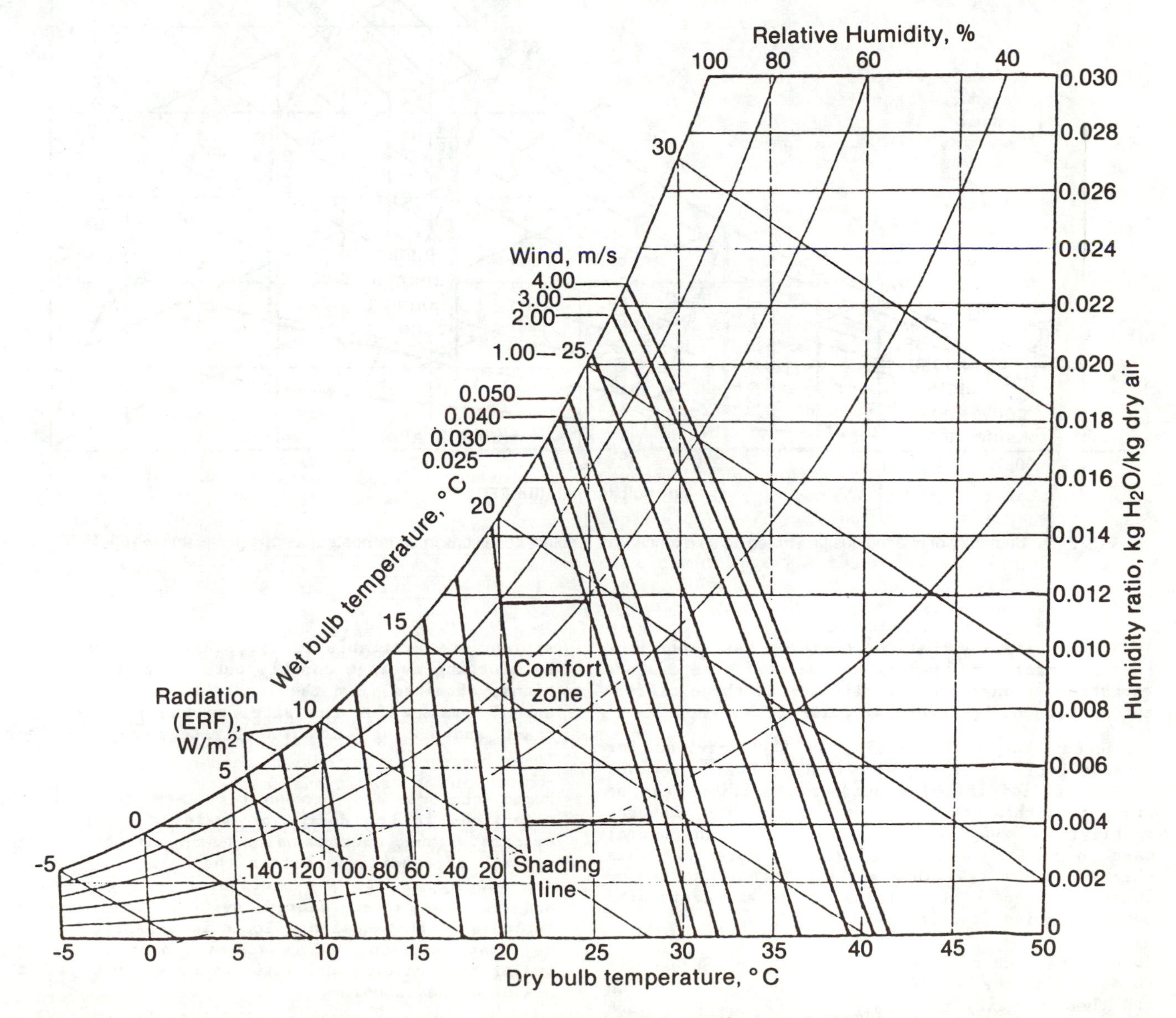

Fig. 3-23. Bioclimatic chart — psychrometric format (Arens et al. 1980)

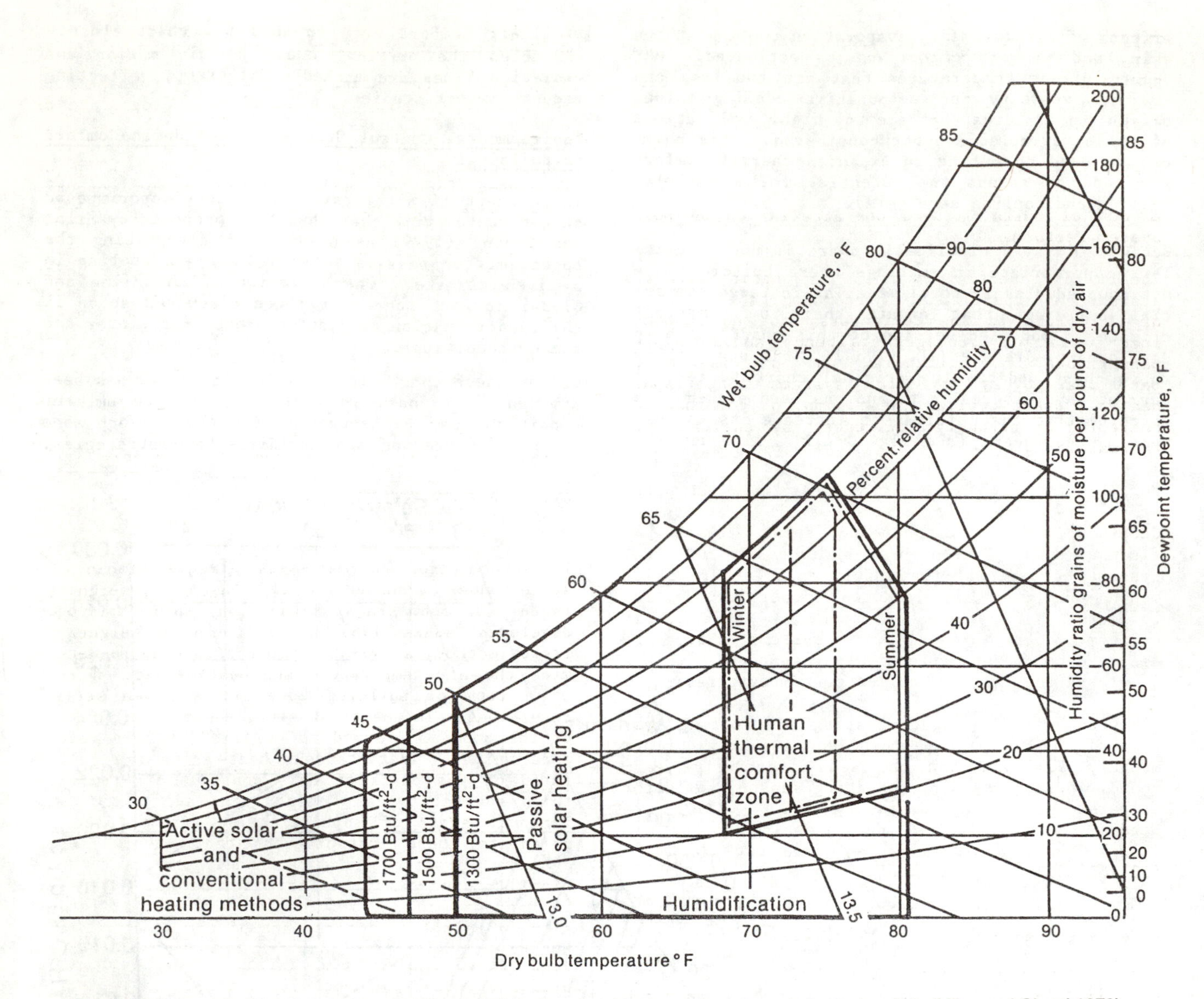

Fig. 3-24. Summary of heating design strategies as a function of ambient conditions in underheated periods (Milne and Givoni 1979)

This chart assumes that the building has sufficient thermal mass to temper the diurnal temperature variation. Zones are established for three different average daily levels of solar irradiation.

A second chart (Fig. 3-25) has been devised for passive cooling strategies. On this chart, regions are identified in which outdoor conditions can be brought within the comfort zone using natural ventilation, evaporative cooling, and high thermal mass to reduce temperature swings. Note that this chart does not take into account latent, solar, or internal source cooling loads; it merely deals with outside ambient conditions.

To use these charts, the general design approach is to determine the ambient climatic data for the site and then identify the design strategies that work to create human comfort conditions with the most appropriate available means. This is accomplished by plotting average monthly outside ambient conditions, obtained from the local climatological data annual summary for the site, on the psychrometric chart and noting in which strategy regions they fall.

Human comfort and productivity are the primary objectives in the design of building conditioning systems. Thus, understanding human comfort and the control of conditions within the built environment are fundamental to any successful building design and to the solar energy systems that may be included in that design. It is important to remember that human comfort is achieved over a range of indoor temperatures and humidities, and not just at a single condition.

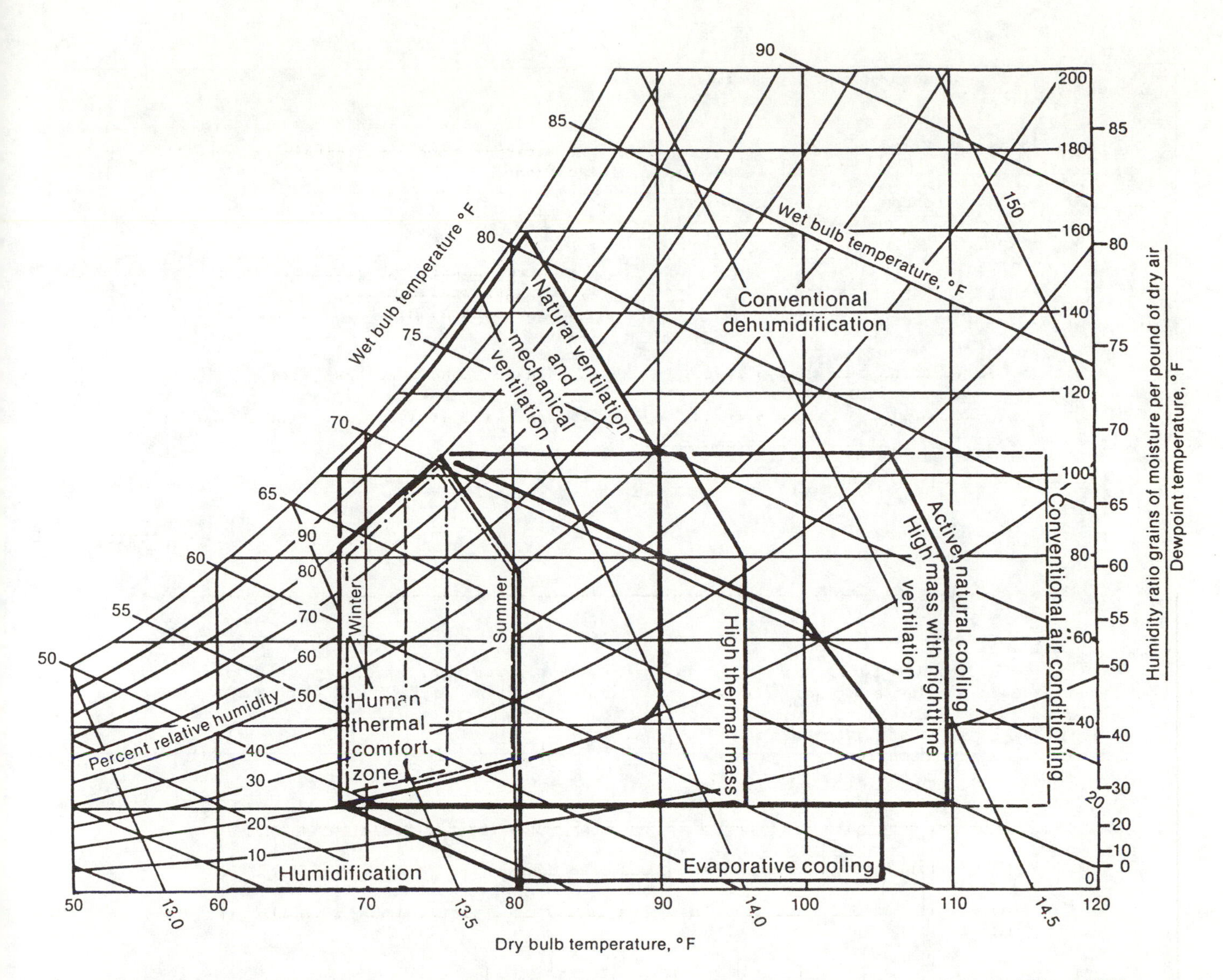

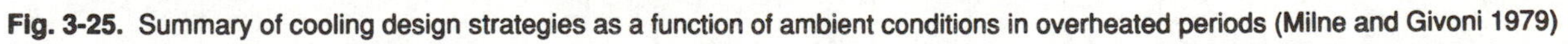

Fig. 3-25. Summary of cooling design strategies as a function of ambient conditions in overheated periods (Milne and Givoni 1979)

Part II
Active Solar System Design and Sizing

Chapter 4

Active Solar Heating and Cooling Systems

INTRODUCTION

The principal applications for active solar systems are service water heating, space heating, and space cooling. For water and space heating, such systems can be combined with heat pumps to improve the range and efficiency of system application. Characteristics of active systems are presented in the following sections.

SERVICE HOT WATER SYSTEMS

Several detailed discussions are available on solar service hot water (domestic or process) systems (e.g., Dickinson and Cheremisinoff 1980b; DOE 1978; Jordan and Liu 1977; Liu and Fanney 1980); only highlights will be presented here. Many features of service hot water systems, particularly in the collector-to-storage loop, are also common to active solar space heating systems. Details of the collector-to-storage configuration, particularly with regard to freeze protection, are included in the discussion of space heating systems.

Separate Systems

If a flat-plate collector is coupled directly to a storage tank such that the same fluid is used for both collection and storage, the system is called a direct system. However, in colder climates a nonfreezing heat-transfer fluid with a liquid-to-water heat exchanger often is used between collection and storage. This configuration is called an indirect system.

A hot water system can be classified as a single- or double-tank type. Each type can be further classified as a direct or indirect system, depending on whether a heat exchanger is placed between collection and load. Service hot water systems can also be classified as open or closed systems, depending on whether the collector fluid is open or closed to the atmosphere. Finally, systems are classified as pumped or thermosiphon, depending on whether a pump or natural convection provides for circulation of the collector fluid. Not all combinations of these system types are used; systems representing the most common types (Fig. 4-1) are described here.

Single-Tank Systems

In a single-tank system, auxiliary heating is done in the solar storage tank. This configuration is not recommended if the auxiliary heat source is located at the bottom of the tank because in such units, the thermostat is located in the tank bottom where water is coolest. Thus, if the temperature

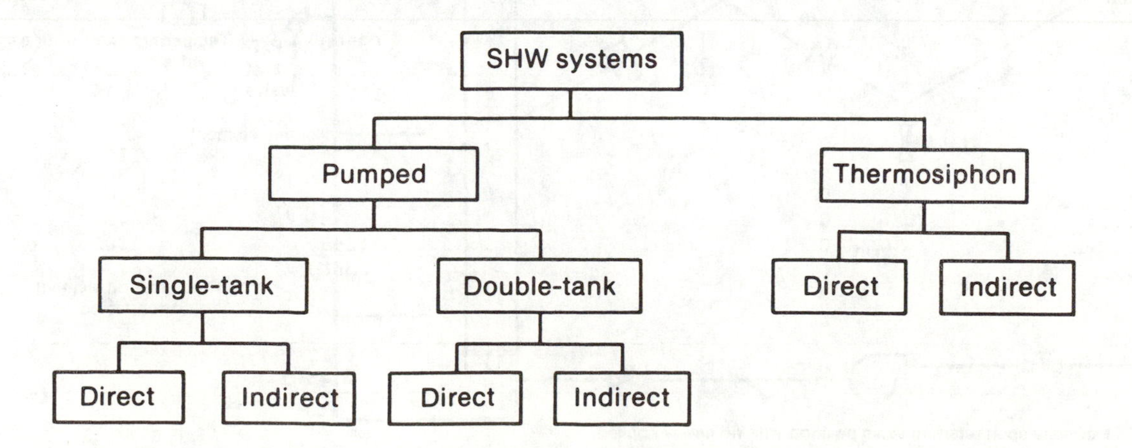

Fig. 4-1. Service hot water system types

of this cool water is less than the thermostat setting, heat would be added to the water until the set-point temperature is reached. This would cause the solar system to operate at a lower efficiency because relatively high-temperature water is delivered to the collectors from the bottom of the tank.

This problem is alleviated by heating only the upper third of the tank, keeping the lower third as cool as possible. Conventional gas water heaters do not operate in the desired manner since the whole tank is heated from the bottom. Electric water heaters, with their lower heating element disconnected, provide the desired auxiliary heating action. However, some gas-fired solar storage tanks are designed to have the heat source located in the upper portion of the tank. Only those tanks that provide for upper-tank auxiliary heating should be used in a single-tank hot water system.

One of three possible freeze-protection strategies, the drain-out option, is shown in Fig. 4-2 for a single-tank, direct system. The collector and storage fluids in this case would be ordinary tap water that mixes with service hot water in the load loop. Freeze protection is accomplished by a series of solenoid valves (or a single multiport drain-out valve) that are opened when near-freezing conditions are detected by the controller, allowing collector fluid (water) to drain onto the ground or to a sump. The collector and storage tank are maintained at line pressure. A vacuum breaker allows air into the line whenever drain valves are opened and line pressure is valved off. A check valve and a solenoid valve prevent loss of storage tank fluid under these circumstances; only collector fluid is drained. The collector loop must be recharged before the system is reactivated. An automatic air vent valve at the high point of the system allows air to escape. Also, piping must be pitched to assure proper drainage. Shut-off, relief, tempering, and other valves are shown in Fig. 4-2.

Use of drain-out systems has decreased because of poor reliability resulting from the failure of drain-out valves and vacuum breakers. To prevent freezing, a possible variation on this system is to pulse the collector pump during periods of marginally freezing ambient temperatures. The system is then drained only during intervals of extremely cold weather.

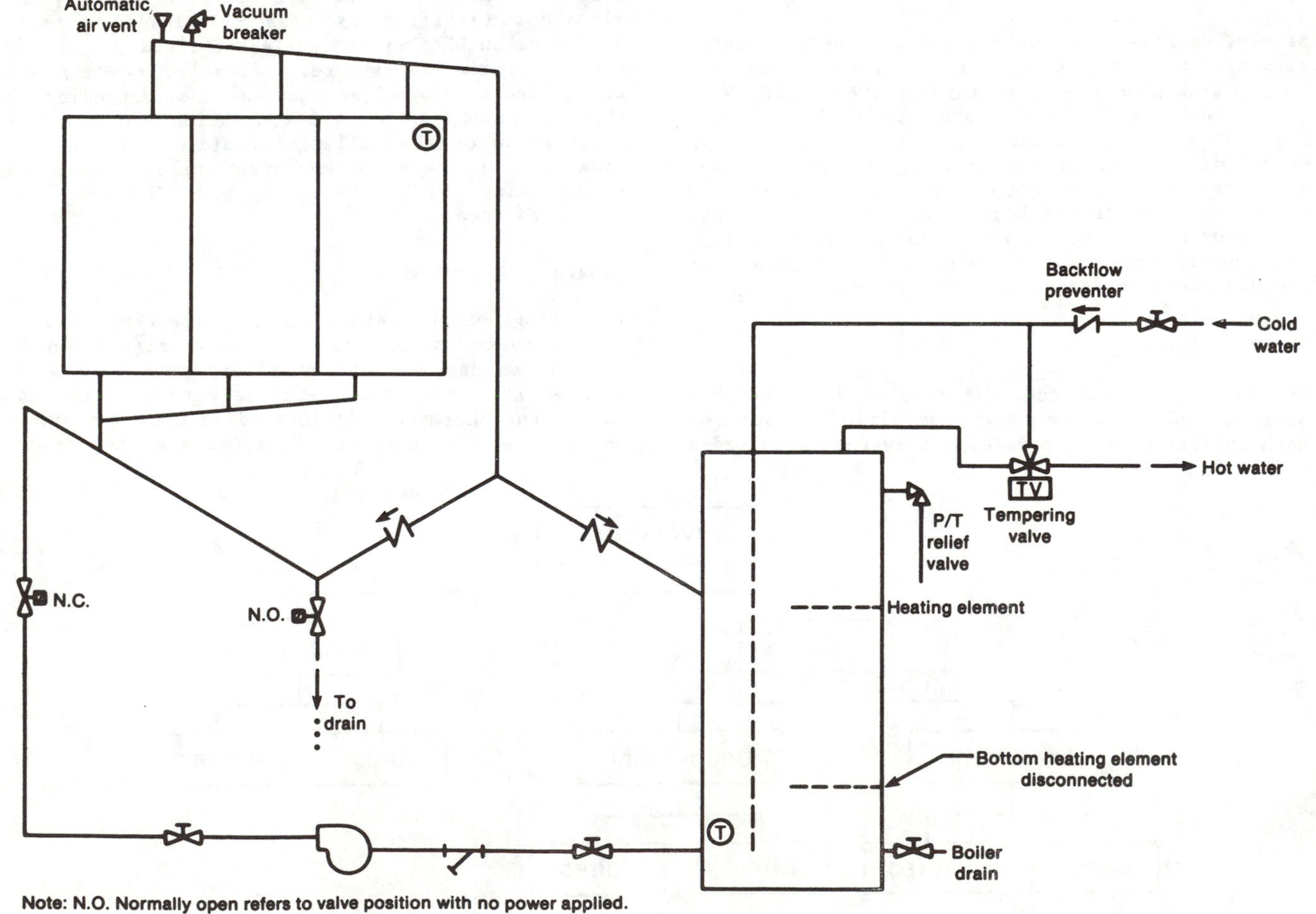

Fig. 4-2. Single-tank, direct, drain-out service hot water system

The second freeze-protection strategy is used in the single-tank, indirect service hot water system shown in Fig. 4-3. An ethylene- (or propylene-) glycol and water antifreeze solution is used in the collector loop with heat transferred to the storage tank via a double-walled, vented, in-tank heat exchanger. (Because water and either ethylene- or propylene-glycol are used as an antifreeze solution, the term glycol/water will refer to either.)

Many manufacturers use coils in tanks or external heat exchangers to accomplish the necessary isolation. A double-walled, jacketed tank (wraparound heat exchanger) arrangement can also be used where the annulus around the tank contains the collector fluid. The tank serves as the thermal storage and is in direct thermal contact with the collector fluid jacket. Single- or multiple-pass shell-and-tube heat exchangers are generally used when the heat exchanger is external to the storage tank; in some cases finned-tube exchangers are used. In this configuration two pumps are required in the collector-storage loop. The high flow rates required for efficient heat exchange in single-pass exchangers increase pumping power cost and reduce storage tank stratification as a result of mixing. However, efficient heat exchange is more important to overall system performance than is establishment of storage tank stratification. Heat exchanger design for solar systems is discussed in detail later.

The toxic glycol/water mixture is double separated from the potable water supply in all of these cases. Glycol solutions can weaken or break down to form acids that corrode system piping. Therefore, fluid condition must be checked regularly to ensure protection and minimize corrosion. (Recent tests at Los Alamos National Laboratory indicate that all-copper systems are not very susceptible to corrosion from degraded glycol; but even in this case, an annual check of the antifreeze is the safest course.) Make-up systems with automatic glycol/water refill systems must be manual since the water can dilute glycol concentrations without the owner's knowledge. The expansion tank should contain a low-level alarm, indicating the need to add antifreeze. Automatic systems should contain a tank of premixed antifreeze. Nonaqueous, nonfreezing fluids can be used instead of glycol/water

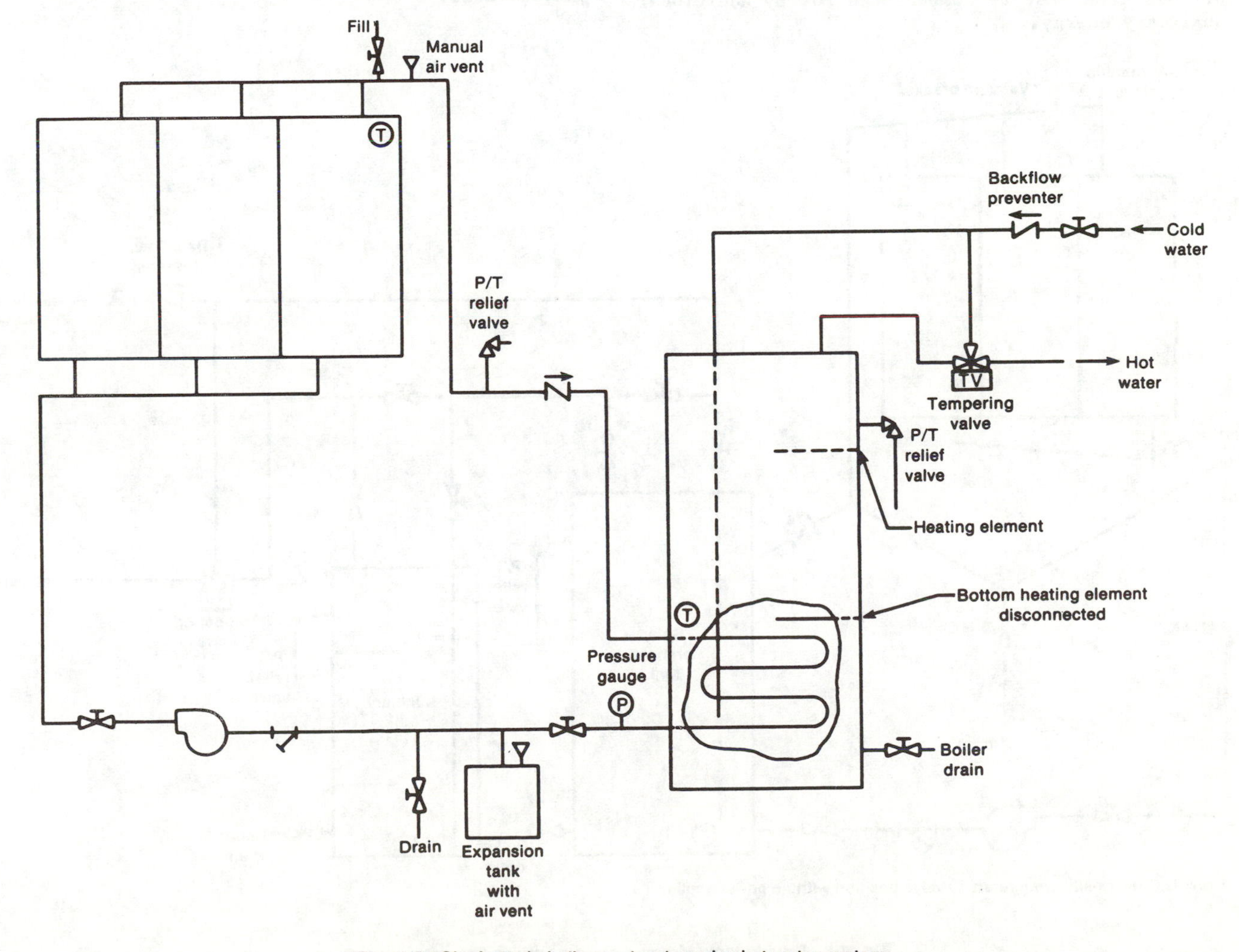

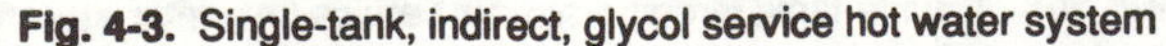
Fig. 4-3. Single-tank, indirect, glycol service hot water system

mixtures. (See the discussion of collector coolants in Ch. 5.) An expansion tank and relief valve are provided to prevent overpressurization of collectors as the heated coolant expands. Performance is less than that of the direct system because the heat exchanger forces the collectors to operate at higher temperatures. The extent of degradation is discussed later. Additional penalties include heat exchanger cost and pump operation cost. These can be weighed against questions of the reliability of a self-draining system and the cost of using a nonfreezing liquid throughout the collector and storage subsystems.

Double-Tank Systems

In a double-tank system the functions of solar storage and auxiliary heating can be separated; the solar tank acts as a preheater for a conventional gas or electric unit. Because auxiliary heat is supplied to the second tank, not to the solar-heated tank, the solar-heated part of the system can operate at lower storage (and hence collector) temperatures and higher collection efficiency. However, overall efficiency is also affected by heat loss from the auxiliary tank. Standby losses of this tank must be compensated for by additional auxiliary energy.

Figure 4-4 shows a double-tank, direct, drain-out service hot water system. As in the single-tank, direct system there is no separation between collector fluid and potable water supply; ordinary tap water flows through the collector. The system is the same as the single-tank, direct system except for interposition of a solar preheat tank. As shown in Fig. 4-4 (and in Fig. 4-5 as well), isolating valves above the auxiliary tank allow bypass operation in the summer when the auxiliary tank is not needed.

A possible variation has warm, solar-heated fluid circulated by a second pump that is actuated by a differential temperature controller between the tanks (i.e., from the solar preheat tank to the auxiliary tank). However, the modified system must be carefully piped and controlled to achieve the desired result, which is net displacement of purchased auxiliary energy. Both thermosiphon action between the tanks and incompatible control of the pump and auxiliary tank could negate any benefit. Normally, the increased cost of additional equipment would not be economical when compared to the marginal energy savings. Such a modified system should be analyzed carefully before it is implemented.

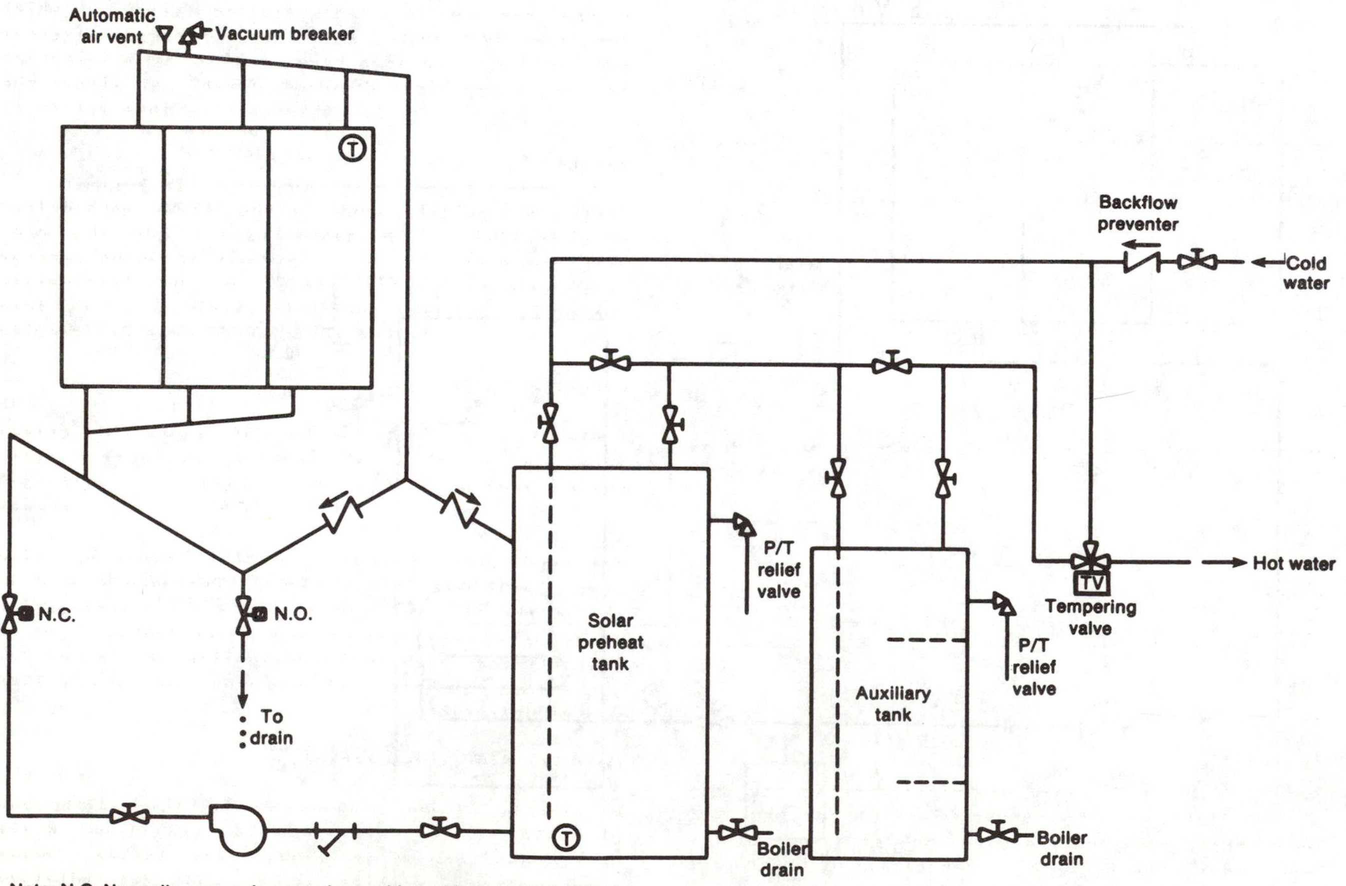

Note: N.O. Normally open refers to valve position with no power applied.

Fig. 4-4. Double-tank, direct, drain-out service hot water system

A double-tank system can be arranged so that the cold water supply passes through a heat exchanger in the solar preheat tank on its way to the auxiliary tank. This double-tank, indirect, drain-back system with load-side heat exchanger is shown in Fig. 4-5. Line pressure is not used in the collector loop. Instead, water is maintained in the collector loop only when the pump is on. Freeze protection is accomplished by allowing the collector and piping to drain by siphon return into the storage tank whenever pumped circulation is stopped. Air that enters the collectors when circulation is stopped must be purged through the collector return line to the space at the top of the preheat tank whenever the circulation pump starts again. As shown in Fig. 4-5, an optional vacuum breaker can be placed in the collector return line to prevent boiling in siphon-return systems that have high static heads.

A final variation on the indirect drain-back system is shown in Fig. 4-6. This system uses a reservoir holding tank in the collector loop that is sized to hold collector fluid when the collector drains at shutoff; it also uses a collector-side heat exchanger that can be of the immersed (internal) type shown in Fig. 4-6 or an external shell-and-tube type. Because hot collector fluid constantly circulates through the heat exchanger whenever the pump is on, water delivered to the auxiliary tank is preheated whenever the collector pump is on. This contrasts with the load-side heat exchanger case shown in Fig. 4-5 where preheating occurs only when water is delivered to the load.

The heat exchanger can be eliminated from drain-back systems if a pressurized air space exists inside the storage tank in a direct, closed system configuration. This air space, maintained by an air compressor, allows city water in the collector loop and drain-back freeze protection. In this system air vents are replaced with an air bleed tube that physically rests against the collector return piping from the uppermost point of the collector array to the top the of the storage tank. This tube ensures that water will drain from the top of the collectors into the tank. This system requires a pressure relief valve to prevent the loop from being overpressurized.

A third type of freeze protection is provided by the pulse, or recirculation, system. A freeze-detection sensor is located on the section of pipe or collector subject to lowest system temperature.

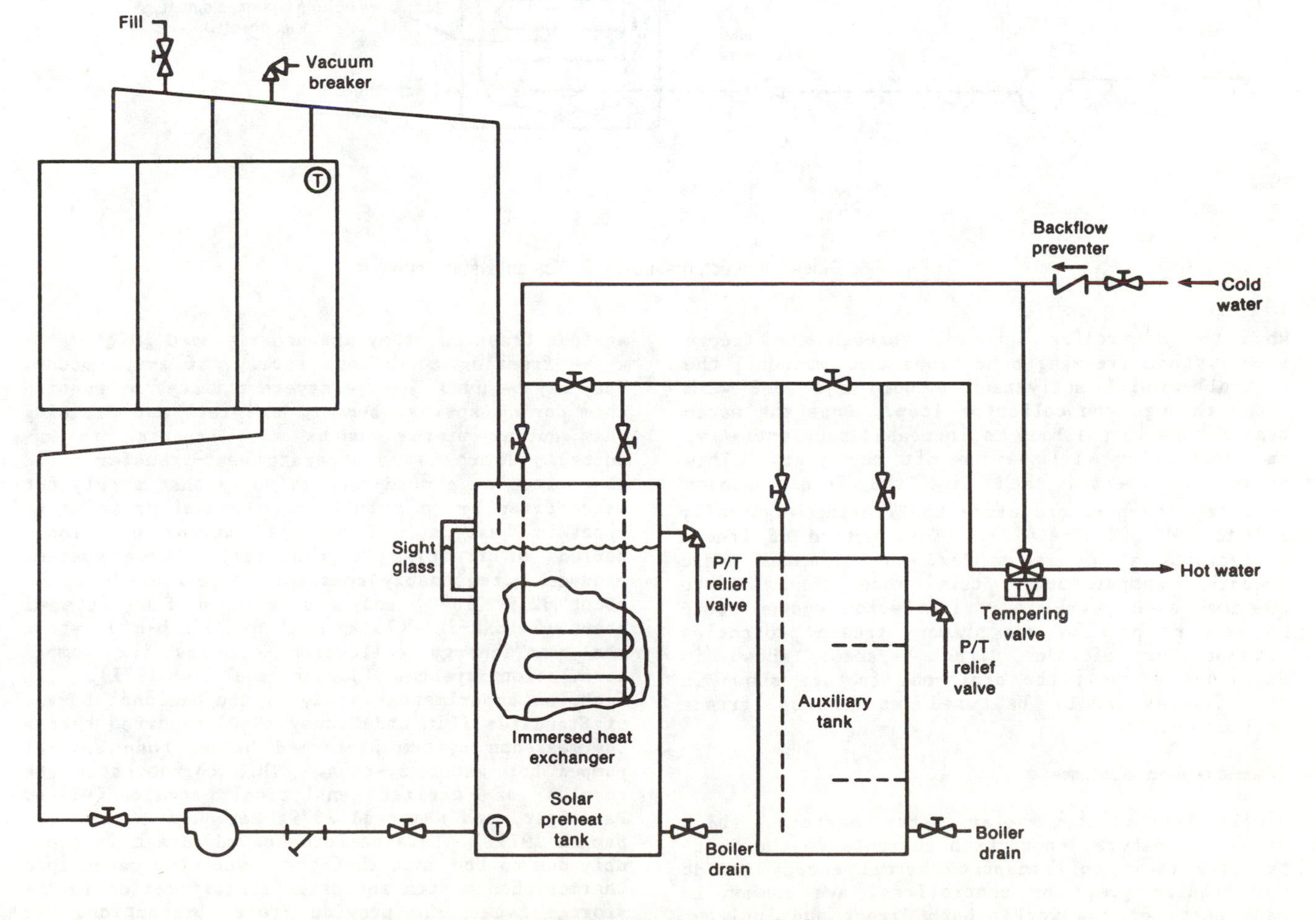

Fig. 4-5. Double-tank, indirect, drain-back service hot water system

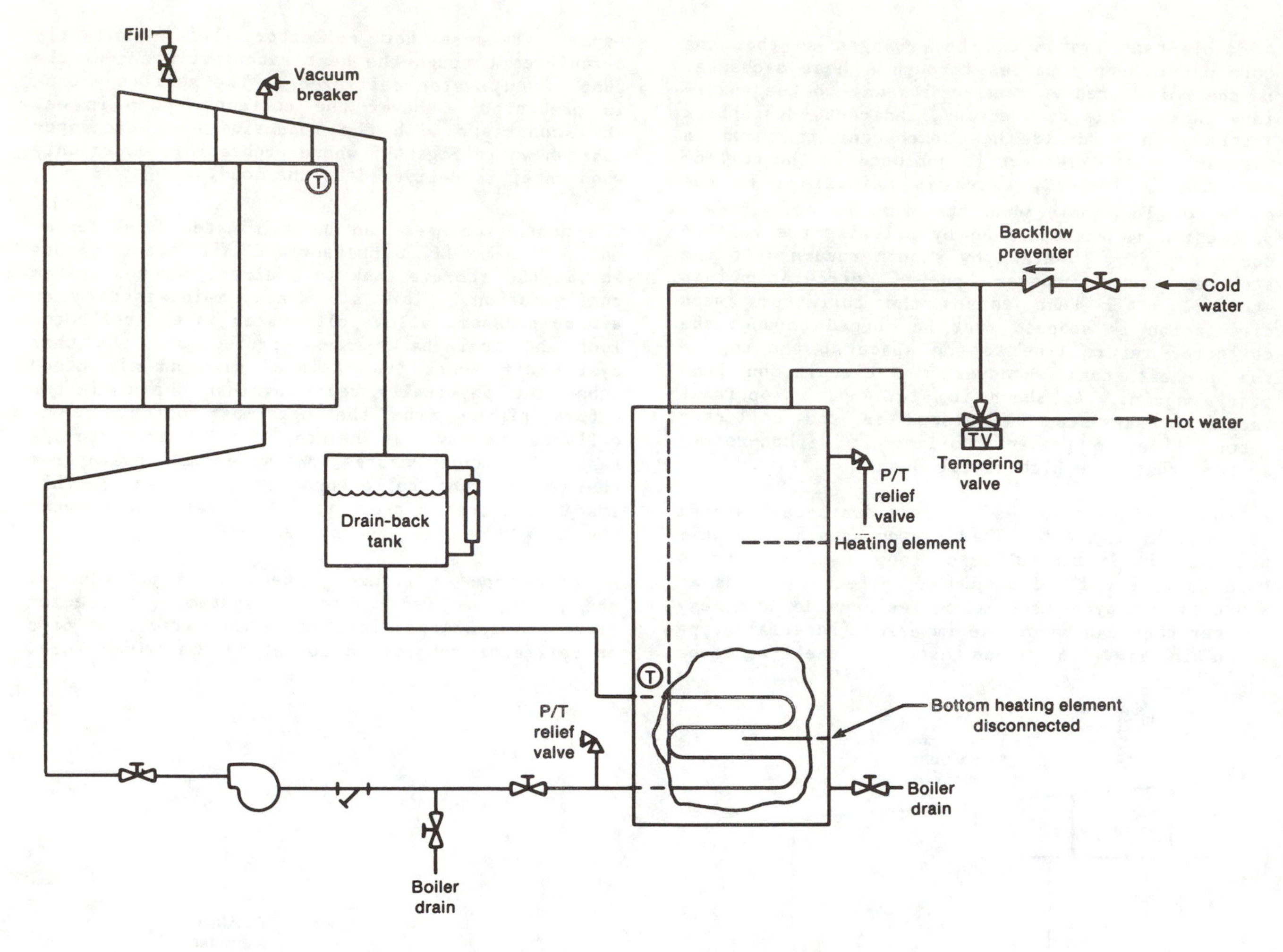

Fig. 4-6. Closed, indirect, drain-back service hot water system

When the controller detects, through the freeze sensor, that freezing conditions are imminent, the controller will activate the pump and send warm water through the collector loop. Once the water temperature in the loop is increased substantially, the controller will switch off the pump. This pulse is repeated each time the freeze sensor detects a temperature close to freezing, typically 38° to 40°F (3.3°-4.4°C). This method of freeze protection is not recommended for climates where freezing temperatures occur more than 20 to 30 times each year. A recirculation system could be used as part of the primary freeze protection with either of the direct systems shown in Figs. 4-2 or 4-4; the drain-out feature shown in the figures could be used as backup freeze protection.

Thermosiphon Systems

Simple thermosiphon solar water heaters, which depend on natural convection currents in the heat-transfer fluid to transport thermal energy and do not require pumps or controllers, are common in many parts of the world; both direct and indirect configurations are used. Because direct thermosiphon systems are inherently difficult to protect against freezing, they are usually used in climates where freezing conditions rarely, if ever, occur. They may be used in more severe climates by running them during spring, summer, and fall and draining them during winter months when freezing is expected. There is no separate heat-transfer fluid in rooftop or ground-mounted units that supply hot water directly to a building. Manual draining or electric heaters can be used during occasional periods of freezing temperatures. These systems produce a reasonably constant temperature rise of about 32°F (18°C) and a peak water flow rate of about 15 $lb/h\text{-}ft^2$ (73 $kg/h\text{-}m^2$ or 73 $L/h\text{-}m^2$), about the same energy collection rate as for pumped circulation systems (Jordan and Liu 1977). In fact, an experimental study by the National Bureau of Standards (Liu and Fanney 1980) reported that a thermosiphon system performed better than several pumped hot water systems. This corroborated the results of earlier analytical studies (Place, Daneshyar, and Kammerud 1979; Bergquam, Young, and Baughn 1979.) This performance advantage is probably due to the fact that the lower flow rates in a thermosiphon system encourage stratification in the storage tank. To provide freeze protection, refrigerant-charged (phase-change) thermosiphon solar water heaters, with heat exchangers placed in the storage tanks, have been the subject of recent research (Mertol et al. 1981; Schreyer 1981) and of field operational studies (Kissner 1980).

Figure 4-7 is a schematic of a direct, drain-out, thermosiphon service hot water system. The direct thermosiphon system resembles in many ways the direct system with forced circulation (Fig. 4-2). Water circulates solely because of the density differences between cold water in the storage tank and heated water in the collector. The storage tank must be located above the collector for the water to circulate. The flow is self-regulating. When energy gain in the collector increases, so does the flow rate; when there is no energy gain, the flow stops. Piping must be continuously sloped upward, with smooth bends to prevent formation of air pockets. To minimize fluid friction losses, collector piping should be at least 3/4 in. (1.9 cm) inside diameter.

An experimental study (Morrison and Sapsford 1983) of the long-term performance of six direct thermosiphon systems has shown that performance is a function of the way the system is operated, and significant variations in performance exist between different system configurations. Specifically, the annual performance of a thermosiphon system, in contrast to a pumped circulation system, was found to improve by about 15% if a morning-peak load profile was used. Also, performance improved when the system was operated with off-peak, as compared to continuous, electric boosting. However, with off-peak boosting the delivery temperature may fall below an acceptable level at times. Morrison and Sapsford (1983) also present a method of predicting the long-term performance of thermosiphon systems.

The direct thermosiphon configuration, one of the simplest and least expensive systems, has the potential for excellent performance. Its greatest disadvantage relative to the direct system with

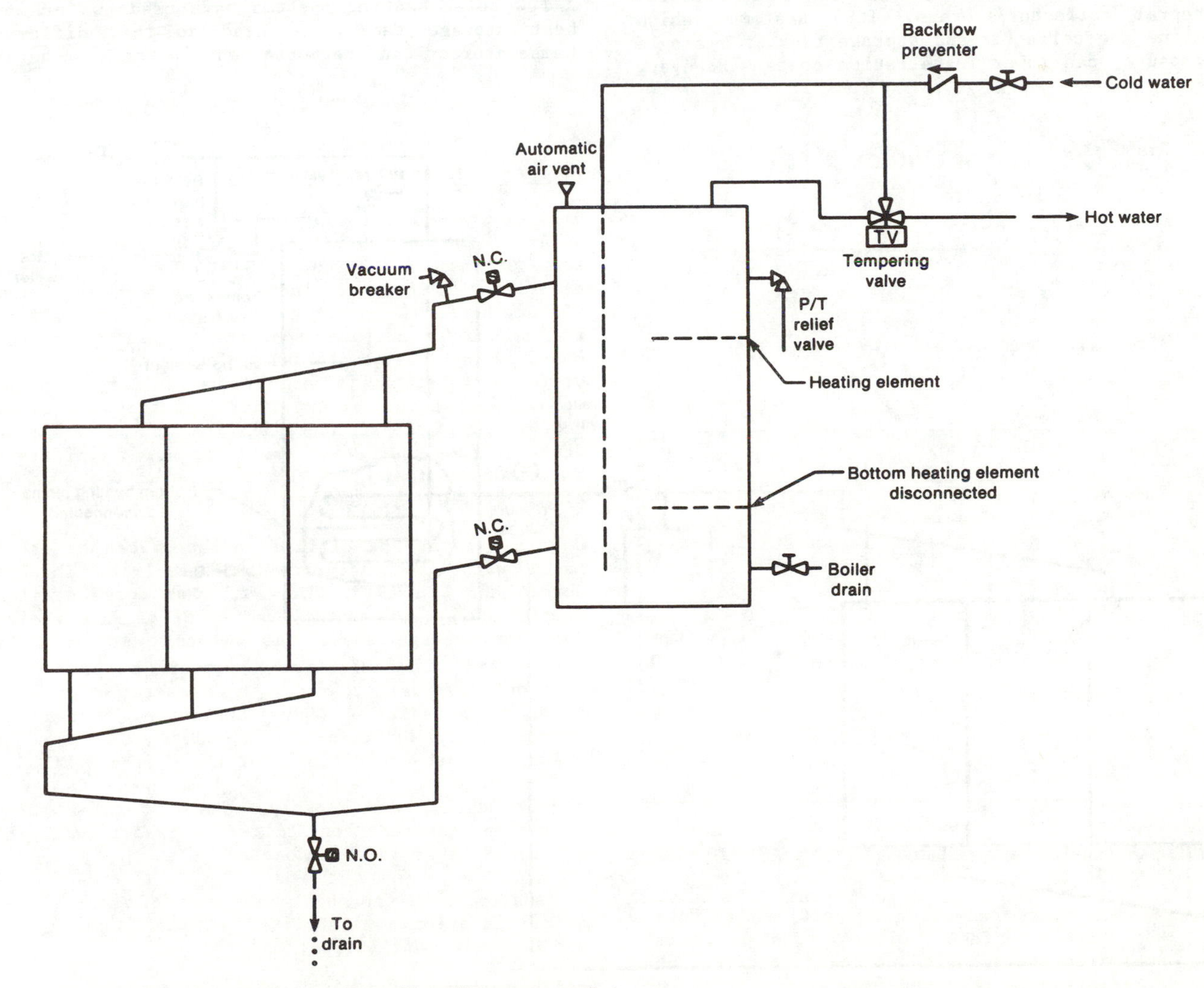

Fig. 4-7. Direct, drain-out thermosiphon service hot water system

forced circulation is the requirement that storage must be located above the collectors and freeze protection must be by drain out.

Figure 4-8 is a schematic of an indirect thermosiphon service hot water system. This system is self-controlling, self-pumping, and reliably freeze protected if a nonfreezing collector coolant is used. It is actually a single-tank active system containing an internal heat exchanger with a modified geometry and pump and controller removed. The requirement for collector-fluid circulation is the same as for the single-tank thermosiphon system, that is, the storage tank must be located above the collector.

Thermosiphon systems located inside a building are not widely used, probably because of the need for storage tank elevation, which is constrained by architectural and structural problems. These systems are not easily used in retrofit but are workable and reliable systems where applicable. Integral collector/storage (ICS) systems, which combine the collector and storage tank in a single enclosure, can lower installation costs. However, because of their small collector size, ICS systems are not usually used in commercial building applications.

Hot Water Systems Using Air Collectors

Another type of hot water system, shown in Fig. 4-9, is composed of air collectors and an air-to-water heat exchanger coil. Freezing must be prevented in an air system just as in a water system. Thermosiphon action can cause air at subfreezing temperatures to pass over the air-water heat exchanger and freeze the water therein. A motorized damper is installed in the return duct from the collectors to prevent this from occurring. This system could be designed as a stand-alone system, but it would be more effective in a combined space heating and hot water system.

Collector/Storage Loop Considerations

A few solar heating systems have used two or more heat storage tanks, arranged so that different temperatures can be obtained in each, thereby

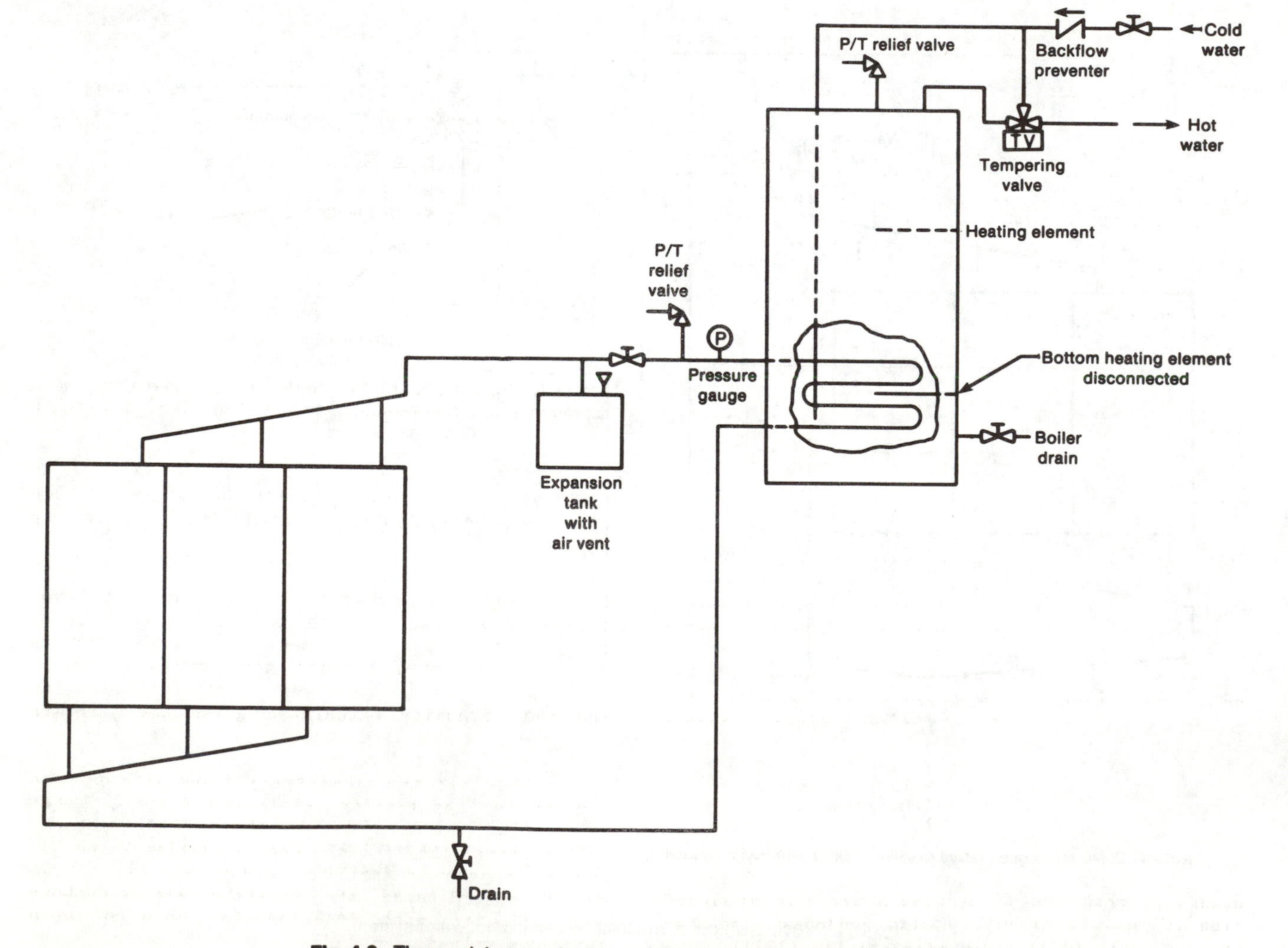

Fig. 4-8. Thermosiphon service hot water system with internal heat exchanger

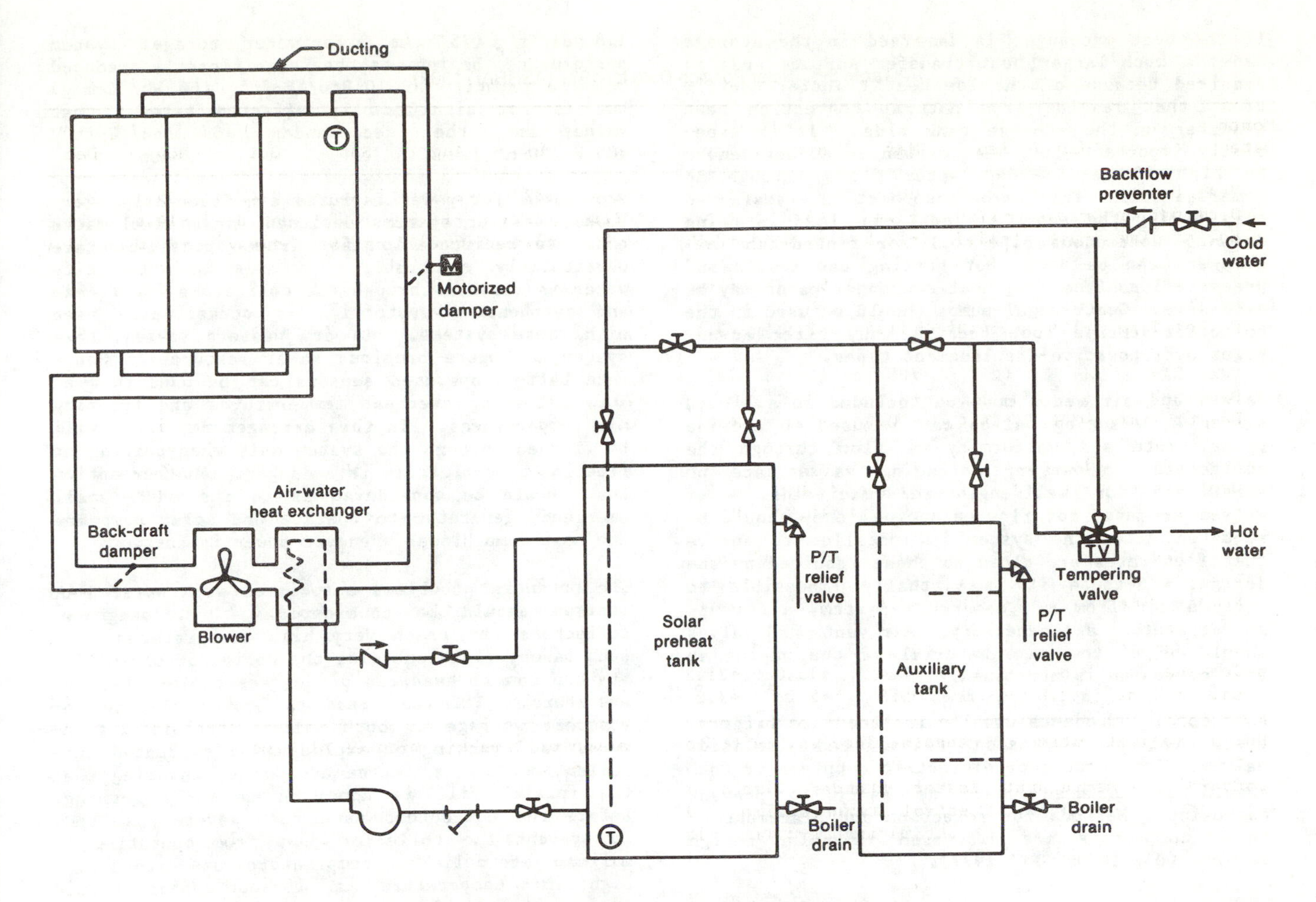

Fig. 4-9. Indirect service hot water system with forced air circulation

providing the advantages of storage stratification (highest temperature to the load and lowest temperature to the collectors). The controller selects the tank to be heated and the tank to meet the building load such that solar energy collection is maximized while the highest practical heat delivery temperatures are maintained. The highest temperature tank is heated when the potential for solar energy collection is high; a cooler tank is heated when solar intensity is low. Water may be supplied to the collector from a cooler tank and returned to a warmer tank, with series flow between them. The load is carried by the tank with a temperature nearest that required to meet the demand. The added cost of a considerably more complex control system and of multiple insulated tanks must be weighed against performance improvement. It is unlikely that the benefit would outweigh the cost in residential-size installations, but there may be economic advantages in large commercial systems.

Stratification in liquid storage tanks should be encouraged because it allows the hotter fluid to be supplied to the load and the colder fluid to be applied to the collector, thus improving collector efficiency. The resultant improvement in system collection is in the range of 5% to 10%. Phillips and Dave (1982) discuss the effects of stratification on system performance in terms of a stratification coefficient and several parameters that affect it.

For maximum efficiency, water from the bottom of the storage tank should be supplied to the collector (or to the collector heat exchanger), and heated water from the collector or exchanger should be delivered to the top of the tank. Similarly, the load loop is supplied with hot water from the top, and return from the load or make-up water is admitted to the bottom. Under normal operating conditions, temperature gain through the collector loop is 10° to 20°F (6°-11°C) near noon on a sunny day, and the load exchange system is designed for approximately the same water temperature decrease. Although the storage temperature could be 10° to 20°F (6°-11°C) higher at the top than at the bottom if effective baffles are used, stratification is difficult to achieve because of comparatively high pumping rates, turbulence at the tank entrance and exit openings, and variations in temperature rise through the collectors during the day. A vertical tank is best for stratification but requires greater height for installation. Practical design, therefore, should normally be based on assumptions of uniform storage tank temperature. Mechanisms to enhance stratification are typically not used in routine design practice because increased performance does not pay for the cost of implementation.

If the heat exchanger is immersed in the storage tank, a much larger heat transfer surface area is required because of the low heat transfer coefficient that results from natural convection heat transfer on the storage tank side. It is especially important that heat exchanger effectiveness be high if service hot water flows through an immersed coil. This is because heat is transferred only during the short periods when load is being drawn. Continuous pipe coils or finned-tube exchangers can be used, but scaling can result and pressure loss from long heat exchanger paths may be excessive. Centrifugal pumps should be used in the collector-storage loop because they offer advantages over positive-displacement types.

Valves and air vents must be included in a liquid system. Balancing valves can be used to provide proper rate and uniformity of flow through the collectors. However, balancing valves are no substitute for well-engineered manifolds. If valves are used for flow balancing, they should be adjusted after the system is installed to ensure that flow rates are close to those required by the design. Because it is virtually impossible to exclude air from a pressurized system, air vents and separators are necessary. Air vents and valves should be of the same material as the piping so that corrosion is minimized.

Additional components usually included are filters, check valves, expansion tanks, and isolation valves. Ion getters or dielectric couplers or both should be included where necessary to prevent corrosion. Methods for selection and placement of these components are discussed in solar design manuals (DOE 1978a, ITT 1977).

The flow capacity of the collector for liquid systems is important because increased flow rate reduces collector temperature rise and, in turn, reduces collector losses at a given inlet fluid temperature. However, higher flow rates increase pumping power, so a compromise must be reached. Simulations have shown that solar heat utilization is relatively insensitive to the product of collector coolant flow rate and specific heat above a value of about 2 Btu/h-ft^2_c°F (11.5 W/m^2_c°C) (Dickinson and Cheremisinoff 1980b). To obtain maximum collection efficiency, a collector coolant capacity rate of 10 Btu/h-ft^2_c°F (57 W/m^2_c°C) is required (Hewitt and Griggs 1976). This corresponds to the typically recommended coolant flow rate of 0.02 gpm/ft^2_c (48.8 kg/h-m^2_c), if water is used. (Recent tests at NBS have indicated that for systems that do not use a heat exchanger, performance can be improved by using much lower flow rates because the lower flow rate encourages tank stratification. The increased stratification can more than make up for decreased heat transfer in the collectors. However, little field experience with these lower flow rates is available, and proper use of this principle will likely require changes in collector design and pipe sizes.)

Many studies have confirmed that the appropriate thermal storage capacity for service hot water systems is 15 Btu/°F-ft^2_c (85 Wh/°C-m^2_c) (Dickinson and Cheremisinoff 1980b); this is equivalent to 1.8 gal/ft^2_c (75 kg/m^2_c) for water storage. System performance begins to be significantly reduced below a capacity of 10 Btu/°F-ft^2_c (60 Wh/°C-m^2_c) but is not significantly affected between that value and the recommended 15 Btu/°F-ft^2_c (85 Wh/°C-m^2_c).

Provisions for power failures are frequently overlooked. Water systems designed for mild climates that use recirculation for freeze protection are particularly vulnerable. A mode in which city water is flushed through the collectors for freeze and overheating protection is occasionally used with these systems. In dry western states, this system can waste precious water resources. Separate battery-operated sensors can be used to measure collector overheat temperatures and freezing air temperatures. In this arrangement, water will be flushed through the system only when boiling or freezing of collectors is a danger. Another option that should be considered is to install a small emergency generator to operate the solar pump and the heat dump blower whenever power is lost.

The potential problems associated with solar loop startup should be considered. Dry, stagnated collectors can reach very high temperatures. In such cases, fluid entering the collector on initial startup can be hundreds of degrees cooler than the absorbers. This can lead to broken glazings or absorber warpage or both with flat-plate collectors, and cracking or exploding of evacuated tube collectors. This can be avoided by ensuring that the initial fill is done during early morning, before the collectors have been severely heated. To prevent the collector pump from operating if maximum safe collector temperatures are exceeded, a high limit temperature pump lockout control should be installed on the absorber.

System performance can be improved during winter morning startup of large, antifreeze-filled collector arrays by adding a bypass warmup mode. On cold startup, a significant volume of very cold fluid is contained in the piping outside the building so that heat is not removed from storage as this cold fluid passes through the heat exchanger. The fluid can be made to bypass the heat exchanger until the entire collector loop has been heated above the storage temperature.

Performance Comparison of Systems

A National Bureau of Standards study (Liu and Fanney 1980) compared measured and simulated performances of several solar hot water systems. Aside from thermosiphon systems, single-tank systems performed better than double-tank systems by about 15%, probably because the double-tank systems had considerably larger surface area-to-volume ratios, thereby increasing storage losses. Direct systems performed about 5% better than indirect systems in both single- and double-tank configurations.

In another study (Guinn, Novell, and Hummer 1981) six commercially available solar hot water systems of various types were tested for several months under laboratory conditions. No consistent trends emerged on the basis of system type, but the performance advantage of single-tank systems was evident.

Finally, the Tennessee Valley Authority tested 32 commercially available solar hot water systems under laboratory conditions according to ASHRAE Standard 95 for service hot water systems (Chinery and Wessling 1981). No conclusions were drawn concerning relative performances of different system types. Field tests of similar systems later revealed that comparable systems performed better under laboratory conditions. However, the performance rankings based on the laboratory tests did not change with the field test results (Wessling 1981). Several studies have compared systems in terms of reliability (ESG, Inc. 1984; Jorgensen 1984). In general, drain-back and recirculation systems are the most reliable, and drain-out the least (due largely to failure of valves and vacuum breakers).

Combined Space and Service Hot Water Systems

The economics of a solar space heating system can be improved if heating is combined with a solar hot water system. A dual-purpose system is usually more cost effective because an otherwise idle collector can be used in the summer to heat water. Note that all collector and storage subsystem design considerations discussed above also apply to combined space and hot water systems. Design considerations pertinent to the space heating-hot water interface will be discussed here.

In a combined system, as shown in Fig. 4-10, heat can be transferred from the main storage tank to a much smaller service hot water tank through a liquid-to-liquid heat exchanger. Pumped circulation of both liquids, as shown in the figure, or thermosiphon circulation can be used. Circulation and heat transfer to the preheat tank occur whenever a positive temperature difference exists between the main solar storage and the solar preheat tank (with a temperature-limit control to avoid overheating). Cold water from the main is supplied to the preheat tank when a hot water tap is opened, and warm water from this tank flows to the auxiliary heater (equipped with a thermostat). The auxiliary unit is a standard water heater (gas-fired or electric).

Another type of auxiliary heater arrangement is an in-line, high-input (demand) heater that has very little storage and simply boosts the temperature of water passing through. A disadvantage of this booster unit is the variable temperature of the solar-heated water supplied to it and the resulting variable delivery temperature from it. Using a water heater with a thermostat avoids this difficulty.

Natural convection in the storage tank results in a low heat-transfer coefficient outside the tube surfaces. The systems described above may also involve use of conventional shell-and-tube, liquid-to-liquid heat exchangers external to the solar storage tank. The total surface area of coil-type exchangers must be considerably larger than the area of external exchangers, but the cost may be lower. Only one pump is required for each heat exchanger, rather than two. However, power requirements may be greater because of the larger pressure drop through the long tubing, a consideration that is negligible in most cases. Comparison of alternatives should also include such considerations as accessibility and maintenance, assurance of leak detection, code compliance, overall thermal performance, and economic feasibility.

Building code restrictions may prohibit use of conventional shell-and-tube exchangers to transfer heat from solar storage to potable water. Possible leakage that might result in contamination of the potable supply can impose a design requirement that eliminates the possibility of back-flow. An exchanger should be designed (unless codes change) so that fluids cannot mix if a tube wall leaks. This implies a double-walled design.

Another option is to immerse the hot water preheat tank in the larger main solar storage tank. With no pumps or heat exchangers, heat can be conducted from main storage through the tank wall into the hot water supply. Again, questions of accessibility and leak detection must be considered. Many building codes will not permit this design because of possible contamination of the potable supply by leakage from the main storage tank. Another problem is the near equality of temperature between the two tanks, but delivery of overheated water to the load can be prevented by using a tempering valve to mix cold water with overheated water to deliver water at the preset temperature.

An air heating solar system can be combined with a service hot water system, involving use of an air-to-liquid heat exchanger at the collector outlet and a one- or two-tank hot water system. It will provide hot water less efficiently than an all-liquid system because of thermal resistance on the air side of the heat exchangers, but it does prevent water system contamination by collector or storage fluids.

In winter, solar heat supplied to service water cannot be supplied to space heating, but it is immaterial to which use the solar heat is applied, assuming the same auxiliary energy source is used for both. However, in summer nearly all service hot water requirements can be supplied by the solar air system. Hot air from the collector, after passing through the service water coil and blower, can be vented externally. Air to the collector can be supplied from the building space itself or from outdoors through an inlet duct and damper. A simple arrangement requires adjustment of a manual damper twice a year to provide summer hot water by drawing building air through the collector, coil, and blower to the discharge vent. An alternative is a closed loop that provides air from the collector through the coil and blower, then back to the collector through a bypass duct rather than through the rock bed (see section on air systems). Several system configurations and control schemes for service hot water are discussed in the <u>Solar Heating Systems Design Manual</u> (ITT 1977).

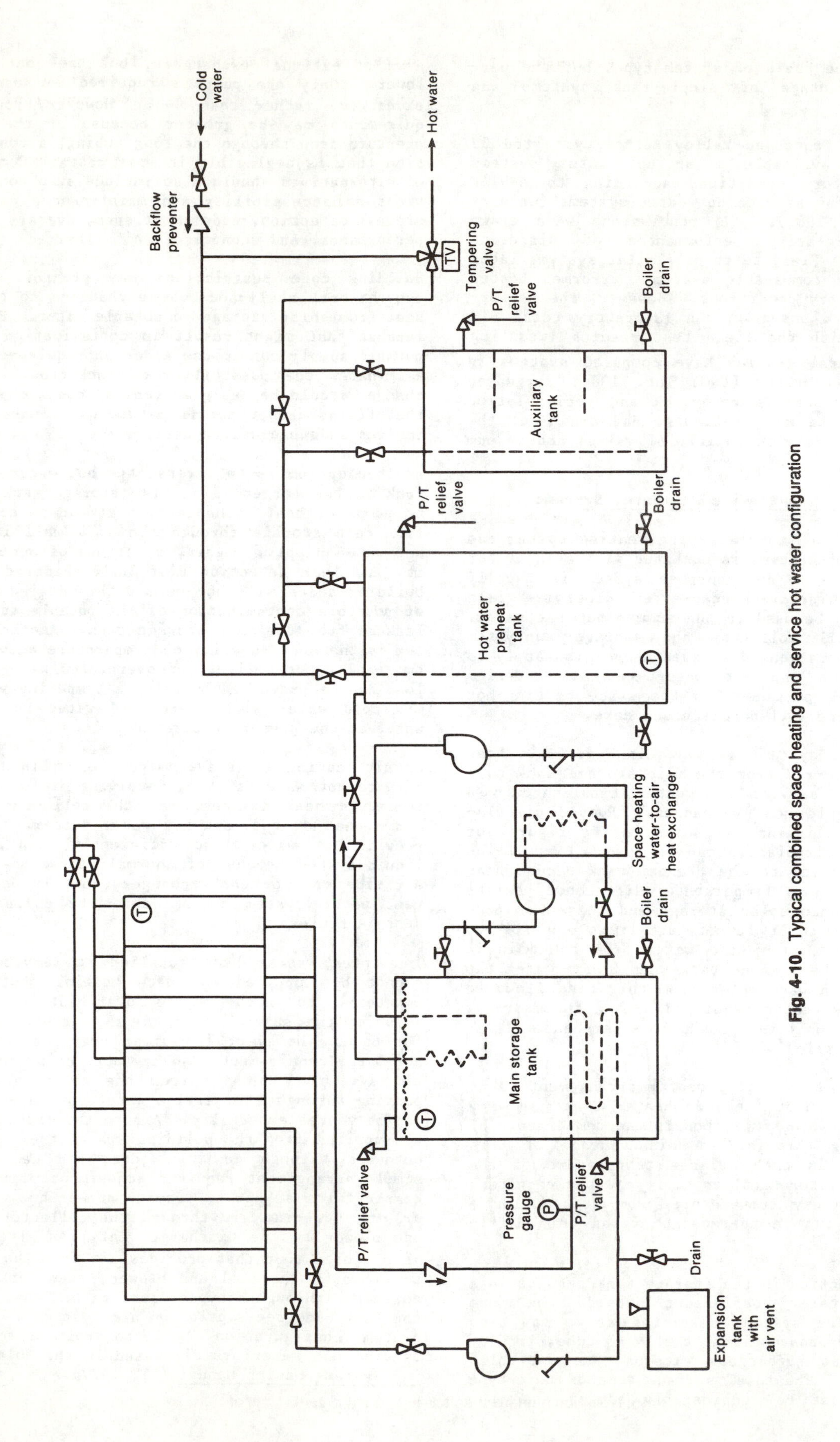

Fig. 4-10. Typical combined space heating and service hot water configuration

ACTIVE SPACE HEATING SYSTEMS*

In active space heating systems, either air or a liquid can be used to collect solar heat and to supply heat to the occupied space. If a liquid collector is used and heat is stored in a hot water tank, the stored hot water can be used as the heat source for a hydronic heating system (e.g., under-window radiators, baseboard heating strips, or radiant floor or ceiling panels). Due, however, to the lower water temperatures supplied by solar collectors, it is usually necessary to use a forced air system in which the air is heated with a central heat exchanger (a water-to-air coil) to transfer heat to the space. For an air collector, warmed air can be delivered directly to the space via a warm-air heating system. Heat can be stored in a rock bed and subsequently delivered to air circulated through the heated rocks to the occupied space. Alternatively, a smaller air collector can be used without storage to supply heat only during daylight hours. Warm air is normally supplied to the load at temperatures of 120° to 150°F (49°-66°C), a comparatively moderate requirement for air collectors. For commercial buildings, liquid systems are usually preferred. Such buildings are almost always heated by terminal boxes, fan coil units, heating coils in air ducts, or radiant hot water units, all requiring hot water. Air systems, however, may be appropriate for small buildings.

The temperature at which the heated fluid must be supplied to the load is dictated primarily by the amount of heat-transfer surface available. With liquid systems, the smaller the heat exchange surface, the higher the fluid temperature must be. Space heating systems that use radiant floor or ceiling heating panels are well suited to solar heating because of the lower temperatures at which these devices operate. Hydronic baseboard units operate most effectively at temperatures of 140°F (60°C) or higher. However, lower temperatures can be used by installing an additional length of baseboard heater to increase the amount of convective heat-transfer surface. Flat-plate collector systems work best in applications that do not require high temperatures. Assuming availability at acceptable cost, concentrating or evacuated tube collectors can be used when high temperatures are required.

Fluid temperature requirements thus depend on the characteristics of a specific space heating system. These requirements affect, in turn, operating conditions in the solar heat supply and storage systems.

Liquid Heating Systems

Virtually all collector and storage subsystem design considerations discussed for service hot water systems also apply to active space heating systems. Additional considerations for space heating systems are discussed below.

Collector and Storage

Recommended storage capacity for a space heating (or combined space and hot water or just a hot water) system is 15 Btu/°F-ft^2_c (85 Wh/°C-m^2_c), enough to provide diurnal storage. This corresponds to 1.8 gal/ft^2_c (75 kg/m^2_c) for water storage (Dickinson and Cheremisinoff 1980b).

Arguments have been advanced for sufficient capacity for seasonal storage, where solar heat collected in summer and fall could be used in winter and spring. Because of economies of scale, seasonal storage systems are advantageous for large apartment complexes or for districts of buildings. However, seasonal storage is not useful in systems that provide fairly constant, nonseasonal loads such as service hot water systems; the gain achieved by storing summer heat is more than offset by lower seasonal storage system efficiency. Also, threshold temperatures for meeting the entire load are higher for hot water systems than for space heating systems. Therefore, for combined space and hot water systems, a two-tank (one large and one small) system has better performance (Sillman 1981).

The analytical study reported by Sillman indicates that, in contrast to diurnal storage systems, seasonal storage systems show only slightly diminishing returns as system size increases. Therefore, seasonal storage systems that provide nearly 100% solar space heat may be economically preferable (on a large scale) to systems sized for about 50% solar space heat. This is especially true for space heating applications with relatively small loads.

Another analytical study reached similar conclusions (Braun, Klein, and Mitchell 1981). Significant reductions in collector area and increases in system performance were shown for seasonal storage systems; Fig. 4-11 illustrates performance trends for Madison, Wisconsin. The effects of load heat exchanger size, tank insulation, and year-to-year weather variations were studied; the optimum collector slope for a seasonal storage system was shown to be approximately equal to the latitude. Furthermore, a simplified design method for large water storage volume-to-collector area ratios is presented. Drew and Selvage (1980) present another sizing procedure, which includes economic optimization.

In regions where freezing does not occur, water can be used both as the collection medium and as storage. In such cases, a direct system is suitable (see Figs. 4-4, 4-2, and related discussion). In regions where freezing seldom occurs, a recirculation system (see service hot water systems section) can be used for freeze protection. In a freezing climate, the drain-out or drain-back options previously discussed can be used.

Alternatively, an unpressurized, vented collector can be designed to drain into storage whenever the pump (centrifugal) stops; this is the drain-back

---

*Jordan and Liu (1977, Ch. XII) discuss space heating methods and solar space heating systems in detail. See also DOE (1978a) and Dickinson and Cheremisinoff (1980b).

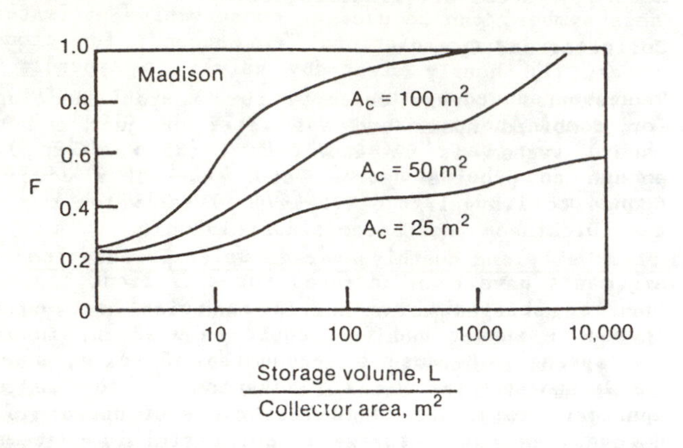

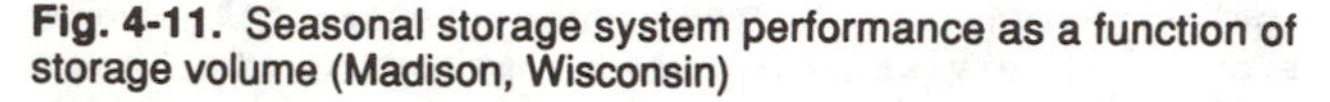

Fig. 4-11. Seasonal storage system performance as a function of storage volume (Madison, Wisconsin)

system discussed previously. If a self-draining collector is not used in a climate with subfreezing temperatures, a nonfreezing liquid may be used in an indirect configuration, with a heat exchanger provided for transfer of heat from the collector loop to the water storage tank. A design using such a system is shown in Fig. 4-10.

In space heating systems not used in summer months, protection against boiling can be provided by draining the collectors, pressurizing the piping system, or "dumping" excess collected heat through an air-cooled fan coil unit. Under no load or no flow conditions, a low-pressure glycol system could boil. Boiling either a propylene- or ethylene-glycol solution can cause the solution to turn acidic. An acidic solution (i.e., a pH of less than 7.0) will increase the probability of corrosion. Also, as the solution boils off, a residue or varnish is deposited inside the collector and degrades its performance. If one suspects that the solution has boiled and has a pH of less than 7.0, the solution should be tested and if necessary, drained; the collector loop should then be flushed, and the fluid replaced as required.

Finally, hybrid active-passive systems have been built that combine liquid-cooled collectors with a massive integral storage component that passively heats the building by radiation and natural convection; these are called active charge-passive discharge systems. They can use a floor slab, a structural wall, or an interior water tank as the storage component. Performance is comparable to active and passive designs over a range of climatic conditions, yet system costs show an economic advantage over all-active systems (Swisher 1981).

Space Heating and Auxiliary Use

A solar space heating system can be designed to supply 100% of the annual energy requirements of a building without an auxiliary system, given sufficient solar collector area and thermal storage capacity. However, such a system is generally not cost effective. A direct solar system is normally designed to meet from 30% to 80% of the annual energy requirements for heating. The backup system should be designed to meet 100% of the heating or cooling requirements. Backup systems can include oil- or gas-fired furnaces or boilers, electric resistance heat, wood furnaces or stoves, central steam or hot water generated off site, or a heat pump. The nature of the backup system will affect solar system configurations and controls.

An auxiliary heater uses conventional energy and conventional means to transfer heat from storage to the occupied space. In most solar heating systems involving liquid collection and storage, heat is supplied to the load by warm air circulated through a central blower and a water-to-air heat exchanger (see Fig. 4-10). Alternatively, warm water from the storage tank can be circulated through heat-exchange surfaces in the occupied space or through reheat coils in HVAC ducts. Practical design temperatures are about 140°F (60°C) for the central air exchange system and baseboard convector units, and about 120°F (49°C) for panel (radiant) heating systems. Solar heat is especially suited for use in reheat coils where usable water supply temperatures may be as low as 100°F (38°C). Unlike conventional systems, the temperature of solar-heated water available to the heating system may vary considerably, so components must be able to provide good heat exchange at lower supply-water temperatures.

Maximum advantage is gained by supplying auxiliary heat to the load loop rather than to the collector loop or storage unit. Such a design minimizes auxiliary energy use by using it only to augment or replace solar heat being supplied to the load. Otherwise, the collector operates at a higher temperature (with corresponding lower efficiency) than necessary, and some heat storage capacity is used for auxiliary heat rather than for solar heat as designed.

Auxiliary heat can be supplied by a conventional hot water boiler in one of two modes. Although not preferred, solar-heated water can be pumped through the auxiliary heater on its way to the load (auxiliary in series), as shown in Fig. 4-12a, with a temperature increase provided by auxiliary energy if needed. The water is then returned to the storage tank. In this mode, auxiliary heat is used as a "booster" so that solar-heated water is further heated in the boiler. The temperature of the return water may be higher than the storage temperature, thereby adding part of the auxiliary energy to storage. Continued operation would gradually drive the solar storage temperature up and thus use storage capacity for auxiliary energy rather than for solar energy, reducing collector efficiency. A storage bypass loop can be provided to avoid this problem.

A preferable arrangement, using minimum auxiliary fuel, has the auxiliary in parallel with the load. Only solar-heated water is used whenever the storage temperature exceeds the required heating coil, convector, or radiant panel temperature (see Fig. 4-12b). When the storage temperature is too low, circulation of solar-heated water is discontinued, and auxiliary heat is used exclusively. Piping and valving must be arranged so that water bypasses the storage tank when auxiliary heat is in use; an automatic valve and single pump provide heat from the appropriate source. Under severe conditions, comparatively warm water in storage may go unused for a time while the auxiliary supply is meeting a high demand. Stored solar heat can be called upon later, when the load is less severe or when the storage temperature has been increased by addition of solar heat. Solar heat is therefore not wasted.

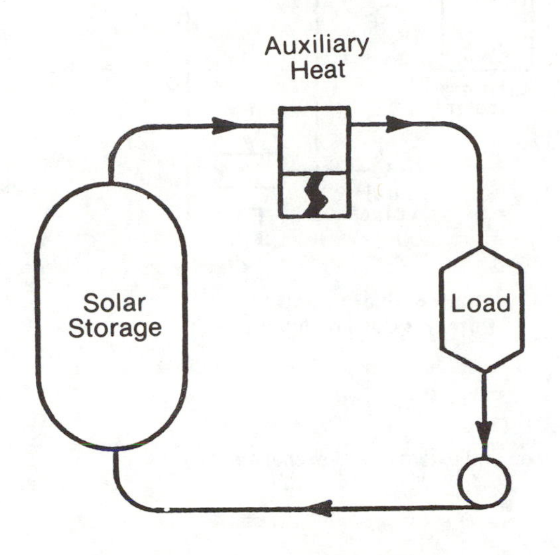

(a) Auxiliary Heat in Series with Storage to Load

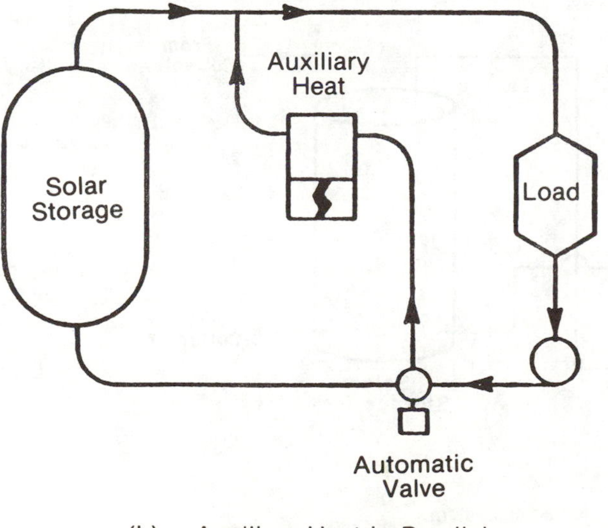

(b) Auxiliary Heat in Parallel with Storage to Load

Fig. 4-12. Auxiliary interfacing with solar storage

If distribution is by air heated by exchange, use of a two-coil arrangement provides a better strategy. In this mode, solar-heated water supplies heat to air in a preheater coil whenever the storage temperature is above a minimum, usually about 80°F (27°C). A second coil, immediately downstream from the solar preheater coil, is heated by auxiliary energy, increasing the air temperature to the required level. This method (Fig. 4-13) ensures maximum use of solar heat and minimum use of fuel, since even low-temperature solar heat is applied usefully whenever it is available. The cost of a second heating coil and increased blower power is incurred, typically a small expense for improved performance.

An alternative design for a warm-air system uses an auxiliary hot water boiler and only one water-to-air heat exchanger. Solar-heated water is circulated to the coil when the water temperature is sufficient to meet heating demand. If not sufficient, the solar supply is discontinued, and hot water from the auxiliary boiler is circulated through the coil.

A variation on this last strategy uses a conventional warm-air furnace just downstream of the solar preheater coil (Fig. 4-14). The furnace, designed to meet peak heating requirements, boosts the air temperature when necessary or furnishes all the heat if none is available in solar storage. In warm-air systems, this is an economical auxiliary supply.

Another location for the furnace is in a bypass air circuit designed for exclusive use with auxiliary energy. In this arrangement either solar heat or auxiliary energy is used. The temperature "boosting" advantage is sacrificed for better furnace efficiency, and an additional motor-operated damper is needed.

System Operation and Control

Controls for a liquid solar heating system must execute functions dictated by the operating requirements previously outlined. Most control systems involve sensors, switches, and motors to (a) transfer solar-heated liquid from collector to storage, directly or indirectly, whenever storage temperature can be increased by this operation; (b) transfer heated water from storage to load whenever heat is needed and solar-heated water is at a usable temperature; and (c) provide auxiliary energy to the load whenever solar storage cannot meet the requirements. Other control elements may be required for safety and other purposes, such as to prevent undesirable high temperatures and to protect from freezing.

A practical method for controlling solar collection uses a sensor to actuate the collector pump (and associated storage pump, if a heat exchanger is used) whenever the fluid temperature at the collector outlet exceeds the lowest temperature in the storage by a preset amount, say 10°F (6°C). (Controlling collector temperature alone is not advisable because, under many conditions, the fluid would be cooled rather than heated.) The collector

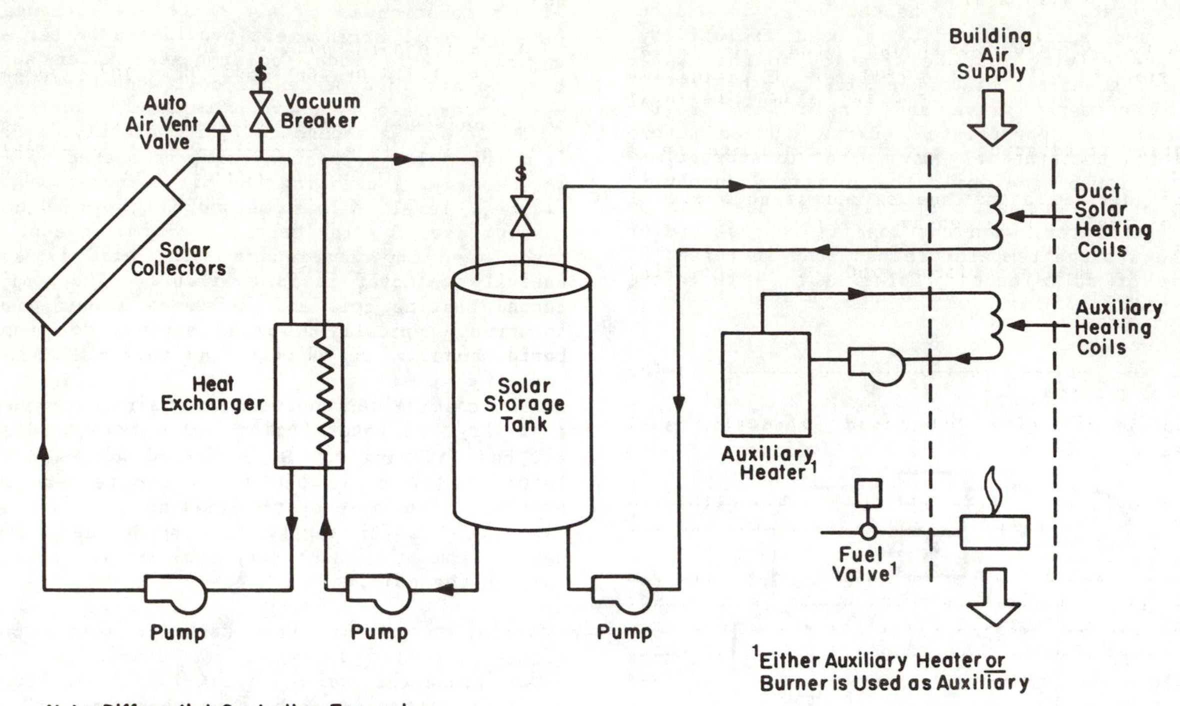

Fig. 4-13. Indirect space heating system with antifreeze loop and external heat exchanger

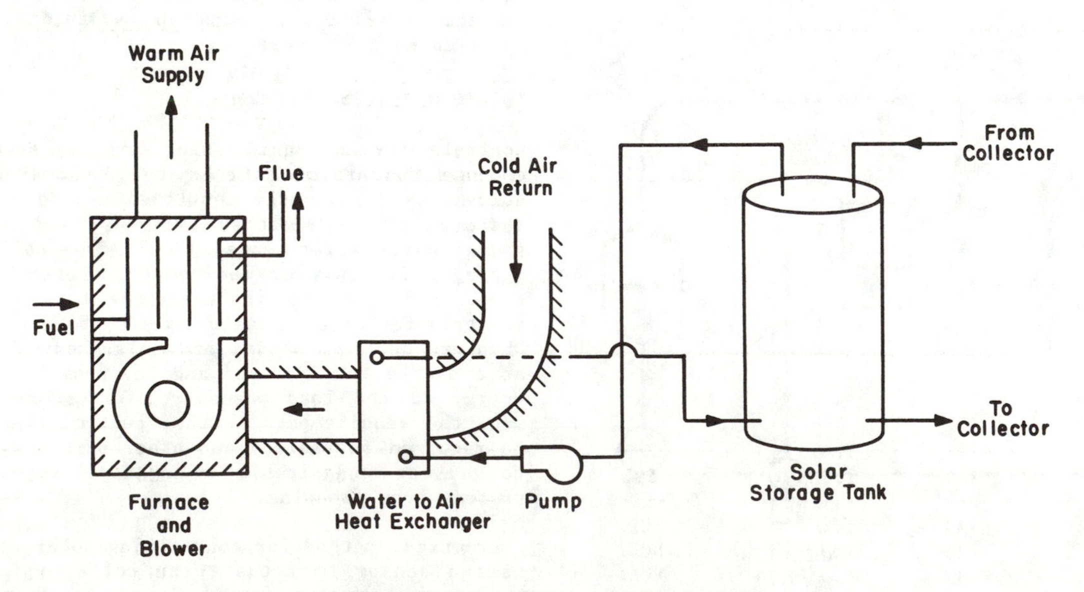

Fig. 4-14. Solar supply in a forced air system

sensor should be located either on the absorber plate near the fluid exit or close enough to the absorber plate to be influenced by collector temperature even though the pump is not running.

Several differential thermostats are commercially available for this service. The on-off control described above is common in residential applications; but, in some commercial installations, use of modulating controls will improve performance. Electric energy can be saved by operating the collector pump at reduced flow (by variable-speed motors or by modulating valves) when solar input is below normal levels. In drain-back systems, a variable speed pump can supply sufficient head to fill the system at the beginning of the day and allow for less pumping power during normal flow. Blower power requirements in air solar systems can be substantial; the payoff potential of the variable flow rate strategy is therefore greater in air systems than in liquid systems. For a constant flow rate, the collector-storage thermostat should be set at the lowest temperature difference, typically 5° to 10°F (3°-6°C), at which the cost of energy required for pumping does not exceed the value of heat collected. A different approach to collector control logic, where the difference between collector stagnation temperature and tank temperature is used to control both on and off functions, is described by Lunde (1982).

A practical control strategy for residential heating involves a room thermostat with double set point, eliminating the need for sensing the storage temperature. Whenever the room requires heat, water from the solar storage tank is pumped to the load, regardless of storage tank temperature. (A low limit near 80°F (27°C) may, however, help reduce temporary decreases in room temperature.) If storage is warm enough, room temperature rises, and heat supply is subsequently discontinued. If, however, the stored water is not hot enough to meet the load, room temperature will continue to decrease until another thermostat contact, set to operate at 1° to 2°F (0.6°-1.1°C) below the initial, or higher, temperature contact, turns on the auxiliary system.

This control system is the most practical if the building is heated by air from a solar water-to-air coil, and if auxiliary heat is supplied to the air either by a hot-water boiler and separate water-to-air coil or by a warm-air furnace. The double-contact thermostat first actuates the air blower and hot water pump for solar heat supply. If that source is not adequate, the second thermostat actuates the auxiliary heat to supplement (boost) the solar. Solar heat, even at low temperature, is thus fully utilized. This control system prevents use of auxiliary heat when solar heat can carry the entire load and provides intermittent deliveries of solar heat when the source is not hot enough to meet the full demand. However, if the room temperature drops slowly to a value between the set points, uncomfortable air delivery temperatures may result before auxiliary heat is supplied.

Control system details are discussed in Ch. 6.

Air Heating Systems

Solar air system development has been much more limited than that of liquid systems. Continuity of effort, however, has placed solar air heating at a fully comparable development stage.

Some important technical and operational factors of air heating systems are

- Solar-heated air can be used to heat space directly and more efficiently without heat exchange.
- Heat storage can be accomplished in a bed of loose solids, typically 3/4 to 1-1/2 in. (19-38 mm) rocks, which also serves as the heat exchanger.
- Both temperature stratification in a rock bed and return of air to the collector directly from the occupied space provide low-temperature 70°F (21°C) air to the collector, resulting in very favorable operating efficiency.
- The combination of air density, specific heat, and practical flow rates provides a considerably higher temperature rise through the collector, typically 70° to 90°F (40°-50°C), than through a liquid collector.

Typical System Design

As with liquid systems, there are numerous options in air systems for integrating the solar collector and rock-bed storage into the complete space heating assembly. A schematic of one of the simplest practical designs is shown in Fig. 4-15. This system uses a solar air collector, rock-bed storage, a single blower, and a conventional gas-fired furnace and distribution duct system. There are three primary modes of operation, depending on both availability of solar energy and space heating requirements: collector to storage, collector to space, and storage to space. To take advantage of rock-bed stratification, flow through the bed is reversed from the charge to discharge mode.

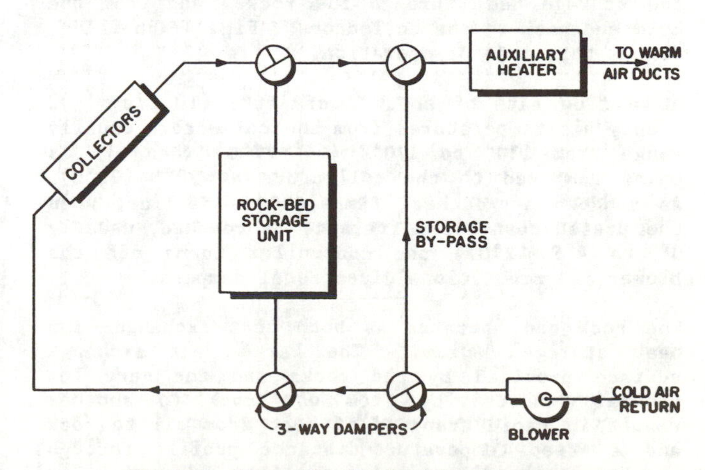

Fig. 4-15. Schematic of basic air-heating system

The two-blower system shown in Fig. 4-16 is more commonly used. In addition to the components in the single-blower system, it has an air-to-water heat exchanger for service water heating. It also has two modes of auxiliary use for space heating and one for summer hot water supply (Fig. 4-16d).

<u>Collector-to-Space Mode: Daytime Space Heating.</u> When heat is needed and the collectors are on, a room thermostat signals the control unit to move dampers to bypass storage and direct the flow of heated air from the collectors directly to the zone requiring heat, as shown in Fig. 4-16a. In this mode, hot air passes from the collectors through the blower, through the furnace, and into the warm-air distribution system. Air from the rooms circulates back to the collectors through conventional cold-air return ducts. Either a motorized damper or a check damper (operated by a slight pressure difference) is in this return duct. When room temperature requirements are satisfied, the thermostat breaks contact, and the storage mode (see below) is used.

When heat demand is high and the air temperature from the collectors is insufficient to meet demand, room temperature will continue to decline. A lower thermostat set point then turns on the auxiliary heater, which increases the air temperature in the distribution system. The full design capacity of the furnace will always provide sufficient heat to meet the demand, so the room temperature will be restored to the preset value.

<u>Collector-to-Storage Mode: Storing Solar Heat.</u> Solar collection and delivery of heat to storage are achieved in the collectors. The collector blower is actuated by a differential thermostat, with a "hot" sensor in the air passage at the collector exit and a "cold" sensor near the cold end of the storage bed. A temperature difference setting of 10° to 20°F (6°-11°C) permits operation whenever the value of collected heat exceeds the cost of blower operation. The controller signal that actuates the blower also positions dampers, so collector discharge air passes to the hot end of the storage bed, through the rocks, and from the cold end back to the collectors. Fig. 4-16b illustrates this mode of operation.

At a flow rate of about 2 cfm/ft^2_c (10 $L/s\text{-}m^2_c$), midday air temperatures from the collectors usually range from 130° to 170°F (54°-77°C) when air is being admitted to the collectors at 70°F (21°C). As sundown approaches, temperatures decline; when the preset turn-off difference is reached, usually 3° to 4°F (2°C), the controller turns off the blower and repositions directional dampers.

The rock bed operates as both heat exchanger and heat storage medium. The large heat exchange surface provided by the rocks and the very low thermal conductivity from one rock to another result in rapid transfer of heat from air to rock and a steep temperature-distance profile through the bed in the direction of air flow. Normally, as the air temperature supplied to the rock bed varies throughout the day, nearly all heat transfer occurs in a zone occupying about 3 ft (1 m) of bed depth

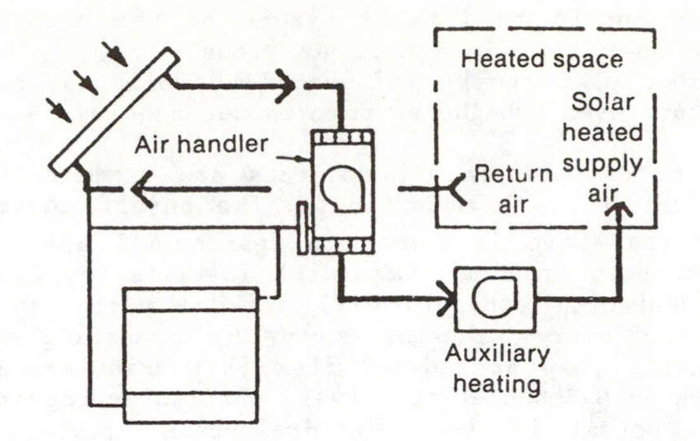

A Heating from collectors

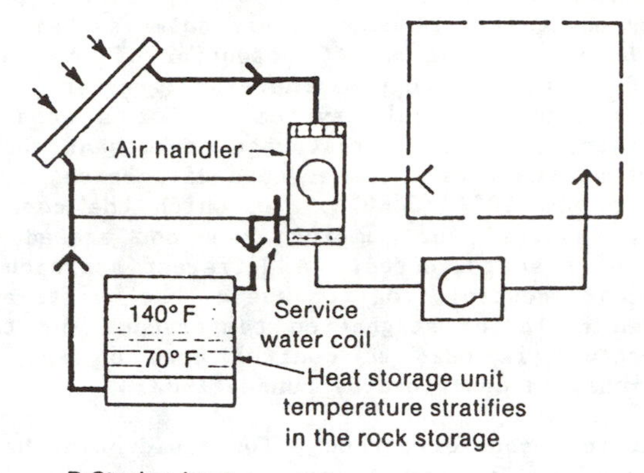

B Storing heat

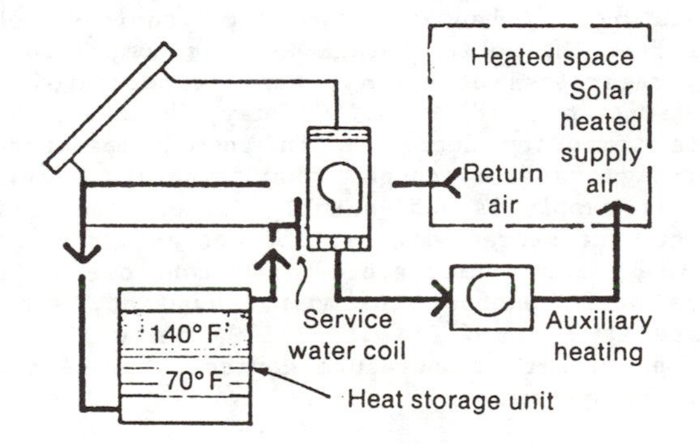

C Heating from storage

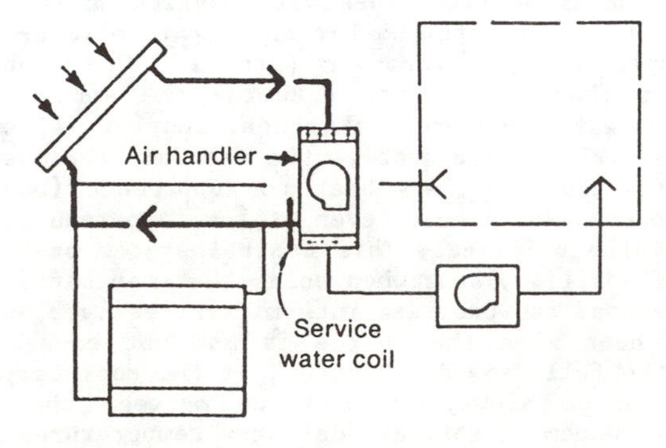

D Summer water heating.

Fig. 4-16. Operating modes of air collectors with rock-bed storage

(Karaki, Armstrong, and Bechtel 1977). Useful heat is completely extracted from air delivered to the bed from the collectors, while the cold end of storage remains essentially at room temperature, 68° to 77°F (20°-25°C). Typical rock bed temperature profiles for charge and discharge cycles are shown in Fig. 4-17.

During mild weather and on most spring and fall days, sufficient heat may be supplied to storage during each day (or during several days when little stored heat has been used) for rock temperatures at the cold end of the bed to be driven upward. The entire rock bed can then be at a temperature substantially higher than room temperature. Under this condition, air returning to the collectors is at a temperature considerably higher than 70°F (21°C). Collector heating can then drive outlet air temperatures to 200° to 250°F (93°-121°C) in mild sunny weather. Collector efficiencies are significantly lower at these conditions, and total energy collection is reduced as the bed becomes fully heated.

<u>Storage-to-Space Mode: Space Heating from Storage.</u> The third mode of operation, illustrated in Fig. 4-16c, is called for when space heat is required but solar collection is not taking place. Under these conditions, a room thermostat signals blowers to operate and dampers to move so that room air flows to the cold end of the storage bed, picks up heat from the rocks as it passes through the bed, and is directed through the furnace to the room via the air distribution system. Heat is thus supplied to room air by transfer from the heated rocks; the air leaving the hot end of the bed is only a few degrees cooler than the rock temperature at the exit point.

If the rock-bed discharge temperature is sufficient, air entering the room will provide enough heat to satisfy the thermostat, and the blower will stop. If, however, the room temperature continues to drop, the auxiliary heat supply will be actuated by the lower thermostat set point, and auxiliary heat will also be supplied. Operation then alternates between stored heat and stored heat plus auxiliary until stored heat is again sufficient to

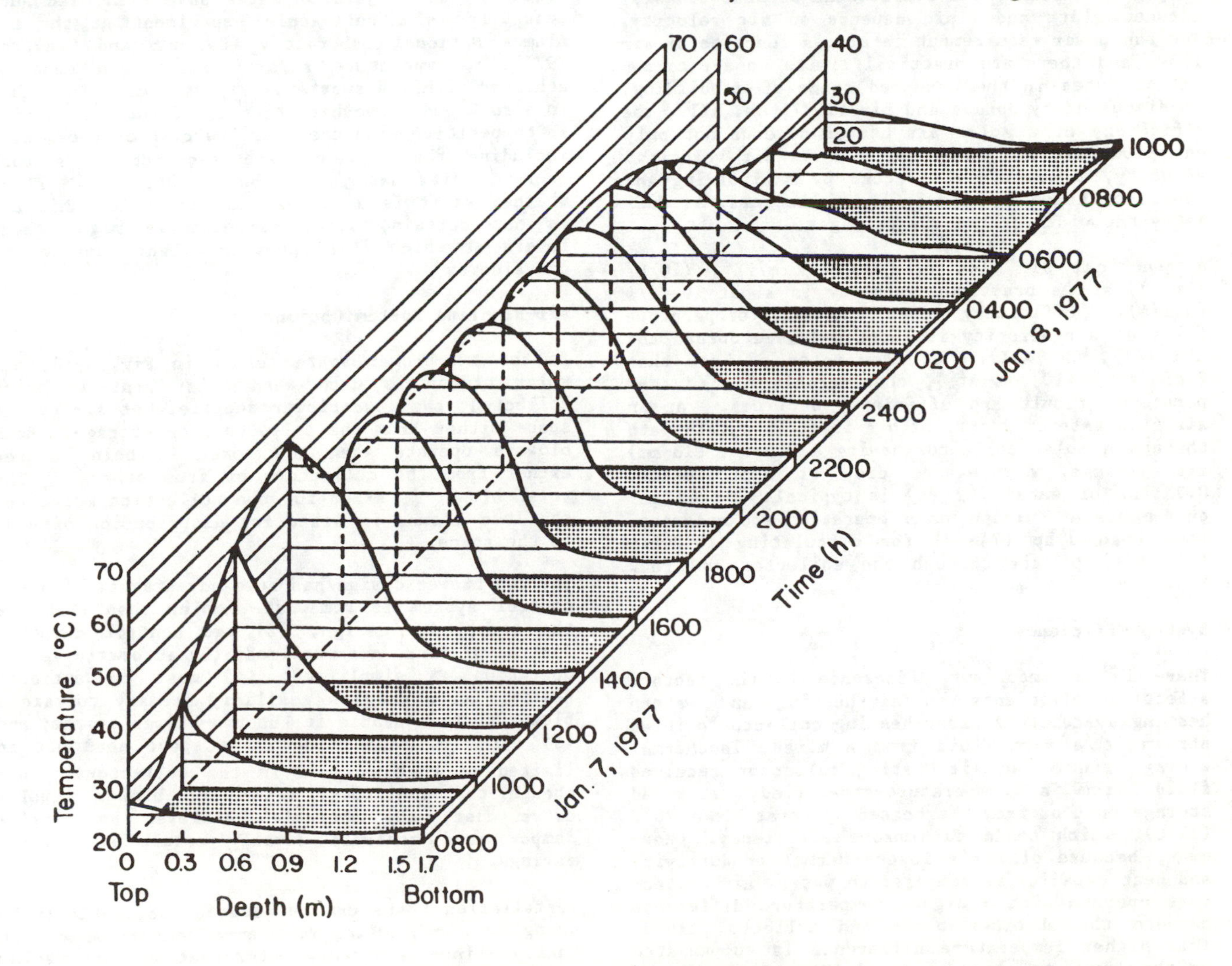

Fig. 4-17. Typical rock-bed temperature profiles

maintain the load by itself or until direct solar heat is available. Alternatively, the system may run until the upper set point is regained and shutoff occurs.

From the descriptions above, we can see that solar heat use is maximized by (a) collecting solar heat whenever moderate temperature delivery of 80° to 90°F (27°-32°C) is possible; (b) using such low-temperature heat, supplemented if necessary with auxiliary; (c) providing, by temperature stratification, high-temperature storage even when the storage unit is only partially heated; (d) bypassing storage when heat is needed during sunny hours; and (e) using auxiliary energy only as a supplement, not as a replacement, for solar energy.

Air Circulation Rates

The designer of a solar air system determines the air flow rate through the collector. Delivery temperature and, hence, absorber plate temperature are strongly dependent on air circulation rate; in a specific design, the coefficient of heat transfer between plate and fluid depends on air velocity. The fan power requirement is also a function of air flow, and there are practical limits to air circulation rates in the occupied space of a building. As discussed by Jordon and Liu (1977, Ch. XII), the efficiency of a solar air heater depends not only on volumetric air rate but also on air velocity. Velocity, in turn, is affected by manifolding and length of travel of air in the collectors, as well as by the width of the air passages.

A practical air flow rate is 2 cfm/$ft^2{}_c$ (10 L/s-$m^2{}_c$), and a practical velocity is about 10 ft/s (3 m/s). Efficiency rises and exit temperature falls as air velocity is increased (see Jordan and Liu 1977, Ch. XII). At flow rates of less than 2 cfm/$ft^2{}_c$ (10 L/s-$m^2{}_c$), considerably higher temperatures result and efficiency declines. At an air flow rate of 2 cfm, with a 13 ft (4 m) air path through a solar collector having a 0.5-in. (13-mm) air passage, a pressure drop of approximately 0.25 in. of water (62 Pa) is typical. Power requirements at this point of operation are moderate, less than 1 hp (746 W) for circulating 1000 cfm (472 L/s) of air through the collector and rock bed.

System Efficiency

There is an important difference in the factors affecting efficiency in air-heating and water-heating systems. A water-heating collector's inlet stream is a warm fluid from a mixed, isothermal storage tank. An air-heating collector receives fluid from a temperature-stratified, rock-bed storage unit or from the heated space at about 70°F (21°C), which tends to improve efficiency. However, because of air's lower thermal conductivity and heat capacity as compared to water, air collectors operate with a higher temperature difference between the absorber plate and collector fluid. This higher temperature difference is compensated by the lower temperature of fluid supplied to the collector. The net result is an average difference between plate temperature and ambient temperature approximately equal for the two systems; collection efficiencies thus are comparable.

Storage Options

Storage in water with a heat exchanger and storage in phase-change materials have been suggested as options to rock-bed storage in solar air systems. System efficiency is reduced, however, with either of these storage media, due to a lack of temperature stratification. Thermal energy storage for building heating and cooling is discussed in the next chapter.

One other type of storage may be comparable to a rock bed in its capacity to store heat and produce low recirculation temperatures. A bin filled with plastic spheres, metal cans, or other small containers of water or other high heat-capacity liquids can be used (Saha 1978). The containers should be small to achieve a high surface-to-volume ratio for effective heat transfer into and out of storage. Small jars of water have been used successfully in a full-scale experiment at the Los Alamos National Laboratory (Balcomb and Hedstrom 1977). Temperature stratification can thus be achieved within a substantially smaller volume than in a rock bed. Whether the cost of such containers is competitive with the very low cost of rock, even including the larger space required by a rock storage unit, has yet to be demonstrated. Also, whether water is a reliable material for this use is not certain, since temperatures might reach levels at which fluid pressure might rupture the containers.

Air-Handling System Options

In the two-blower system shown in Fig. 4-18, the solar blower is used when solar heat is being collected; the load blower supplies hot air to the space either from the collectors or storage. Both blowers operate when the space is being heated either from the collectors or from storage. The solar blower is sized for the collection mode, and the load blower is sized for distribution of heat to the space.

The two-blower design has some advantages: (1) the control system is somewhat simpler than that for the one-blower design; (2) two control dampers, instead of four, are required; (3) automatic damper operation is simplified; (4) most conventional heating units used as auxiliary already contain a blower that is usable in the two-blower system; and (5) air flow rate through the space need not be limited to that desired in the collectors. Although the additional motor and blower involve extra cost, the economies achieved by simpler dampers and controls usually result in overall savings.

Installation costs can be reduced substantially by using an air-handling module--a factory-made unit that combines a blower, water-heating coil, solar collector damper, solar storage damper, and summer vent damper in a cabinet that can be used along with a conventional furnace. Figures 4-19 and 4-20 show commercially manufactured air handlers for use in single-blower and two-blower systems, respectively.

Though air systems would appear to be freeze-proof, freezing has occurred when air dampers to air-to-water heat exchangers leaked. This encourages thermosiphoning of the air in the collector, which may freeze the water in the heat exchanger, and cause the heat exchanger to rupture. Locating the hot water heat exchanger as shown in Fig. 4-18 eliminates this problem.

Commentary on Air and Liquid Systems

The relative virtues of solar air collection and liquid collection systems have been a subject of considerable attention. There are advantages and disadvantages to each that bear on types of use, most suitable geographic locations, and future potential for widespread application.

Historically, liquid systems have had more attention and use. Conduits for transfer of heating fluid between collector and storage are conveniently small, and its high specific heat makes water a compact storage medium. Also, hot liquid systems are needed to drive solar absorption air-conditioning systems. Disadvantages of liquid systems are possible freezing of collector water, the corrosive effects of water in the presence of air on many common and inexpensive metals, damage that can result from accidental leakage from a solar system, and problems associated with boiling under special conditions. All these problems and hazards can be satisfactorily handled, but at the cost of disadvantages imposed by heat exchangers, self-draining and self-venting collector arrangements, pumping energy increases, corrosion-resistant metals, leakproof fittings, and control requirements.

Air system advantages and disadvantages are essentially the reverse of those associated with liquid systems. Drawbacks include bulkier conduits to move heated fluid between the collector and storage, a storage volume about three times that required by water storage, and increased parasitic power consumption. Air system advantages include freedom from hazards associated with corrosion, freezing, boiling, and liquid leakage; direct association with conventional warm-air heating systems; and heat supplied at temperatures usable in such systems.

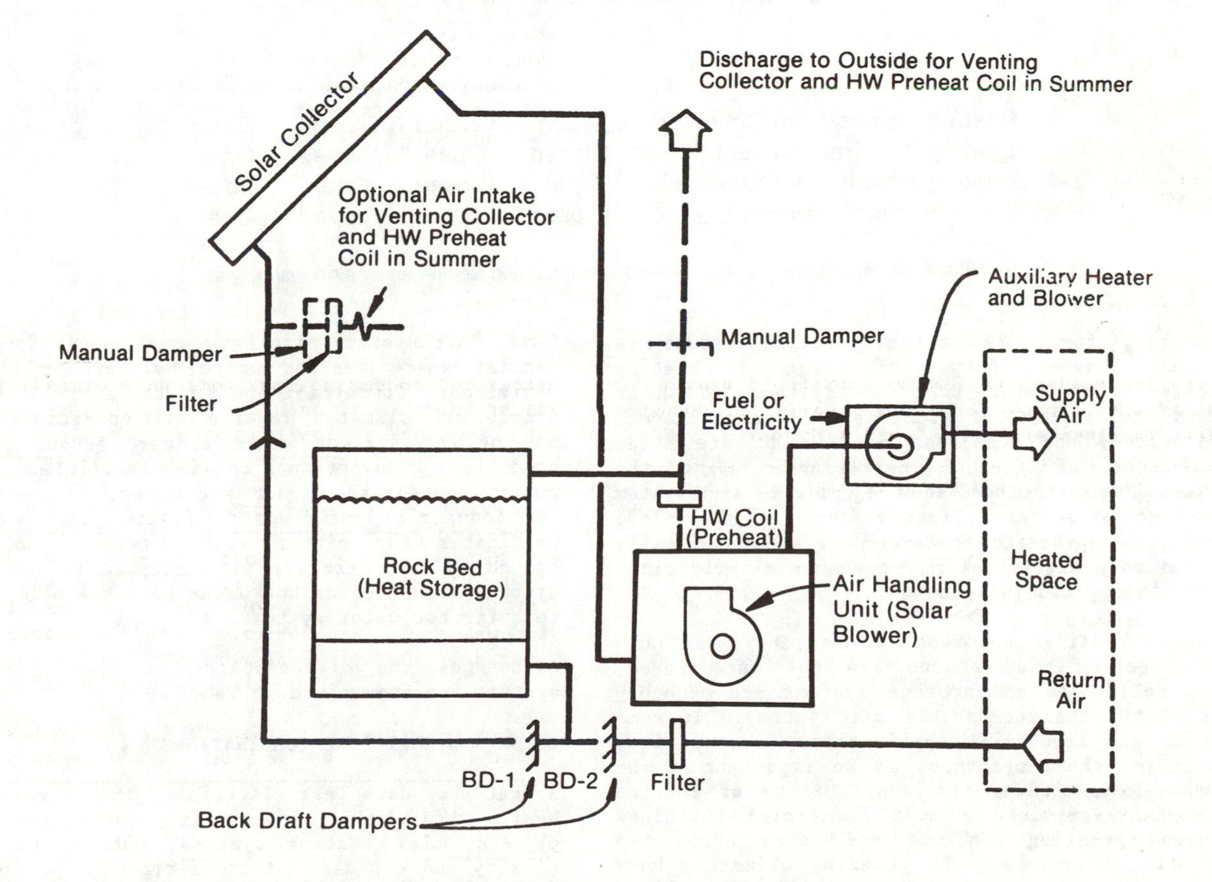

Fig. 4-18. Two-blower air-heating system schematic

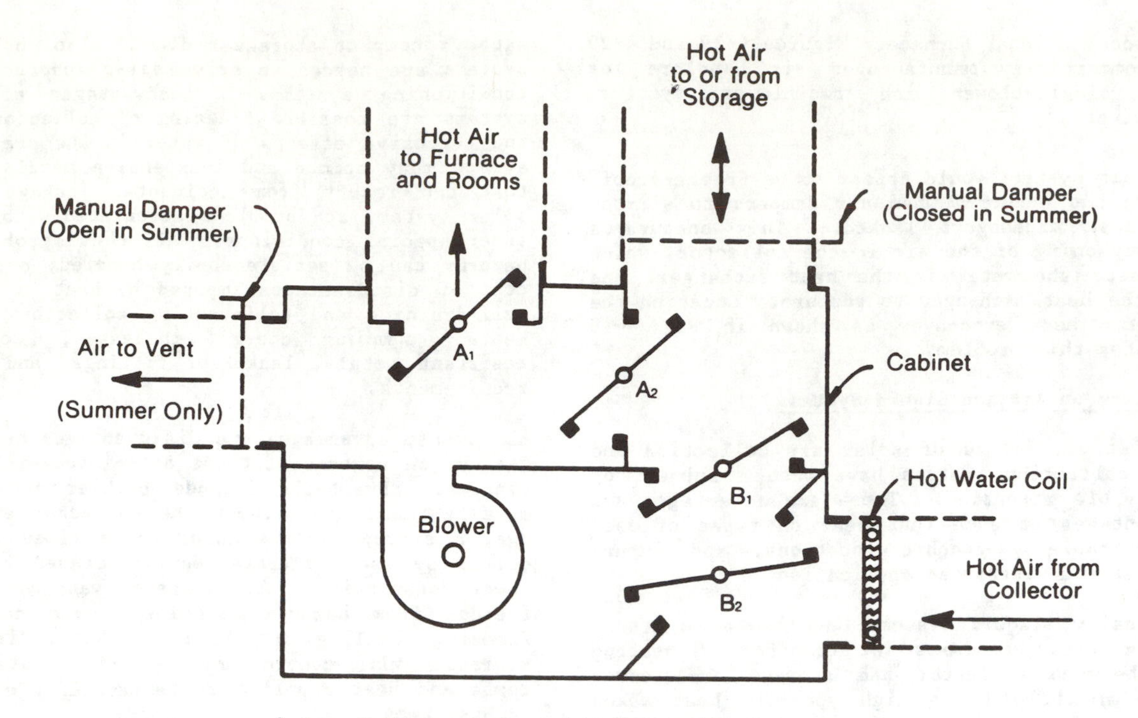

	A_1	A_2	B_1	B_2
Heating Rooms from Collector	Open	Closed	Closed	Open
Storing Heat from Collector	Closed	Open	Closed	Open
Heating Rooms from Storage	Open	Closed	Open	Closed
Summer Water Heating	Closed	Closed	Closed	Open

Fig. 4-19. Air handling unit for air-type collectors and single-blower air-heating system

Although performance data for operating systems are limited, seasonal output of solar air heating systems approximates that of systems in which liquid collectors of comparable transmittance, absorptance, and heat-loss coèfficient are used. Comparisons of typical performance using the F-Chart design method show a slight air system advantage in certain climates (Oonk et al. 1976). However, in practice, solar air systems typically consume more electrical power than similarly sized heat-delivery liquid systems.

If solar cooling equipment is used and/or if outdoor freezing temperatures are not encountered, water collection and storage systems are probably less costly than comparable air systems. In commercial and industrial applications, system sizes are large; the compactness is an important advantage. In addition, the availability of routine maintenance service in most commercial buildings reduces operating problems and costs associated with liquid systems. In freezing climates, however, and particularly in new residential applications, solar air systems avoid use of expensive measures required by liquid systems. In these circumstances, air systems, with rock-bed storage, appear to be more cost effective than liquid systems. Air systems also have more appeal for residential space heating where maintenance must be minimized. Commercial and industrial buildings in nearly all climates require air conditioning, so use of air systems would require, under present conditions, conventional cooling facilities. Solar cooling potential is not yet clear, so commercial and industrial buildings in freezing climates are candidates for either system type. Buildings heated by warm air are well adapted to solar air systems; those using hot water heat are well suited to solar hot water systems.

Advantages and disadvantages of liquid and air systems are summarized in Table 4-1.

SOLAR AND HEAT PUMP COMBINATIONS

A heat pump uses less electricity than a resistance heater when used as an auxiliary source for liquid or air solar heating systems, but it adds complexity and expense. A coefficient of performance (COP = heat delivered ÷ thermal equivalent electrical energy input) for heating of about 2 to 4 is usually obtained. However, COP depends strongly on heat supply temperature, which is usually the outdoor air temperature. The heat pump capacity,

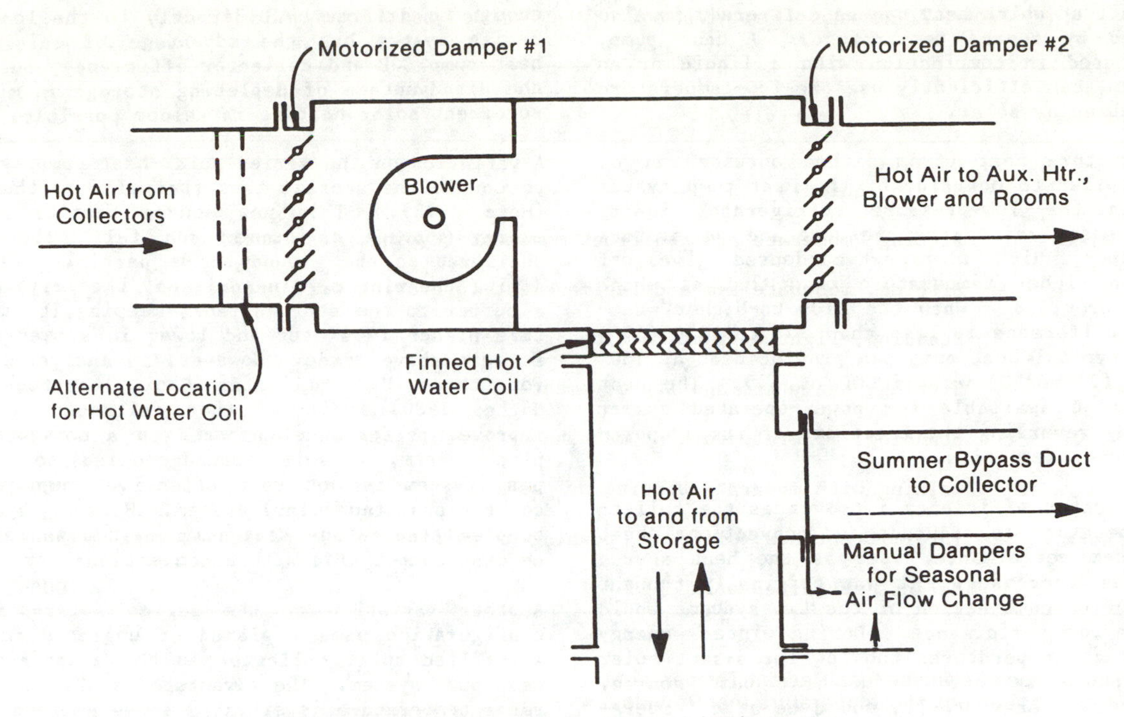

Fig. 4-20. Air handling unit for two-blower air-heating system

Table 4-1. Advantages and Disadvantages of Liquid and Air Systems

Characteristic	Liquid	Air
Efficiency	Collectors generally more efficient for a given temperature difference.	Collectors generally operate at slightly lower temperature.
System Configuration	Can be readily combined with service hot water and cooling systems.	Space heat can be supplied directly but does not adapt easily to cooling. Can preheat hot water.
Freeze Protection	Freeze protection may require antifreeze and heat exchangers that add cost and reduce efficiency.	No freeze protection needed.
Maintenance	Precautions must be taken against leakage, corrosion, and boiling.	Low maintenance requirements. Leaks repaired readily with duct tape, but leaks may be difficult to find.
Space Requirements	Insulated pipes take up only nominal space and are more convenient to install in existing buildings.	Duct work and rock storage units are bulky, but ducting is a standard HVAC installation technique.
Operation	Less energy required to pump liquids.	More energy required to drive blowers to move air; noisier operation.
State of the Art	Has received considerable attention from solar industry.	Has received less attention from solar industry.

or the rate at which heat can be delivered, is also influenced by source temperature. A heat pump, properly used in combination with a liquid or an air system, can efficiently use the low-temperature heat supplied by solar.

Low-temperature heat (from either outside air or water) applied to one side of the heat pump system evaporates the low-pressure refrigerant liquid. The compressor raises the pressure and temperature of the vapor, which, when next condensed, gives off heat at a higher temperature than that at which heat was provided. When the "low-to-higher" temperature difference is less than 30° to 40°F (17°-22°C), a typical heat pump can provide heat at 160° to 180°F (71°-82°C) with a COP of 3.5. The heat pump is also available for power-operated summer cooling by reversing the functions of its evaporator and condenser coils.

Solar collector efficiency improves as the collection temperature is reduced, and conventional heat pump system performance rises as the heat source temperature increases. It was originally thought that a series combination of the two systems would have superior performance. During winter, energy collected at temperatures too low for direct solar heating could be used as a heat pump source. Because this energy usually would be at a temperature above that of ambient outdoor air, heat pump capacity and COP would increase over those for the heat pump alone. At these low temperatures, collector heat losses would decrease, and higher collection efficiency would result. Alternatively, less expensive collectors might be used without sacrificing efficiency. However, use of an air-to-air heat pump in parallel with the solar system, with outdoor air rather than solar heat as the heat pump source, usually requires less electric energy than does a series system.

System Descriptions

Combined solar and heat pump systems are of three basic types: parallel, series, and dual source (Mitchell, Freeman, and Beckman 1978; Freeman, Mitchell, and Audit 1979). The simplest is a conventional solar system, either liquid or air, with an air-to-air heat pump as an auxiliary energy source. This arrangement is called a parallel system (Fig. 4-21). Direct solar heating is used whenever possible; the heat pump operates whenever solar energy is insufficient. Electric resistance heat may be used when neither source can meet the load. This arrangement does not benefit from the use of solar energy as a heat pump source.

In a series solar and heat pump system, the heat pump is placed between the solar system and load, such that it "boosts" the solar energy temperature by using solar (from either a liquid or air system) as the heat pump source (Fig. 4-22). The evaporator is in the storage tank (or in a storage tank loop), and the condenser supplies heat to the fluid being circulated to the space. The heat pump uses stored solar energy whenever storage is above a set minimum temperature (usually just above the freezing point of the liquid in the tank). Provision is also made for direct solar heating by bypassing the heat pump when the storage temperature is high enough to deliver heat directly to the load. The series system has the advantage of raising both heat pump COP and collector efficiency, but it has the disadvantage of depleting storage in midwinter so direct solar heating is seldom possible.

A variation on the series solar-heat pump system is to couple the storage tank thermally to the ground (Metz 1978). This new source permits seasonal energy storage of summer and fall solar energy delivered to the ground to be partially retrieved during the winter. In addition, the earth acts as a buffer to the storage tank, keeping its temperature higher in winter and lower in summer than if it were above grade. However, an analytical study for three U.S. climates (Choi, Morehouse, and Hughes 1980) indicated that, although it showed improved performance compared to a nonground coupled system, a solar/ground-coupled series heat pump system is not cost effective compared to a conventional (nonsolar) system. However, a ground-coupled stand-alone heat pump system was shown to be cost competitive with a conventional system.

Another variation on the series solar-heat pump configuration uses a glazed or unglazed refrigerant-filled solar collector as the evaporator in a heat pump system. The advantage is that the evaporator temperature is elevated above ambient because of solar energy. With an appropriate collector area, these systems can have a 10% to 15% performance advantage over conventional heat pumps (O'Dell, Mitchell, and Beckman 1983). A related variation is to place the evaporator of a conventional air-to-air heat pump in an attached sunspace, using the sunspace as a collector (Kinloch, Hinsey, and Tichy 1981). This concept may be an attractive retrofit. A simulation of this system for a northeastern U.S. climate (Kinloch, Hinsey, and Tichy 1981) showed this use of passive collection saves less energy, relative to a conventional heat pump, than active series collection but has a superior payback period due to lower initial cost.

In the dual-source system (Fig. 4-23), the heat pump has two evaporators, one in the storage tank and the other outdoors. This arrangement allows the heat pump to use either collected solar energy or ambient air as the source (but not both simultaneously), depending on which results in a higher COP. Direct heating is possible when the tank is at high temperature, so the heat pump is off. When the tank temperature drops below the control point, the heating mode is the same as in the series system. When the tank temperature is either below a minimum (usually just above the freezing point of the liquid in the tank) or less than the outdoor air temperature, operation is like that of the parallel system. Thus the dual-source system appears to take advantage of the best features of both systems, but the equipment is more expensive and control is more complex. In all three systems the preheat coils for service hot water are located on the solar side and can be used when there is no space heating load.

A variation on the usual operating mode of the dual-source system is to place the heat pump condenser coil downstream of the solar coil in the

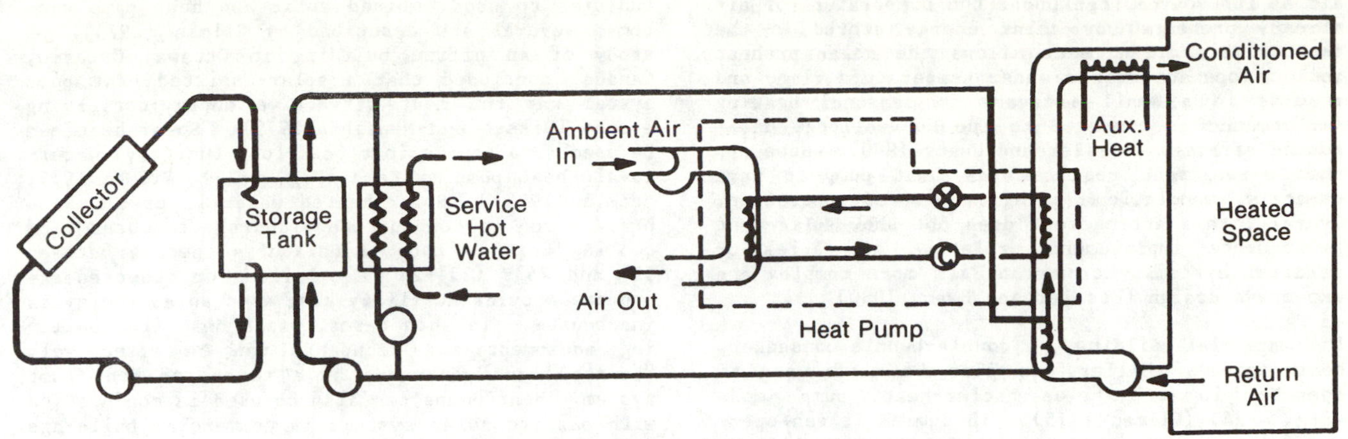

Fig. 4-21. Parallel solar-heat pump system

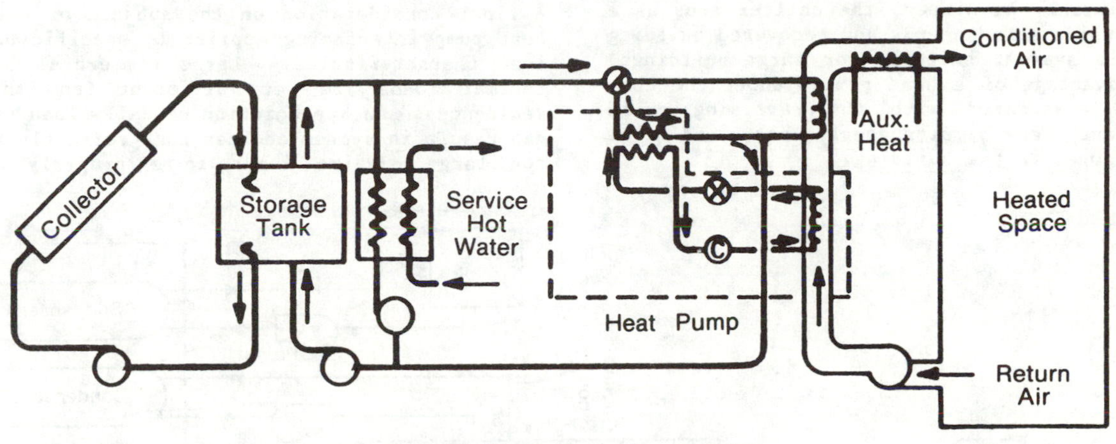

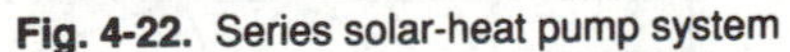

Fig. 4-22. Series solar-heat pump system

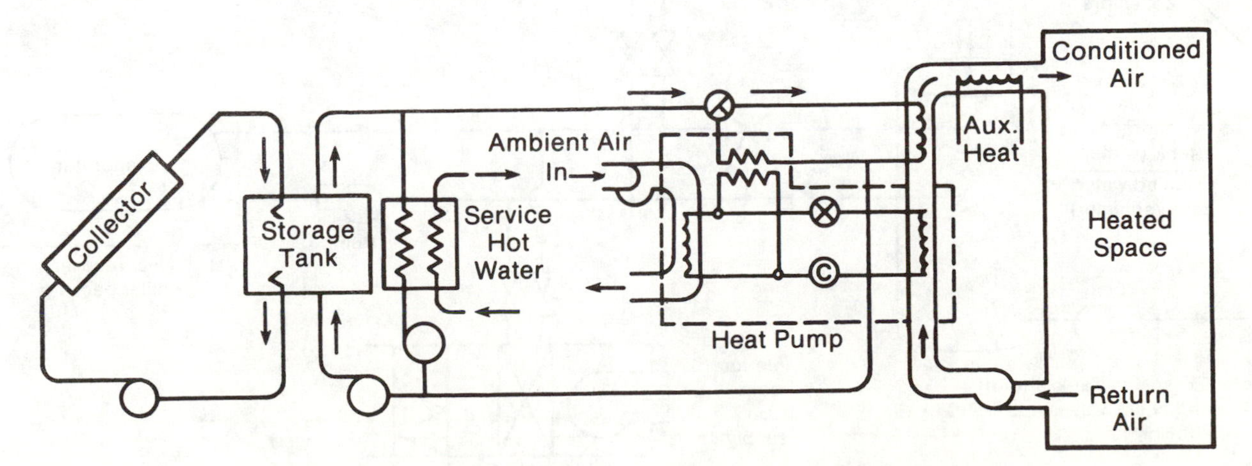

Fig. 4-23. Dual-source solar heat pump system (shown in direct heating mode)

supply air stream. The heat pump, using ambient air as its source, can boost the temperature of air already preheated by solar energy stored in the tank. This arrangement allows the solar preheat coil to operate over a wider temperature range and results in a small increase in seasonal heating performance as compared to the conventional dual-source strategy (Bessler and Hwang 1980). However, this arrangement requires the heat pump to have capacity modulation using two-speed operation. Overall, this arrangement does not show sufficient performance improvement, relative to series or parallel systems, to warrant its more complex and expensive design (Bessler and Hwang 1980).

In commercial buildings, a double-bundle condenser, heat-recovery chiller can also be configured to operate in a series solar-heat pump mode (Fig. 4-24) (Gilman 1975). In summer it can operate as a conventional, central-station, chilled-water, air-conditioning system by day and take advantage of favorable ambient temperatures and offpeak power rates at night to produce chilled water in the storage tank; the solar system is not used in summer. In winter, the chiller acts as a heat pump using solar energy and recovered building heat. This system, suitable for large buildings, has the advantage of a heat pump without the complications associated with the reversing cycle features and can simultaneously heat and cool different zones in the building.

Commercial building applications offer other opportunities to use combined solar and heat pump systems; several are described by Gilman (1975). A study of an office building in Ottawa, Ontario, Canada, concluded that a solar-assisted heat pump system was the most attractive energy-conserving system (Bisset and Monaghan 1979). Solar heat can be used as a source in closed-loop (unitary) water-to-air heat pump systems (Fig. 4-25) (Gilman 1975; Meckler 1978). Solar-heated water is used as the heat source in combination with rejected heat from cooling units. Loop temperatures operate between 55° and 75°F (13°-24°C). Boilers or other energy sources provide auxiliary heat when solar energy is inadequate. In such cases, waste heat from building equipment and occupants can be effectively "boosted" and recovered by a solar and heat pump system. Heat pumps can also be used in conjunction with passive solar systems in commercial buildings (Bridgers, Stoltys, and Broughton 1979). Finally, a solar pond has been used as the source for a series solar and heat pump system to heat a college campus (Ahmed and Scanlon 1980).

A final consideration on the subject of solar and heat pump interfacing applies to specific building load characteristics. Large commercial building thermal loads are very different from those of residences. An appreciation of daily load profiles can result in system choices that correctly account for large daytime loads in a properly managed

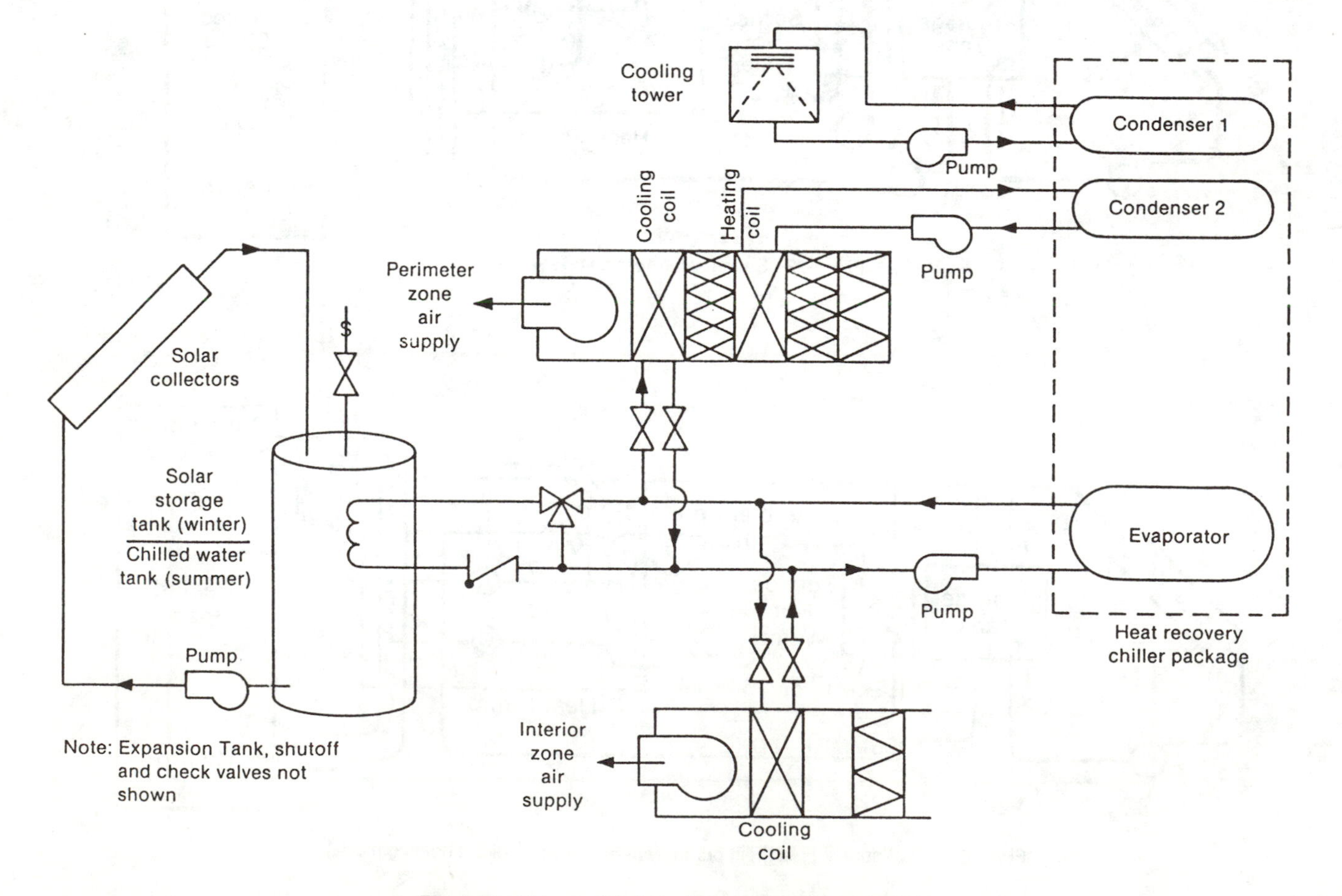

Fig. 4-24. Solar-assisted, double-bundle condenser chiller

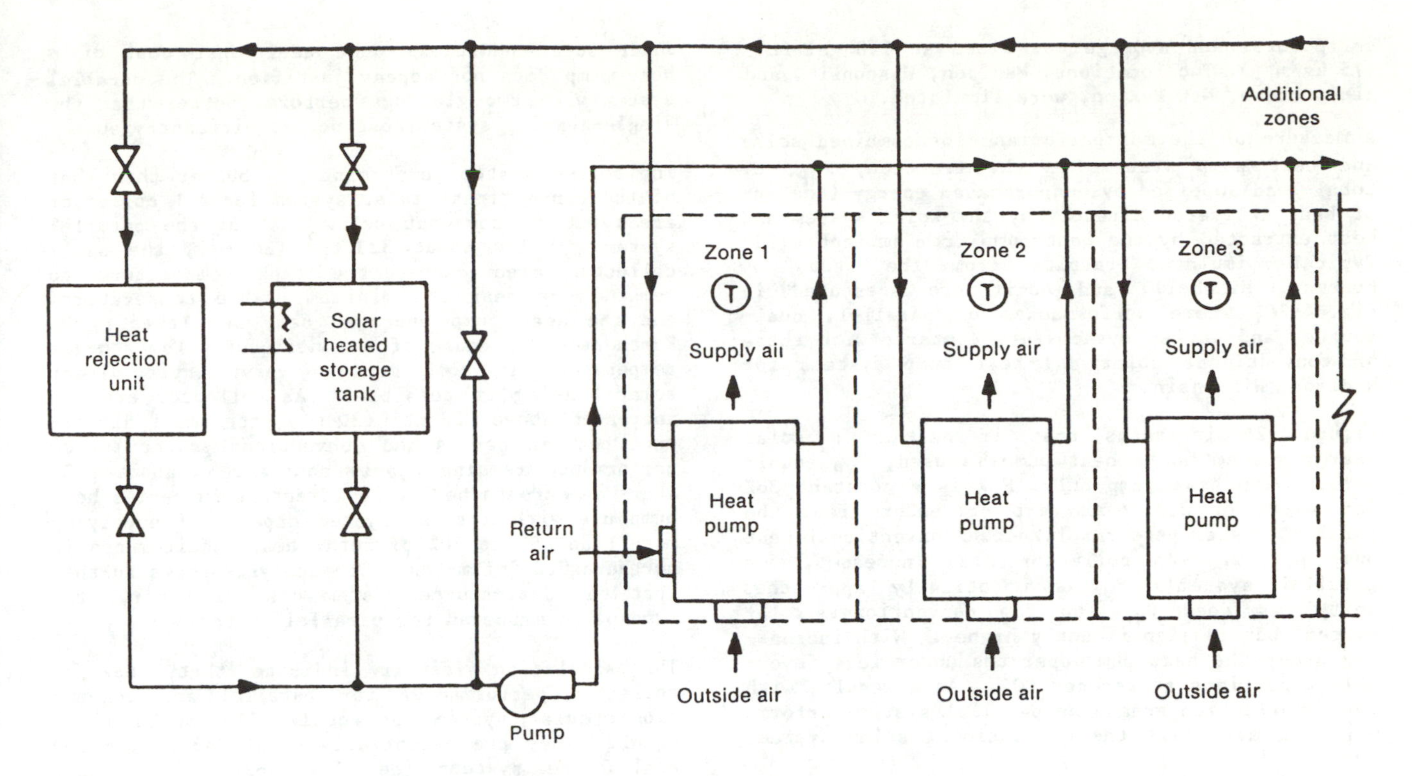

Fig. 4-25. Solar unitary hydronic heat pump system

commercial building, for example. In this case, short-term storage (one day) may not be as important because nighttime loads are low.

Component Considerations

Several types of heat pumps can be used in solar and heat pump systems. By far the most commonly used heat pumps, and the only types commercially available, are those that operate on a vapor-compression refrigeration cycle. These usually use electrically driven compressors, but some use fuel-powered engines to drive the compressor. Experimental or unconventional heat pumps (such as absorption, chemical, or thermoelectric types) have been suggested for use with solar systems. However, these are the subject of current research and are not yet recommended for solar and heat pump applications.

The heat pump most widely available has a single-speed compressor. It has a COP curve that rises with evaporator temperature, peaks, then falls off, thus restricting the temperature range over which solar heat can be efficiently used. Small, variable-capacity heat pumps with variable-speed compressors are being developed. Capacity varies with load, and they have the advantage of higher COPs over a wider load range. Capacity-modulated heat pumps for larger systems that use screw machines or centrifugal compressors are commercially available.

Selecting components appropriate to the task is imperative in system design. Medium- or high-temperature, concentrating collectors should be restricted to direct space heating or to parallel solar and heat pump applications where efficiencies at higher temperatures are better, and higher quality heat is matched to the load temperature.

System Performance

Numerous analytical studies have been conducted on the performance of combined solar and heat pump systems (Freeman, Mitchell, and Audit 1979; Mitchell, Freeman, and Beckman 1978; Metz 1978; Choi, Morehouse, and Hughes 1980; O'Dell, Mitchell, and Beckman 1983; Oonk, Shaw, and Hopkinson 1978; Andrews 1978; Morehouse and Hughes 1979; Manton and Mitchell 1981; White, Morehouse, and Swanson 1980). Most have shown that a properly designed combination of solar and heat pump systems will result in substantial energy savings over either a conventional (direct) solar or a stand-alone heat pump system. However, combining the two systems does not always produce a synergistic effect on energy savings, and economic savings are not always realized.

Three studies summarize the energy and economic performance of solar and heat pump systems and their performance sensitivity to design parameters (Freeman, Mitchell, and Audit 1979; Morehouse and Hughes 1979; Manton and Mitchell 1981). The first study used a TRNSYS thermal analysis and a model of a standard commercially available 3-ton (10.6 kW) heat pump. A conventional liquid solar system was associated with space and water heating load characteristics of a single-family residence with 1290 ft^2 (120 m^2) of floor area. The collector model had parameters characteristic of typical zero-, single-, or double-glazed flat-plate collectors, and storage was sized at 1.8 gal/ft^2_c (75 kg/m^2_c). Two locations, Madison, Wisconsin, and Albuquerque, New Mexico, were simulated.

A measure of thermal performance of combined solar and heat pump systems is the fraction, F_{NP}, of total load supplied by nonpurchased energy (the sum of heat delivered directly by the solar system and heat extracted by the heat pump from ambient air). Typical simulated results from the study by Freeman, Mitchell, and Audit are presented in Fig. 4-26, where performances of parallel, dual-source, and series systems are compared with those of conventional solar and heat pump systems for Madison, Wisconsin.

Figure 4-26 indicates that if neither a solar energy system nor a heat pump is used, F_{NP} equals zero. With heat pump only, F_{NP} is a constant 36% for a COP of 2.8. At zero collector area, the parallel system performs like the conventional heat pump system. As collector area increases, the parallel system's F_{NP} asymptotically approaches unity, as does F_{NP} for the conventional solar system, but is significantly higher. With increasing area, the heat pump operates under less favorable conditions at reduced COP. As a result, with large collection areas the parallel system performs more and more like the conventional solar system. Under such conditions, the additional cost of a heat pump does not appear justified. The parallel system with two glazings performs better than the single-glazing system, but not significantly so.

The series system performance is better than that of the conventional solar system for all collector areas but is somewhat below that of the parallel system. Below about 323 ft^2 (30 m^2), the small collector area causes the tank temperature to remain very near the minimum usable temperature, and the heat pump operates near its lowest COP. Furthermore, because of a continuously low storage temperature in cold months, very little direct solar heating is possible. As collector area is increased above 323 ft^2 (30 m^2), the small difference between series and conventional solar system performance remains nearly constant at about 10%. There appears to be little advantage in series heat pump use with a solar system capable of supplying more than 60% to 70% of total heat requirements in northern U.S. climates. The analysis shows further that the dual-source system has very nearly the same performance as the parallel system.

The simulation results indicate that seasonal collector performances for parallel and conventional solar systems of equal collector area are equal. They are essentially equal for series and dual-source systems (see Fig. 4-27). Collector performances for series and dual-source systems are

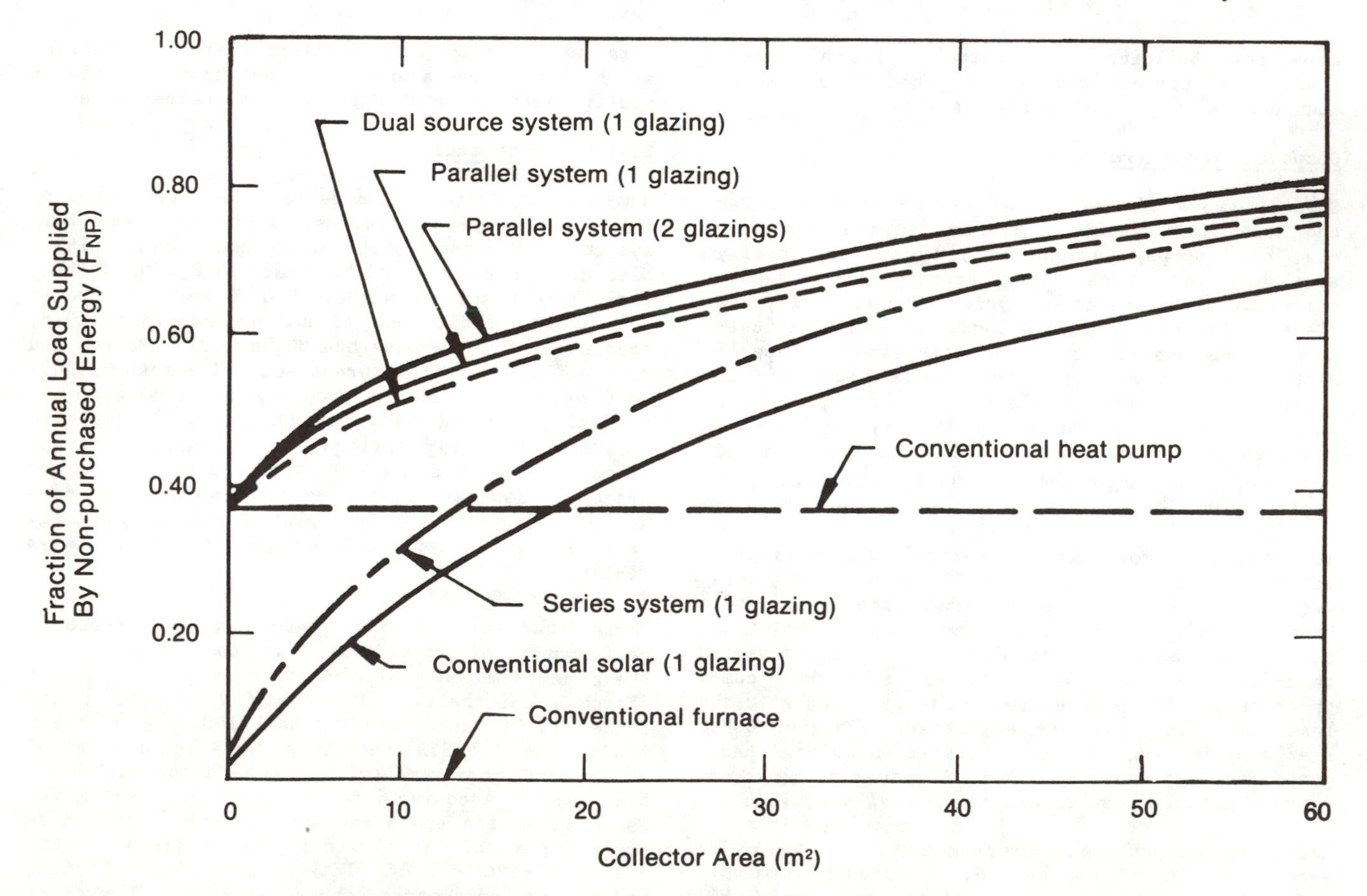

Fig. 4-26. Performance comparisons for solar and heat pump options in Madison, Wisconsin

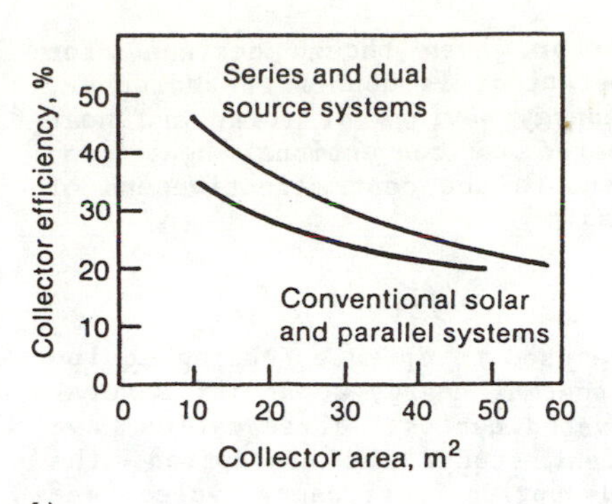

Fig. 4-27. Annual collector efficiencies for single-glazing solar-heat pump systems in Madison, Wisconsin

better than for parallel and conventional solar systems because solar source heat pump capability maintains lower storage temperatures and hence higher collector efficiencies.

Series, dual-source, and conventional solar systems were also investigated for larger storage sizes (Fig. 4-28), with improved heat pump performance, and for a coverless collector (Freeman, Mitchell, and Audit 1979). Even when storage size was increased by a factor of 10 over the nominal 1.8 gal/ft^2_c (75 kg/m^2_c), system performance improved only slightly (about 7%); seasonal COPs showed little or no increase with storage size in all systems investigated. Improvements in heat

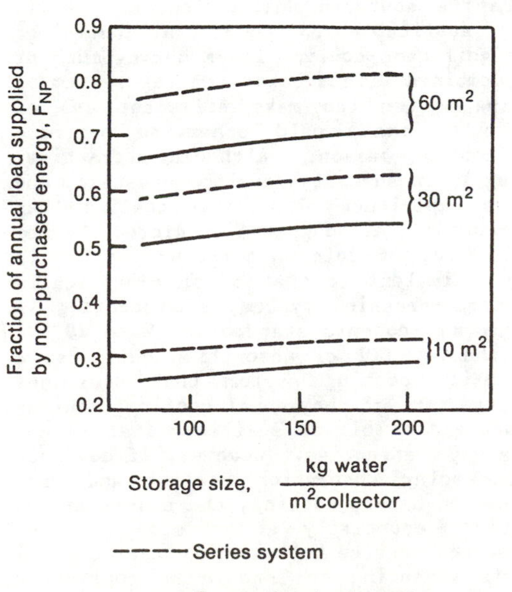

Fig. 4-28. Effect of storage size on series and conventional solar system performance in Madison, Wisconsin

pump COP by a factor of two improved parallel and dual-source system performance moderately (about 15%), but series system performance improved very little. Dual-source system performance for a coverless collector was only slightly better than that for the conventional heat pump. The findings of Freeman, Mitchell, and Audit (1979) were corroborated by other studies (Morehouse and Hughes 1979; Manton and Mitchell 1980). It was also found that climate does not appear to alter the relative performance rankings.

In summary, a properly designed combined solar and heat pump system should require less auxiliary energy than a conventional heat pump system. Parallel systems require less auxiliary energy than series systems in all locations studied and therefore appear to be the best combined system. Energy saved by both conventional solar and conventional heat pump systems is relatively large, but combining the two systems does not produce great additional energy savings. These conclusions are consistent with those of Oonk, Shaw, and Hopkinson (1978), who conclude that a parallel configuration, air-source solar and heat pump system saves the most auxiliary energy. However, development of special liquid-source heat pumps combined with ground-coupled storage and special low-cost collectors may make series systems the best competitor for conventional heat pumps (Andrews 1978). Also, using a series system with a refrigerant-filled collector, as reported by O'Dell, Mitchell, and Beckman (1983), may achieve a performance advantage of 10% to 15% with large collector areas.

It should be noted that under particular design criteria one system type may appear to be superior to another. However, many factors, including relative storage size, collector characteristics, and the relationship between site ambient temperature and solar radiation, play important roles in system performance. Daily temperature profiles are a factor; for climates with identical heating degree-days, the minimum daily temperature profile will alter the relative values of system benefits.

Many solar and heat pump systems are in operation throughout the world. However, only a few are adequately instrumented so their performance can be assessed. Even so, monitored systems generally show that combined solar and heat pump systems perform reliably, save energy compared to conventional solar and heat pump systems, and reduce peak demands. Data, however, are not sufficient to confirm results of the reported performance simulations.

One of the earliest solar and heat pump applications to a commercial building involved use of a solar-source water-to-water heat pump that operated in both heating and cooling modes with a thermal storage tank. Measured performance of this 5000 ft^2 (464.5 m^2) office building in Albuquerque, New Mexico, is reported by Gilman, McLaughlin, and Wildin (1976). The 750 ft^2 (69.7 m^2) liquid solar collectors provided energy to storage. The energy was used for space heating at a seasonal heat pump COP of 3.29; peak collector efficiencies of 38% were measured.

A large number of residential solar and heat pump systems have been monitored in several parts of the United States. Examples include monitoring of a series (liquid) system in an occupied residence in Madison, Wisconsin (Terrell 1979), a series (liquid) system in a test house in Knoxville, Tennessee (McGraw, Bedinger, and Reid 1981), and a parallel (air) system in a test facility, also in Knoxville, Tennessee (Bedinger et al. 1981).

Economic Considerations

Several studies of life-cycle costs of solar and heat pump systems have been conducted (Bessler and Hwang 1980; Morehouse and Hughes 1979; Manton and Mitchell 1981; White, Morehouse, and Swanson 1980). An extensive study of residential systems (Morehouse and Hughes 1979) concluded that combined solar and heat pump systems have initial costs 5 to 10 times higher than conventional heat pump systems. Using best estimates of present collector costs for series and parallel systems, these systems were shown to have similar life-cycle costs (within 15% of each other) but were at best marginally competitive with conventional oil and electric furnace systems. Payback periods ranged from 15 to 25 years. Combined solar and heat pump systems were not economically competitive with conventional gas furnace systems or conventional heat pump systems. Liquid series systems come closest but are still not life-cycle cost competitive. These conclusions are in general agreement with those of Bessler and Hwang (1980) and Andrews (1978) although Bessler and Hwang concluded that parallel systems are more cost effective than series systems, and both are more cost effective than direct solar systems. Collector cost sensitivity analysis did not offer encouragement regarding significant system cost reductions. In a related sensitivity analysis, White, Morehouse, and Swanson (1980), similarly concluded that none of the engineering design parameter variation studies significantly reduced life-cycle costs.

One economic study (Manton and Mitchell 1981), which assumed the existence of a 40% federal tax credit for solar systems, concluded that very low-cost collectors (at least a factor of two reduction in single-cover collector costs and about a factor of three reduction in unglazed collector costs below present single-cover collector costs) would be required for series systems to be competitive with parallel or conventional solar systems. However, it was shown that the best alternative solar or solar and heat pump system is economically better, over an assumed 20-yr system lifetime, than electric resistance heat in all 60 locations studied, with electricity prices of 3¢-7¢/kWh. Gas furnaces, with typical present fuel prices, are more economic than the best solar alternative in all regions but the Southwest. Oil furnaces, with typical present fuel prices, are less economic than the best solar alternative in virtually all U.S. regions.

In summary, if low-temperature collectors could be produced at low cost, series solar and heat pump systems could be used with them to economic advantage. However, availability of a durable, low-cost trouble-free collector, even though designed for low-temperature operation, is doubtful. Moreover, the considerable energy savings of solar and heat pump systems compared to conventional heat pump systems does not ensure the cost effectiveness of the combined systems.

SOLAR COOLING

Solar energy can be used to operate cooling equipment by supplying thermal energy to any of several types of heat-activated cycles. Three cycles have received the greatest study: the absorption, the Rankine, and the adsorption (desiccant) cycles [see Dickinson and Cheremisinoff (1980b) and Jordan and Liu (1977, Ch. XIII)]. For conventional cooling cycles, the figure of merit for rating systems is the COP (ratio of cooling produced to energy required to produce that cooling). For solar operation, a COP definition is appropriate that includes terms for all nonsolar energy additions to the cycle (auxiliary fuel and parasitic power use converted to primary energy source units) (Curran 1977). Also, temperatures required for solar collection and system COP (ratio of cooling produced to solar energy incident on the collector) are also important performance measures. Because of the transient operation of solar cooling systems, seasonal COP is an appropriate measure of performance.

Because solar cooling systems require higher temperatures than solar heating systems, evacuated tube collectors are often used in cooling applications. However, many difficulties and problems were experienced in their early use. To prevent many of these problems, design considerations for evacuated tube heating and cooling systems are discussed by Ward and Ward (1979).

Solar cooling is an attractive idea, applicable to buildings in the southern United States. It is particularly applicable to commercial buildings that have significant cooling loads during much of the year. Combined with a solar heating system, a solar cooling system can make efficient use of solar collectors that would otherwise be idle during the cooling season. Although attractive, solar cooling is in an early development stage, and we have less experience with solar cooling than with solar heating. Although the predicted thermodynamic efficiency of solar absorption cooling is very nearly equivalent to that of an electrically driven, vapor-compression system, such cooling is marginal from an economic standpoint (Ward 1979). Because of the low COP of absorption chillers, a solar absorption cooling system that does not replace a large part of the annual cooling requirement and uses gas or oil as auxiliary fuel is not likely to be cost effective. However, if designed with limited solar absorption cooling and with vapor-compression backup cooling, the system may be cost effective, especially if it also supplies space heating and service hot water. Solar cooling technology is advancing and should be considered for some buildings. Dickinson and Cheremisinoff (1980b), Jordan and Liu (1977), deWinter and deWinter (1976), Curran (1975), and Clark and deWinter (1978) all present detailed information on solar cooling systems. An extensive solar cooling bibliography is given in Clark and deWinter (1978).

Absorption Cooling Systems

Absorption cooling systems are heat-operated thermodynamic cycles based on absorption of refrigerant in liquid absorbent solutions. They can be operated by supplying solar energy to accomplish generation or regeneration. Operation of absorption air conditioners using energy supplied by solar collectors, combined with thermal storage, is the most common approach to active solar cooling today. Cooling is accomplished as the absorption cooler generator is supplied with heat by a fluid pumped from the collector-storage system or from auxiliary. Heat is supplied to a refrigerant/absorbent solution in the generator, where refrigerant is distilled off. Refrigerant is condensed and goes through a pressure-reducing valve to the evaporator where it evaporates and cools air or water. The refrigerant vapor then goes to the absorber where it comes in contact with the solution that is weak in refrigerant and that flows from the generator. The vapor is absorbed in the solution and returned to the generator. A heat exchanger is used for sensible heat recovery and greatly improves cooler COP. Pumping to move the absorbent solution can be by mechanical means or by vapor-lift in the generator for low-pressure systems (Dickinson and Cheremisinoff 1980b).

Figure 4-29 shows a combined solar heating and cooling system with an absorption chiller. In such combined systems, several storage tanks can be provided. Two or three tanks can be used--one or two for hot water and one for chilled water. Figure 4-30 shows how the absorption cycle works.

Low-pressure lithium bromide-water vapor absorption cooling units, originally designed for gas, steam, or hot water heat supply, have been developed to operate at generator temperatures as low as 170°F (77°C) with typical COPs of 0.6 to 0.8 (Dickinson and Cheremisinoff 1980b) for single-effect machines and 1.0 to 1.2 for double-effect machines (i.e., units that use either ambient or solar-heated air). Evaporator and absorber pressures are controlled by cooling fluid temperature; water cooling is essential because lithium bromide crystallizes. Flat-plate collectors and cooling towers are usually used, although evacuated tube collectors are sometimes used (Ward and Ward 1979). These units are commercially available in limited sizes and have been used successfully in both residential and commercial installations. Several 3-ton residential units, optimized for solar applications, have been demonstrated (Auh 1978). Manufacturers of large-capacity absorption chillers have examined their potential use in solar-powered systems; 25-ton capacities are available. These units can be used without hardware modification or with slight modification in the generator heat-transfer surface. Absorption cooling is recommended for cooling with solar energy because it has been

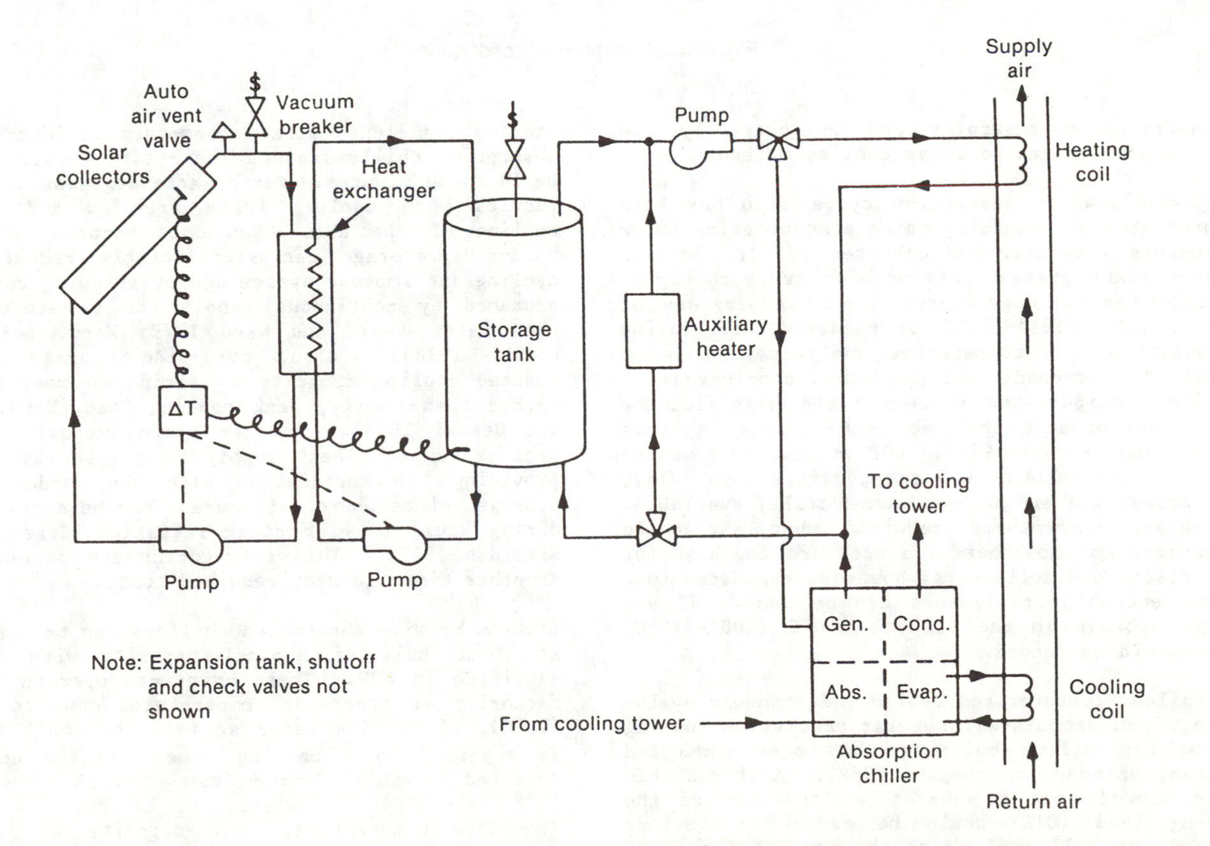

Fig. 4-29. Typical solar absorption cooling system

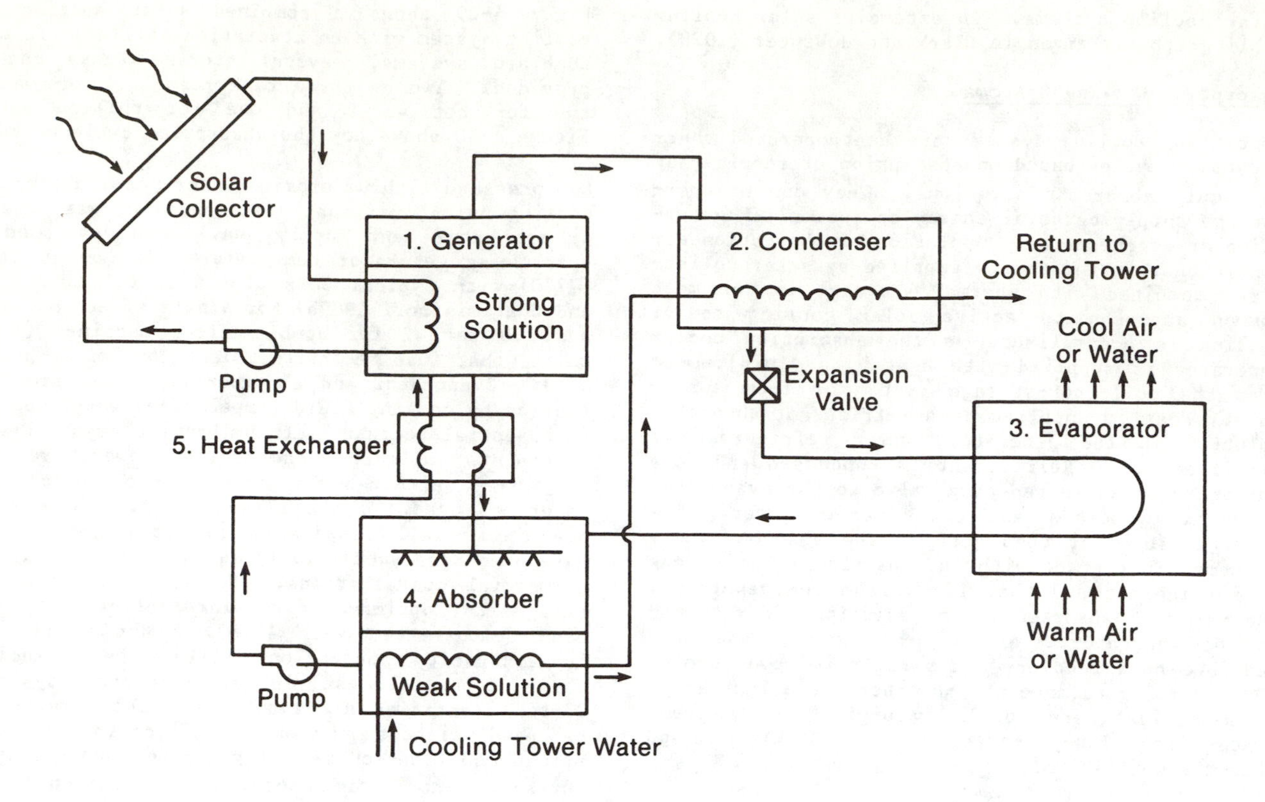

Fig. 4-30. Absorption air conditioner

demonstrated successfully and is currently the least expensive active solar cooling method.

The ammonia-water absorption cycle also has been demonstrated successfully using concentrating solar collectors (deWinter and deWinter 1976). Because pressures and pressure differences are much higher in this cycle, a generator input temperature of 250° to 350°F (121°-177°C) is required; air cooling is possible. A comparative analytical study of absorption air-conditioning systems concluded that lithium bromide-water systems yield significantly better performance than do ammonia-water systems (Wilbur and Mancini 1976). COP values for ammonia-water units of 0.4 to 0.5 are possible (Auh 1978), but these units are not yet commercially available. Generator temperatures required when air-cooled condensers and absorbers are used are too high for most flat-plate collectors; however, evacuated tube or concentrating collectors can be used. If so, energy storage in the 212° to 350°F (100°-175°C) range would be needed.

A detailed, computerized design and economic evaluation of an ammonia-water solar absorption cooling system for office buildings has been conducted (Shiran, Shitzer, and Degani 1982). Although this study showed that a substantial fraction of the cooling load (81%) could be met if collectors covered the full roof area, the projected cost of energy supply under optimal system conditions was not cost effective.

Startup transients affect long-term performance of absorption chillers since their time constants may be 10 to 30 minutes. Performance degradation often results when cooling loads are low and on-off cycling of the absorption unit occurs. Use of cold-side storage can significantly reduce unit cycling and improve system COP by allowing coolness produced by continuous running to be stored and used later (Ward and Ward 1979; Ward, Löf, and Uesaki 1978). Also, cold-side storage allows reduced cooling capacity by sizing to meet daily, rather than hourly, peak cooling load (Ward, Löf, and Uesaki 1978). Another cycle improvement addresses variable heat supply in a solar system by providing, in association with the condenser, a storage volume where refrigerant can be accumulated during hours of high solar radiation (Grassie and Sheridan 1977). This refrigerant can be expanded at other times to meet required loads.

Lithium bromide absorption chillers can be operated at about half of nominal capacity with little sacrifice in COP. These units can operate satisfactorily at generator temperatures down to 165°F (74°C), if cooling water at less than 85°F (29°C) is supplied by a cooling tower and the unit is operated at chilled-water temperatures of 40° to 50°F (4°-10°C). However, at or near this low generator temperature, unit capacity is greatly derated (see Fig. 4-31). The steady-state COP, as a function of solar thermal source temperature for

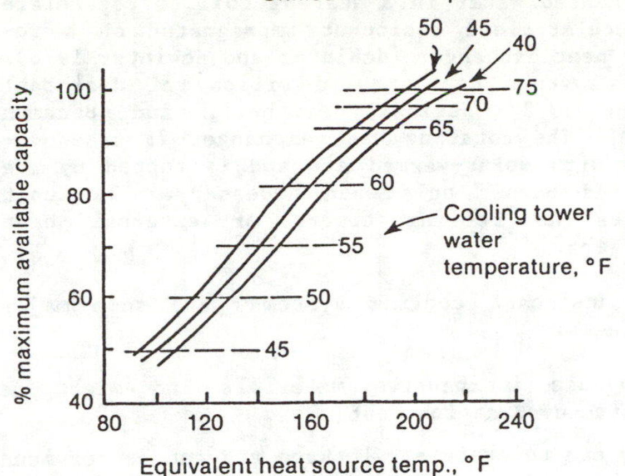

Fig. 4-31. Performance map of typical commercial absorption units (adapted from Jordan and Liu 1977)

both single- and double-effect lithium bromide-water machines, is shown in Fig. 4-32 for typical cooling water temperatures and a 44°F (7°C) chilled-water temperature.

Current research to increase absorption machine COP, is aimed at double-effect and ammonia-water units, advanced and hybrid cycles, new working fluids, and absorption heat pumps. System studies are focusing on system configuration, subsystem options, control and storage strategies, and parasitic power requirements.

Several experimental and demonstration solar absorption cooling systems have been operating successfully for a number of years. Performance data have been collected on a residential-scale lithium bromide cooling system at Colorado State University using both flat-plate and evacuated tube collectors (Duff et al. 1978). A large lithium bromide absorption system has also been in successful operation for several years on a library and conference

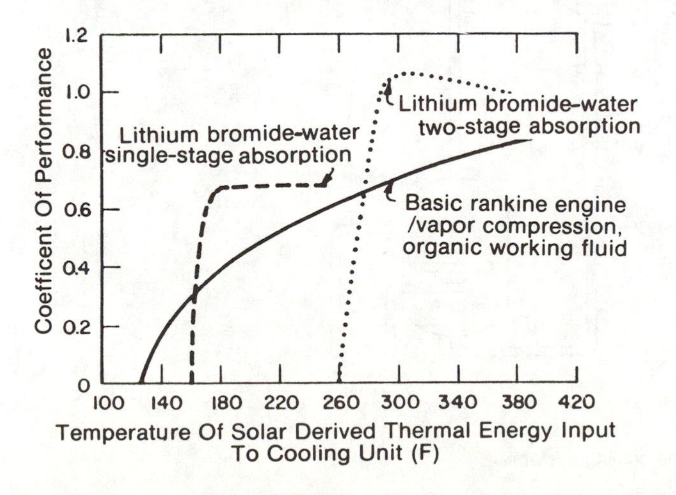

Fig. 4-32. Source temperature dependency for solar cooling systems

center at the Los Alamos National Laboratory (Murray, Hedstrom, and Balcomb 1978). Considerable performance data and operating experience have been obtained from these systems. Although not cost effective, they have worked well. Ward and Oberoi (1981) present a summary of recent results, including many problems encountered.

Rankine Cooling Systems

Solar-powered Rankine-cycle turbine engines have been used to drive conventional vapor compression cooling systems. These engines use a fluorocarbon refrigerant, such as Freon®, as the working fluid because of its ease of evaporation at temperatures available in advanced flat-plate [200°F (93°C)] or moderate concentration [320°F (160°C)] collectors. For a condensing temperature of 90°F (32°C), cycle efficiencies of about 10% are attainable at COPs of 0.3 to 0.7. A schematic of a solar Rankine-cycle engine is shown in Fig. 4-33. Thermal energy is converted to mechanical work by pumping the liquid working fluid into a boiler where it is evaporated by addition of heat from solar-heated water. The vapor is expanded through an expander, usually of the turbine or rotary-vane type, which lowers the vapor temperature and pressure and drives a shaft. The vapor then flows to a condenser where it returns to the liquid phase by rejecting heat to a cooling tower or ambient air. The liquid is pumped to the boiler, and the cycle is repeated (Curran 1978). Operating data for a 3-ton Rankine-cycle prototype are reported by Prigmore and Barber (1976).

The other component shown in Fig. 4-33 is a standard vapor compression refrigeration machine. Its energy efficiency ratio (thermal energy removed from building to shaft energy supplied to machine) averages about 5 for water-cooled machines and 3 for air-cooled machines (Curran 1978). Thus, the cycle COP is as shown in Fig. 4-32.

Two basic problems are associated with solar Rankine systems: generation of mechanical energy from solar energy and coupling of air conditioning components to the solar energy source under part-load operation. The efficiency of solar collection decreases as operating temperature increases; efficiency of the Rankine-cycle heat engine increases with operating temperature. Therefore, overall system efficiency peaks at an optimum operating temperature. Because operating temperature varies throughout the day, control to meet time-varying loads is complicated. If excess capacity is available, energy supplied to the engine can be throttled, or the engine can be coupled to an electric generator. The system could be designed to operate at variable speed, but the air conditioning system would be operating in an off-design condition, with reduced efficiency as well as output.

Parametric studies of solar Rankine systems have been conducted using flat-plate collectors (Olson et al. 1977) and evacuated tube collectors with shaped reflectors (Mehta and Lavan 1982). Optimal storage sizes were found to be less than 1.8 gal/$ft^2{}_c$ (75 kg/$m^2{}_c$) (recommended for solar heating and service hot water systems) because higher tank temperatures were needed. For a fixed collector area, an optimal engine size was found (Olson et al. 1977), and operation at 500°F (260°C) was superior to operation at temperatures typical of flat-plate collectors (Mehta and Lavan 1982).

Although these systems have been demonstrated successfully (deWinter and deWinter 1976; Clark and deWinter 1976; Curran 1978), they are experimental and prohibitively expensive. They are therefore not yet considered practical (Olson et al. 1977; Choi, Hughes, and Morehouse 1982). Several solar Rankine cooling field tests are in progress; system capacities range from 3 to 77 tons (Clark and deWinter 1976; Ward and Oberoi 1981).

Desiccant (Adsorption) Cooling Systems

A recently introduced prototype desiccant cooling system (Fig. 4-34), to be powered by solar energy and natural gas, performs particularly well in dry climates, but its COP in humid regions may be lower than that of absorption chillers. The system uses solar-heated water in a heating coil to regenerate a molecular-sieve desiccant impregnated on a rotating heat exchanger (deWinter and deWinter 1976). [Other systems use a solid silica gel desiccant (Curran 1978; Jurinak, Mitchell, and Beckman 1984).] The rotating heat exchanger is preconditioned with solar-warmed air and is topped by the gas-fired burner only when necessary. The unit requires no cooling towers or external heat exchangers.

Solar desiccant cooling systems have some major advantages:

- They use inexpensive materials and might be manufactured at low cost.
- They can tolerate air leakage and can be serviced easily.
- They can tolerate wide variation in solar input and still generate usable output.

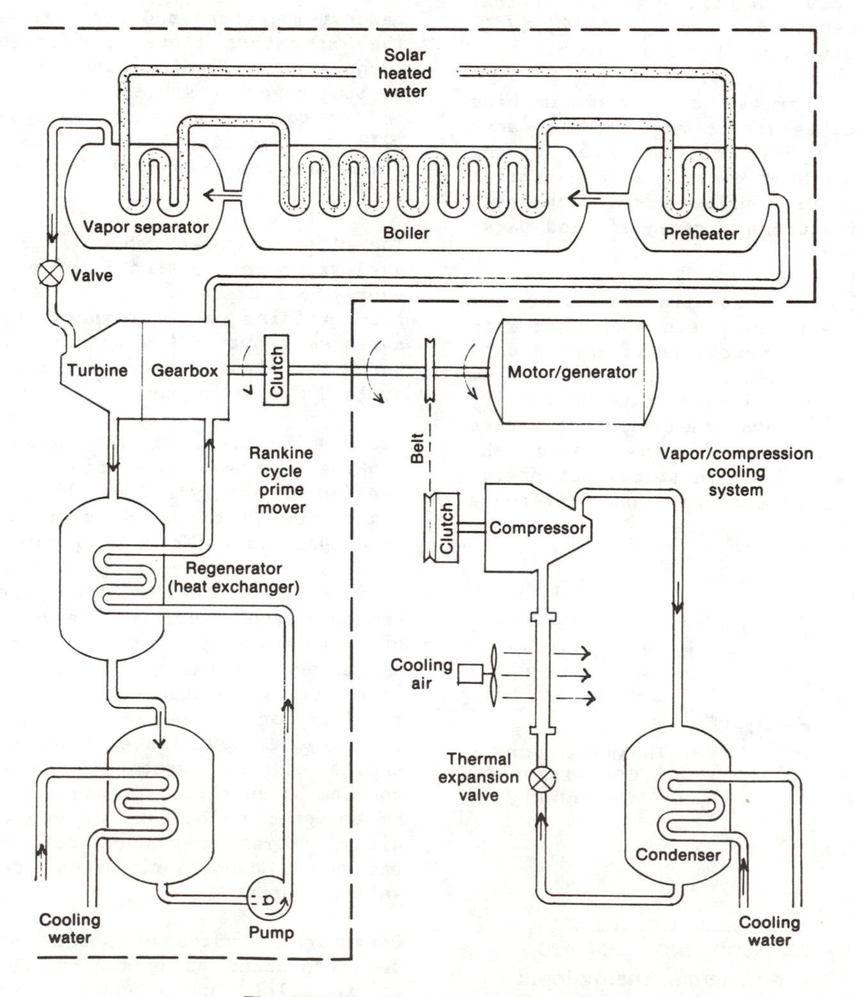

Fig. 4-33. Rankine-cycle cooling system

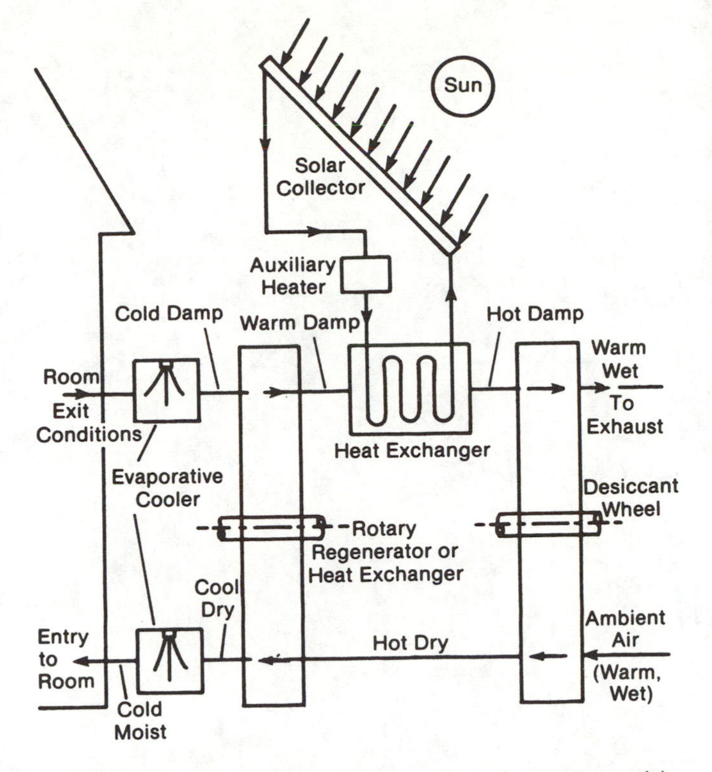

Fig. 4-34. Desiccant cooling system schematic (ventilation mode) (Kettleborough 1983)

An advanced desiccant system with a COP of about 1.0 to 1.2 can be achieved at solar collector temperatures of about 160°F (71°C) (Jurinak, Mitchell, and Beckman 1984). The system uses a parallel-channel desiccant wheel consisting of a spiral-wound Mylar® strip coated with silica gel. Problems to be surmounted include reduction in cost or amount of adsorbent material used, development of cost-effective heat exchangers, and development of an optimized system control strategy and hardware. Further work is under way to develop desiccants that exhibit less hysteresis on adsorption/desorption cycling. Several units are commercially available and continue under development (Clark and deWinter 1978; Kettleborough 1983).

Other Nonsolar Cooling Strategies

Heat pumps have long been used for cooling as well as for heating. Heat may be recovered from the vapor compression heat pump cycle so that simultaneous heating and cooling can be accomplished, as is often needed in commercial buildings.

An effective load management approach is to use the heat pump to produce ice that is stored for later cooling use. An icemaker heat pump approach has the advantage of smaller storage volumes for the same heat storage capacity and therefore lower standby losses than chilled water storage systems. Several demonstration projects using this concept are in operation (Hubbell et al. 1982).

EFFECTS OF COMPONENT SIZING ON ACTIVE SOLAR SYSTEMS

The need for correct sizing of equipment is obvious in any heating and cooling system. However, with solar systems, the impact of incorrect sizing is substantial, because of the relatively large initial investment required for the solar array.

An oversized solar array will be costly, and it may also require a method of heat rejection. If the array is undersized, the solar contribution is poor.

Oversized pumps consume excess energy; turbulence and flow balance problems can also result. Undersized pumps have failed, requiring the installation of new, larger pumps. Higher collector temperatures and lower collector efficiency also result if the pump flow rate is too low for the collector array.

Correct sizing of heat exchangers is critical. If they are undersized, the solar system capacity is choked off at the heat exchanger, resulting in reduced system efficiency. An investment in proper heat exchanger capacity is well worth the initial cost, considering the high relative cost of unused collector array capacity.

Storage tanks that are oversized often result in unusable low-grade heat, as well as adding both cost and space penalties. Undersizing storage can prevent the use of available solar energy once the maximum allowable storage temperature is reached. Heat rejection may also be required at this point. To achieve optimum efficiency during system operation, a thorough engineering analysis must be made during the design phase regarding building loads and optimum sizing of components.

Chapter 5
Principal Components of Active Systems

SOLAR COLLECTORS

Solar collectors are basically heat exchangers that transfer the energy of incident solar radiation to sensible heat in a working fluid--liquid or air. In space or water heating systems, the working fluid transports heat to the building interior where it is used to supply service hot water and space heat or is stored for later use. In most solar cooling applications, the working fluid heats the generator section of an absorption chiller.

Major types of collectors used in building heating and cooling applications are flat plates, evacuated tubes, and various linear concentrator types. Flat plates are by far the most common type for space and water heating; the others often supply higher temperatures needed for process heating or solar cooling.

Flat-Plate Collectors

A flat-plate collector has the following basic components (see Fig. 5-1):

- Absorber plate--usually copper, steel, aluminum, or plastic; surface covered with flat black paint or a special selective coating to maximize absorption of solar radiation and minimize emission of thermal radiation.
- Flow passages--with liquids, flow is usually through tubes attached to, or integrated with, the absorber plate. In some collectors, flow trickles down open passages formed by corrugated aluminum troughs (see Fig. 5-2, Thomason system); evaporation/condensation processes inherent in this configuration lead to reduced collector efficiencies at high temperatures (Beard et al. 1976). With air, flow occurs above and/or below the plate; heat-transfer surface area is maximized by means of fins, slots, or metal screening.
- Cover plate(s)--one, two, or three transparent covers used to reduce convective and radiative heat losses to outside air. Tempered glass and plastic materials are commonly used.
- Insulation--used to reduce heat losses from back and sides of the collector (e.g., low-binder fiberglass or isocyanurate polyurethane foam).
- Enclosure--a box to hold collector components together and protect them from weather.

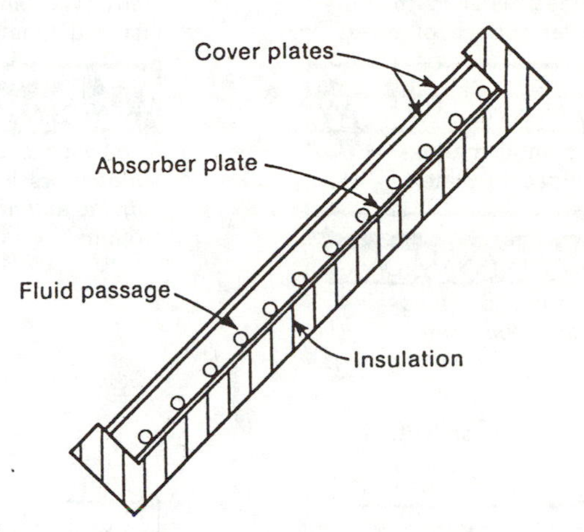

Fig. 5-1. Cross section of typical flat-plate collector

Absorber Plate, Flow Passages, and Coatings

In a liquid collector, tubes typically are spaced several inches apart, with absorber surfaces between acting as fins that absorb heat and conduct it to the tubes. Tube spacing is determined by fin efficiency versus cost tradeoff. For 1/2 in. (12.7 mm) tubes spaced 6 in. (152 mm) apart in good thermal contact with a black copper plate 0.02 in. (0.5 mm) thick, heat collection is 97% of that for a completely water-cooled black sheet (Anderson 1977).

Copper tubes are usually welded, soldered, or clamped to copper plates or clamped to aluminum plates. The thermal conductance of the bond is critical and can vary from 1000 Btu/h-ft-°F (1730 W/m-°C) for a well-soldered tube to less than 5 Btu/h-ft-°F (8.65 W/m-°C) for a poorly clamped one (ASHRAE 1982, Ch. 58). From a heat transfer standpoint, it is better to extrude or otherwise form a tube pattern into the plate during manufacture. Aluminum and steel absorber plates can be less expensive than copper but must be protected against corrosion. Aluminum flow passages are not compatible with copper piping, and steel flow passages may corrode in a drain-back or drain-out system unless nitrogen is used as a refill gas when the system is drained. Plastic absorbers are less expensive still but are less durable and can degrade due to high stagnation temperatures.

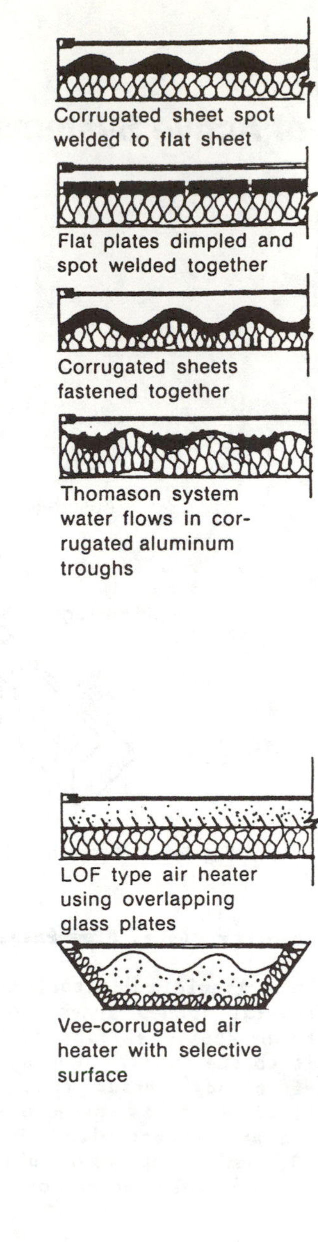

Fig. 5-2. Common collector configurations

Tubes can be routed through the collector in parallel paths from inlet to outlet headers, or a single tube can be routed in a serpentine fashion. The latter technique eliminates the possibility of header leaks and assures uniform flow but increases pressure drop. If a drain-back or drain-out freeze protection system is used, flow passages must be easy to drain.

Air collector absorber plates need not have high thermal conductivity because the air comes in contact with the entire surface. Flow can be above or below the plate, but the latter results in lower heat losses through the glazing. The heat transfer from absorber to air can be enhanced by creating turbulent flow and by increasing the surface area with fins, corrugations, etc. This, however, involves a tradeoff between heat transfer, fan power consumption, and cost.

Absorber coating type will determine the fraction of incident solar energy absorbed. Flat black paint has an absorptivity, α, between 0.92 and 0.96; that is, it absorbs between 92% and 96% of incident short-wave solar radiation. However, a hot absorber plate will radiate long-wave energy to the cover plate and cooler environments to an extent that depends on its emissivity, ε. Flat black paint has a high α, but also a high ε--on the order of 0.88. So-called selective surfaces have high short-wave absorptivities and low long-wave emissivities; i.e., they have a high α/ε ratio, and thus retain heat better.

Ordinarily, a metal absorber plate is coated with a metal having a low ε; a thin layer [on the order of 10^{-5} in. (2.5×10^{-4} mm)] of a high α material is laid on top (Meinel and Meinel 1976). Black chrome, one of the more popular selective surfaces, has an α of about 0.95 and an ε of 0.1. Selective surfaces improve collector performance, but they also increase cost. An important characteristic of any absorptive coating is that it have good adhesion and not peel or otherwise deteriorate during high-temperature stagnation conditions. It should also have a high vaporization temperature so as not to outgas. A list of absorptive coatings is given in Table 5-1.

Cover Plates

Glass is the most commonly used cover material. A 1/8 in. (3.2 mm) sheet of window glass (0.12% iron content) has a transmittance for solar radiation (at normal incidence) of 85% (τ = 0.85). Water-white glass (0.01% iron) has a τ of 0.92. Glass is practically opaque to long-wave radiation given off by the absorber plate; if tempered, it has high durability as well. Deterioration is negligible, even over very long periods of exposure to intense ultraviolet radiation.

Plastic materials used for collector glazings are cheaper and lighter in weight than tempered glass. Because they are used in thin sheets, they often have a higher transmittance than glass. However, they do not trap thermal radiation as well as does glass and generally are not as durable. Degradation due to ultraviolet radiation, high temperature, and wind action can be severe. Tedlar®, for example, should not be used as an inner glazing in a two-cover collector because it is susceptible to heat degradation. The high thermal expansion of plastic materials requires special mounting. Special supports within the enclosure are often required because plastics lack rigidity.

The number of glazings used depends on the application and on a cost versus performance tradeoff. Generally, the higher the temperature difference between load temperatures and ambient temperatures, the more covers are needed. A single cover is suitable for service hot water applications in most climates; two covers may be needed to supply space heat. However, a single-glazed collector with a

Table 5-1. Characteristics of Absorptive Coatings (HUD 1977)

Property or Material	Absorptance[1] α	Emittance ε	$\frac{\alpha}{\varepsilon}$	Breakdown Temperature °F (°C)	Comments
Black Chrome	~.93	.1	~9		
Alkyd Enamel	.9	.9	1		Durability limited at high temperatures
Black Acrylic Paint	.92-.97	.84-.90	~1		
Black Inorganic Paint	.89-.96	.86-.93	~1		
Black Silicone Paint	.86-.94	.83-.89	1		Silicone binder
PbS/Silicone Paint	.94	.4	2.5	662(350)	Has a high emittance for thicknesses >10 μm
Flat Black Paint	~.95	.89-.97	~1		
Ceramic Enamel	.9	.5	1.8		Stable at high temperatures
Black Zinc	.9	.1	9		
Copper Oxide over Aluminum	.93	.11	8.5	392(200)	
Black Copper over Copper	.85-.90	.08-.12	7-11	842(450)	Patinates with moisture
Black Chrome over Nickel	.92-.94	.07-.12	8-13	842(450)	Stable at high temperatures
Black Nickel over Nickel	.93	.06	15	842(450)	May be influenced by moisture at elevated temperatures
Ni-Zn-S over Nickel	.96	.07	14	536(280)	
Black Iron over Steel	.90	.10	9		

[1]Dependent on thickness and vehicle-to-binder ratio.

selective coating will often outperform a double-glazed one with a nonselective coating. A list of cover plate materials and their properties is given Table 5-2. Unglazed collectors have application where working fluid and ambient temperature differences are small, as in summer pool heating.

Insulation

Several types of insulation are used in collectors to prevent heat loss from the back and sides. Insulation is also needed to protect wooden roof members from collector stagnation temperatures that can be well above 250°F (121°C) even in single-glazed collectors. It is important that insulation not outgas under stagnation conditions since such gases can coat glazing surfaces and greatly reduce transmittance; resin-free binders should be used to avoid outgassing from the insulation. Insulation materials and their characteristics are listed in Table 5-3.

Enclosure

Collector housings are generally made of steel, aluminum, or fiberglass reinforced plastic (FRP) and support the absorber plate and covers. (Wood is not generally recommended because of fire hazard). Adequate clearance (around a glass cover, for example) and proper use of gaskets must be provided for differential expansion between collector components. The frame should cause little shading of the absorber plate, and aperture area should be at least 85% of gross area (DOE 1978a). The box should be well sealed to keep water out; however, provision should be made for drainage of any rainwater or condensed moisture present. Vents or weepholes at the bottom of the collector enclosure will allow moisture to escape. A desiccant can help prevent condensation on the inner surface of the cover. External pipe connections should receive particular attention. Sealing compounds and gaskets should be capable of withstanding stagnation temperatures without outgassing (should have high vaporization temperatures) and must be capable of withstanding thermal cycling. Commonly used flat-plate collector configurations are shown in Fig. 5-2.

Improving Flat-Plate Collectors

A number of techniques have been used to improve the performance of basic flat-plate collectors. Glass treatment to reduce reflection can increase performance by as much as 4% (Anderson 1977). Inner glass surface coatings have been used to reflect long-wave radiation emitted from the absorber plate (which would otherwise heat the glass), though these also reduce transmittance. A honeycomb located between the inner cover and

Table 5-2. Characteristics of Cover Plate Materials (DOE 1978c)

Test	Polyvinyl Fluoride[a]	Polyethylene Terephthalate or Polyester[b]	Polycarbonate[c]	Fiberglass Reinforced Plastics[d]	Methyl Methacrylate[e]	Fluorinated Ethylene-Propylene[f]	Ordinary Clear Lime Glass (Float)(0.10-0.13% iron)	Sheet Lime Glass (0.05-0.06% iron)	Water White Glass (0.01% iron)
Solar Transmission (%)	92-94	85	82-89	77-90	89	97	85	87	85-91
Thickness (in.)	0.004	0.001	0.125	0.040	0.125	0.002	0.125	0.125	0.125
Maximum Operating Temperature (°F)	227	220	250-270	200° produces 10% transmission loss	180-190	248	400	400	400
Tensile Strength (psi)	13000	24000	9500	15000-17000	10500	2700-3100	1600 annealed 6400 tempered	1600 annealed 6400 tempered	1600 annealed 6400 tempered
Thermal Expansion Coefficient (in./in./°F × 10^6)	24	15	37.5	18-22	41.0	8.3-10.5	4.8	5.0	4.7-8.6
Elastic Modulus (psi × 10^{-6})	0.26	0.55	0.345	1.1	0.45	0.5	10.5	10.5	10.5
Weight (lb/ft^2) for given thickness	0.028	0.007	0.77	0.30	0.75	0.002	1.63	1.63	1.63
Length of Life (yr)	In 5 years retains 95% of total transmission	4	--	7-20	--	--	--	--	--

[a]e.g., Tedlar®
[b]e.g., Mylar®
[c]e.g., Lexan, Merlon®
[d]e.g., Kalwall's Sunlite®
[e]e.g., Lucite, Plexiglas, Acrylite®
[f]e.g., Teflon®

Table 5-3. Characteristics of Insulation Materials (HUD 1977)

Material	Density (lb/ft^3)	Thermal Conductivity at 200°F (Btu/h-ft^2-°F-in.)	Temperature Limits (°F)
Fiberglass with Organic Binder	0.6	0.41	350
	1.0	0.35	350
	1.5	0.31	350
	3.0	0.30	350
Fiberglass with Low Binder	1.5	0.31	850
Ceramic Fiber Blanket	3.0	0.4 at 400°F	2300
Mineral Fiber Blanket	10.0	0.31	1200
Calcium Silicate	13.0	0.38	1200
Urea-Formaldehyde Foam	0.7	0.20 at 75°F	210
Urethane Foam	2-4	0.20	250-400

absorber plate can reduce both convective and radiative heat losses (Hollands 1976). Parallel strips running up and down the collector have also been shown to be an effective convection suppressor. Planar reflectors in front of a row of flat-plate collectors can increase energy collection by as much as 40% over a heating season (Kaehn et al. 1978). Most of these techniques, which offer various advantages, have not yet been widely applied.

Numerous attempts have been made to reduce collector costs either by using cheaper materials or by decreasing the amount of material used (e.g., thin-film absorbers). To date such designs have been lacking either in performance or durability, but research is continuing.

Advanced Collector Designs

As temperature requirements increase, flat-plate collector efficiencies rapidly fall off. Other collector alternatives are often chosen, particularly to supply 195°F (91°C) water to operate an absorption chiller. Examples of more commonly used advanced collector designs are discussed below.

Evacuated Tube

Convective heat loss can be reduced by use of a vacuum between glazing and absorber surfaces. Because a vacuum would cause a typical flat-plate collector to collapse, the technique is used with a tubular design. There are several types of evacuated tube collectors on the market.

One design, shown in Fig. 5-3, uses three concentric glass tubes. Fluid flows into the annular space between the inner and second tubes and out the inner tube. The annulus between the second and third (outer) tubes is evacuated, and the outside of the second tube is coated with a selective surface.

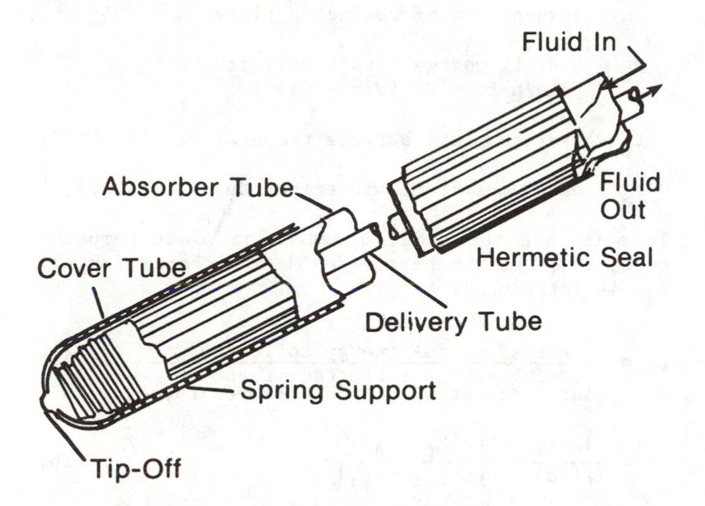

Fig. 5-3. Concentric glass evacuated tube collector

Another design (Fig. 5-4) uses only two concentric glass tubes with the space between evacuated; the outer surface of the inner tube contains an absorptive coating. A metal fin conforms to the inside surface of the inner tube; a U-tube is attached to carry the fluid. The U-tube readily accommodates thermal expansion, and glass breakage will not result in a leak.

Various reflector shapes are used behind evacuated tubes to provide a small amount of concentration (1.1 to 1.2 concentration ratio). Evacuated tubes can collect both direct and diffuse solar radiation and do not require tracking. Due to the extremely high stagnation temperatures [>600°F (316°C)] that evacuated tube collectors can reach, thermal shocking with cold water and boiling during power failure can be problems.

Parabolic Trough

Concentrating solar energy onto a smaller absorber surface can also reduce heat losses that impair efficiency. The most commonly used concentrator is

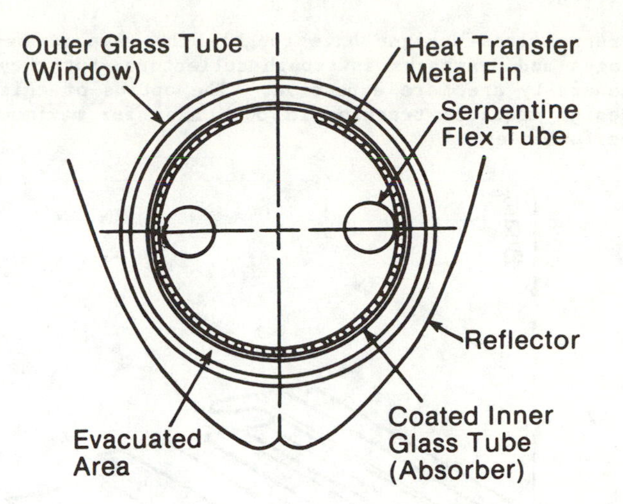

Fig. 5-4. Evacuated tube collector with copper flow tube

a parabolic reflector trough with an absorber pipe located along the focal line, as shown in Fig. 5-5. Collectors are typically oriented east-west or north-south, though any orientation is possible. In most cases, concentration ratios (ratio of aperture to absorber area) are high enough (i.e., greater than 2:1) to require continuous single-axis tracking even with east-west troughs. Since most diffuse radiation falls outside concentrator acceptance angles, only direct radiation can be collected.

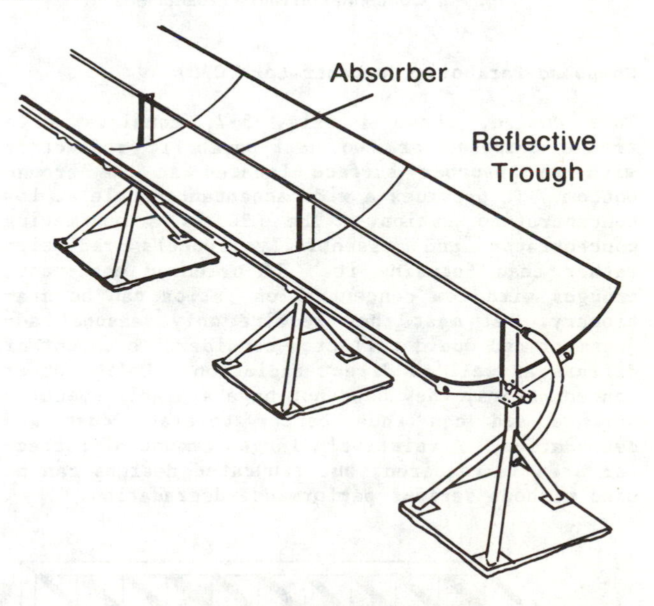

Fig. 5-5. Parabolic trough collector

Fresnel Lens

Rather than being reflected, sunlight can instead be refracted onto an absorber pipe, as shown in Fig. 5-6. Although this can be done with a convex lens, a Fresnel lens (consisting of many small refractive surfaces) is ordinarily used because it is cheaper, lighter, and absorbs less solar energy.

Fresnel lens designs have roughly the same advantages and drawbacks as trough collectors, but they generally are more expensive. The optics of this design requires tracking in both axes for maximum performance.

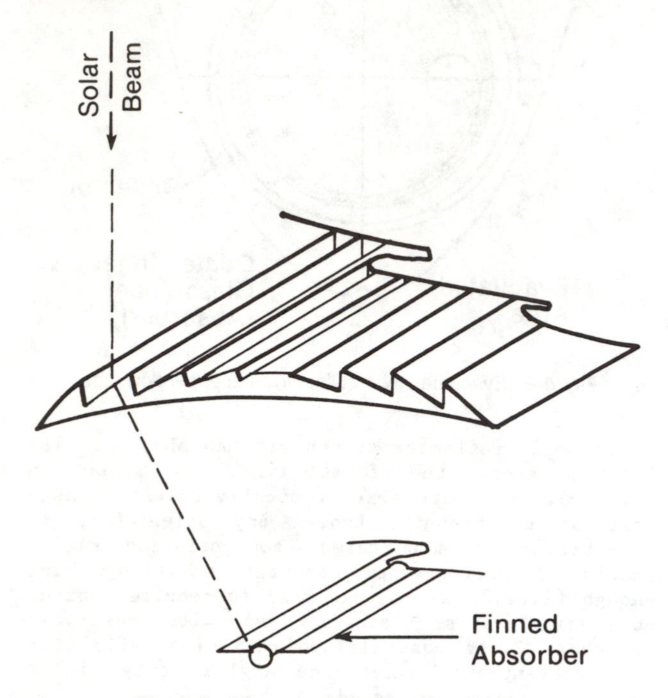

Fig. 5-6. Concentration with a Fresnel lens

Compound Parabolic Concentrator (CPC)

This design, shown in Fig. 5-7, consists of a trough composed of two half-parabolic reflectors with an absorber surface located at the trough bottom. It provides a wide acceptance angle at low concentration ratios. The CPC is a nonimaging concentrator and essentially funnels radiation rather than focusing it. If oriented east-west, troughs with low concentration ratios can be stationary. At most they require only seasonal adjustment and would collect a considerable amount of diffuse as well as direct radiation. Unlike other concentrators, they need not have a highly specular surface and can thus better tolerate dust and degradation. A relatively large amount of reflector area is required, but truncated designs can be used without serious performance degradation.

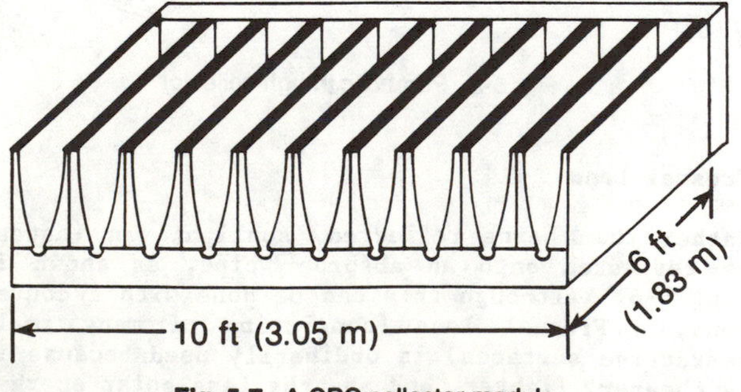

Fig. 5-7. A CPC collector module

Collector Efficiency

Consider an energy balance on a simple flat-plate collector, as illustrated in Fig. 5-8.

$$\begin{bmatrix}\text{useful energy}\\ \text{collected}\end{bmatrix} = \begin{bmatrix}\text{energy absorbed}\\ \text{by plate}\end{bmatrix} - \begin{bmatrix}\text{energy lost}\\ \text{to surroundings}\end{bmatrix}$$

$$\frac{Q}{A} = I_t(\tau\alpha) - U_L(t_p - t_a) \qquad (5\text{-}1)$$

where

Q = rate of useful energy gain [Btu/h, (W)]

A = gross collector area [ft^2, (m^2)]

I_t = global solar radiation incident on collector plane [Btu/h-ft^2, (W/m^2)]

τ = net transmittance of glazings

α = absorptance of collector plate

U_L = overall energy loss coefficient [Btu/h-ft^2-°F, (W/m^2-°C)]

t_p = average plate surface temperature [°F, (°C)]

t_a = outside ambient air temperature [°F, (°C)]

To eliminate the need to determine plate temperature, a collector heat removal efficiency factor, F_R, is introduced.

$$F_R = \frac{\text{Actual useful energy collected}}{\text{Useful energy collected if entire plate were at inlet fluid temperature}} \qquad (5\text{-}2a)$$

$$= \frac{I_t(\tau\alpha) - U_L(t_p - t_a)}{I_t(\tau\alpha) - U_L(t_{f,i} - t_a)} \qquad (5\text{-}2b)$$

Then,

$$\frac{Q}{A} = F_R[I_t(\tau\alpha) - U_L(t_{f,i} - t_a)] \qquad (5\text{-}3)$$

where

$t_{f,i}$ = inlet fluid temperature [°F, (°C)]

Useful energy gained by the collector heats the collector fluid, so,

$$\frac{Q}{A} = \frac{\dot{m}C_p(t_{f,e} - t_{f,i})}{A} \qquad (5\text{-}4)$$

where

$\dot{m}$ = mass flow rate of fluid [lb/h, (kg/s)]

C_p = heat capacity of fluid [Btu/lb-°F, (J/kg-°C)]

$t_{f,e}$ = fluid exit temperature [°F, (°C)]

Thus,

$$\frac{Q}{A} = F_R[I_t(\tau\alpha) - U_L(t_{f,i} - t_a)] \qquad (5\text{-}5)$$

$$= \frac{\dot{m}C_p\,(t_{f,e} - t_{f,i})}{A}$$

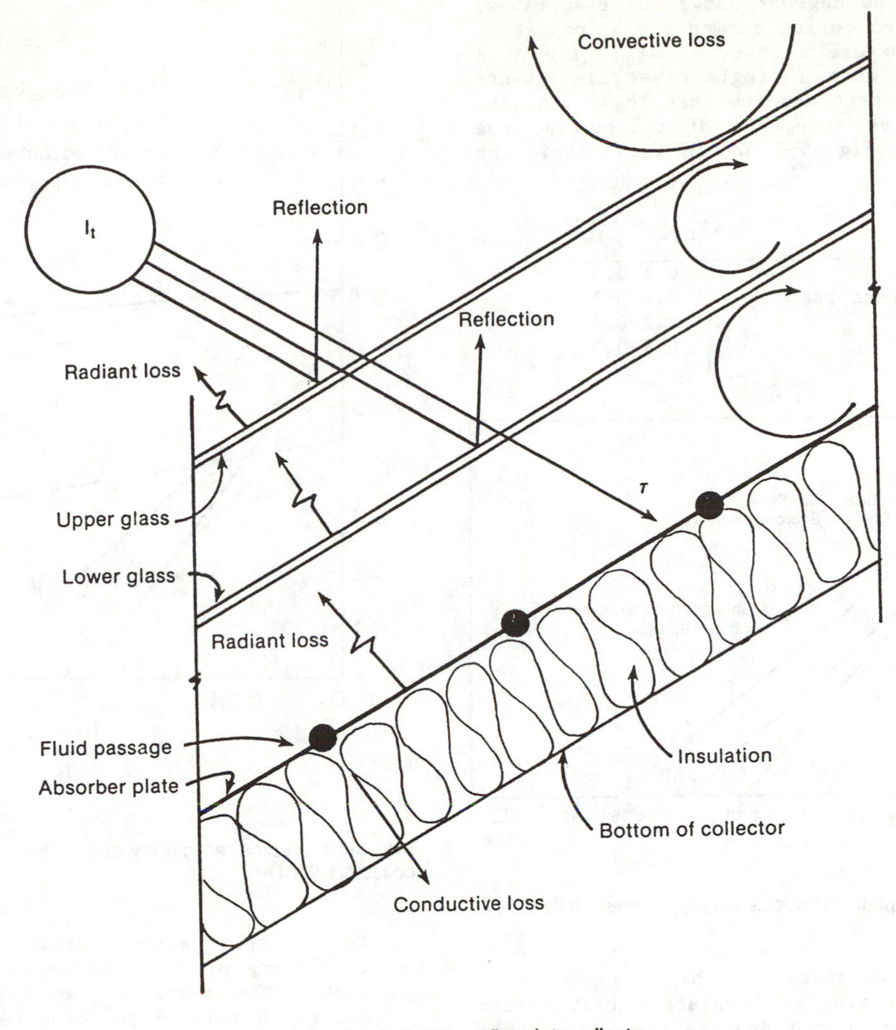

Fig. 5-8. Energy balance on flat-plate collector

The collector efficiency is defined as

$$\eta = \frac{\text{useful energy collected}}{\text{solar radiation incident on collector}} \quad (5\text{-}6)$$

$$= \frac{Q/A}{I_t}$$

So,

$$\eta = F_R(\tau\alpha) - F_R U_L \frac{(t_{f,i} - t_a)}{I_t} \quad (5\text{-}7)$$

$$= \frac{\dot{m}C_p\,(t_{f,e} - t_{f,i})}{AI_t}$$

The term on the far right in Eq. (5-7) is used to experimentally measure collector efficiency. Although the U_L term in the other part of the equation is a function of windspeed and temperature, it is usually considered constant for a flat-plate collector. Thus, when test measurements are made, and η is plotted against $(t_{f,i} - t_a)/I_t$, a straight line is ordinarily fitted to the data points, with:

$$\text{slope} = -F_R U_L$$

$$\text{intercept} = F_R\,(\tau\alpha)$$

Given a collector's efficiency curve, the F_R and U_L parameters can easily be calculated.

Figure 5-9 shows efficiency curves for three flat-plate collectors: two with flat black absorbers (one has a single cover; the other, two covers) and one with a single cover and a black chrome selective surface. Consider the flat black collectors. Note that, for large values of $(t_{f,i} - t_a)/I_t$ (corresponding to the collector supplying water at a temperature much higher than ambient or to very low radiation levels), the two-cover collector has a higher efficiency. The second cover reduces heat loss, resulting in smaller U_L and a flatter slope. At small values of $(t_{f,i} - t_a)/I_t$, however, the single-cover collector outperforms the two-cover model because the former has a higher transmittance, τ, and thus a larger value of $F_R(\tau\alpha)$, the intercept. It is easy to go one step further and see that for a swimming pool collector, which heats water to only a few degrees above (or even below) ambient temperature during summer operation, it is often best not to use a cover at all. Use of a selective surface with a single cover has advantages of high intercept and low heat loss. For the most cost-effective selection of collectors, the loss parameter on Fig. 5-9 should fall within the following ranges:

	$(t_{f,i} - t_a)/I_t$
Service hot water	0.1-0.3
Series solar-heat pump	0.1-0.3
Space heating	0.2-0.6
Cooling	0.5-0.9

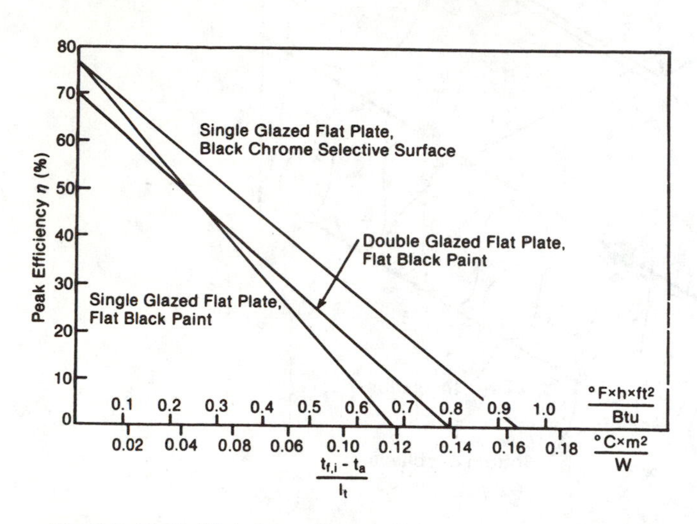

Fig. 5-9. Typical flat-plate collector efficiency curves (HUD 1977)

For collectors that operate at high temperatures, such as parabolic troughs, radiation heat losses make the efficiency curve deviate from a straight line. A second-order equation is thus often used (Duffie and Beckman 1980):

$$\eta = F_R\eta_o - U_{L,1}\frac{t_{f,i} - t_a}{I_a} - U_{L,2}\frac{(t_{f,i} - t_a)^2}{I_a} \qquad (5\text{-}8)$$

where

η_o = optical efficiency that accounts for $\tau\alpha$ as well as reflectance of a reflector and accuracy of the reflector

$U_{L,1}$ and $U_{L,2}$ = heat loss coefficients

I_a = solar radiation in the aperture

Typical efficiency curves for a variety of high-temperature collector types are given in Fig. 5-10.

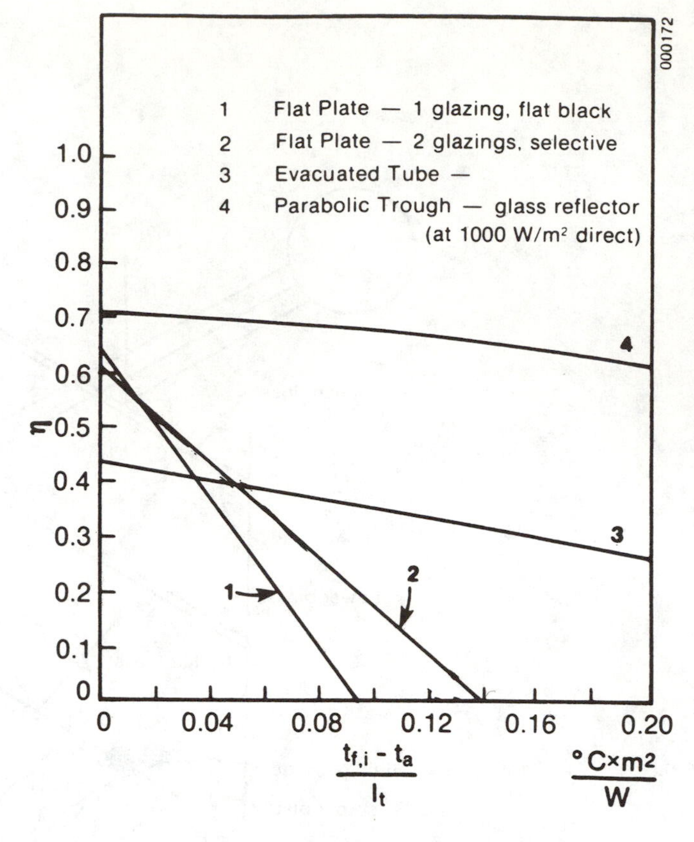

Fig. 5-10. Typical efficiency curves for high-temperature liquid collectors (HUD 1977)

Collector Testing

Two basic test methods have been used to determine solar collector efficiencies (Hill and Streed 1977):

1. Calorimetric method--uses a closed system consisting of a collector and a small storage tank. The system's time rate of temperature change is related to solar radiation and efficiency as

$$\eta = \frac{Q/A}{I_t} = \frac{m_c C_p \, dt/dT}{I_t} \qquad (5\text{-}9)$$

m_c = mass of the medium in the calorimeter per unit area of collector [lb/ft^2, (kg/m^2)]

C_p = specific heat of the medium in the calorimeter [Btu/lb-°F, (J/kg-°C)]

t = average system temperature of the medium [°F, (°C)]

T = time (s)

This is a good method for determining the daylong efficiency of a collector; only I_t and dt/dT must be measured during the day. However, it is difficult to extrapolate results from one day to another, and the procedure is sensitive to thermal losses. In addition, it is difficult to test an air collector by this method because of the low heat capacity of air.

2. Instantaneous method--uses an open system with collector isolated. Fluid mass flow rate, inlet and outlet temperatures, and solar

radiation are measured at solar noon under steady-state conditions. Efficiency is then determined as shown earlier:

$$\eta = \frac{Q/A}{I_t} = \frac{\dot{m}C_p(t_{f,e} - t_{f,i})}{AI_t} \qquad (5\text{-}10)$$

This method requires accurate measurement of $\dot{m}$, C_p, Δt, and I_t and is best suited to determination of liquid or air collector instantaneous efficiency.

Based on a procedure developed by the National Bureau of Standards, the American Society of Heating, Refrigerating and Air-Conditioning Engineers (ASHRAE) adopted Standard 93-77, "Methods of Testing to Determine the Thermal Performance of Solar Collectors" (ASHRAE 1977). This standard uses the instantaneous method to determine collector efficiency, based on gross collector area. A test configuration for liquid collectors is shown in Fig. 5-11. Three basic tests are called for, and these determine the following: collector time constant, instantaneous efficiency, and incident angle modifier.

Collector Time Constant

The collector time constant, τ_c, is the time required for the difference between fluid inlet and outlet temperatures to drop to 36.8% of its initial value after a step decrease in solar radiation or inlet fluid temperature. In a typical test the incident solar flux is abruptly reduced from a steady-state value of at least 250 Btu/h-ft^2 (790 W/m^2) to zero. At time $T = \tau_c$ after such a step change, with inlet collector temperature $t_{f,i}$ held constant,

$$\frac{t_{f,e}(\tau_c) - t_{f,i}}{t_{f,e}(o) - t_{f,i}} = \frac{1}{e} = 0.368 \qquad (5\text{-}11)$$

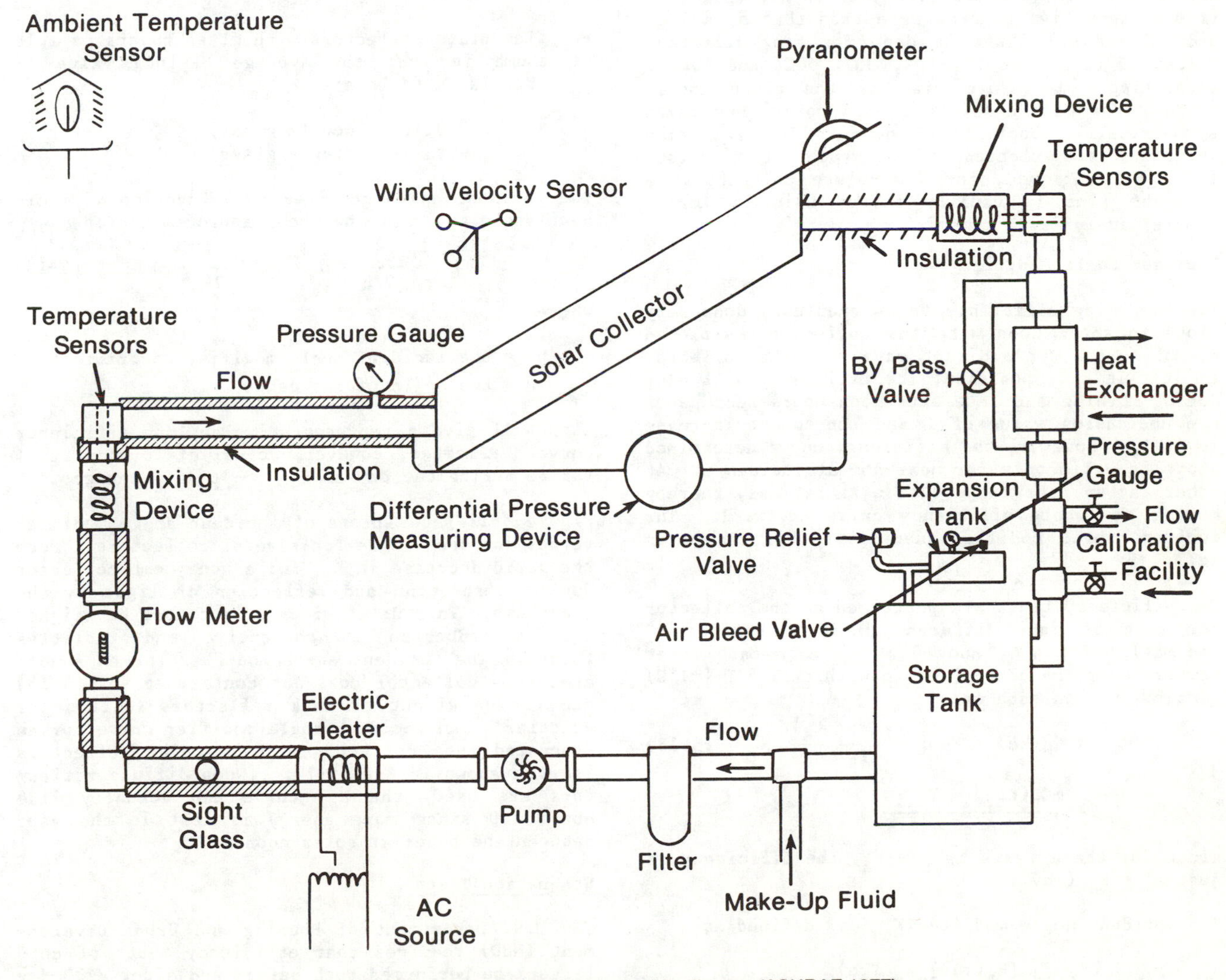

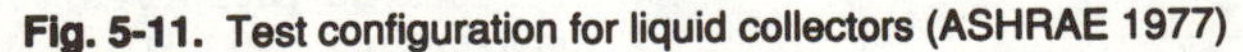

Fig. 5-11. Test configuration for liquid collectors (ASHRAE 1977)

The smaller the time constant, the more rapidly a collector will respond to changes in solar radiation. The time constant is needed to perform the instantaneous efficiency test described next.

Efficiency Test

Over a time period T_1 to T_2 the efficiency is determined as

$$\eta = \frac{\dot{m}C_p}{A} \frac{\int_{T_1}^{T_2} (t_{f,e} - t_{f,i})dT}{\int_{T_1}^{T_2} I_t dT} \quad (5\text{-}12)$$

For each data point, the efficiency is determined as the average over a period of 5 minutes or τ_c (whichever is larger) during which solar radiation is steady and flow is maintained at 14.7 lbm/h-ft^2 (0.02 kg/s-m^2) for a liquid collector and at both 1.96 and 6.0 cfm/ft^2 (0.01 and 0.03 m^3/s-m^2) for an air collector. (The efficiency of an air collector is more sensitive to flow rate than that of a liquid collector.) This is done for four different values of $(t_{f,i} - t_a)$: 10%, 30%, 50%, and 70% of stagnation temperature rise at the given conditions. For each case, four data points are taken symmetrically about solar noon. The resulting 16 points are plotted on a graph of η versus $(t_{f,i} - t_a)/I_t$ and, for flat-plate collectors, a straight line is usually fitted using a least-squares analysis.

Incident Angle Modifier

The preceding efficiency tests should be done very close to solar noon with the collector positioned normal to the sun's direct rays. If the collector is left in this position, its efficiency will drop during morning and late afternoon hours because of the decreasing value of I and consequent increase in $\Delta t/I_t$. However, the efficiency curve determined above is valid only for near-normal incidence. At other angles, the $\tau\alpha$ product will be less, thereby moving the whole efficiency curve downward. The incident angle modifier quantifies this change in $\tau\alpha$.

Two efficiency tests are performed on the collector for each of four different incident angles: 0° (normal), 30°, 45°, and 60°. In each case inlet temperatures are controlled to within ±1.8°F (±1°C) of ambient. In each case,

$$\eta = F_R(\tau\alpha) - F_R U_L \frac{(t_{f,i} - t_a)}{I_t} \quad (5\text{-}13)$$

$$= \frac{\dot{m}C_p(t_{f,e} - T_{f,i})}{AI_t} ;$$

since for these tests $t_{f,i} = t_a$, the efficiency is just $\eta' = F_R (\tau\alpha)_\theta$.

The incident angle modifier $K_{\alpha\tau}$, is defined as

$$K_{\alpha\tau} \equiv \frac{[F_R(\tau\alpha)]_\theta}{[F_R(\tau\alpha)]_n} = \frac{(\tau\alpha)_\theta}{(\tau\alpha)_n}$$

where

$(\tau\alpha)_n = (\tau\alpha)_\theta$ for the normal incidence ($\theta = 0°$) case

Since at any given incident angle θ, $\eta' = F_R(\tau\alpha)_\theta$ for these tests, $(\tau\alpha)_\theta = \frac{\eta'}{F_R}$, and

$$K_{\alpha\tau} = \frac{\eta'}{F_R(\tau\alpha)_n} . \quad (5\text{-}14)$$

η' values are determined for each angle, and $F_R(\tau\alpha)_n$ is the efficiency value for the 0° case. The $K_{\alpha\tau}$ values allow one to take the normal incidence efficiency curve and determine the appropriate curve for other angles (i.e., morning and afternoon hours). These other curves would have the same slope but a lower y-intercept, $F_R(\tau\alpha)$. The collector efficiency curve corrected for incident angles thus becomes

$$\eta = K_{\alpha\tau} F_R(\tau\alpha)_n - F_R U_L \frac{(t_{f,i} - T_a)}{I_t} .$$

For flat-plate collectors with glass covers, a rule of thumb is that the average daylong value of $K_{\alpha\tau}$, $\bar{K}_{\alpha\tau}$ is as follows:

$\bar{K}_{\alpha\tau}$ = 0.91 (for double glass)
= 0.93 (for single glass)

The incident angle modifier as a function of incident angle typically has been approximated as

$$K_{\alpha\tau}(\theta) = \frac{(\tau\alpha)_\theta}{(\tau\alpha)_n} = 1 + b_o \left(\frac{1}{\cos\theta} - 1\right) \quad (5\text{-}15)$$

where

b_o = the incident angle modifier coefficient
θ = the angle of incidence

This will give a representative daylong efficiency curve. Remember, however, actual efficiency still varies during the day as $(t_{f,i} - t_a)/I_t$ changes.

Figure 5-12 shows plots of incident angle modifier versus incident angle for several collectors. Note the rapid decrease in $K_{\alpha\tau}$ for a honeycomb collector due to absorption and reflection of light by the honeycomb. This deleterious effect must be weighed against reduction in convective and radiative losses. The incident angle modifier for an evacuated tube collector does not conform to Eq. (5-15) due to the effect of back reflectors (diffuse or specular). An incident angle modifier curve for an evacuated tube collector with specular reflectors is also shown in Fig. 5-12. (When diffuse reflectors are used, the $K_{\alpha\tau}$ curve can actually rise above 1.0 since more energy is lost in the gaps between the tubes at solar noon.)

Stagnation Tests

The U.S. Department of Housing and Urban Development (HUD) requires that efficiency tests of collectors be performed both before and after a 30-day stagnation test to reveal signs of deterioration (HUD 1977; DOC 1976).

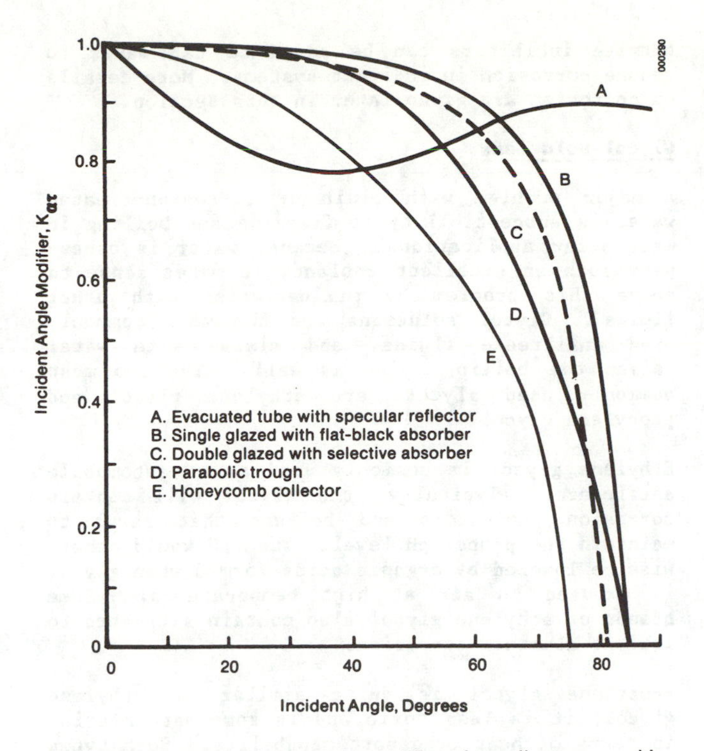

Fig. 5-12. Typical incident-angle modifiers for collectors used in IPH applications

Prior to the test, liquid collectors are filled with clean tap water, except those made exclusively for drain-back or drain-out systems, which are gravity drained first. (If freezing temperatures are possible, an antifreeze solution is used in place of pure tap water). The collectors are sealed and capped with pressure relief valves set within 10% of the collector manufacturer's maximum recommended operating pressure. Flat-plate collectors are set at a tilt angle that provides solar radiation within 10° of normal incidence at solar noon. Collectors are then exposed to 30 days of cumulative exposure to a radiation flux of at least 1500 Btu/ft^2-day (17 MJ/m^2-day) in their plane. This exposure must include at least 4 consecutive hours in a flux of 300 Btu/h-ft^2 (946 W/m^2), with outside ambient temperature of at least 80°F (27°C). This so-called 30-day stagnation test often requires several months to complete according to the above criteria.

Performance Ratings

In 1978 DOE initiated the Interim Solar Collector Testing (ISCT) Program. More than 100 solar collectors were tested at approved laboratories, and the results were published by Kirkpatrick et al. (1983). A number of state collector certification programs were also developed, notably by Florida (Florida Solar Energy Center 1982) and California (Trenschel and Goetze 1981). Eventually a national certification program, the Solar Rating and Certification Corporation (SRCC), was incorporated under the auspices of the Solar Energy Industries Association and the Interstate Solar Coordination Council. The SRCC publishes collector performance ratings (SRCC 1982) based on tests conducted at independent accredited laboratories. These include ASHRAE 93-77 tests and others, such as waterspray tests used to simulate the thermal shock of rainwater on collectors heated by stagnation. Design engineers should consult these publications before choosing a particular brand of collector.

Collector Selection and Use

Based on past government demonstration programs, the following precautions are recommended (DOE 1978b, 1978c; Cash 1978):

- Thoroughly review collector efficiency and stagnation test results. Pre- and post-stagnation test results should show a decrease in collector efficiency of <10%.
- Consider maintenance costs, such as replacement of plastic glazings every several years. Be sure the collector array installation provides access for maintenance.
- If collectors are to be used at city water pressure, make sure that they are specified for such application. Collectors must be capable of withstanding design pressures without leaking.
- Make sure freeze protection is adequate in the event of power failure.
- When using tracking collectors, talk with others or review DOE project reports regarding problems encountered with trackers and pipe joints.
- The best collector area is that which minimizes total life-cycle cost. Procedures for determining collector area are discussed in Ch. 7. However, the following very approximate rules of thumb will serve as a check on a given design:
  - Space heating and cooling--collector area of 1/5 to 1/2 ft^2 per square foot of floor area
  - Service hot water--10 to 20 ft^2_c per person (residential only)
  - Space cooling only--1/3 to 2/3 ft^2_c per square foot of floor area.

COLLECTOR COOLANTS

Air

Air used for heat transfer is readily available at no cost. Leaks will not cause damage (although they can definitely degrade system performance). Air cannot freeze or boil, but its low density and specific heat dictate high flow rates that can result in use of significant fan power, require large ducts, and generate noise. Some care must be taken to avoid corrosion; aluminum, for example, should be avoided in an air system located in the salty atmosphere near the ocean.

Liquids

A heat-transfer fluid chosen for use in liquid collectors must be compatible with the application. A liquid coolant should have good heat transport capabilities, require low pumping power, remain in a liquid state over all possible temperatures and pressures, not promote corrosion, be safe to use, and be inexpensive. The coolant should be selected for freezing point and boiling point, considering the minimum and maximum temperatures expected. The heat transfer fluid chosen must be compatible with the collectors, piping, pipes, and seals. For example, aluminum should never be exposed to the chlorinated water used in swimming pools.

The first two characteristics listed above can be combined to yield a heat-transfer efficiency factor (HTEF), the ratio of heat-transfer coefficient to pumping power (Fried 1973). Values of HTEF for a variety of coolants are given in Fig. 5-13. The superiority of water can be clearly seen. Water is also inexpensive, safe, and, in most cases, readily available. If used in a warm climate, or in conjunction with a drain-back or drain-out freeze protection system, and if kept below the boiling point, only corrosion inhibitor additives might be needed.

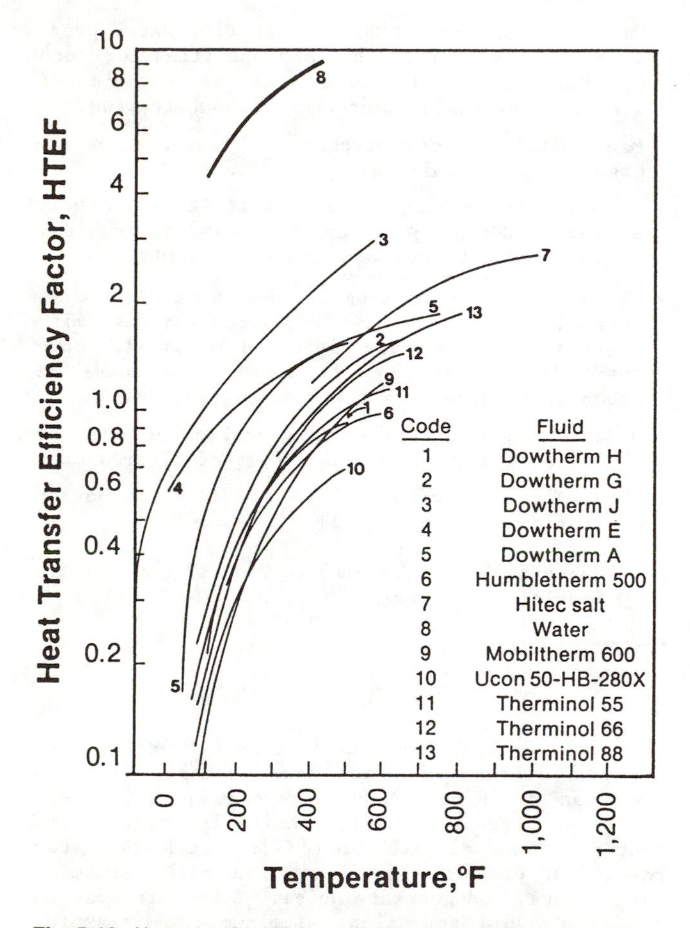

Fig. 5-13. Heat-transfer efficiency factors for 13 fluids (Fried 1973)

The likelihood of corrosion depends on water quality and system materials. If copper pipe is used throughout, plain tap water can ordinarily be used. Many direct solar hot water systems, in fact, circulate pressurized potable water through the collector loop. If aluminum or steel is used, attention must be paid to corrosion control. Nitrite inhibitors can be added to tap water to reduce corrosion in indirect systems. More details on corrosion are given later in this section.

Glycol Solutions

A major problem with plain or corrosion-treated water is susceptibility to freezing and boiling in many solar applications. Because water is otherwise such an excellent coolant, it makes sense to solve this problem by mixing water with other fluids. Glycol solutions are the most commonly used antifreeze fluids, and mixing with water raises the boiling point as well. The two most commonly used glycols are ethylene glycol and propylene glycol.

Ethylene glycol is commonly used as an automobile antifreeze. Typically, the glycol will contain corrosion inhibitors and buffers that serve to maintain the proper pH level. The pH would otherwise be lowered by organic acids formed when glycol is exposed to air at high temperatures. Some brands of ethylene glycol also contain silicates to inhibit aluminum corrosion.

Propylene glycol is quite similar to ethylene glycol; it is less toxic and is somewhat inferior in terms of heat transport capability. Both types of glycol-water mixtures require periodic maintenance to prevent corrosion. They should be replaced at least once a year or, preferably, be monitored regularly for pH. It is also advisable to check the collector fluid after any stagnation condition. Recent experiments at Los Alamos National Laboratory have indicated that a properly sealed all-copper system can withstand degraded glycol over long periods. However, an annual check of the solution is still the safest course of action.

A system using glycols should be cleaned and flushed before being filled. No galvanized pipe should be used because glycol inhibitors react with zinc to form sludge. Boiling should be avoided to prevent sludge formation. No chromate treatment should ever be added to glycol solutions. (Chromates are generally not recommended in any case due to high toxicity.)

It is important to note that physical characteristics of glycol-water solutions differ somewhat from those of pure water. Typical 50% water-ethylene glycol mixtures have a slightly higher density but lower specific heat than water. This means that a higher volumetric flow rate is required, as shown in Table 5-4. Pumping power requirements are also greater because of increased flow rate and differences in viscosity. (See Table 5-5.) Finally, since glycol mixtures have an expansion rate 20% greater than that of water, an expansion or compression tank must have a volume at least 20% greater than for a water system (Liu and Fanney 1980).

Aqueous Salt Solutions

Aqueous salt solutions have been suggested as an alternative to glycols. According to Kauffman

Table 5-4. Increased Flow Requirement for Same Heat Conveyance: 50% Glycol

Fluid Temperature (°F)	Flow Increase Needed for 50% Glycol as Compared with Water
40	1.22
100	1.16
140	1.15
180	1.14
220	1.14

Table 5-5. Pressure Drop Correction Factors: 50% Glycol

Fluid Temperature (°F)	Pressure Drop Correction Flow Rates Equal	Combined Pressure Drop Correction: 50% Glycol Flow Increased Per Table 5-4
40	1.45	2.14
100	1.1	1.49
140	1.0	1.32
180	0.94	1.23
220	0.9	1.18

(1977), 23% sodium acetate and 38% sodium nitrite with phosphate and copper inhibitor additives are the preferred solutions. Their advantages are low toxicity, low pumping power, long-term stability, and no oxidation at high temperatures like glycol. Thus far, however, they have not been widely used.

Nonaqueous Fluids

Nonaqueous heat-transfer fluids have low vapor pressure at high temperatures, low corrosion, and long-term stability. Some examples are paraffinic oils, aromatic oils, and silicone fluids.

Paraffinic oils are generally low in toxicity and are often used in high-temperature industrial applications. Example trade names are Exxon Caloria and Dowtherm HP. Exposure to air at high temperatures can result in corrosive byproducts, and pumping power is high, particularly at low temperature. Due to their fairly low flash point [typically 374°F (190°C)], most paraffinic oils will fail to meet HUD and NBS requirements that flash points be at least 100°F (55°C) higher than the highest possible system (stagnation) temperature.

Aromatic oils flow better at low temperatures than paraffinic oils but have higher toxicity and higher cost. Examples are Dowtherm J, Monsanto Therminol 55, and Therminol 60. The flash point [typically 140°F (60°C)] is even lower than that of paraffinics.

Manufactured primarily by Dow Corning and General Electric, silicone fluids have low freezing and pour points, low vapor pressure, low toxicity, low corrosion, relatively high flash point, and excellent long-term stability. However, they are also high in viscosity (resulting in high pumping power), have low specific heats, are conducive to leakage (due to low surface tension), and are high in cost.

More details on available heat-transfer fluids can be found in reports issued by Argonne National Laboratory (Cole et al. 1979; ANL 1981). Table 5-6 lists major characteristics of many of the heat-transfer fluids discussed above.

Coolant Characteristics

Toxicity

The main hazard of use of a heat transfer-fluid is that it may contaminate potable water. Table 5-7 lists the LD_{50} values for various heat-transfer fluids. LD_{50} refers to the quantity of substance that kills 50% of dosed test animals within 14 days. Thus the higher the LD_{50} value, the lower the toxicity.

For service hot water applications, HUD and NBS require a double-walled heat exchanger whenever a toxic collector fluid is used. Some fluids may qualify as nontoxic, but they can deteriorate with time. Since it is possible that someone will inadvertently replace the fluid with a toxic substitute at some future date, it is safest to use a double-walled heat exchanger unless city water is supplied directly to the collectors. It is advisable in any case to place a tag at the fluid fill location that contains information regarding the proper coolant to be used.

Corrosion

The importance of preventing corrosion when using water and water-glycol mixtures (by far the most commonly used fluids in solar heating and cooling applications) has already been mentioned in describing these fluids. Corrosion is a complex phenomenon involving many parameters, including composition of metals and fluids, temperature, flow rate, system design (use of multiple metals, effects of air in open systems), and presence of additives (HUD 1977). Popplewell (1975) has summarized the four types of internal corrosion caused by a heat-transfer fluid.

Galvanic corrosion is caused by the junction of two dissimilar metals in an electrolytic solution. Insulating couplings should be used between dissimilar metals to prevent the flow of electric current. Care must be taken to ensure that no path for electric flow (e.g., flange bolts) exists. Pitting corrosion involves rapid local metal loss. Aluminum and steel are susceptible to the presence of chloride ions. Aluminum is susceptible to the presence of metal ions. **Crevice corrosion** is similar to pitting corrosion but occurs at crevices in the system, such as those at rivets and spot-welded joints. **Erosion corrosion** can be produced on any metal by high local fluid velocity.

Corrosion is more of a problem with aqueous fluids than with nonaqueous fluids. High chloride content

Table 5-6. Commercially Available Heat-Transfer Fluids (Kauffman 1977)

Medium	Density kg/m^3	Viscosity centipoise	Heat Capacity cal/gm °C	Freezing Point °C	Anti-corrosion Protection[b]	High Temperature Reactions	Toxicity Requires Double Wall[c]
AQUEOUS FLUIDS							
Water	1000	0.5-0.9	1	0	Nitrites or chromates and a pH buffer in a closed loop for steel or aluminum	Boils at 100°C	No[e]
50% wt.% water-ethylene glycol	1050	1.2-4.4	0.83	-36	Silicates, pH buffer[d]	Boils at 110°C; degrades slowly slowly to formic acid above 150°C	Yes
50% wt.% water-propylene glycol	1022	1.4-7.0	0.87	-31	Phosphates, pH buffer[d]	Boils at 110°C; not recommended above 150°C	No
NONAQUEOUS FLUIDS				(pour pt.)			
Paraffinic Oils (Exxon Caloria or Dowtherm HP)	821	11	0.51	-7	Not needed	Flash point 190°C; oxidizes slowly if exposed to air above above 110°C	
Aromatic Oils (Dowtherm J or Monsanto Therminol 55)	830	0.6	0.46	-70	Not needed	Flash point 63°C; oxidizes slowly on exposure to air above 150°C	Yes
Silicone Oils (GE SF-96 or Dow	970	50	0.36	-85	Not needed	No known problems; flash point is above 260°C	No
Corning Q2-1132)	960	15	0.38	-120			

[a]Properties are average values between 30°C and 65°C. Thermal conductivities are not presented because of a lack of data.
[b]Inhibitors are added primarily to protect aluminum and steel. Special copper anticorrosion agents such as MBT (2-mercaptobenzotriazole), BTZ (benzotriazole), or TTZ (tolyltriazole) may be added to prevent galvanic corrosion from copper ions.
[c]The double wall requirement applies only to potable water systems.
[d]Glycol solutions should not be used with zinc galvanized plumbing.
[e]Double wall required if chromate or nitrite inhibitors are added.

Table 5-7. LD_{50} Values for Various Heat-Transfer Fluids (Cole et al. 1979)

Fluid	LD_{50}
Water	--
100% Ethylene Glycol (No inhibitors)	8.0
100% Propylene Glycol (No inhibitors)	34.6
100% Diethylene Glycol (No inhibitors)	30
100% Triethylene Glycol (No inhibitors)	30
100% Dowtherm SR-1	4
SF-96(50) (Silicone)	50
Q2-1132 (Silicone)	50
Dowtherm J	1.1
Therminol 44	13.5
Therminol 55	15.8
Therminol 60	13.0

will accelerate corrosion, as will improper pH. Corrosion usually increases with temperature and oxygen concentration. Table 5-8 shows acceptable and unacceptable use conditions for metals in contact with aqueous fluids in an open system (i.e., one in which the fluid is constantly exposed to air, such as drain-out systems). Table 5-9 shows similar conditions for a closed system. More details on corrosion are given in the storage section, below. Further information is also found in reports from HUD (1977), NBS (DOC 1976), and Argonne National Laboratory (Cole et al. 1979).

THERMAL ENERGY STORAGE FOR BUILDING HEATING AND COOLING

Building heating and cooling loads are often poorly matched to the time profile of available energy supply. Solar energy, unlike energy from fossil fuels, is not available for heat at night or on cloudy days. Even cooling loads, which more nearly coincide with maximum levels of solar radiation, often are present after sunset. Energy storage can be an economical means of correcting this availability mismatch.

A building designer requires basic information about energy storage before proceeding with a given project. What types of storage are available? How much storage is required? How will inclusion of storage affect system performance, reliability, and cost? What storage systems or designs are available? This section provides answers to these questions or directs the reader to pertinent literature. Although thermal energy can be stored by a variety of chemical, electrochemical, or mechanical means, only low-temperature sensible storage and latent thermal energy storage are considered here.

Duration of Storage

Solar applications require storage of thermal energy from very short durations (e.g., buffer storage of minutes for solar thermal power plants) to annual-cycle time scales. Most solar systems use diurnal storage, with energy stored for at most a day or two. Annual-cycle storage offers a number of advantages.

Table 5-8. Generally Acceptable and Unacceptable Use Conditions for Metals in Direct Contact with Heat-Transfer Liquids in Open Systems (HUD 1977)

Generally Acceptable Use Conditions[a]	Generally Unacceptable Use Conditions
ALUMINUM	
• When in direct contact with distilled or deionized water that contains appropriate inhibitors and does not contact copper or iron. • When in direct contact with distilled or deionized water that contains appropriate inhibitors and a means of removing heavy metal ions obtained from contact with copper or iron. • When in direct contact with stable anhydrous organic liquids.	• When in direct contact with untreated tap water with pH <5 or >9. • When in direct contact with aqueous liquid containing less electropositive metal ions such as copper or iron or halide ions. • When specific data regarding the behavior of a particular alloy are not available, the velocity of aqueous liquid shall not exceed 4 ft/s. • When in direct contact with a liquid that is in contact with corrosive fluxes.
COPPER	
• When in direct contact with distilled, deionized, or low-chloride, low-sulfate, and low-sulfide tap water. • When in direct contact with stable anhydrous organic liquids.	• When in direct contact with aqueous liquid containing high concentrations of chlorides, sulfates, or liquid containing hydrogen sulfide. • When in direct contact with chemicals that can form copper complexes such as ammonium compounds. • When in direct contact with an aqueous liquid having a velocity greater than 4 ft/s.[b] • When in direct contact with a liquid that is in contact with corrosive fluxes. • When in contact with an aqueous liquid with a pH <5. • When the copper surface is initially locally covered with a copper oxide film or a carbonaceous film. • When operating under conditions conducive to water line corrosion.
STEEL	
• When in direct contact with distilled, deionized, or low-salt content water containing appropriate corrosion inhibitors. • When in direct contact with stable anhydrous organic liquids. • When adequate cathodic protection of the steel is used (practical only for storage tanks).	• When in direct contact with untreated tap, distilled, or deionized water with pH <5 or >12. • When in direct contact with a liquid that is in contact with corrosive fluxes. • When in direct contact with an aqueous liquid having a velocity greater than 6 ft/s[b]. • When operating under conditions conducive to water line corrosion.
STAINLESS STEEL	
• When the grade of stainless steel selected is resistant to pitting, crevice corrosion, intergranular attack, and stress corrosion cracking in the anticipated use conditions. • When in direct contact with stable anhydrous organic liquids.	• When the grade of stainless steel selected is not corrosion resistant in the anticipated heat transfer liquid. • When in direct contact with a liquid that is in contact with corrosive fluxes
GALVANIZED STEEL	
• When adequate cathodic protection of the galvanized parts is used (practical only for storage tanks.) • When in contact with stable anhydrous organic liquids.	• When in direct contact with aqueous liquid containing copper ions. • When in direct contact with aqueous liquid with pH <7 or >12. • When in direct contact with aqueous liquid with a temperature >55°C.
BRASS AND OTHER COPPER ALLOYS	
Binary copper-zinc brass alloys (CDA 2XXX series) exhibit generally the same behavior as copper when exposed to the same conditions. However, the brass selected shall resist dezincification in the operating conditions anticipated. At the zinc contents of 15% and greater, these alloys become increasingly susceptible to stress corrosion. Selection of brass with a zinc content below 15% is advised. There is a variety of other copper alloys available, notably copper-nickel alloys, which have been developed to provide improved corrosion performance in aqueous environments.	

[a]The use of suitable antifreeze agents and buffers is acceptable provided they do not promote corrosion of the metallic liquid containment system. The use of suitable corrosion inhibitors for specific metals is acceptable provided they do not promote corrosion of other metals present in the system. If thermal or chemical degradation of these compounds occurs, the degradation products shall not promote corrosion.

[b]The flow rates at which erosion/corrosion becomes significant will vary with the conditions of operation. Accordingly, the value listed is approximate.

Table 5-9. Generally Acceptable and Unacceptable Use Conditions for Metals in Direct Contact with Heat-Transfer Liquids in Closed Systems (HUD 1977)

Generally Acceptable Use Conditions[a]	Generally Unacceptable Use Conditions
ALUMINUM	
• When in direct contact with distilled or deionized water that contains appropriate corrosion inhibitors.	• When in direct contact with untreated tap water with pH <5 or >9.
• When in direct contact with stable anhydrous organic liquids.	• When in direct contact with liquid containing copper, iron, or halide ions.
	• When specified data regarding the behavior of a particular alloy are not available, the velocity of aqueous liquids shall not exceed 4 ft/s.
COPPER	
• When in direct contact with untreated tap, distilled, or deionized water.	• When in direct contact with an aqueous liquid having a velocity greater than 4 ft/s[b].
• When in direct contact with stable anhydrous organic liquids.	• When in contact with chemicals that can form copper complexes such as ammonium compounds.
• When in direct contact with aqueous liquids that do not form complexes with copper.	
STEEL	
• When in direct contact with untreated tap, distilled, or deionized water.	• When in direct contact with liquid having a velocity greater than 6 ft/s[b].
• When in direct contact with stable anhydrous organic liquids.	• When in direct contact with untreated tap, distilled, or deionized water with pH <5 or >12.
• When in direct contact with aqueous liquids of 5 <pH <12.	
STAINLESS STEEL	
• When the grade of stainless steel selected is resistant to pitting, crevice corrosion, intergranular attack and stress corrosion cracking in the anticipated use conditions.	• When the grade of stainless steel selected is not corrosion resistant in the anticipated heat transfer liquid.
• When in direct contact with stable anhydrous organic liquids.	• When in direct contact with a liquid that is in contact with corrosive fluxes.
GALVANIZED STEEL	
• When in contact with water of pH >5 but <12.	• When in direct contact with water with pH <7 or >12.
	• When in direct contact with an aqueous liquid with a temperature >55°C.
BRASS AND OTHER COPPER ALLOYS	
Binary copper-zinc brass alloys (CDA 2XXX series) exhibit generally the same behavior as copper when exposed to the same conditions. However, the brass selected shall resist dezincification in the operating conditions anticipated. At zinc contents of 15% and greater, these alloys become increasingly susceptible to stress corrosion. Selection of brass with a zinc content below 15% is advised. There is a variety of other copper alloys available, notably copper-nickel alloys, which have been developed to provide improved corrosion performance in aqueous environments.	

[a]The use of suitable antifreeze agents and buffers is acceptable provided they do not promote corrosion of the metallic containment system. The use of suitable corrosion inhibitors for specific metals is acceptable provided they do not promote corrosion of other metals present in the system. If thermal or chemical degradation of these compounds occurs, the degradation products shall not promote corrosion.

[b]The flow rates at which erosion/corrosion becomes significant will vary with the conditions of operation. Accordingly, the value listed is approximate.

Energy collected on summer days can be used on winter days. Larger storage has lower unit heat loss because of lower surface-to-volume ratios. The need for backup systems can be eliminated, since periods of adverse weather have little effect on thermal energy availability. Collector areas can be reduced, and collector stagnation in summer is minimized. Also, annual-cycle energy systems are a natural match for well-designed energy management systems in which excess heat or coolness from the environment or adjacent structures is saved for later use. It may be economical only in multi-dwelling designs, where expensive energy distribution systems are required, and novel institutional ownership and financing may have to be devised.

Advantages of diurnal systems include: capital investments for storage and energy loss are usually low, devices are smaller and can be easily manufactured off-site, and sizing of daily storage for each application is not as critical as for larger annual storage. Additional information on storage duration is given in Ch. 4; the following section will address only diurnal systems.

Storage Technologies

Options for storing low-temperature thermal energy are numerous. Figure 5-14 lists a classification of available storage devices. Each is briefly described below. More complete descriptions can be found in the referenced literature.

Sensible Heat Storage

The amount of energy stored by sensible heat devices is proportional to the differences between storage input and output temperatures times the medium's heat capacity. Water has approximately twice the heat capacity [1 Btu/lb-°F (2326 J/kg-°C)] of rock and earth. Sensible heat storage

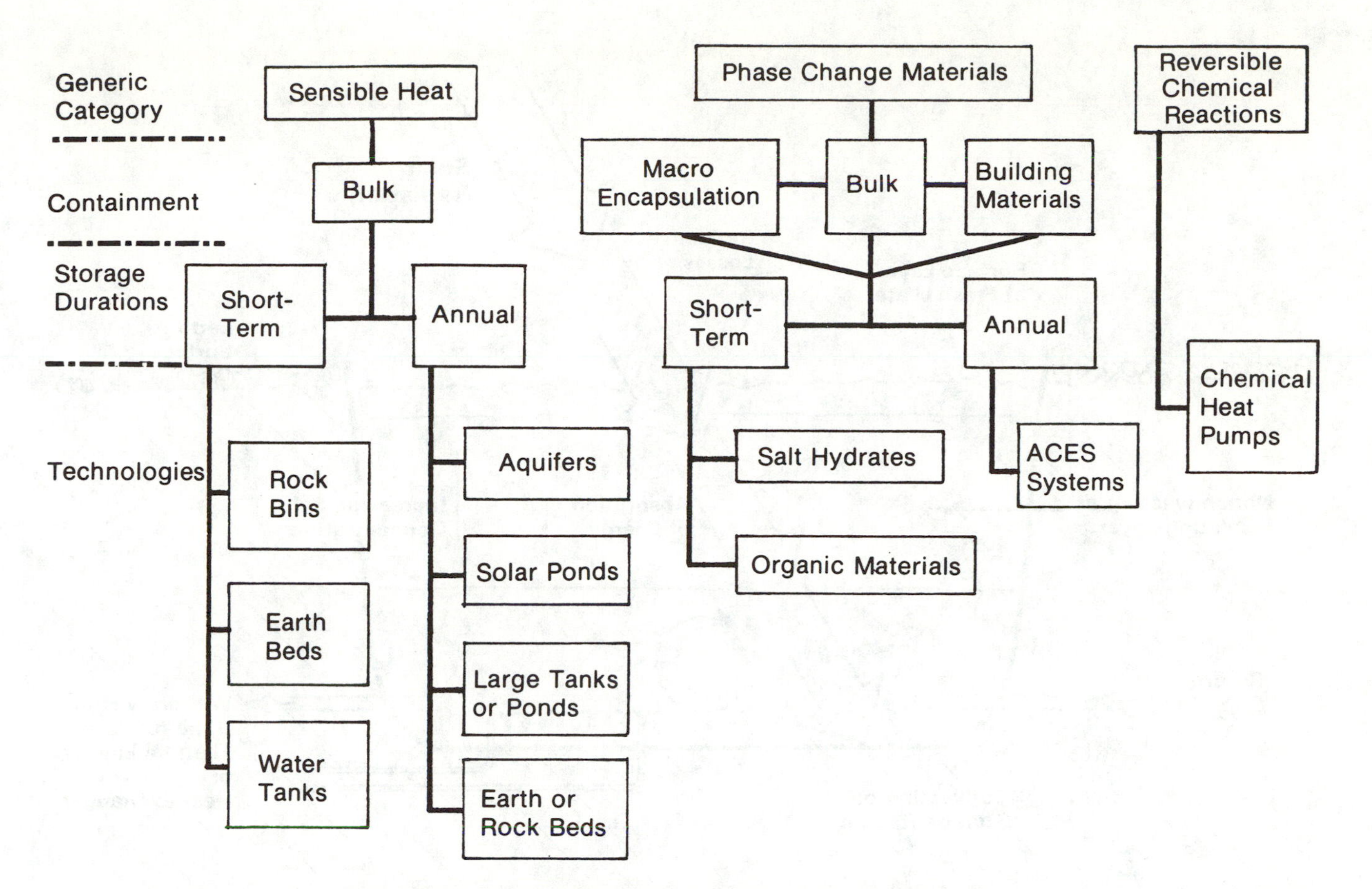

Fig. 5-14. Available devices for storing low-temperature thermal energy

consists of a storage medium, a container, and input-output devices. Containers must both retain storage material and prevent loss of thermal energy. As discussed previously, thermal stratification, a thermal gradient across storage, is desirable. Maintaining stratification is much simpler with solid storage media, such as rocks or earth, than with water or other fluids.

A number of novel technologies are among possible annual storage systems (Givoni 1977). Aquifers are large geological formations that can be tapped by wells. Although some technical problems, such as biofouling and clogging of pipes and wells, remain to be solved, aquifers offer potentially very low-cost storage for larger scale systems. Except for the cost of access wells, the aquifer container is essentially free. However, rather low round-trip aquifer storage efficiencies (70%) may mandate the use of aquifers only in conjunction with such low-cost energy supplies as industrial waste heat.

Solar ponds are both low-temperature solar collectors and long-term storage devices. Solar energy is trapped and stored by the bottom layer (see Fig. 5-15) because there is no convective loss through the salt gradient part of the pond. Salt concentration increases with depth because temperature increases with pond depth. This establishes a density gradient, which counteracts any tendency for convection in the gradient layer. Salt-gradient solar ponds may be economically attractive in climates with little snow and in areas where land is readily available.

Latent Heat Storge

Use of the latent heat capacity of storage materials allows smaller storage volumes. The relatively constant temperature of storage can maximize collector efficiency and minimize storage heat loss. An ideal phase-change material (PCM) should have the following characteristics:

- appropriate phase-change temperature
- high latent heat
- low cost
- ready availability
- nontoxicity and nonflammability
- uniform phase-change characteristics (no subcooling or separation)
- long life under repeated phase change.

Two basic material types have been used for short-term phase change storage--organic compounds and salt hydrates. A good example of an organic compound is paraffin. Like other organics, it suffers certain limitations--flammability, low thermal

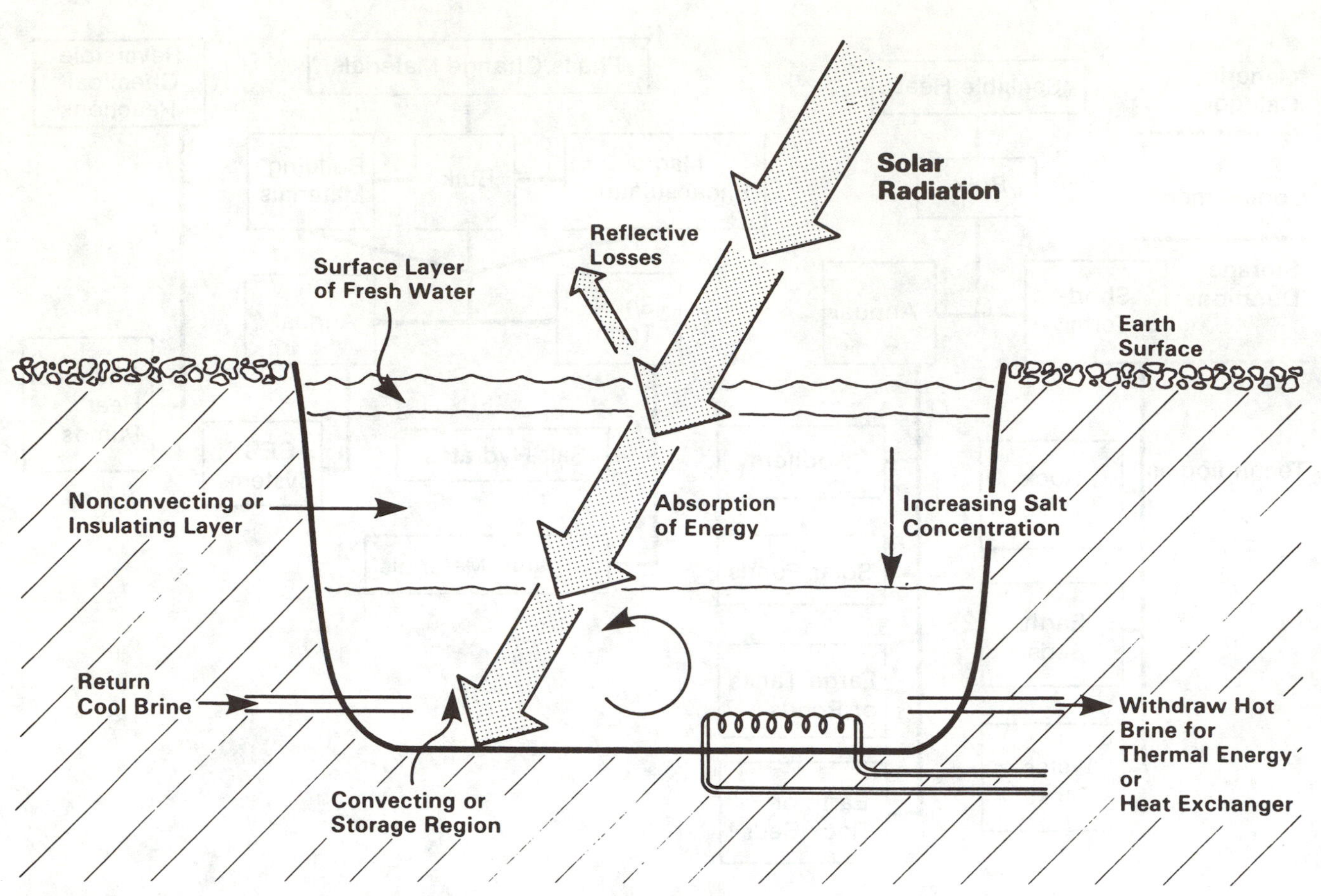

Fig. 5-15. Solar pond

conductivity, and volumetric change during phase change (causing separation from container walls). The latter two problems make it difficult to get heat into and out of the material. As a result of such limitations, most work in phase-change materials has centered on salt hydrates. Maria Telkes, who has done much pioneering work in this field, has concentrated on one salt hydrate in particular because of its low cost--sodium sulfate decahydrate, or Glauber's salt. The chemical reaction involved in the phase change of Glauber's salt ($Na_2SO_4 \bullet 10\ H_2O$) is

$$Na_2SO_4 \bullet 10\ H_2O + 108\ Btu/lb_m\ (251\ kJ/kg) \underset{Cooling}{\overset{Heating}{\rightleftarrows}} Na_2SO_4 + 10\ H_2O$$

When heated, Na_2SO_4 dissolves in the water of hydration; the result is a liquid. When cooled, the hydrate is reformed and 108 Btu are released for each pound (251 kJ/kg) solidified. To illustrate the volumetric storage capacity of Glauber's salt, consider 220 lb (100 kg) undergoing a 54°F (30 K) temperature swing that includes a phase change at 89°F (32°C):

$$Q = Q_{sensible} + Q_{latent}$$

$$Q = Mc_p\Delta T + Mh_{fg}$$

where

h_{fg} = latent heat of fusion, 108 Btu/lb_m (251 kJ/kg)

c_p = specific heat at constant pressure, 0.5 Btu/lb_m-°F (2.1 kJ/kg-K)

ΔT = temperature change, 54°F (30 K)

Substituting numerical values, we have

$$Q = 220\ lb\ (0.5\ Btu/lb_m\text{-}°F)\ 54°F + 220\ lb\ (108\ Btu/lb_m)$$

$$Q = 27{,}208\ Btu\ (28{,}700\ kJ)$$

Note that the latent heat term is a constant, independent of temperature. The smaller the temperature swing, the less the contribution from sensible heat. In this example of a 54°F (30 K) change, 87% of the heat stored is latent and 13% is sensible.

In Table 5-10, the advantage in both mass and volume inherent in Glauber's salt over rocks or water is plainly seen. The comparison is based on each material sized for a heat storage capacity of 1 GJ over a 54°F (30 K) temperature swing (typical for a solar space heating system). Only 7670 lb_m (3480 kg) of Glauber's salt is needed, compared with 17,640 lb_m (8000 kg) of water and 84,000 lb_m (38,100 kg) of rock.

Table 5-10. Storage Size Comparison of Different Media Based on 1 GJ of Storage and a 54°F (30 K) Temperature Swing

Medium	Weight		Volume	
	(lb_m)	(kg)	(ft^3)	(m^3)
Rocks	84,000	38,100	218.6	20.3
Water	17,640	8,000	283.0	26.3
Glauber's Salt	7,670	3,480	22.6	2.1

Unfortunately, commercialization of Glauber's salt storage has faced a number of obstacles that arise from its phase-change behavior. Glauber's salt is considered an incongruent material because, upon melting, two distinct phases result: a saturated solution of Na_2SO_4 in water and an excess amount of insoluble Na_2SO_4 that precipitates out. Once the precipitate settles to the bottom, it no longer participates in the reaction and results in a decrease in overall heat capacity with each melting and freezing cycle. Various approaches have been used to address this problem, such as use of thickeners like paper, pulp, and thixotropic agents (similar to those used in thickening paints), active mixing, and thin containers. Glauber's salt also tends to supercool below its normal freezing point. To prevent this, nucleating agents (such as borax) have been added, or a cold rod has been inserted in the salt hydrate to serve as a nucleation site.

Although storage devices containing Glauber's salt have been marketed, the problems discussed above have not been completely solved. As a result, some manufacturers have examined other salt hydrates--calcium chloride hexahydrate, for example. It is considered a semicongruent material and does not exhibit the same degree of separation as does Glauber's salt. According to one manufacturer, calcium chloride hexahydrate only supercools for the first several cycles [to 20°F (11°C) below its freezing point]. Unfortunately, it is expensive in packaged form [>$1/lb ($2/kg)] and has a low melting point [81°F (27°C)], which makes it suitable only for very low-temperature storage applications.

Several characteristics of phase change materials make them attractive for solar applications. First, they store heat over a narrow temperature range. As such, thermal stratification is not a major design consideration. In addition, the isothermal nature of the storage make PCMs attractive for passive designs. Second, the amount of energy stored per unit volume can be substantially higher than in sensible heat storage. Therefore, using a PCM can facilitate building retrofits with solar systems where system volume and/or weight is a consideration. However, heat exchange into and out of storage is difficult because heat transfer is more difficult from solid surfaces than through fluids. In general, phase change materials must still be considered experimental and are not yet cost effective for building applications.

Reversible Chemical Reaction Storage

Thermal energy can be used to drive a chemical reaction, and the products can be stored indefinitely at room temperature with no loss of capacity. When required, they are recombined in a chemical reaction that releases thermal energy.

Chemical heat pumps are still in the experimental stage and cannot be considered practical yet for building applications. The chemical heat pump (CHP) is the primary device that uses reversible chemical reactions for low-temperature thermal energy storage. These pumps can be used for both heating and cooling applications, and, like standard electrical heat pumps, they can use environmental energy. COPs of 1.6 for heating and 0.7 for cooling applications have been predicted. Currently, six basic systems are in different early stages of development: ammoniated salt, hydrated salt, methanolated salt, dilute/concentrated sulfuric acid, hydrated zeolite, and hydrogenated metal systems.

Rock-Bed and Water-Tank Storage Characteristics

Commercially available storage devices consist of a well-insulated container, a storage medium, and provision for adding and withdrawing heat. Design parameters for water and rock-bed storage, by far the most widely used and least expensive options, are outlined below; more detailed information is found in Design and Installation Manual for Thermal Energy Storage (Cole et al. 1979).

Types of Containers

Water storage requires insulated, leakproof tanks that are the major cost component of such systems. In 1977, Solar Engineering magazine listed manufacturers of steel tanks and 17 manufacturers of plastic or fiberglass-reinforced plastic tanks of different sizes and shapes (Solar Engineering 1977). Tanks are manufactured of four major materials: steel, fiberglass-reinforced plastics, concrete, and wood. Characteristics and costs of each are listed in Table 5-11.

Cost ranges listed in Table 5-11 are very approximate. Costs depend on tank size (see Fig. 5-16), location, installation requirements, temperature ranges, and insulation. It must be stressed that costs for very large tanks for long-duration storage can be much lower, perhaps as low as $0.05/gal ($0.19/L).

Rock beds are simpler to construct than water tanks (see Fig. 5-17). Common materials can be used, and excessive insulation is not required because rock has a rather low thermal conductivity. In addition, corrosion problems are eliminated, and small leaks cause little problem. A detailed discussion of rock-bed design is given below.

Table 5-11. Storage Tank Characteristics

Type	Advantages	Disadvantages	Temperature Limitations	Cost[a] Range ($/gal)
Steel	• Can be pressurized • Much field experience • Easy plumbing connections	• Complete tanks difficult to install indoors • Subject to rust and corrosion	None	0.70-0.96
Fiberglass-Reinforced Plastic	• Factory insulated tanks available • Much field experience • No corrosion	• Maximum temperature is limited • Usually not made to be pressurized • Complete tanks difficult to install indoors	<150°-200°F	1.29-2.42
Concrete	• May be precast or cast in place	• Possibility of cracks and leaks • Cannot be pressurized • Difficult to make leak-tight plumbing connections	<210°F	0.69-0.92
Wooden	• Easy indoor installation	• Maximum temperature is limited • Cannot be pressurized • Not suitable for underground installation	<160°F	0.62-1.78

[a]These cost estimates are based upon 1975 costs inflated to 1979 prices. Costs are for small tanks for daily storages only.

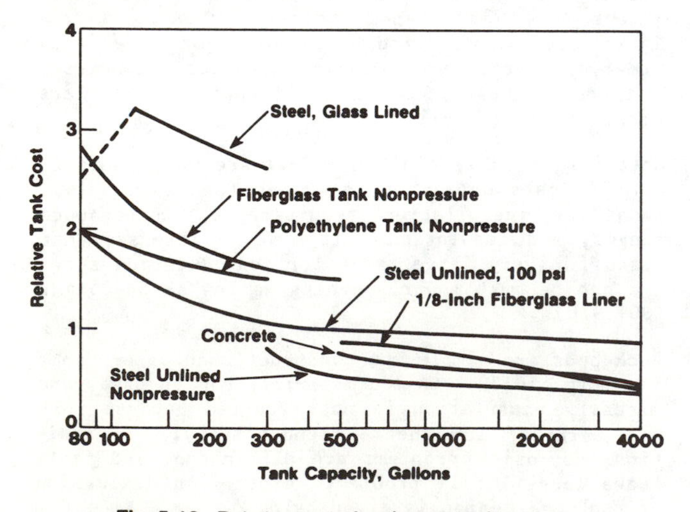

Fig. 5-16. Relative cost of various container types

Storage Size

A given collector area will supply a given portion of building load. Undersizing thermal energy storage will result in lower system efficiency. Short-duration storage costs much less than collectors do, so undersizing is an uneconomical practice. Storage must be sized correctly.

Storage quantity required depends on heat capacity and, for latent heat storage, the latent heat of fusion. Volumetric heat capacity is determined by multiplying heat capacity by density: it defines the material volume required for a given amount of energy storage. Water has a volumetric heat capacity of 62.4 Btu/ft^3-°F (4184 kJ/m^3-°C); rock (with allowances for 30% voids) has a volumetric heat capacity of about 20 Btu/ft^3-°F (1341 kJ/m^3-°C). The quantity of heat stored also depends on the daily temperature range experienced by the material. The larger the range, the smaller the storage can be.

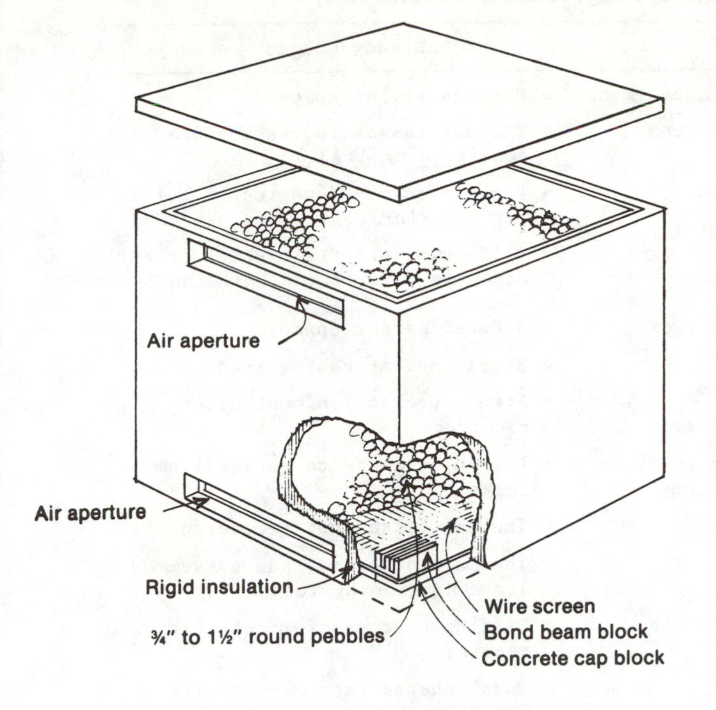

Fig. 5-17. Typical rock-bed heat storage unit

A number of design tools have been developed to aid sizing of solar system components. These are discussed in Ch. 7, as are rules of thumb for storage sizing. Recommended storage sizes for different applications were also presented in Ch. 4. It is important to remember that a sophisticated design tool such as TRNSYS may give very different results than rules of thumb for a particular application. For example, a passive home with integral storage in structural mass and a lower-than-normal heating load may require much less than 15 Btu of storage capacity per °F per ft^2_c = 1.8 gal/ft^2_c (305 kJ/°C-m^2_c). A 100% solar house built in Saskatchewan (Besant, Dumont, and Schoenau 1979) has 14.6 gal/ft^2_c (2433 kJ/°C-m^2_c) of collectors.

Thermal Stratification

Thermal stratification in storage tanks improves system performance and should be encouraged. Stratification can be facilitated somewhat by baffles or antiblending headers at the top and bottom of water storage tanks, and proper separation of supply and return pipes. However, rapid input of heat can upset equilibrium conditions. Stratification is much easier to ensure in rock beds than in water tanks because rocks are immobile. Rock beds need adequate inlet and outlet plenum volumes to ensure uniform flow throughout the bed.

Location

Storage can be located inside or outside a building. Advantages and disadvantages of various options are listed in Table 5-12. Note that latent heat storage devices have much smaller volumes for a given capacity and reduce storage weight and size problems.

Insulation and Storage Efficiency

Storage efficiency (ratio of stored energy withdrawn to energy placed in storage, over one cycle) and thermal energy loss are opposite sides of the same coin. Heat loss can be reduced by insulation but at a cost. The Sheet Metal and Air Conditioning Contractors' National Association recommends a standard of less than 2% loss in 12 hours. The HUD Intermediate Minimum Property Standards (HUD 1977) suggest at most a 10% energy loss in 24 hours. When thermal loss supplements winter heating load and the loss is easily vented in summer, lower storage efficiencies are acceptable. Such considerations also apply to ducts and pipes that carry thermal energy coming from collectors to load or storage and from storage to load.

Rock-Bed Storage Design

Depending on storage temperature and on the timing and control of heat release to the space, rock-bed storage systems can be either thermally coupled to heated spaces or isolated from them. By intentionally coupling a rock bed with the heated space, stored heat will be conducted to a surface within the space and radiated and convected to the space. Figure 5-18 illustrates a space-coupled rock bed arrangement. The conductance of the material separating bed and space (usually a floor or wall panel) determines the rate and delay of heat release to the space. Floor or wall surface temperatures will be only a few degrees above room temperature. If the design is correct, the net result will be comfortable, with slow heating from a large radiant panel. Such an arrangement is usually desirable with low-temperature [65°-90°F (18°-32°C)] rock beds, characteristic of passive or hybrid systems. With a rock bed isolated from the space, one has greater control over heat delivery. In isolated rock beds, discharge is with air flow. Design of space-coupled rock beds for passive and hybrid applications is described in Passive Solar Design Analysis (Balcomb et al. 1980).

Isolated Rock-Bed Flow Configurations

Vertical and horizontal flow rock beds are most common. A U-shaped rock-bed design has been tried, but it has significant disadvantages, and little performance data are available; its use is not recommended.

Vertical flow rock beds (Fig. 5-19) provide the best thermal characteristics, since they take advantage of warm air's natural tendency to rise. Air flows vertically through the bed, downward when charging and upward when heat is removed. The main disadvantage is the need for considerable vertical space, and rock-bed design must fit available space.

Horizontal flow rock beds (Fig. 5-20) require less vertical space than vertical flow rock beds. However, if the rocks settle with time, an air gap can be formed between the top of the rocks and the

Table 5-12. Advantages and Disadvantages of Storage Locations (Cole et al. 1979)

Location	Advantages	Disadvantages
Utility Room or Basement	• Minimal insulation requirement • Insulation protected from weather • Leaks easily detected • Easy access for repairs	• Reduced living space • Thermal losses increase summer air conditioning load • Leaks possibly damaging to building interior • Difficult to install steel or FRP tanks in an existing building
Unheated Garage	• Insulation protected from weather • Leaks easily detected • Easy access for repairs • Easy installation of steel or FRP tanks in an existing garage	• Reduced garage space • Extra insulation required • Freeze protection most often required • Possible damage to garage from leaks • Thermal losses not recovered • Thermal losses may add to summer air conditioning load • Difficult access for retrofit or repairs • Usual shaped tank may require extra insulation
Outdoors, above grade	• Easy access • No increased air conditioning load from thermal losses • No reduction in living space	• Extra insulation required • Weather protection required • No recovery of thermal losses • Freeze protection usually required • Possibility of vermin attacking insulation
Outdoors, below grade	• No increased air conditioning load from thermal losses • No living space reduction	• Access for repairs difficult • Several problems can be caused by contact with ground water • Thermal losses not recovered • Possibility of vermin attacking insulation • Careful design required to ensure sufficient net positive suction head for the pump • Cost for excavation

rock-bed cover, unless precautions are taken. If rocks are tamped and allowed to settle for a few weeks before the top is installed, the air gap will be minimized. Baffles installed from the top of the box into the rocks, perpendicular to the air flow, can also minimize the air gaps.

As warm air is distributed in the inlet plenum in a horizontal flow rock bed, the air rises and heats rocks at the top of the bin more than those on the bottom. When flow is reversed for space heating, part of the air passes through warmer top rocks and the rest passes through cooler bottom rocks. Mixed warm and cool air may be delivered to the space at a temperature significantly lower than that of the warmest rocks. Design of rock-bed plena to avoid flow channeling is discussed later in this chapter.

Research on improved designs for horizontal rock beds is being conducted in Japan and the United States. An experimental design is shown in Fig. 5-21. However, the most effective way to ensure adequate and even flow through the bed is to use a vertical flow arrangement.

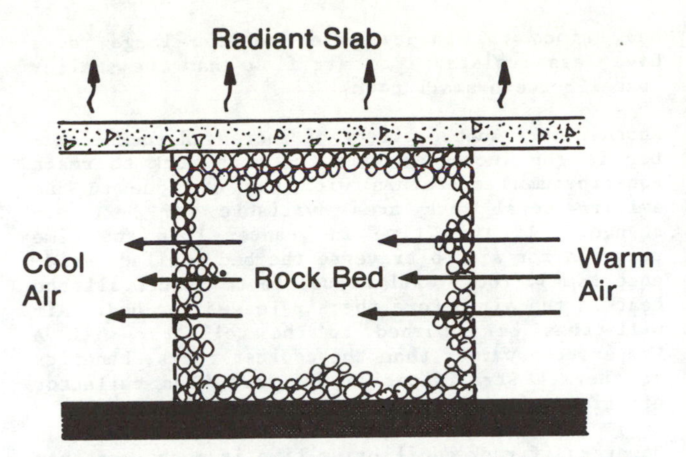

Fig. 5-18. Space-coupled rock bed arrangement (Cole et al. 1979)

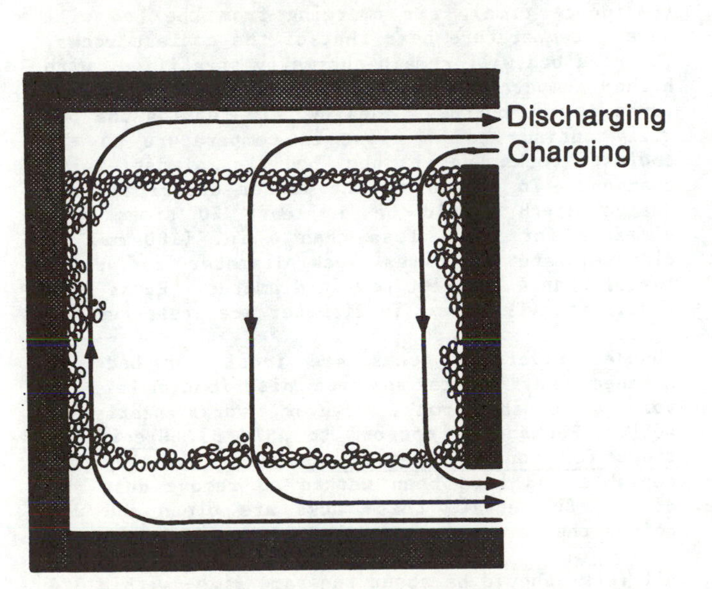

Fig. 5-19. Vertical flow rock bed (Cole et al. 1979)

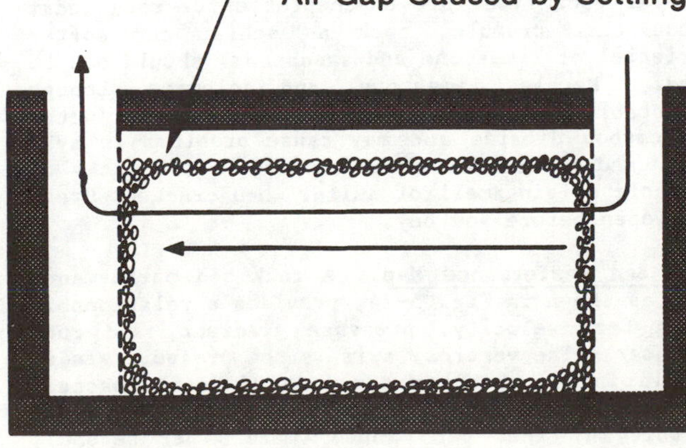

Fig. 5-20. Horizontal flow rock bed (Cole et al. 1979)

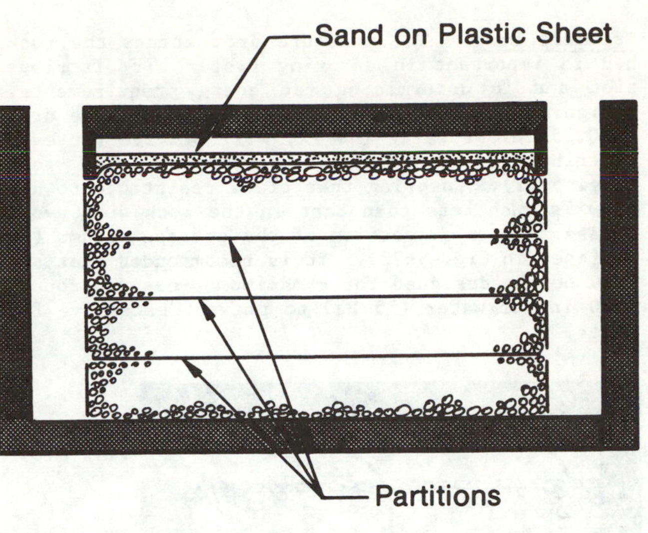

Fig. 5-21. Horizontal flow rock bed with partitions and floating cover (Cole et al. 1979)

Rock-Bed Design Guidelines

The following guidelines are summarized from the Design and Installation Manual for Thermal Energy Storage (Cole et al. 1979). Rock-bed performance depends strongly on bed volume, face velocity, pressure drop across the bed, and rock size in relation to bed depth.

Heat Storage Capacity. Storage volumes ranging from 0.50 to 0.75 ft^3/ft^2_c (0.15-0.20 m^3/m^2_c) are recommended. With rock density of 160 lb/ft^3 (2600 kg/m^3), specific heat of 0.22 Btu/lb-°F (0.90 kJ/kg-°C), and typical void fraction of 0.4, a nominal midwinter daily collection of 800 Btu/ft^2_c (9 MJ/m^2_c) will result in an average bed temperature rise of

$$\Delta T = \frac{Q}{mC_p}$$

$$\Delta T = \frac{800\ Btu/ft^2_c}{\left(\frac{160\ lb}{ft^3\ rock}\right)\left(0.6\frac{ft^3\ rock}{ft^3\ bed}\right)\left(0.75\frac{ft^3\ bed}{ft^2_c}\right)\left(0.2\frac{Btu}{lb^oF}\right)}$$

$$\Delta T = 55°F$$

This would raise the average bed temperature from 85°F to 140°F (assuming collectors can deliver 140°F air), if all heat collected in a day was stored. However, during the winter in most practical systems, half to two-thirds of collected heat is placed in storage; the balance is used directly during the day.

Face Velocity. Face velocity is the volumetric air flow rate divided by the cross-sectional flow area of the rock bed. It measures air velocity just before it enters the rocks. Increased face velocity increases the pressure drop across the rock bed.

Pressure Drop. The pressure drop across the rock bed is important in ensuring proper air distribution and in determining fan energy requirements. Designing the rock bed for a minimum pressure drop of 0.15 in. of water (38 Pa) will provide for even distribution of air flow through the rock (Fig. 5-22). Ensuring that plena resistance to air flow is much less than that in the rock will avoid bypassing a major portion of the storage volume (as depicted in Fig. 5-23). It is recommended that the rock bed be designed for a maximum pressure drop of 0.30 in. of water (75 Pa) to prevent excessive fan power.

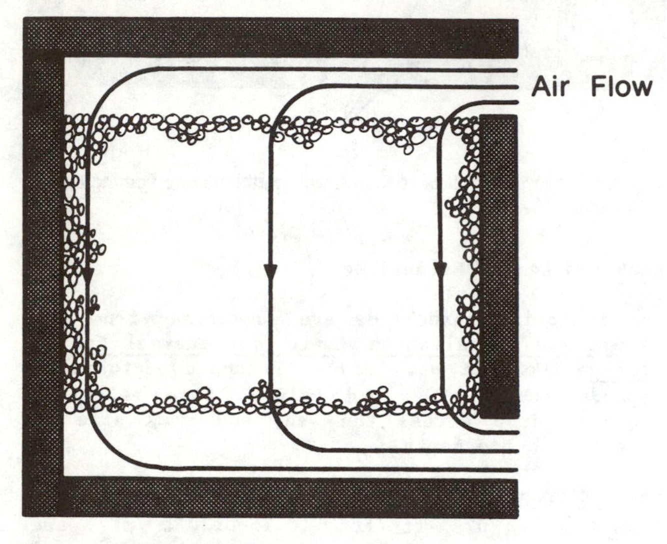

Fig. 5-22. Air flow pattern when pressure drop and plenum size are adequate (Cole et al. 1979)

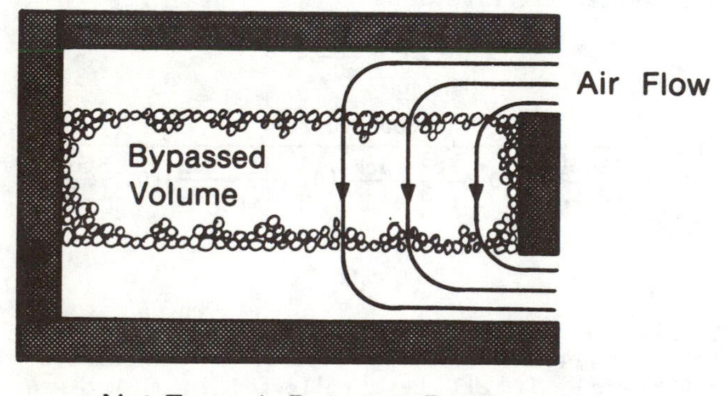

Fig. 5-23. Air flow pattern when pressure drop is too low or plena too small (Cole et al. 1979)

Dividing pressure drop by bed depth (for a vertical-flow rock bed) yields the pressure gradient, which will be used in later calculations.

Rock Size and Type. The fourth parameter affecting performance is rock diameter. Increasing rock diameter decreases the pressure drop across the bed, since the large spaces between large rocks have less resistance to air flow than the smaller spaces between small rocks.

Another more subtle effect of increased rock diameter is the increased time for each rock to reach equilibrium temperature with the air due to the smaller total rock area available for heat exchange. If this time is greater than the time required for air to traverse the bed (called residence time), rocks will be unable to absorb all the heat in the air before the air leaves the bed. Air will thus be returned to the collector with a temperature warmer than the coolest rocks, benefits of thermal stratification are lost, and collector efficiency is lowered.

However, if rock equilibrium time is much less than the time required for air to traverse the bed (residence time), air emerging from the bed will have a temperature near that of the coolest rocks. The rock bed will remain thermally stratified, with higher temperatures at the top and cooler temperatures at the bottom. Ideally, air leaving the bed will continue to be close in temperature to the coolest rocks until the bed is almost fully charged. To behave in the manner described, bed design depth should be at least 20 times rock diameter for rocks less than 4 in. (100 mm) in diameter and 30 times rock diameter for rocks larger than 4 in. (100 mm) in diameter. Rocks 0.75 to 1.5 in. (19-38 mm) in diameter are preferred.

Rounded river-bed rocks are ideal for bed use because they create an even distribution of air voids. Crushed rock, however, works nearly as well. Rocks that conform to ASTM 33, Specifications for Concrete Aggregates, are generally acceptable and have been washed to remove dust and dirt. Be certain that rocks are clean and dry before they are put in the bed.

All rocks should be about the same size--within 3/4 to 1-1/2 times the average rock diameter. This prevents nonuniform settling that can cause flow channeling. Rocks are usually sorted by size with screen meshes.

Several rock types are unacceptable for rock beds. Those that crumble, such as schist and softer varieties of limestone and sandstone, should not be used. Marble, limestone, and dolomite, though acceptable for heating systems, react with water and carbon dioxide and may cause problems in systems used for summer night cooling. Some rocks of volcanic origin smell of sulfur when cracked; break some open before you buy.

Rock Bed Performance Map. A rock bed performance map, as shown in Fig. 5-24, provides a relationship among face velocity, pressure gradient, and rock diameter. The vertical axis is the pressure gradient in inches of water per foot of depth (pascals per meter) and the horizontal axis is the face velocity in feet per minute (meters per second); curved lines show relationships between pressure gradient and face velocity for various rock diameters. Two typical calculations using a rock bed performance map are:

- For given air flow rate and rock bin dimensions, determine rock size needed for adequate performance.
- For given air flow rate and available rock size, determine necessary dimensions of rock bin.

Use of a performance map will be demonstrated in the design example that follows.

Rock Bin. Rock bins must be strong enough to withstand the outward pressure of the rocks. This pressure, already great when the bin is first filled, increases as rocks expand, contract, and settle with bed heating and cooling. Typical rock bin materials include wood, concrete, and cement blocks. Figure 5-17 shows a typical design. The Design and Installation Manual for Thermal Energy Storage (Cole et al. 1979, Appendix F) shows construction details for a wooden bin. Construction of a poured concrete bin is similar to that of the cast-in-place water tank (Cole et al. 1979, Appendix H), except that no lining is needed, and air inlet and outlet openings must be provided.

Plena. Plena distribute air uniformly over the top and bottom of the rock bed. They must have much

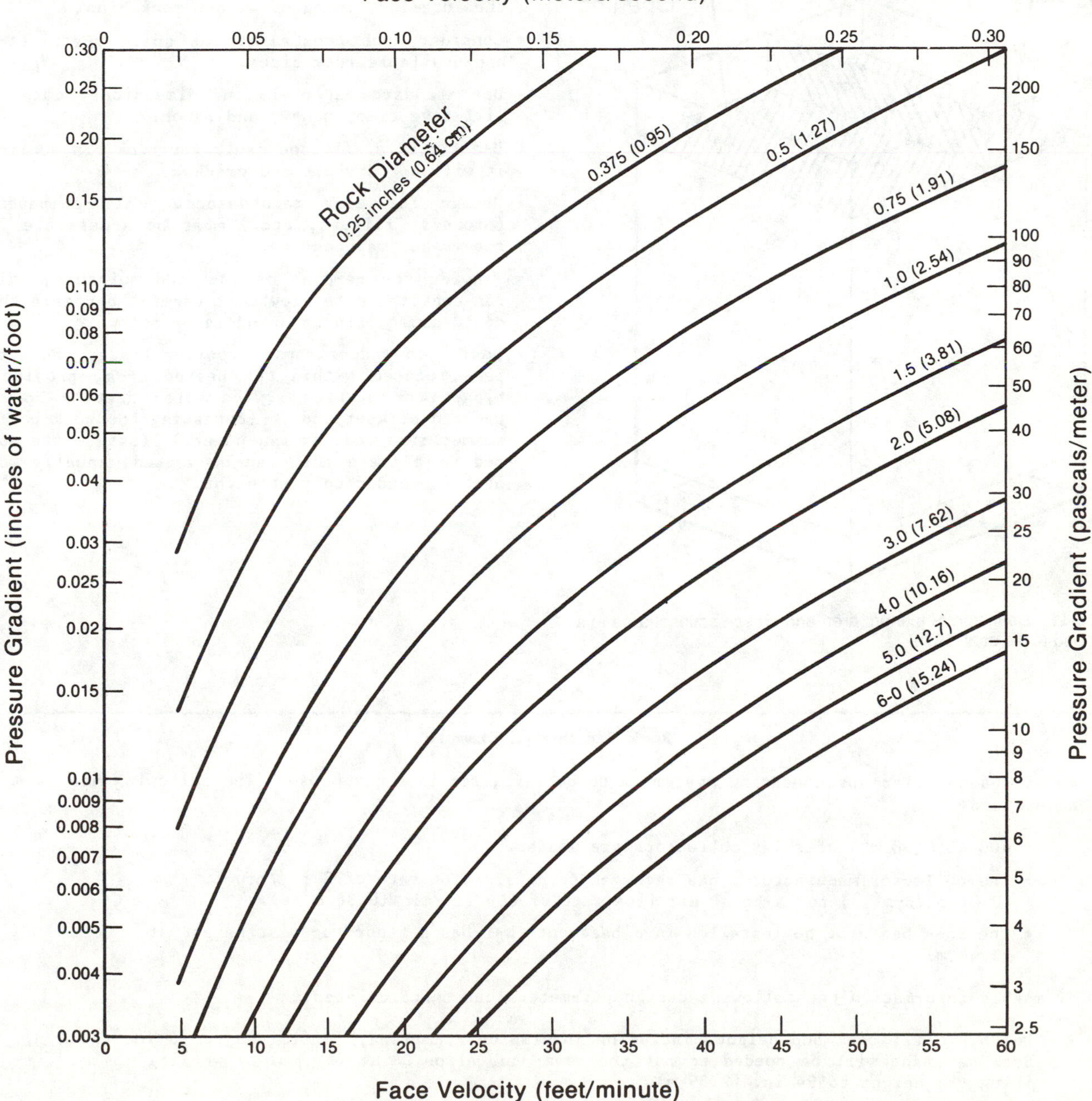

Fig. 5-24. Rock bed performance map (Cole et al. 1979)

lower air flow resistance than the rocks. It is recommended that the cross-sectional area of each plenum, as shown in Fig. 5-25, be at least 8% of the cross-sectional area of the rock bed. That is, plenum height times width should be at least 8% of rock bed width times length. If a plenum is partially obstructed by such bed supports as bond beam blocks, the plenum area should be increased to 12% of bed cross-sectional area.

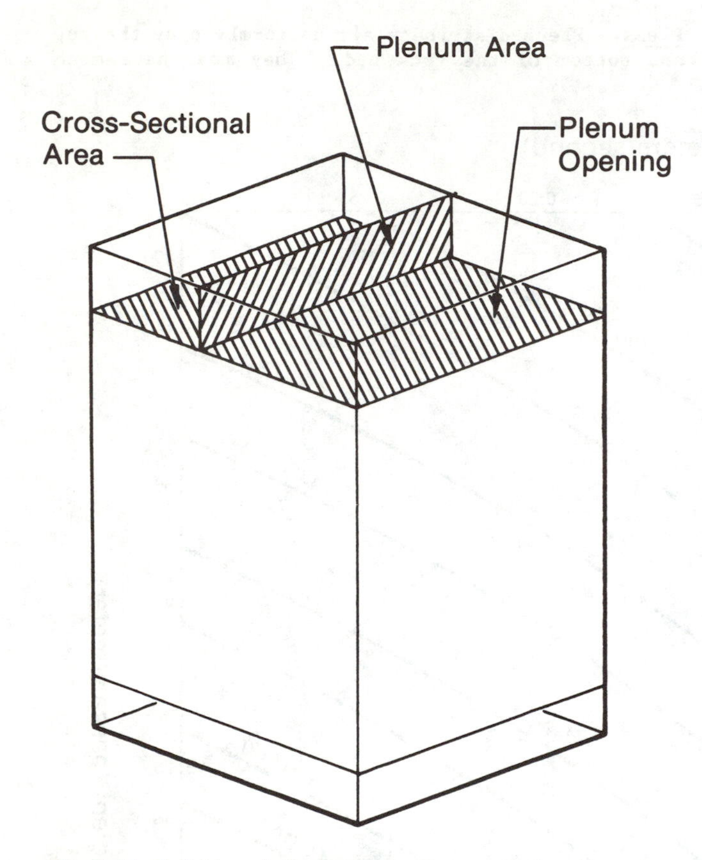

Fig. 5-25. Definition of plenum area and cross-sectional area in a rock bed (Cole et al. 1979)

Other Considerations. Common sense and building codes dictate several considerations that a designer should keep in mind when designing a rock bed.

- Do a thorough design job on the first few systems; many details can be incorporated into later systems.
- Have a structural engineer design a rock-bed foundation that will meet building code requirements and ensure building safety.
- Choose materials that can withstand high temperatures for the life of the system. Fire codes generally prohibit placing flammable materials in contact with the hot air system, so fire-retardant gypsum board is usually specified as the interior lining of wooden rock bins.
- Consider such constraints as space restrictions and available rock sizes.
- Use standard materials and dimensions; doing so will save time, money, and labor.
- Use flexible silicone caulk for rock bin sealing; it will not dry out and crack.
- Design for easy maintenance. All components (blowers, filters, etc.) must be accessible for repair and maintenance.
- Provide for removal of rock-bed moisture, which can contribute to growth of harmful bacteria that could cause illness or odors or both.
- Design to reduce summer cooling loads. For rock beds located within the heated area, provide a bypass for summer service water heating so the bed can be kept cool. (Bypassing the rock bed in summer also reduces fan power.) Install the rock bed in a space that can be vented manually when heat is needed only at night.

Rock Bed Design Example

An air-based solar space heating system is to be installed in a new house. The following has been determined:

- 400 ft^2 (38 m^2) of solar collectors are needed.
- The collector manufacturer has recommended an air flow rate of 2 $ft^3/min\text{-}ft^2_c$ (0.01 $m^3/s\text{-}m^2_c$), for a total air flow rate of 800 ft^3/min (0.38 m^3/s).
- The rock bed will be installed in a basement that has a floor-to-joist height of 8 ft (2.44 m).

From this information the following design parameters can be calculated:

A. Depth. Total rock bed height, including insulation and plena, is limited to 96 in. (2.44 m). Because space will be needed to put the cover on, allow 2 in. (50 mm) of working space, limiting the height to 94 in. (2.39 m).

Example (continued on next page)

Example (continued)

The cover is to be made of 2 × 6 studs (to allow room for insulation) with 3/8-in. plywood sheathing, so it will be 6 1/4 in. (155 mm) thick.

Because the bond beam blocks used to support the rocks come in a standard 7-5/8-in. (193 mm) size, it is assumed that the bottom plenum, which will contain the bond beam blocks, will be 7-5/8 in. (193 mm) high and the top plenum will be 5-1/8 in. (130 mm) high.

The bond beam blocks will be placed on a piece of 2-in. (51-mm) rigid fiberglass insulation.

The total height of these components is 21 in. (0.53 m), which leaves 94 in. minus 21 in. = 73 in. (1.85 m) for rocks. For convenience, it is assumed that 72 in., or 6 ft (1.83 m), of rocks will be used.

B. **Volume Limits.** From the rule of thumb for rock bed volume, for 400 ft^2 of collector, minimum rock volume is

$$400 \times 0.5 = 200 \text{ ft}^3 \ (5.66 \text{ m}^2)$$

and maximum rock volume is

$$400 \times 0.75 = 300 \text{ ft}^3 \ (8.50 \text{ m}^2) \ .$$

Therefore, any volume between 200 and 300 ft^3 will satisfy the rule of thumb requirement.

If F-Chart, or some other design tool, is available, it should now be used to calculate optimum storage volume before proceeding with the following calculations.

C. **Cross-sectional Area.** The cross-sectional area of the rock bed is volume divided by height. Using the volume limits from Step B and the height from Step A, the minimum cross-sectional area is

$$\frac{200 \text{ ft}^3}{6 \text{ ft}} = 33.3 \text{ ft}^2 \ (3.09 \text{ m}^2)$$

and the maximum cross-sectional area is

$$\frac{300 \text{ ft}^3}{6 \text{ ft}} = 50 \text{ ft}^2 \ (4.65 \text{ m}^2) \ .$$

To obtain the length, the plenum sizing rule of thumb is used. The bottom plenum (with the bond beam blocks) rule says the plenum area must equal at least 12% of cross-sectional area. Minimum plenum area is

$$\text{Minimum Plenum Area} = 0.12 \times \text{Cross-sectional Area} \ ,$$

which can be written as

$$\text{Plenum Height} \times \text{Width} = 0.12 \times \text{Maximum Length} \times \text{Width} \ .$$

Since "Width" is common to both sides of the equation,

$$\text{Plenum Height} = 0.12 \times \text{Maximum Length} \ .$$

Therefore, since a plenum height of 7-5/8 in., or 7.62 in. (194 mm), has been specified, a maximum length can be computed as

$$\text{Maximum Length} = \frac{7.62}{0.12} = 63.5 \text{ in. } (1.61 \text{ m}) \ .$$

A 63 in. (1.60 m) length is chosen, less than the maximum length permitted by the rule of thumb. The top plenum height is to be about two-thirds (8%/12%) of the bottom plenum height that satisfies the rule of thumb.

Anticipating that walls will be made from 2 × 6 studs with 1-in. plywood outside and 1/2-in. gypsum board inside, the other dimension will be 82 in. (2.08 m).

Example (continued on next page)

Example (continued)

The cross-sectional area is then

$$\frac{63 \times 82}{144} = 35.9\ ft^2\ (3.33\ m^2)$$

which is within the limits of 33.3 to 50.0 ft^2, as calculated above.

D. **Face Velocity.** Face velocity is the air velocity just before it enters the rocks. It is calculated by dividing air flow rate by rock bin cross-sectional area. In this case, face velocity will be

$$\frac{800\ ft^3/min}{35.9\ ft^2} = 22.3\ ft/min\ (0.11\ m/s)$$

This value will be used later to help determine rock size and rock bed pressure drop.

E. **Pressure Gradient Limits.** The pressure gradient is the pressure drop per unit length through the rock bin. It is calculated by dividing total pressure drop by rock depth. From the second rule of thumb given earlier, the minimum pressure gradient is

$$\frac{0.15\ in.\ of\ water}{6\ ft} = 0.025\ in.\ of\ water/ft\ (20.4\ Pa/m)$$

and the maximum pressure gradient is

$$\frac{0.30\ in.\ of\ water}{6\ ft} = 0.05\ in.\ of\ water/ft\ (40.8\ Pa/m).$$

These limits will be used to determine rock size and bed pressure drop.

F. **Rock Size.** The rock bed performance map in Fig. 5-26 and the values calculated in Steps D and E will be used to determine a suitable rock size.

A vertical line is first drawn at a face velocity of 22.3 ft/min (0.11 m/s). Horizontal lines are then drawn at the pressure gradient limits of 0.025 and 0.05 in. of water/ft (20.4-40.8 Pa/m).

Any curved line, for various rock sizes, that crosses the face velocity line between the pressure gradient limit lines is suitable. In this case, only the curve that corresponds to 3/4-in.-diameter (19 mm-diameter) rock is indicated, so that rock size will be used. Only washed rocks with average diameter between 3/4 to 1-1/2 times the nominal diameter, or from 5/8 in. to 1-1/8 in. (16 to 28 mm) will be accepted.

To check the fourth rule of thumb, note that 20 × 3/4 in. = 15 in. (0.38 m) is much less than the rock bed depth of 6 ft (1.70 m). Thermal stratification in the rock bed will be good.

G. **Pressure Drop.** The rock bed design point is the point on the rock bed performance map where the pressure gradient line crosses the rock size curve. In this case, the design point is at a pressure gradient of 0.032 in. of water/ft (26.1 Pa/m). Multiplying this value by rock bed depth gives a total pressure drop of

$$0.032 \times 6 = 0.19\ in.\ of\ water\ (47.3\ Pa)\ .$$

This number will be used along with other system pressure drops to size the blower.

H. **Volume and Floor Loading.** Total rock volume will be

$$6\ ft \times 35.9\ ft^2 = 215\ ft^3\ (6.09\ m^3)\ .$$

If the rocks have a 30% void fraction, they will weigh

$$0.7 \times 167\ lb/ft^3 \times 215\ ft^3 = 25{,}234\ lb\ (11{,}470\ kg)\ ,$$

or almost 13 tons. Floor loading will be

$$\frac{25{,}134\ lb}{35.9\ ft^2} = 700\ lb/ft^2\ (3425\ kg/m^2)\ .$$

Example (continued on next page)

Example (continued)

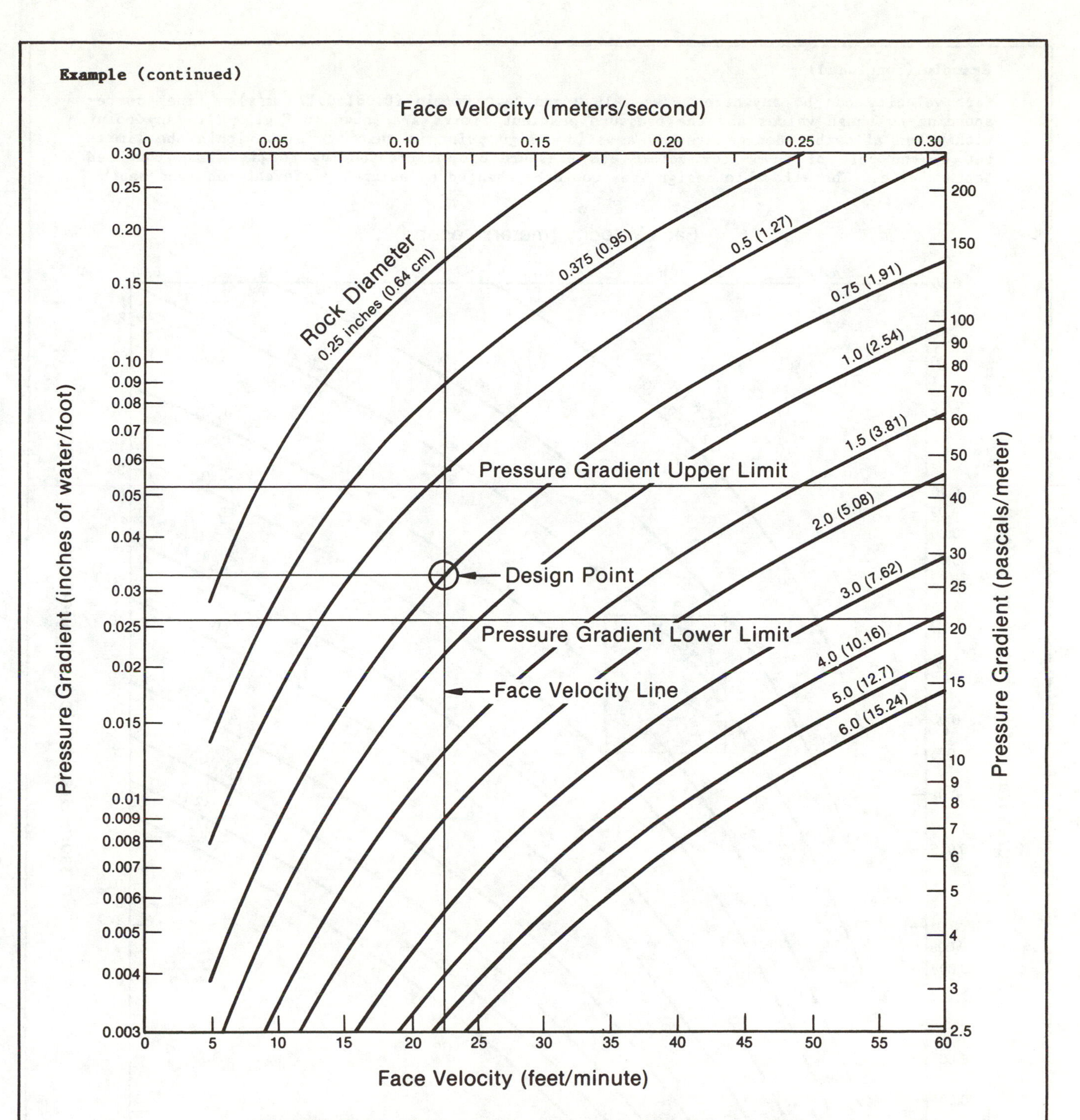

Fig. 5-26. Rock bed performance map for design example (Cole et al. 1979)

Since most basement floors can support only 150 to 400 lb/ft^2 (734-1957 kg/m^2), a structural engineer must determine the load capacity and whether design reinforcements are necessary. Failure to do this may lead to a building code violation and structural damage to the building.

Note that the design point selected is not unique. If maximum and minimum face velocity values are calculated by using minimum and maximum area values, respectively, from Step C, the

Example (continued on next page)

Example (concluded)

face velocity can be anywhere between 16.0 and 24.0 ft/min (0.081-0.122 m/s). Lines corresponding to these values and the pressure gradient limits are shown in Fig. 5-27. Any point within the allowable design area is a valid design point in that it falls within the limits set by the rule of thumb for volume and pressure drop corresponding to the first rock-bed depth choice. The allowable design area could be changed by using a different rock-bed depth.

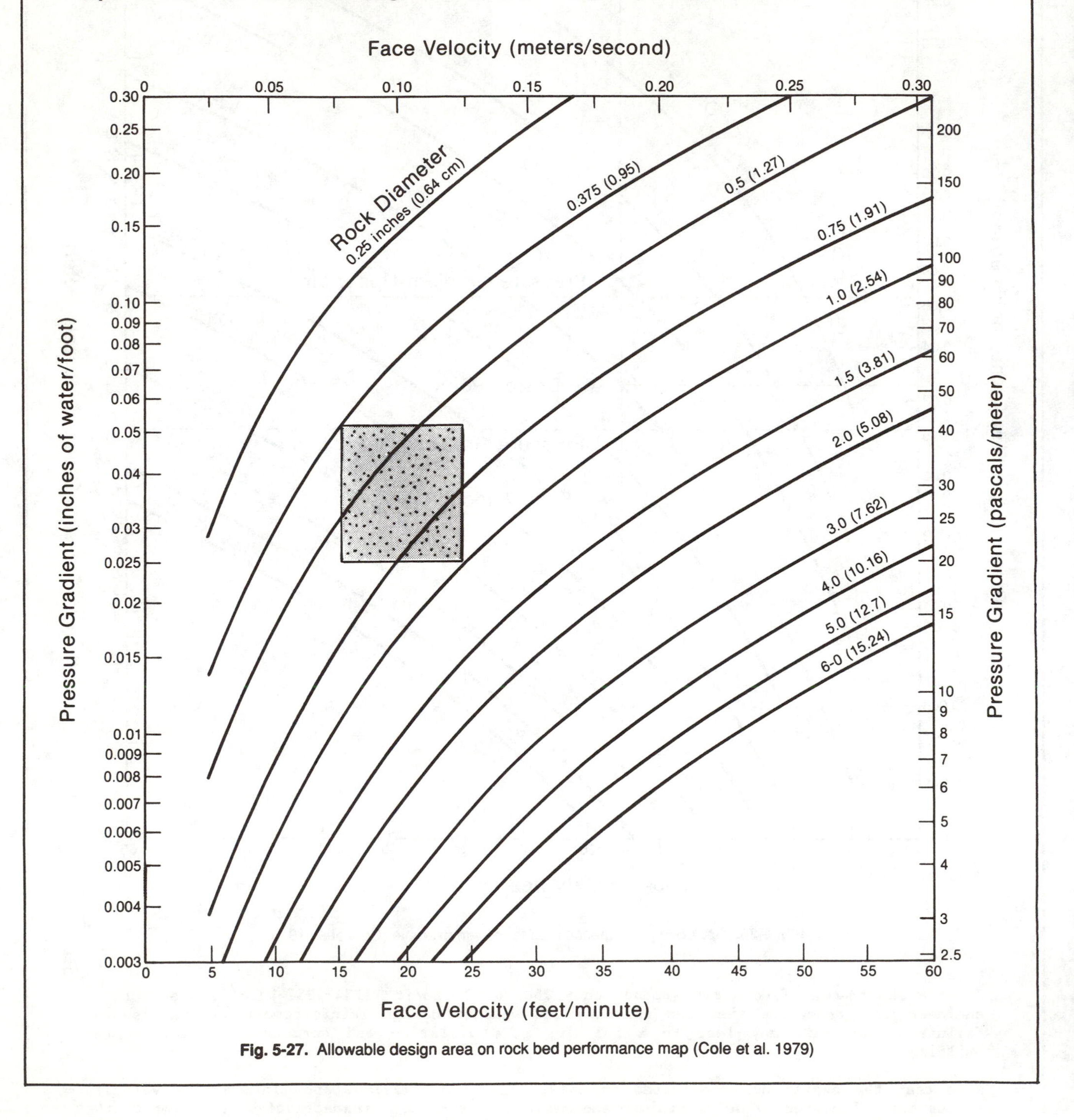

Fig. 5-27. Allowable design area on rock bed performance map (Cole et al. 1979)

Liquid-Based Thermal Energy Storage Design

Before selecting a particular type of liquid storage tank, one must consider a number of parameters including size and shape, material, location, insulation, corrosion, cost, leak protection, protective coating, installation, and pressure and temperature limits.

Each liquid-based storage system has particular requirements. Temperatures listed below are subject to variation and are intended only as a general guide. Direct space heating systems usually require storage temperatures no higher than 160°F (71°C), although some hydronic heating systems require higher temperatures, and most use unpressurized tanks. Solar-assisted heat pump systems typically operate at less than 100°F (38°C), which dramatically reduces insulation requirements of the tank. Absorption air conditioning systems use hot storage at temperatures above 170°F (77°C) or cold storage at temperatures below 55°F (13°C); a pressure vessel may be required for hot storage.

Load management storage may resemble storage for direct space heating if the tank is heated electrically at off-peak rates. If a load management system uses heat pumps to move heat from one part of the building to another or to air condition during the day and heat at night, temperature requirements will be similar to temperature requirements for a solar-assisted heat pump.

Direct heating of potable water usually requires temperatures of less than 140°F (60°C) and a pressurized tank. Water preheating usually requires temperatures under 120°F (49°C). The preheat tank may be pressurized or unpressurized, depending on system configuration. It is not unusual for service hot water systems to operate at higher temperatures than those listed here.

Various tank materials that meet these temperature and pressure requirements are available. Use of tested materials, such as steel, FRP, concrete, or wood with plastic lining, will avoid risks inherent in using unproven materials. Advantages and disadvantages of each tank type were listed earlier. All can be purchased or constructed in any size likely to be used in storage systems.

Only one storage tank is normally used in space heating systems. Where a properly sized tank is unavailable, where space restrictions dictate use of smaller tanks in place of a larger one, or where multiple tanks are used to achieve stratification, two or more may be used. Two small tanks cost more than one large one, however, and a multitank system requires more insulation because more surface area is exposed.

Tank Types

Steel Tanks

The principal advantages of steel tanks are relative ease of fabrication to ASME Pressure Vessel Code requirements, ease of attaching pipes and fittings, and much experience with their use. Corrosion is the main disadvantage, but difficulties in installation in enclosed buildings and cost are also problems.

Four types of corrosion threaten steel tanks: electrochemical corrosion, oxidation (rusting), galvanic corrosion, and pitting. Six protective methods, which can be used singly or in combination, are system sealing, addition of chemical corrosion inhibitors (see Table 5-13), use of protective coatings or liners, cathodic protection, increased metal thickness, and use of noncorroding alloys. Since corrosion rates double with each

Table 5-13. Corrosion Inhibitors

Generalized Categories	Metals Protected	Specific Compounds	Acute Oral Toxicity[a]
Nitrate salts	Iron, aluminum	Lithium nitrate, sodium nitrate	High
Sulfate salts	All metals	Sodium sulfate	?
Sulfite salts	All metals	Sodium sulfite	Moderate
Borate salts	Iron and iron alloys	Sodium borate	Moderate
Phosphate salts	Iron, aluminum	Potassium hydrogen phosphate	Moderate
		Trisodium phosphate	Low
Silicate salts	Copper, iron, aluminum	Sodium metasilicate	Moderate
Triazoles	Primary cuprous metals	Benzoltriazole	?
Benzoate salts	Iron	Sodium benzoate	Moderate

[a]Source: N. Irving Sax. Dangerous Properties of Industrial Materials. Van Nostrand Reinhold Company, New York, 1979.

20°F (10°C) increase in temperature, limiting maximum temperature prolongs tank life.

Electrochemical Corrosion. Electrochemical corrosion is governed primarily by two conditions: liquid (electrolyte) pH and metal electric potential. pH is a measure of acidity or alkalinity; an acidic solution has a pH lower than 7, an alkaline solution has a pH higher than 7, and a neutral solution has a pH of 7. The metal electric potential is usually given relative to that of the hydrogen always present in water.

Marcel Pourbaix has produced diagrams showing corrosion conditions for various metals as functions of pH and electric potential (Cole et al. 1979). Figure 5-28 is the Pourbaix diagram for iron at 77°F (25°C). In addition to corrosion conditions for the metal (i.e., anode reactions), the figure shows two lines, a and b, representing cathodic reactions. Below line a, water decomposition in the form of reduction of H^+ ions, with evolution of hydrogen, is possible; above line b, water decomposition in the form of oxidation of OH^- ions, with evolution of oxygen, is possible.

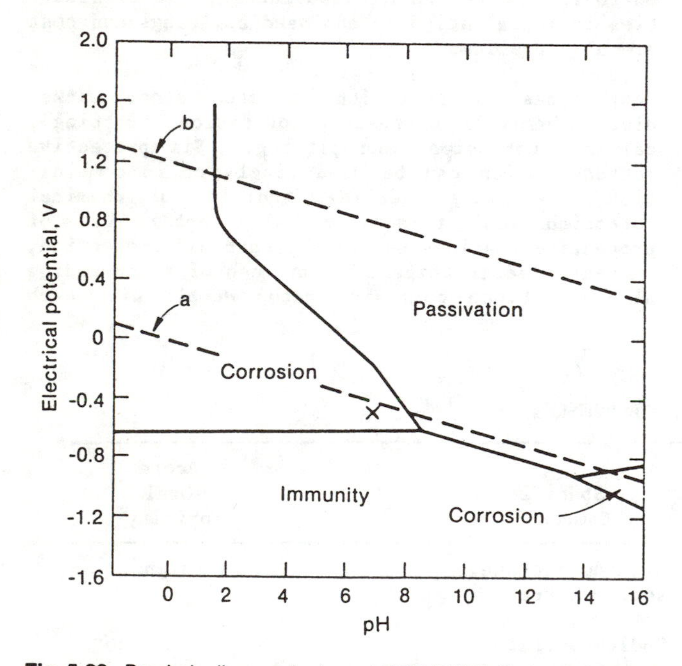

Fig. 5-28. Pourbaix diagram for iron at 77°F (25°C) (Cole et al. 1979)

Under normal conditions, solution pH is 7 and iron has an electric potential of -0.44 volts relative to hydrogen. This state is shown as Point X in Fig. 5-28. Point X is in a region where corrosion can take place.

Three methods to prevent corrosion can move Point X. Cathodic protection can move it down, alkalinization can move it to the right, and anodic protection can move it up. Cathodic protection is most reliable and can be used effectively to protect either the inside or the outside of a tank. A negative electric potential is applied by connection either to a direct current source or to a bar of a more reactive metal.

If direct current is used, the tank must be connected to the negative (-) terminal; an electrode in contact with the water but insulated from the tank must be connected to the positive (+) terminal. Electrodes should be of an inert material. Pourbaix recommends use of the following voltages:

- For $pH < 10$, $E = -0.62$ V
- For $10 < pH < 13$, $E = -0.08 - (pH \times 0.059)$ V

Cathodic protection by direct current has the disadvantage of requiring electrical equipment and a power source that must be checked periodically.

A more common method of cathodic protection, which does not require external power, is use of a sacrificial anode. The anode must be submerged in the water and electrically connected to the tank. Protection ends when the anode has completely dissolved; it should be inspected annually and replaced if necessary.

The metal chosen as a sacrificial anode must be more reactive than the steel tank, as shown in Table 5-14. The most commonly used anodes are magnesium, aluminum, or zinc. We recommend a magnesium bar for most storage tanks. Aluminum has a tendency to form a protective coating over itself, reducing the effectiveness of protection. Zinc, whether as a metal bar, as galvanizing, or as an additive to epoxy or paint, is not recommended for most thermal energy storage applications. At temperatures above approximately 155°F (68°C), zinc and steel reverse roles; steel is then sacrificed instead of protected. However, zinc can provide effective protection if the maximum tank temperature is kept below about 140°F (60°C).

Table 5-14. Electromotive Force Series on Metals (Baumeister 1967)

MOST REACTIVE	Magnesium
	Beryllium
	Aluminum
	Manganese
	Zinc
	Chromium
	Iron
	Cadmium
	Nickel
	Tin
	Lead
	Hydrogen
	Copper
	Mercury
	Silver
	Palladium
	Platinum
LEAST REACTIVE	Gold

Either cathodic protection method works particularly well if the tank is lined with epoxy, glass, or hydraulic stone. The coating protects most of the steel and leaves only a few small uncoated areas. Since current required for cathodic protection is proportional to exposed steel area, coating allows use of a smaller current, or the anode lasts longer.

Cathodic protection should not be combined with strong alkalinization in most situations. Alkalinization may cause a protective coating to form over the anode, or it may cause rapid consumption of the anode. In either case the effectiveness of protection will be decreased.

Steel is relatively inactive in an alkaline environment; adding lime, caustic soda, or trisodium phosphate to maintain a pH of 9.5 to 12.0 will protect steel from electrochemical corrosion. Increasing solution pH also moves iron into the region of passivation, with a protective film formed over the surface. Water pH must be tested periodically to maintain required alkalinity. Alkalinization cannot ordinarily be used to protect service hot water tanks, but it works well for vented or sealed tanks in other applications. Do not use alkalinization if aluminum parts will be exposed to the water; aluminum corrodes quickly in an alkaline environment.

Anodic protection is rarely used because it does not stop corrosion completely. Increased corrosion rates can occur if steel potential is raised too high, and imperfections in the passivated surface can lead to pit corrosion. For these reasons, anodic protection is not recommended.

Oxidation. Oxygen can enter a tank in two ways: it can be dissolved in water that enters the tank, or it can enter through the tank's air vent. Besides causing rust, oxygen tends to catalyze other types of corrosion.

One way to prevent oxidation is to seal the system so that no air or water can enter. Oxygen initially in the system will be quickly removed via reaction with system parts. Although minor rusting may occur at that time, no further rusting will take place. Sealing the system requires that all components be able to withstand pressures generated when the system is heated. Tanks must be designed to the American Society of Mechanical Engineers (ASME) Pressure Vessel Code or local code requirements and be provided with a pressure relief valve. An expansion tank may be required by some systems. The main disadvantage of a sealed system is cost, but if the system must operate close to or above the normal boiling point, sealing provides effective oxidation protection.

Oxygen dissolved in water flows continuously into service hot water tanks. Thus, even though a tank is pressurized, oxygen cannot be excluded. Glass or hydraulic stone linings limit steel and oxygen contact and thus limit rusting. In tanks larger than 120 gal (454 L), glass or hydraulic stone is expensive. For these tanks, four interior coatings of baked-on phenolic epoxy are recommended.

Vented tanks are continuously exposed to oxygen. Four coats of baked-on phenolic epoxy are recommended to protect the tank interior. If a baked-on treatment is unavailable, four coats of two-part epoxy can be specified, although it is less effective than the baked-on coating. Sodium sulfite added to water will scavenge oxygen; however, the sodium sulfite must be periodically tested and replenished.

If the tank is to be located underground, two coats of coal-tar epoxy on the outside are recommended. For indoor or aboveground locations, a coat of primer followed by two coats of enamel provides adequate external protection.

Galvanic Corrosion. Galvanic corrosion occurs when dissimilar metals in an electrolyte are in electrical contact with each other. This often occurs when a copper fitting is screwed into a steel tank; tank water serves as the electrolyte. The more reactive metal (steel) dissolves in the vicinity of the less reactive metal (copper), and the eventual result is a leak in the system.

Galvanic corrosion can be minimized by electrically insulating dissimilar metals from each other. Dielectric bushings are used to connect pipes to tanks, and gaskets or pads are used to insulate other components. Be careful to eliminate other electrical connections between dissimilar metals. A common ground, for example, would defeat the effort to insulate dissimilar metals from each other. Use a volt-ohmmeter (VOM) to measure the electrical resistance between components before the system is filled with water. A resistance of more than 1000 ohms indicates adequate insulation, but a resistance of less than 100 ohms means an electrical connection still exists. Because dissimilar metals plus electrolyte form a battery, testing after system filling will give false resistance measurements.

Dissimilar metal combinations encountered most frequently in solar systems are iron-copper (or iron-brass), aluminum-iron, aluminum-copper (or aluminum-brass), and zinc-iron. The more reactive metal is listed first, except for the zinc-iron combination. As previously mentioned, zinc is more reactive than iron below about 155°F (68°C) but less reactive above that temperature.

Pitting. Pitting is a localized form of corrosion in which small-diameter holes penetrate the base metal. One type of pitting is believed to be caused by ions or particles of a less reactive metal plating onto a more reactive metal. Localized galvanic corrosion produces a pit that can quickly penetrate the more reactive metal. In a storage system, a typical source of less reactive metal ions is a copper pipe. Copper ions circulate in the water until they plate out on steel and cause a pit.

Another type of pit corrosion is believed to be caused by small imperfections in a passivated surface. As the pit grows, solution chemistry changes locally so the passivating agents become deficient within the pit, accelerating pit growth.

A similar type of corrosion occurs in crevices and screw threads. Cathodic protection is more effective against pitting than alkalinization or anodic protection. The presence of some ions, especially chloride ions, tends to encourage this form of pitting.

Other Methods of Protection. Occasionally the thickness of the wall in steel tanks is increased for corrosion protection. Although thicker metal does not stop corrosion, it does increase the time before corrosion causes system failure. Thicker metal use is often combined with alkalinization.

Stainless steel alloys are rarely encountered in solar energy storage systems because of high cost. Stainless steels form a passivated surface that protects the metal from corrosion. Cathodic protection will destroy the surface condition and, therefore, should not be used with stainless steel.

Recommendations for Steel Tanks. Following are recommendations for most steel tanks used in space heating applications:

- Use four coats of baked-on phenolic epoxy on the inside.
- Use a magnesium bar as cathodic protection.
- Electrically insulate dissimilar metals, except the magnesium anode, with dielectric bushings, gaskets, or pads.
- Use sodium sulfite to scavenge oxygen.
- Do not use chromate-type corrosion inhibitors. They are highly toxic and carcinogenic, cause damage to a common type of pump seal, and are difficult to dispose of properly. Since chromates act by passivating a steel surface, their effectiveness is much reduced when combined with cathodic protection.
- If possible, limit maximum tank temperature to 160°F (71°C).
- Protect the outside of a buried tank with two coats of coal-tar epoxy and a magnesium anode. Use a coat of primer and two coats of enamel to protect the outside of aboveground tanks.
- If the tank will be pressurized, it must comply with Section VIII of the ASME Boiler and Pressure Vessel Code or local codes.

Fiberglass-Reinforced Plastic Tanks

Both factory-insulated and on-site-insulated fiberglass-reinforced plastic (FRP) tanks are available and have been used successfully in solar energy installations. Their main advantage is they do not corrode.

It is recommended that the designer use factory-insulated FRP tanks, designed specifically for solar energy storage and available in convenient sizes and shapes. A typical factory-insulated FRP tank consists of an inner FRP shell covered with 2 to 4 in. (5-10 cm) of urethane insulation. An outer FRP shell protects the insulation. Factory-insulated FRP tanks can be used outdoors, above or below grade, or in a garage.

Nearly all FRP tanks have two limitations:

- They must not be pressurized or subjected to a vacuum inside unless they are specifically designed for it. A vent will keep these conditions from occurring.
- Operating temperatures must never exceed the manufacturer-specified limit. Exceeding it will void the warranty and damage the tank. Since the limit is nearly always below the boiling point, the system controller must stop heat addition to the tank before the limit is reached. Adjust the cutoff point on the controller to 5°F (3°C) below this limit. Since many controllers do not provide for limiting temperature, the designer must select the controller carefully.

The temperature limitation is determined by the type of resin used to make the tank. Ordinary polyester resins have a limitation of 160°F (71°C)--suitable only for low-temperature tanks. With premium quality resins, the temperature limitation can be raised to 180° to 200°F (82°-93°C); consult the tank manufacturer for details.

Concrete Tanks

Concrete tanks for solar energy storage are of two types: cast-in-place and precast, including tanks designed primarily for use as septic tanks and utility vaults. Such tanks are relatively inexpensive so long as their shape is kept simple, the mass becomes part of the storage system, and concrete is a readily available construction material. Concrete also has considerable resistance to underground loads. Because it can be cast into almost any shape, it is good for retrofit installations. It is also fireproof and corrosion resistant.

Concrete does have several disadvantages, however. It is subject to capillary action, so water can seep through cracks and joints unless the tank is lined. Leakproof connections through tank walls are often difficult to make. Plaster coatings, adequate for cisterns, have a tendency to crack with the fluctuating temperatures of thermal storage materials. Seepage tolerable in a cistern or swimming pool will degrade the tank insulation. Use of either a spray-on butyl rubber coating, 30 to 50 mils (0.75-1.25 mm) thick, on the inside of the tank or a replaceable liner of the type used in wooden tanks is recommended.

Tank weight may be either an advantage or a disadvantage, depending on the situation. Special footings may be required to carry the weight, particularly if the tank is near a load-bearing wall. A tank can be designed into a corner of a basement wall if the foundation and walls are designed to carry the added load. Proper installation techniques are essential in such cases (see Ch. 8). This technique is more amenable to new construction than to retrofitting. The weight of concrete is an advantage in underground storage containers, since it helps prevent the tank from being buoyed out of the ground by high groundwater.

Wooden or Multicomponent Tanks with Plastic Liners

Wooden or multicomponent tanks can be purchased as kits or can be custom designed. Low cost and easy indoor installation are major advantages. But plastic liners have temperature limitations, and tanks are usually intended for indoor locations only.

One available kit provides a vinyl-lined, 2000-gal (7600-L) cylindrical tank. According to the manufacturer's instructions, the tank, made of 3/8 in. (9.5 mm) CDX plywood and reinforced with steel bands, can be installed with simple hand tools. Insulation for bottom, sides, and cover is included, as is a 1-in. (2.54-cm) PVC compression fitting. Maximum allowable temperature inside the tank is 160°F (71°C).

Another kit uses lock-together panels of 4-in. (10-cm) thick urethane foam sandwiched between steel facings. The plastic liner is rated for 180°F (82°C) continuous service, and an aluminum roof is available for outdoor containers. Sizes from 500 to 2000 gal (1900-7600 L) are available.

A wooden tank could be constructed by modifying the rock bin described in (Cole et al. 1979, Appendix F). The gypsum board or sheet metal lining, tie rods, caulking, and openings for air ducts can be deleted, but otherwise, construction of a water tank is similar to construction of a rock bin. Custom fabricators of plastic linings can be found in most major U.S. cities; the technology is similar to that used for water beds and swimming pools.

Liners

Liners should be about 1% to 3% larger than the inside of the container to avoid stressing the seams. It may be difficult to fabricate corners with some plastics; however, corners can be folded from a flat sheet, if necessary. Care must be taken to remove all sharp edges, burrs, splinters, and debris that might puncture the liner. Avoid working inside the tank with the liner in place; if work must be done, remove anything that could cause a puncture. Properties of several plastics are listed below.

Polyvinyl Chloride (PVC). PVC is one of the least expensive and easiest liner materials to work with; there is considerable experience with it in solar systems, industrial hot water processes, swimming pools, and water beds. Seams can be made by dielectric sealing, a more reliable process than heat sealing or cementing. Repairs can be made with patches and adhesives.

A thickness of 30 to 60 mils (0.75-1.50 mm) should be used. Maximum tank temperature should be limited to 160°F (71°C), although some special compositions can tolerate a water temperature of 180°F (82°C). Lifetime varies from 6 to 15 years, with about 8 years typical. Failure is finally caused by leaching of plasticizers. The plastic then becomes brittle and cracks appear, usually at corners.

Ethylene Propylene Diene Monomer (EPDM). EPDM is a rubber-like material that can withstand boiling water. It is more expensive than PVC and more difficult to fabricate, since dielectric sealing cannot be used. There may be difficulty in finding a fabricator who will make corners, although flat sheet is available. It can be patched with adhesives, but patching is more difficult than for PVC. A thickness of 30 to 60 mils (0.75-1.50 mm) should be specified.

Butyl Rubber. Butyl rubber is a reasonably durable material that is less expensive than EPDM but more expensive than PVC. Seams can be vulcanized, but the process is more difficult than dielectric sealing. Butyl rubber can be patched with adhesives. The sheet form is more durable than the spray-on form.

Polyethylene and Polypropylene. Polyethylene and polypropylene are inexpensive in flat sheets, but joints are difficult to fabricate, and patching is difficult at best. Adhesives apply poorly since polyethylene and polypropylene are unaffected by most ordinary solvents. The material tends to be stiff in thicknesses greater than 30 mils (0.75 mm). Polypropylene can withstand higher temperatures than polyethylene, but there is little experience with use of these materials in solar systems.

Chlorosulfonated Polyethylene. Chlorosulfonated polyethylene is usually laminated with a scrim (a mesh fabric) of another material to give it dimensional stability. Flat sheets are generally available, but fabricated corners may be difficult to obtain. Some users have reported separation of the material from the scrim.

Chlorinated Polyvinyl Chloride (CPVC). Like PVC, CPVC is normally rigid, but it can be made flexible by adding plasticizers. Although CPVC can withstand slightly higher temperatures than PVC, its plasticizers are also subject to leaching.

Tank Cost and Size

Factors that affect tank installation costs include size, whether installation is in a new building or an existing one, location, temperature requirements, insulation requirements, and materials used. Tank size is the most important factor affecting cost. Cost per gallon generally decreases as tank size increases. Because system performance is not very sensitive to tank size (unless the tank is considerably undersized), selection of a standard size close to the optimum is the best approach.

New Building or Existing Building

Whether the tank is to be installed in a new or an existing building limits the choice of tank material and location. Access to a steel, FRP, or precast concrete tank may be designed into a new building, but lack of access in an existing building may require choosing an outdoor location or a different material.

Foundation reinforcement can be specified before a new building is built, but part of the basement floor in an existing building may have to be removed before a reinforced section of floor can be installed. Greater flexibility in choice of material, location, and foundation reinforcement generally gives a solar system in a new building a cost advantage over one in an existing building.

Location

Storage tank location will have an impact on the requirements that the tank is designed to meet. Each special requirement adds to system cost. Some of these requirements are waterproof or extra-thick insulation, freeze protection, long-life components, material limitations, and groundwater protection (i.e., tie-down straps, exterior corrosion protection, provisions for groundwater drainage, etc.).

Basement locations generally impose few special requirements on a storage tank. Weatherproof and extra-thick insulation, freeze protection, and protection from groundwater are not needed when the tank is indoors. Steel and fiberglass tanks are not recommended for existing buildings because they ordinarily do not fit through doors. If a steel or fiberglass tank is to be installed in a new building, it should last the lifetime of the building. Both wooden and cast-in-place concrete tanks are suitable for basement installation. Requirements for tanks in basements also apply to crawl spaces. Since most crawl spaces are unheated, extra insulation and a means of freeze protection are needed.

Garages are an excellent location for steel or fiberglass tanks. The large door allows easy installation or replacement of the complete tank. Since most garages are unheated, extra insulation and freeze protection are needed.

Insulation Requirements

Insulation requirements are primarily determined by location. Indoor tanks in heated areas require the least insulation, and protection of the insulation can consist of a simple cover. Tanks in unheated indoor locations need extra insulation thickness, but a simple cover is sufficient insulation protection. Outdoor tanks have the most severe insulation requirements. Aboveground tanks require extra insulation thickness and weather protection. Underground tanks require waterproof insulation and extra thickness to compensate for groundwater. The cost of insulating an outdoor tank is about two to four times the cost of insulating an indoor tank. Material costs and labor frequently make a factory-insulated tank less expensive than an on-site-insulated one.

Tanks should be insulated so that thermal losses do not exceed 5% of maximum thermal energy capacity over a 24-h winter design day. This generally requires about 3 to 4 in. (7.5-10.1 cm) of fiberglass batts or 2 to 3 in. (5-7.5 cm) of rigid Styrofoam® insulation (at least R-14) for indoor installation. At least R-28 is recommended for outdoor installations. The entire perimeter of the storage, including top and bottom, should be insulated. Heat losses through concrete foundations are probably high since concrete acts as a thermal bridge when inadequately insulated; the bottom, therefore, should be more heavily insulated.

Materials

Costs for various tank types are shown in Table 5-15.

Flow Channeling and Stratification

Solar systems work most efficiently when the coldest liquid flows to the collectors and the hottest liquid is supplied to the load. This is accomplished most effectively with tank stratification. The taller the tank, the greater the stratification (up to a point). Flow channeling, or short circuiting, is undesirable when it occurs across a storage tank. For example, if the inlet from the collectors in a horizontal tank is directly above the tank's outlet to the collectors, flow channeling may occur between them, with hot water from the collectors flowing directly to the collector supply and bypassing the bulk of the storage. Flow channeling reduces tank stratification and reduces overall system efficiency. It is minimized by separating supplies and returns for both collector and load loops. Returns from collectors and load should be at one end of the tank, and supplies to collectors and load should be located at the opposite end. Stratification is enhanced in a horizontal set up if a perforated discharge pipe is installed on the collector return pipe (Fig. 5-29).

However, flow channeling is desirable between two different thermal loops, such as a solar loop and a building heating loop. If the supply of solar-heated liquid is located adjacent to the point where water is drawn from storage to heat the building, the hottest water in storage is delivered directly to the load whenever solar heat is collected.

Although perfect stratification cannot be achieved in a liquid-based system, partial stratification can be encouraged by methods shown in Fig. 5-30a, b, and c. Use of horizontal inlets and outlets and of low-velocity flows (Fig. 5-30a and b) is so easy that it should be standard practice in liquid-based systems. Baffles are most easily installed by the manufacturer.

Additional Considerations for Liquid Storage Systems

- Liquid storage tanks should be leak tested at 1.5 times design pressure. Automatic relief valves should be incorporated to protect tanks against overpressurization (some codes require underpressurization protection).
- It is better to boil storage tank water to dump excess heat than to vaporize collector fluid. Boiling collector fluid could cause its loss or degrade a glycol-water mixture.
- A vented storage tank can create pumping difficulties when operating temperatures approach the

Table 5-15. Tank Costs

Tank Type	Size Range	Installed Cost	Comments
Steel [rated for 100 psi (690 kPa)]	Indoor: 1000-2000 gal (3800-7600 L)	$2.50/gal ($0.66/L)	Includes tank, lining, insulation, and sheath. Cost represents rough guide, not a precise estimate.
Premium for a pressure-rated tank		$0.50/gal ($0.13/L)	
FRP tank (without a high-temperature resin for direct space heating)	1000-2000 gal (3800-7600 L)	$1.75/gal ($0.53/L)	On-site-insulated and factory-insulated tanks are similar in cost, since the more difficult and labor-intensive installation procedures for the on-site-insulated tank tend to bring its installed cost up to the installed cost of a factory-insulated tank.
Cost of high-temperature resin		$0.25-$1.24/gal ($0.07-$0.33/L)	Shipping costs can add a significant amount if the distance to the manufacturer is more than 100 miles (160 km). Due to the difficulty of shipping, installed costs are highly variable.
Precast concrete tank	1000-2000 gal (3800-7600 L)	$1.50/gal ($0.40/L)	If the tank can be designed into a corner of a building and the tank is poured at the same time as the building walls, lower costs are possible. Some liners require periodic replacement, which increases cost.
Plastic-lined wooden or multicomponent tanks installed in sheltered locations	1000-2000 gal (3800-7600 L)	$1.00-$1.50/gal ($0.26-$0.40/L)	

storage water boiling point. Cavitation can occur at the pump inlet, especially if storage is below ground, unless the net positive suction head required by the pump is maintained.

- Unsheltered, aboveground storage tanks should be designed to resist snow, wind, hail, and seismic loads. Completely enclosed tanks need only be designed to resist seismic loads.
- Storage units should be accessible for routine maintenance and repair.
- Tanks should be provided with means for removing the liquid. Those above grade or floor level should have a valve at the lowest point, and buried tanks should have provision for pumping or siphoning fluid from the tank.
- Tanks should have liquid-level indicators--preferably sight glasses--to indicate whether fluid is at the appropriate level. Tanks that require makeup water from the potable water system but contain nonpotable water should be filled using an air gap, a back-flow preventer, or an acceptable means of separation.
- Flammable materials used as storage media or in the storage container must conform to existing standards and fire codes.
- Thermal storage system materials should be chemically compatible to prevent corrosion and deterioration.
- Heat losses from thermal storage containers placed in basements or equipment rooms must be accounted for when determining thermal loading of these spaces.
- Concrete tanks are generally not usable in pressurized systems. The lining of a concrete tank must be leakproof and must be able to withstand temperatures of 180° to 200°F (84°-94°C) over long periods of time.
- Where water storage is vented to the atmosphere, auto-fill or low-water level indicators (sight glasses) are essential. Vented storage may well evaporate excessive moisture into living or equipment areas.

• Major heat losses from tanks can occur due to convection currents in attached pipes. Closed loop pipes should contain check valves to present reverse thermosiphoning. In addition, connecting pipes should have a vertical U-bend at least 12 in. (30 cm) long to provide a heat trap. This prevents small convection cells in the pipes from transporting heat out of storage and making the pipes behave like heat rejection fins.

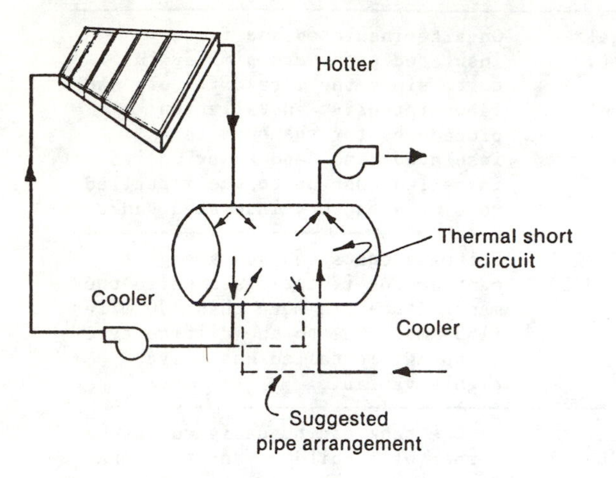

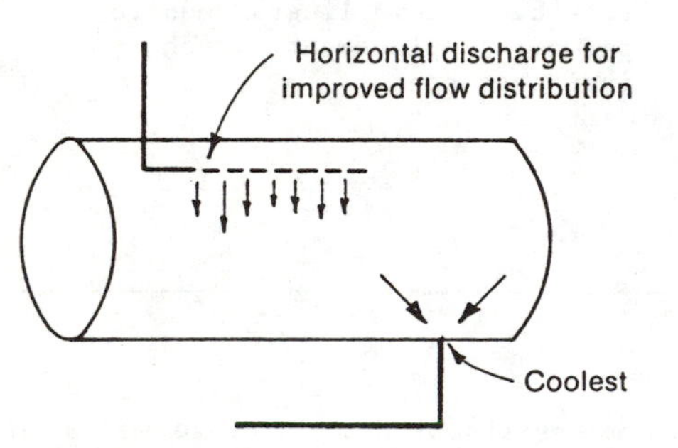

Fig. 5-29. Storage piping to promote stratification

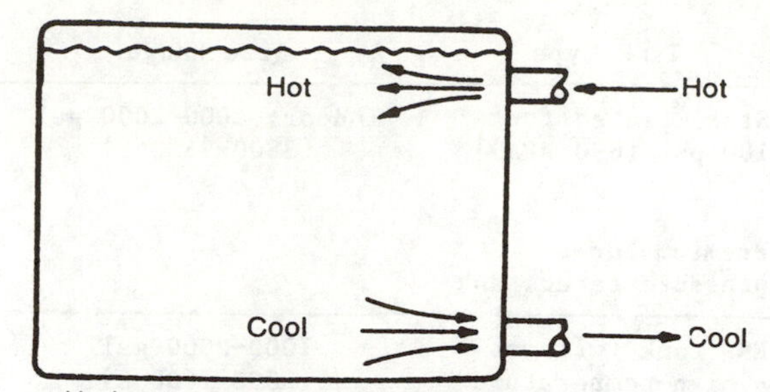

a. Horizontal flow. Hot water should enter or leave at the top; cold water should enter or leave at the bottom

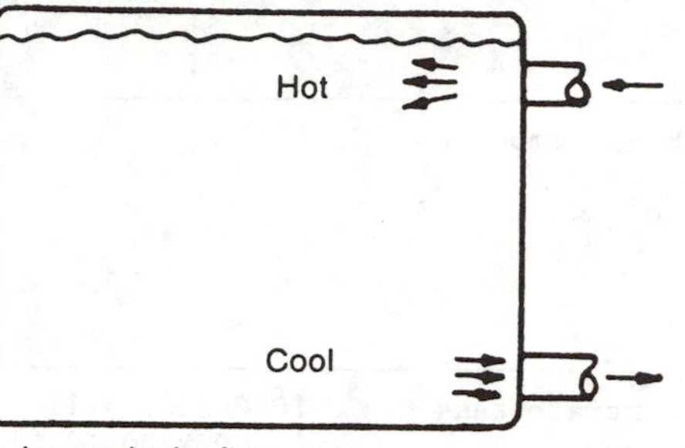

b. Low-velocity flow

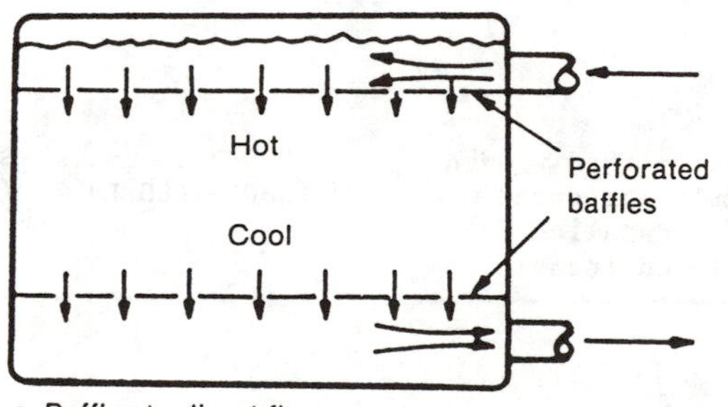

c. Baffles to direct flow

Fig. 5-30. Methods for promoting thermal stratification in water tanks (Cole et al. 1979)

Chapter 6

Active Subsystem Components

INTRODUCTION

Heat exchangers, ducts, pumps, piping, insulation, expansion tanks, and controls, as subsystem components of active and hybrid solar energy systems, are described in this chapter. All are interrelated in functional ways, each influencing and being influenced by the size, type, and performance of the others. Each subsystem component requires a harmonious relationship with the four other major elements (solar collectors, transport fluid, storage system, and terminal load devices) that make up a complete solar energy system. Selection and sizing of one subcomponent must be made with careful regard for all the others.

HEAT EXCHANGERS

A heat exchanger is a device that transfers thermal energy from one fluid to another. In some solar energy systems, a heat exchanger is required between the collector loop and the storage or load loop. They are used to isolate collector piping loops containing antifreeze or nonaqueous solutions from storage and space heating or service hot water loops. Heat exchangers can also be used to reduce hydrostatic head on storage tanks when collectors are mounted at a considerable height above the water storage tank; low pressure tanks are considerably less expensive than pressure vessels.

The information presented here will serve as a general guide for selecting the heat exchange surface area during early design. During final design, heat exchanger selection is generally made in consultation with equipment suppliers who have access to computerized analysis techniques.

Types Most Widely Used

Coil in Storage Tank

This heat exchanger can be used to transfer heat either from collector to storage or from storage to end use. Finned coils provide a large surface area to increase effectiveness. Figure 6-1 illustrates use of a coil in a storage tank that preheats water in a closed-loop collector system. Though some building codes allow a design in which an antifreeze solution circulates through the collector loop, a single separation between a possibly toxic liquid and potable water can be dangerous and should not be specified. The coil should be located in the lower (cooler) section of the storage tank where heat is transferred from collectors to storage. When heat is transferred from storage to load, the coil should be located in the upper (hotter) section of the tank. If the coil is double-walled (tube in tube), double separation is provided between transfer coolant and water supply; this is especially suitable for antifreeze systems.

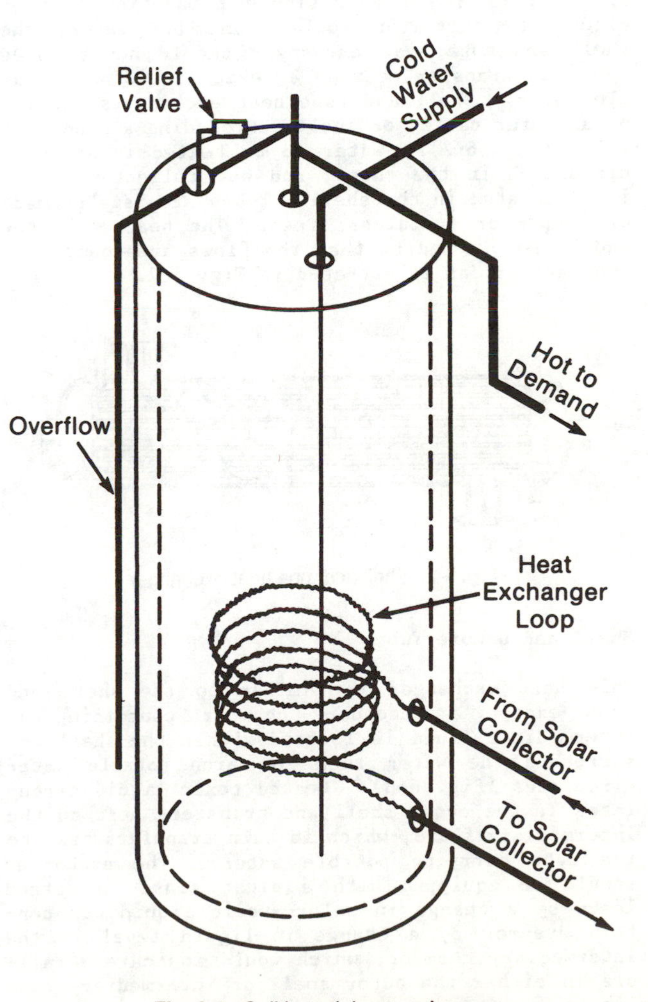

Fig. 6-1. Coil in tank heat exchanger

Coil or Plate-Coil Wrapped Around and Bonded to Storage Tank

This provides a double separation between collector coolant and the heated, potable water. Insulation enclosing the tank and heat exchanger should have an R-value of at least 7.5 [Btu/h-ft^2-°F]$^{-1}$ (1.32 [W/m^2-°C]$^{-1}$). In the simplest design, a serpentine tube, or a plate, is wrapped around the tank in intimate thermal contact with it. The collector coolant flows through this tube or plate. However, the most effective form is a counterflow design in which two tubes are wrapped in a serpentine pattern around the hot water tank; hot collector fluid circulates through one tube. Water admitted to the hot water tank circulates in the opposite direction through the other tube before returning, preheated, to the hot water tank.

Tank in Tank

In this type, an uninsulated service hot water tank, constructed to withstand system pressure, is placed inside a larger tank, sized for the space heating system.

Shell and Tube

Since there is only a single separation between circulating transfer coolant and hot water, the shell and tube heat exchanger should not be used for heat transfer between a toxic liquid and potable water. Shell and tube heat exchangers consist of an outer casing or shell surrounding a bundle of tubes (Fig. 6-2). Water to be heated is normally circulated in the tubes, and hot collector liquid is circulated in the shell. Tubes are usually made of copper or stainless steel. The heat exchanger should be plumbed so that the flows are counter to one another, as illustrated in Fig. 6-2.

Fig. 6-2. Shell and tube heat exchanger

Shell and Double Tube

This heat exchanger is similar to the shell and tube except a secondary chamber containing an intermediate fluid is located within the shell and surrounds the water tube in which potable water circulates (Fig. 6-3). Heated toxic liquid circulates in the outer shell and transfers heat to the intermediary fluid, which in turn transfers heat to the tube carrying potable water. The exchanger should be equipped with a sight glass to detect leaks by a change in color--toxic liquid can contain dye--or by a change in liquid level in the intermediary chamber, which would indicate a failure in either the outer shell or intermediary tube lining.

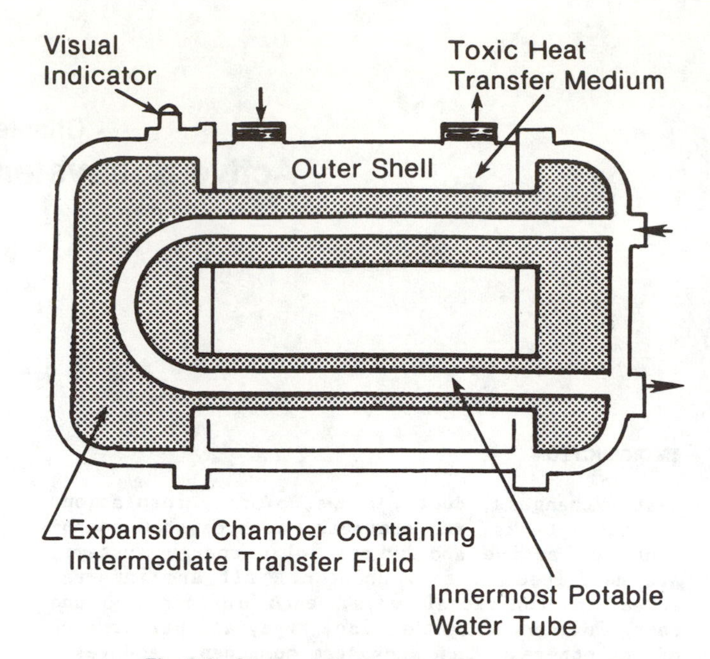

Fig. 6-3. Shell and double tube heat exchanger

Two Separate Heat Exchangers

Double separation can be provided by using two separate heat exchangers between collector and storage. The two may be of the same or different types but are generally of the shell and tube type. A pump is required between the two exchangers.

Rock-Bed Storage

This heat exchanger can be used in an air system. It should contain washed rocks 1/2 to 3 in. (15-80 mm) in diameter, and air flow preferably should be directed vertically rather than horizontally through the rock bed. Rock-bed design is discussed in detail in Ch. 5 and in the Design and Installation Manual for Thermal Energy Storage (Cole et al. 1979).

Hot Water Coil in Air Handler

This exchanger is used to preheat service hot water in a solar energy system with air collectors. It is located in the duct downstream of the solar collector (see Fig. 6-4).

Heat Exchangers in Antifreeze Systems

To avoid potable water freezing in the heat exchanger, precautions (specification of check valves, interlocks with pumps, and other) must be taken to prevent either thermosiphon or forced flow of collector fluid through the heat exchanger when the fluid is below 35°F (2°C).

Antifreeze solutions generally have lower specific heat than water. An increased flow rate is required to obtain the same degree of heat transfer as with water. For example, with a 50% ethylene glycol solution between 100° and 140°F (38°-60°C), the flow rate should be approximately 15% greater than with water (see Table 5-4).

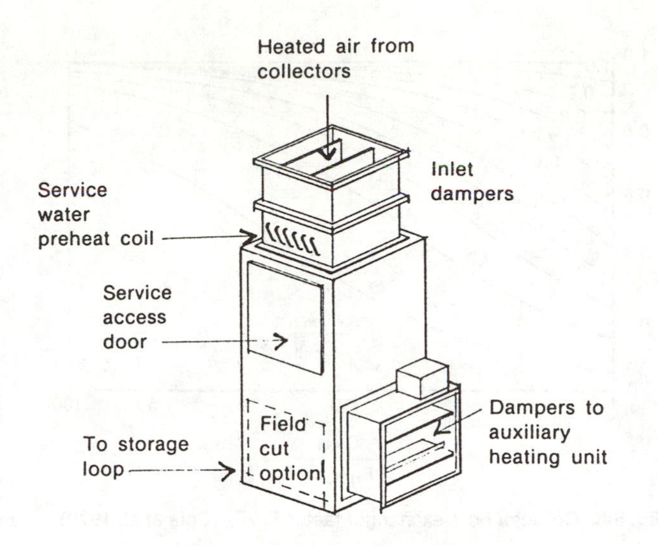

Fig. 6-4. Typical all-air solar air handler

The pressure drop across the heat exchanger will be greater with an antifreeze solution than with water because the flow rate is increased and the antifreeze viscosity is, on average, higher than that of water. (Note that the viscosity depends on the temperature.) Heat exchangers should be designed for different flow rates and pressure drops when antifreeze solutions are used. The pump too must be sized to accommodate increased pumping head.

Sizing

A system heat exchanger should be chosen to satisfy design criteria for that system and should satisfy thermal performance, materials, and construction specifications and requirements on pressure drop and fluid velocity in the tubes. It should afford protection from high pressure and temperature and be reasonably priced. It is very important that the heat exchanger be sized adequately; an undersized heat exchanger restricts transfer of collected solar energy to the load. Consequently, because a higher solar collector operating temperature is required, collector efficiency decreases.

Heat Exchanger Component Design

Nomenclature:

C = capacitance rate, Btu/h-°F (W/°C)

C_p = specific heat at constant pressure, Btu/lb-°F (kJ/kg-°C)

D = inside or outside tube diameter, ft (m)

g = acceleration of gravity, 32.2 ft/s^2 (9.8 m/s^2)

h = convective film coefficient, Btu/h-ft^2-°F (W/m^2-°C)

K = thermal conductivity, Btu-ft/h-ft^2-°F (W-m/m^2-°C)

t = temperature, °F (°C)

Δt = temperature difference, °F (°C)

V = linear velocity, ft/s (m/s)

β = thermal coefficient of expansion, °F^{-1} (°C^{-1})

μ = absolute viscosity, lb/h-ft (kg/s-m)

ν = kinematic viscosity, ft^2/h (m^2/s)

ρ = density, lb/ft^3 (kg/m^3)

U = overall coefficient of heat transfer, Btu/h-ft^2-°F (W/m^2-°C)

Dimensionless Groups:

Reynolds Number $$Re = \frac{\rho VD}{\mu} = \frac{VD}{\nu} \quad (6\text{-}1)$$

Grashof Number $$Gr = \frac{D^3\rho^2\beta g\Delta t}{\mu^2} \quad (6\text{-}2)$$

Prandtl Number $$Pr = \frac{\mu C_p}{K} \quad (6\text{-}3)$$

Nusselt Number $$Nu = \frac{hD}{K} \quad (6\text{-}4)$$

Provision of sufficient heat exchanger surface area is the first concern of preliminary heat exchanger sizing. Heat exchangers require a temperature differential to effect heat transfer; the magnitude of this temperature differential affects system operating temperature. Heat exchanger sizing thus has a direct bearing on overall collection efficiency.

There are two main types of heat exchanger systems: double-loop and single-loop. A double-loop system, illustrated in Fig. 6-5, requires two pumps (forced convection) to maintain positive control of flow on both sides of the heat exchanger. A single-loop system has only one pump, and typically features either a coil inside the tank or a coil fastened to the outside of the tank. Single-loop systems rely on heated water buoyancy to maintain flow on the tank side of the heat exchanger (natural convection). Forced convection is maintained on the other side of a single-loop system by a pump.

Using a heat exchanger leads to a collection penalty, as shown in Fig. 6-5. Collection efficiency decreases with increasing collector temperature, as shown in the curve in the lower part of the figure. The heat exchanger increases collector temperature and hence results in greater collector heat loss to the environment (Cole et al. 1979).

Double-Loop Heat Exchanger Systems

For a double-loop heat exchanger system, if the capacity rates in the two loops are such that:

$$C_{coll} < C_{sto} \quad (6\text{-}5)$$

where C_{coll} is the fluid capacitance rate (Btu/h-°F) in the collector loop and C_{sto} is that in the storage loop. As shown in Fig. 6-5, heat collected by the collector-heat exchanger combination is reduced (Cole et al. 1979); the factor is

$$\frac{F_R'}{F_R} = \frac{1}{1 + \frac{F_R U_c A_c}{C_{coll}} \left[\frac{1}{\varepsilon} - 1\right]} \quad (6\text{-}6)$$

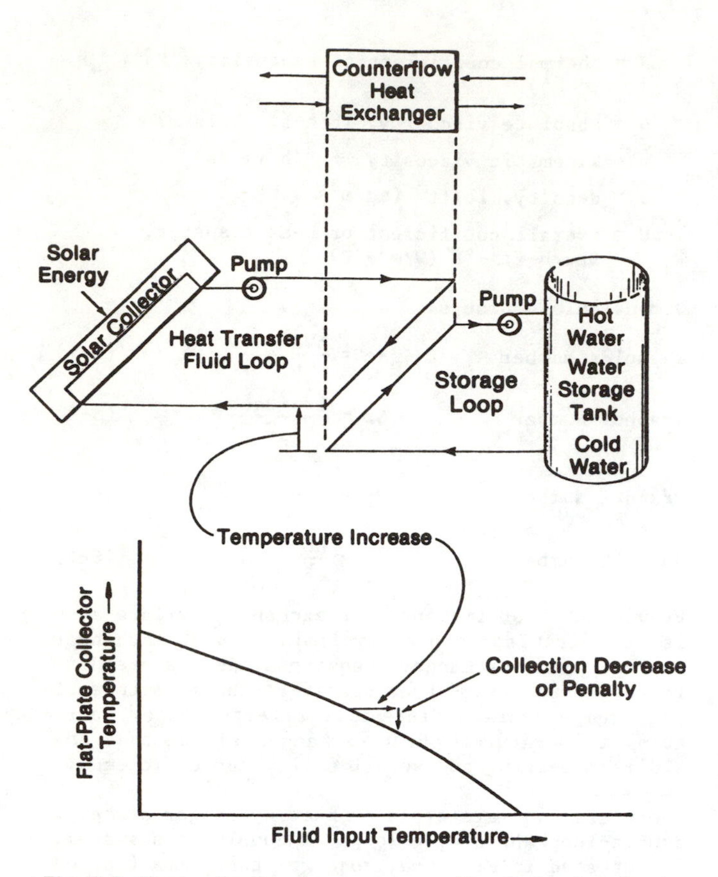

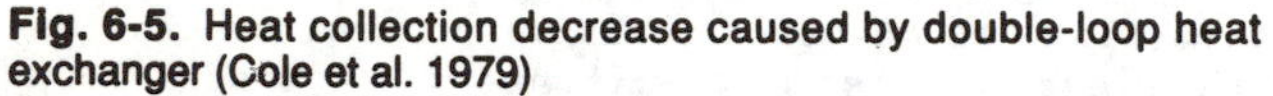

Fig. 6-5. Heat collection decrease caused by double-loop heat exchanger (Cole et al. 1979)

where

F_R = standard collector efficiency factor of the linear flat plate collector model

F_R' = same factor, modified by the heat exchanger effect

A_c = collector area

U_c = collector heat loss coefficient

ε = heat exchanger effectiveness

Where Eq. (6-5) does not hold, a more general case is

$$\frac{F_R'}{F_R} = \frac{1}{1 + \frac{F_R U_c A_c}{C_{coll}}\left[\frac{C_{coll}}{\varepsilon C_{min}} - 1\right]} \tag{6-7}$$

This completely general equation is plotted in Fig. 6-6.

In the general case, the effectiveness of a counterflow heat exchanger is an exponential function of the number of transfer units, NTU = $(U_x A_x)/C_{min}$ and C_{max}, as shown below. C_{min} and C_{max} are capacitance rates of the fluids in the heat exchanger. A_x is the heat exchanger heat transfer area, and U_x is the associated overall heat-transfer coefficient.

$$\varepsilon = \frac{1 - e^{-N}}{1 - (C_{min}/C_{max})e^{-N}} \tag{6-8}$$

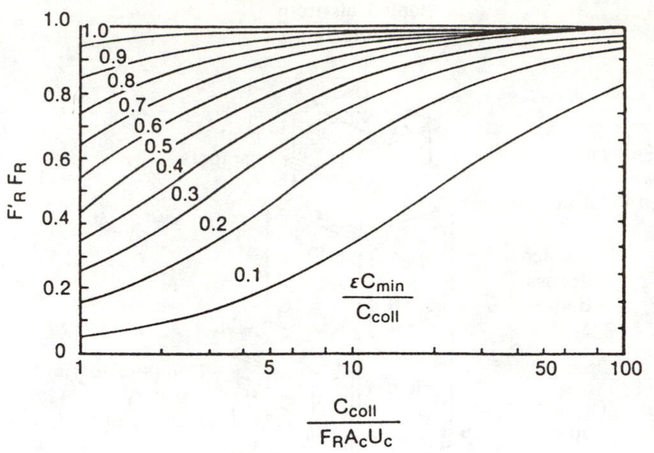

Fig. 6-6. Collector heat exchanger factor F_R'/F_R (Cole et al. 1979)

with

$$N = \text{NTU}\left[1 - C_{min}/C_{max}\right] \tag{6-9}$$

Thus, effectiveness increases as heat-transfer area increases. This reduces the collection penalty--it increases F_R'/F_R, bringing it closer to 1--so heat collection is increased. However, increasing heat exchanger size increases system cost. By means of a computer simulation the designer can balance these considerations to find an optimum heat exchanger size, as illustrated in Fig. 6-7.

If

$$C_{coll} = C_{sto} = C_{min} = C_{max} \tag{6-10}$$

then taking the limit of Eq. (6-8) as $(C_{min}/C_{max}) \rightarrow 1$ yields

$$\varepsilon = \frac{\text{NTU}}{1 + \text{NTU}} = \frac{1}{1 + \frac{C_{coll}}{U_x A_x}} \tag{6-11}$$

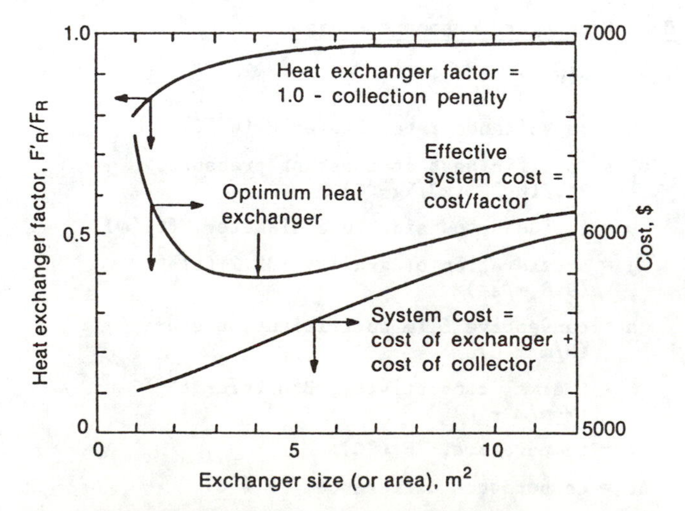

Fig. 6-7. Typical heat exchanger optimization plot (Cole et al. 1979)

Substituting this into Eq. (6-7):

$$\frac{F_R'}{F_R} = \frac{1}{1 + \frac{F_R U_c A_c}{U_x A_x}} \qquad (6\text{-}12)$$

When cost per unit collector area (c_c) and cost per unit heat exchanger area (c_x) are constant, it has been found that if the heat-transfer coefficient U_x does not vary with area A_x, optimum heat exchanger area A_x may be calculated from

$$A_x = A_c \left[\frac{F_R U_c c_c}{U_x c_x}\right]^{1/2} \qquad (6\text{-}13)$$

For typical values of collector capacity rate C_{coll}, C_{coll}/C_{sto} ranges from 0.5 to 0.6 and, for all practical purposes, the above equation can be used to find the optimum heat exchanger area, since this is only about 1% different from that found for the optimum (ummatched capacity rate) case.

EXAMPLE

Select a counterflow heat exchanger for a liquid system with a collector area of 500 ft^2 (46.45 m^2). The collector has the following characteristic parameters:

$$F_R U_c = 0.93 \text{ Btu/h-ft}^2\text{-}^\circ\text{F } (5.28 \text{ W/m}^2\text{-}^\circ\text{C})$$

Assume a particular time of day when the intensity of solar radiation on the tilted collector is at a near-maximum level of 300 Btu/h-ft^2 (946 W/m^2), inlet fluid temperature to the collector, t_i, is 130°F (54°C), storage water temperature is 128°F (53°C), and ambient temperature, t_a, is 30°F (-1°C). The collector fluid is a 50% mixture of ethylene glycol and water. Assume an acceptable penalty in F_R of $F_R'/F_R = 0.95$.

To size the heat exchanger in the collector loop, flow rates for collector and storage liquids are needed. Desired collector flow rate is 0.02 gpm/ft^2 (0.014 L/m^2_c-s), so total flow is 500 × 0.02 = 10 gpm (0.6 L/s). The rate of storage water flow through the heat exchanger is typically twice the rate through the collector, or 20 gpm (1.30 L/s). Temperature rise through the collector is determined as $t_o = t_i + Q_c/C_{coll}$. From Table 6-4, C_p = 0.835 Btu/lb-°F (3.49 kJ/kg-°C) and density is 65.15 lb/ft^3 (1046 kg/m^3). Thus

$$C_{coll} = \frac{10\left[\frac{gal}{min}\right] \times 65.15\left[\frac{lb}{ft^3}\right] \times 60\left[\frac{min}{h}\right] \times 0.835\left[\frac{Btu}{lb\text{-}^\circ F}\right]}{7.48 \text{ gal/ft}^3}$$

$$= 4364 \text{ Btu/h-}^\circ\text{F } (2309 \text{ W/}^\circ\text{C})$$

The fluid capacitance rate of the storage water at a temperature of about 130°F (54°C) is

$$C_{sto} = 20\left[\frac{gal}{min}\right] \times 8.34\left[\frac{lb}{gal}\right] \times 0.985^* \times 60\left[\frac{min}{h}\right] \times 1\left[\frac{Btu}{lb\text{-}^\circ F}\right]$$

$$= 9858 \text{ Btu/h-}^\circ\text{F } (5205 \text{ W/}^\circ\text{C})$$

Clearly, $C_{coll} < C_{sto}$, and C_{min}/C_{max} = 4364/9858 = 0.44. To choose a heat exchanger, calculate

$$\varepsilon = \frac{A_c F_R U_c}{C_{min}\left[\frac{A_c F_R U_c}{C_{coll}} + \left(\frac{F_R'}{F_R}\right)^{-1} - 1\right]} = 0.67$$

From Eq. 6-8 and 6-9 we introduce the intermediate term, the number of transfer units (NTU):

$$NTU = \frac{1}{(1-z)} \ln\left(\frac{1-\varepsilon z}{1-\varepsilon}\right) \text{ for } z \neq 1 \qquad (6\text{-}14)$$

$$= \frac{\varepsilon}{1-\varepsilon} \text{ for } z = 1$$

where

$$z = \frac{C_{min}}{C_{max}}$$

*0.985 is the specific gravity of water at 130°F (54°C) and is applied as a correction to the density value taken at 40°F (4°C) of 8.34 lb/gal.

Example (continued on next page)

Example (concluded)

For this example

$$z = 0.44$$
$$NTU = 1.36$$
$$U_x A_x = (NTU)\ C_{min} = 5935$$

Assume that

$$U_x = 250\ \text{Btu/h-ft}^2\text{-°F}\ (1420\ \text{W/m}^2\text{-°C});$$

then

$$A_x \approx 24\ \text{ft}^2.$$

The calculation should be repeated for other F_R'/F_R values and the cost effectiveness evaluated for the different resultant heat exchangers and performances. For more information see (Cole et al. 1979) and Leflar, Hull, and Winn (1978).

Overall Heat-Transfer Coefficient

The overall heat-transfer coefficient, U_x (Btu/h-ft^2-°F) (W/m^2-°C) of the heat exchanger tubes can be expressed in two ways:

Based on inside area:

$$U_i = \frac{1}{\frac{1}{h_i} + \frac{W_t}{K} + \frac{D_i}{h_o D_o}} \tag{6-15}$$

Based on outside area:

$$U_o = \frac{1}{\frac{1}{h_o} + \frac{W_t}{K} + \frac{D_o}{h_i D_i}} \tag{6-16}$$

where

U_i and U_o = Coefficients based on inside and outside tube areas, respectively, Btu/h-ft^2-°F (W/m^2-°C)

h_i and h_o = Film coefficients for inside and outside tube surfaces, respectively, (Btu/h-ft^2-°F) (W/m^2-°C)

D_i and D_o = Inside and outside diameter of tubes, ft (m)

W_t = Thickness of tube wall, ft (m)

K = thermal conductivity of the fluid tube wall material, Btu/h-ft^2-°F/ft (W/m^2-°C/m)

For practical purposes, in early design stages the U-factor equation for thin-wall tubes free from scale may be approximated as

$$U_x = \frac{1}{\frac{1}{h_i} + \frac{1}{h_o}} \tag{6-17}$$

However, during normal heat exchanger operation, fluid impurities are deposited on the tubes as a thin film. This film is quantified by a fouling factor, ξ, which is tabulated for various applications. For solar systems with relatively clean heat-transfer fluids, the overall U-factor equation becomes

$$U_x = \frac{1}{\frac{1}{h_i} + \frac{1}{h_o} + \frac{1}{\xi}} \tag{6-18}$$

For normal solar system design, an acceptable ξ factor is 1000 Btu/h-ft^2-°F (5682 W/m^2-°C) or $1/\xi$ = 0.001 (Btu/h-ft^2-°F)$^{-1}$ [1.76×10^{-4} (W/m^2-°C)$^{-1}$] as normally published in handbooks (ASHRAE 1985).

Three methods are presented here to solve Eq. (6-18) for the overall heat-transfer coefficient U_x of the heat exchanger.

Method 1: This approach uses fundamental empirical formulas to determine the film coefficient for both heat-transfer fluids.

a. Tube Side - Turbulent Flow (Re > 2100)

$$\frac{hD}{K} = 0.023\ (Re)^{0.8}\ (Pr)^n \tag{6-19}$$

where

n = 0.4 for heating
  = 0.3 for cooling

b. Tube Side - Laminar Flow (Re ≤ 2100)

$$\frac{hD}{K} = 1.86\ (Re \times Pr \times \frac{D}{L})^{0.33} \tag{6-20}$$

where L is the tube length

Note: Physical properties are taken at average fluid temperatures from the ASHRAE Handbook of Fundamentals (ASHRAE 1985).

c. Shell Side - Turbulent Flow (Re > 2100)

$$\frac{hD}{K} = 0.36\ (Re)^{0.55}\ (Pr)^{0.33} \tag{6-21}$$

In computing Re, equivalent diameters of various tube sizes and configurations are needed; these are given in Table 6-1.

Table 6-1. Equivalent Tube Diameters (Kern 1950)

Tube OD, in.	Tube Placement Pattern	d_e		D_e	
		in.	mm	ft	m
3/4	1" square	0.95	0.24	0.079	0.024
1	1-1/4" square	0.99	0.25	0.083	0.025
1-1/4	1-9/16" square	1.23	0.31	0.103	0.031
1-1/2	1-7/8" square	1.48	0.38	0.123	0.038
3/4	15/16" triangle	0.55	0.14	0.046	0.014
3/4	1" triangle	0.55	0.14	0.061	0.014
1	1-1/4" triangle	0.72	0.18	0.06	0.018
1-1/4	1-9/16" triangle	0.91	0.23	0.076	0.023
1-1/2	1-7/8" triangle	1.08	0.27	0.09	0.027

Note: Physical properties are taken at average fluid temperatures.

d. Shell Side - Laminar Flow ($Re \le 2100$) (applicable to tube-bundle type heat exchangers)*

$$\frac{hD}{K} = 0.59\ (Gr)^{0.25}\ (Pr)^{0.25} \qquad (6\text{-}22)$$

when

$Gr \times Pr = 10^4$ to 10^8, and

$$\frac{hD}{K} = 0.13\ (Gr)^{0.33}\ (Pr)^{0.33}$$

Fluid physical properties for use in these equations, for water, 50% propylene glycol, 50% ethylene glycol, and silicone, are tabulated in Tables 6-2 through 6-5.

Method 2: This approach uses tables to solve for fluid-film coefficients. It is simpler and more adaptable for use in early conceptual design stages. The tables (6-6, 6-7, 6-8) are from Stoever (1941).

CASE 1. To find the inside film coefficient, h_i, for turbulent flow in the tubes use Table 6-6. For selection purposes an average tube fluid velocity of 4 fps (1.2 m/s) is adequate. The heat-transfer coefficient equation used for this chart is

$$\frac{hD}{K} = 0.225\ (Re)^{0.8}(Pr)^{0.4} \qquad (6\text{-}23)$$

Each physical property is taken at the average fluid temperature.

CASE 2. To select outside film coefficient, h_o, in the shell side of the heat exchanger with turbulent flow use Table 6-7. An average shell fluid velocity of 2 fps (0.6 m/s) is recommended. The chart is based on the following equation:

$$\frac{hD}{K} = 0.385\ (Re)^{0.56}(Pr)^{0.3} \qquad (6\text{-}24)$$

Each physical fluid property is taken at the average film temperature.

*The assumption has been made that forced laminar flow for $Re \le 2100$ can be suitably approximated by free-convection laminar flow.

CASE 3. For a tube-bundle heat exchanger (a shell and tube heat exchanger without the shell) with natural convective heat transfer on the outside tube surface area use Table 6-8. The chart is used to select the outside film coefficient, h_o, for the tubes. The tube inside film coefficient would use Case 1 previously discussed. The heat-transfer equation for this chart is

$$\frac{hD}{K} = 0.525(Gr)^{0.25}(Pr)^{0.25} \qquad (6\text{-}25)$$

Each physical fluid property is taken at the average film temperature.

Method 3: This approach is the easiest since an overall U-factor is assumed for different heat exchanger applications. For example:

shell & tube heat exchanger	U-factor
water/water	300 Btu/h-ft^2-°F (1705 W/m^2-°C)
glycol/water	250 Btu/h-ft^2-°F (1420 W/m^2-°C)

tube bundle heat exchanger	U-factor
water/water	50 Btu/h-ft^2-°F (284 W/m^2-°C)

Use of assumed U-factors obviously simplifies the previous calculational procedures. Similar U-factors can be developed from experience for other common solar system configurations.

Table 6-2. Physical Properties: Water

Temperature t °F	Density ρ lb_m/ft^3	Dynamic Viscosity lb_m/hr ft	Specific Heat C_p B/lb_m-F	Thermal Conductivity K B/hr-ft^2-F/ft	Prandtl Number Pr	Thermal Coef of Exp β
40	62.4	3.73	1.005	0.326	11.50	.00020
50	4	.16	1.0	32	9.51	49
60	4	.71	1.0	37	8.04	85
70	62.3	2.36	1.0	0.343	6.88	.0012
80	2	.08	1.0	49	5.95	15
90	1	1.85	1.0	54	5.23	18
100	62.0	.66	1.0	0.359	4.63	.0020
110	61.9	.49	1.0	63	4.11	
120	7	1.36	1.0	67	3.70	
130	6	24	1.0	0.372	3.33	
140	61.4	14	1.0	76	3.04	
150	2	1.05	1.0	80	2.77	.0031
160	0	0.97	1.0	0.383	2.53	
170	60.8	0.90	1.005	86	2.34	
180	6	0.84	1.005	88	2.18	
190	4	0.786	1.005	0.390	2.02	
200	60.1	0.738	1.01	92	1.90	0.004
210	59.9	0.693	1.015	93	1.80	
220	7	54	1.015	0.394	1.69	
230	4	18	1.01	95	1.58	
240	59.1	0.585	1.01	96	1.49	
250	58.8	0.555	1.015	0.396	1.42	0.0048

Table 6-3. Physical Properties: 50% Propylene Glycol

Temperature t °F	Density ρ lb_m/ft^3	Dynamic Viscosity lb_m/hr ft	Specific Heat C_p B/lb_m-F	Thermal Conductivity K B/hr-ft^2-F/ft	Prandtl Number Pr
40			0.84	0.223	
50	65.4	22.39	0.845	0.221	85.61
60					
70				0.22	
80				2	
90				2	
100	64.06	7.5	0.865	2	29.5
110	63.77	6.29	0.87	2	24.9
120	63.52	5.57	0.875	2	22.2
130	63.27	4.84	0.88	2	19.4
140	63.02	4.11	0.885	0.219	16.6
150	62.77	3.51	0.89	0.219	14.3
160	62.52	3.03	0.895	0.218	12.4
170	62.28	2.66	0.90	0.217	11.0
180	62.03	2.42	0.905	0.217	10.1
190	61.78	2.06	0.91	0.2165	8.7
200	61.5	1.79	0.915	0.216	7.6
210					
220					
230					
240					
250	60.0	1.17	0.94	0.215	5.12

Reference: GLYCOLS, published by Union Carbide corp;
270 Park Avenue, NY NY 10017

Table 6-4. Physical Properties: 50% Ethylene Glycol

Temperature t ^oF	Density ρ lb_m/ft^3	Dynamic Viscosity lb_m/hr ft	Specific Heat C_p B/lb_m-F	Thermal Conductivity K B/hr-ft^2-F/ft	Prandtl Number Pr
40	66.83			0.243	
50		13.31	0.78	2	42.9
60				2	
70			0.79	0.241	
80			0.80	1	
90			0.81	0.24	
100	65.83	5.57	0.815	4	18.91
110	65.58	5.04	0.82	4	17.22
120	65.33	4.5	0.83	4	15.56
130	65.15	3.97	0.835	4	13.81
140	65.15	3.43	0.84	4	12.01
150	64.65	2.90	0.845	4	10.21
160	64.40	2.54	0.85	4	9.0
170	64.15	2.3	0.86	0.24	8.24
180	63.90	2.18	0.865	0.239	7.89
190	63.65	2.03	0.87	0.2385	7.40
200	63.46	1.82	0.875	0.238	6.69
210			0.88	0.237	
220				0.235	
230				0.232	
240			0.89	0.231	
250	62.03	1.21	0.895	0.23	4.71

Reference: GLYCOLS, published by Union Carbide Corp; 270 Park Avenue, NY, NY 10017

Table 6-5. Physical Properties: Silicone [GE SF 81 (50)]

Temperature t ^oF	Density ρ lb_m/ft^3	Dynamic Viscosity lb_m/hr ft	Specific Heat C_p B/lb_m-F	Thermal Conductivity K B/hr-ft^2-F/ft	Prandtl Number Pr
40					
50	61.5	154.9	0.356	0.0882	625.38
60					
70					
80					
90					
100	60.0	93.0	0.378	0.086	408.86
110	59.7	85.6	0.384	0.0855	384.45
120	59.3	78.2	0.388	0.085	356.96
130	59.0	68.7	0.392	0.0846	318.33
140	58.7	66.0	0.396	0.0842	310.40
150	58.5	61.2	0.40	0.0837	292.42
160	58.1	57.5	0.405	0.0833	279.56
170	57.8	51.6	0.41	0.0828	255.51
180	57.5	46.9	0.414	0.0824	235.64
190	57.2	44.4	0.418	0.082	226.33
200	57.0	42.0	0.422	0.0815	217.47
210					
220					
230					
240					
250	55.35	30.05	0.445	0.0792	168.84

Reference: General Electric Corp. Data

$$h = h_0 \times F_t \times F_d,$$

where h = the film coefficient, B.t.u./(sq. ft.)(hr.)(°F.); h_0 = the base value of the film coefficient, from Fig. F_t = the temperature-correction factor, from Table 8; F_d = the diameter-correction factor, from Table 9.

TABLE 8.—TEMPERATURE-CORRECTION FACTOR FOR CASE 1

Liquid	Ave. temp. of liquid					
	0°F.	50°F.	100°F.	150°F.	200°F.	250°F.
	F_t*					
Acetic acid (100%)	....	....	1.00	1.04	*1.08*	*1.12*
Acetic acid (50%)	....	0.75	1.00	1.15	1.30	*1.46*
Acetone	*0.78*	0.91	1.00	1.02	*1.04*	*1.06*
Ammonia	0.69	0.84	1.00	*1.18*	*1.36*	*1.56*
Amyl acetate	0.96	0.98	1.00	1.06	*1.16*	*1.27*
Amyl alcohol (iso)	*0.41*	0.68	1.00	1.39	1.82	*2.25*
Aniline	....	0.75	1.00	1.31	1.70	*2.10*
Benzene	....	0.80	1.00	1.14	1.28	*1.42*
Brine (CaCl₂) (25%)	0.54	0.74	1.00	1.29	1.63	*2.01*
Butyl alcohol (n)	*0.50*	0.73	1.00	1.33	*1.72*	*2.13*
Carbon disulfide	0.91	0.95	1.00	1.03	*1.05*	*1.06*
Carbon tetrachloride	*0.73*	0.88	1.00	1.05	*1.09*	*1.12*
Chlorobenzene	*0.82*	0.93	1.00	1.02	1.04	*1.05*
Chloroform	*0.72*	0.86	1.00	*1.14*	*1.28*	*1.42*
Ethyl acetate	*1.01*	1.01	1.00	0.99	*0.98*	*0.97*
Ethyl alcohol (100%)	0.63	0.80	1.00	1.21	1.43	*1.64*
Ethyl alcohol (40%)	*0.38*	0.64	1.00	*1.41*	*1.80*	*2.40*
Ethyl bromide	*0.89*	0.95	1.00	1.04	*1.08*	*1.11*
Ethylene glycol (50%)	*0.45*	0.66	1.00	1.40	1.89	*2.40*
Ethyl ether	*0.82*	0.94	1.00	1.06	*1.12*	*1.18*
Ethyl iodide	*0.70*	0.85	1.00	1.14	1.29	*1.44*
Glycerol (50%)	*0.45*	0.69	1.00	*1.39*	*1.85*	*2.30*
Heptane	*0.86*	0.92	1.00	1.08	*1.18*	*1.29*
Hexane	*0.84*	0.92	1.00	1.07	1.12	*1.15*
Methyl alcohol (100%)	*0.66*	0.87	1.00	*1.10*	*1.19*	*1.28*
Methyl alcohol (90%)	*0.63*	0.84	1.00	*1.13*	*1.26*	*1.38*
Methyl alcohol (40%)	*0.39*	0.67	1.00	1.30	*1.61*	*1.91*
Octane (n)	*0.84*	0.92	1.00	*1.07*	*1.13*	*1.19*
Pentane (n)	0.88	0.95	1.00	*1.04*	1.07	*1.10*
Propyl alcohol (iso)	*0.36*	0.69	1.00	1.32	*1.64*	*1.95*
Sulfur dioxide	0.96	0.98	1.00	*1.03*	*1.07*	*1.11*
Sulfuric acid (60%)	....	0.83	1.00	1.19	*1.39*	*1.62*
Toluene	0.80	0.90	1.00	1.08	1.16	*1.23*
Water	....	0.70	1.00	1.22	1.41	*1.58*

* Values in italics are based on extrapolated values of the physical properties of the fluids.

TABLE 9.—DIAMETER-CORRECTION FACTOR FOR CASE 1

Inside tube dia., in.	F_d	Inside tube dia., in.	F_d
0.20	1.38	1.20	0.97
0.30	1.27	1.30	0.95
0.40	1.20	1.40	0.94
0.50	1.15	1.50	0.92
0.60	1.11		
0.70	1.08	2.00	0.87
0.80	1.05	2.50	0.83
0.90	1.02	3.00	0.80
1.00	1.00	3.50	0.78
1.10	0.98	4.00	0.76

TABLE 10.—MINIMUM PERMISSIBLE VALUES OF $V \times d$ FOR CASE 1

V = velocity, ft. per sec.; d = inside tube dia., in.

Liquid*	Ave. temp. of liquid				
	50°F.	100°F.	150°F.	200°F.	250°F.
	$(V \times d)_{min.}$				
Acetic acid (100%)	...	1.2	0.9	0.8	0.7
Acetic acid (50%)	3.5	1.7	1.0	0.7	0.6
Amyl acetate	1.6	1.0	0.6	0.4	0.3
Amyl alcohol (iso)	9.5	4.1	2.0	1.1	0.8
Aniline	7.6	3.7	1.9	1.0	0.7
Brine (CaCl₂) (25%)	1.9	1.1	0.7	0.4	0.2
Butyl alcohol (n)	6.2	3.2	1.7	0.9	0.6
Ethyl alcohol (100%)	2.4	1.5	1.0	0.6	0.5
Ethyl alcohol (40%)	5.8	2.3	1.2	0.8	0.7
Ethylene glycol (50%)	6.3	2.6	1.2	0.7	0.4
Glycerol (50%)	9.7	4.2	2.3	1.3	0.8
Methyl alcohol (40%)	3.6	1.5	0.8	0.6	0.4
Propyl alcohol (iso)	5.4	2.3	1.2	0.8	0.6
Sulfuric acid (60%)	6.5	3.4	2.2	1.8	1.7

* Only those liquids for which $(V \times d)_{min.}$ at 100°F. is greater than 0.2 are listed.

Liquid	Curve No.
Acetic acid (100%)*	13
Acetic acid (50%)*	6
Acetone	7
Ammonia	1
Amyl acetate*	18
Amyl alcohol (Iso)*	21
Aniline*	20
Benzene	14
Brine (CaCl₂)(25%)*	3
Butyl alcohol (N)*	19
Carbon disulfide	9
Carbon tetrachloride	16
Chlorobenzene	16
Chloroform	10
Ethyl acetate	9
Ethyl alcohol (100%)*	14
Ethyl alcohol (40%)*	5
Ethyl bromide	11
Ethylene glycol (50%)*	6
Ethyl ether	9
Ethyl iodide	10
Glycerol (50%)*	8
Heptane	14
Hexane	12
Methyl alcohol (100%)	9
Methyl alcohol (90%)	7
Methyl alcohol (40%)*	5
Octane (N)	15
Pentane (N)	11
Propyl alcohol (Iso)*	17
Sulfur dioxide	4
Sulfuric acid (60%)*	16
Toluene	13
Water	2

*Chart not valid if V x d is less than minimum value given in Table 10

h_o = Base value of the film coefficient, B.t.u. per (sq.ft.)(hr.)(°F.)

V = Velocity, ft. per sec.

Table 6-7. Liquids Heated or Cooled Outside Single Tubes, Direction of Flow Normal to Tube

Temperature-correction Factor for Case 2

Liquid	Ave. temp. of the film* 0°F.	50°F.	100°F.	150°F.	200°F.	250°F.
	F_t†					
Acetic acid (100%)	....	....	1.00	0.96	*0.92*	*0.88*
Acetic acid (50%)	....	0.82	1.00	1.12	1.19	*1.23*
Acetone	*0.90*	0.95	1.00	1.01	*1.02*	*1.03*
Ammonia	0.79	0.89	1.00	*1.11*	*1.23*	*1.34*
Amyl acetate	1.16	1.08	1.00	0.93	*0.86*	*0.78*
Amyl alcohol (iso)	*0.57*	0.77	1.00	1.24	1.48	*1.72*
Aniline	....	0.84	1.00	1.20	1.42	*1.68*
Benzene	....	0.88	1.00	1.08	1.16	*1.24*
Brine ($CaCl_2$) (25%)	0.63	0.80	1.00	1.21	1.47	*1.75*
Butyl alcohol (n)	*0.83*	0.89	1.00	1.21	*1.48*	*1.83*
Carbon disulfide	0.97	0.98	1.00	1.01	*1.02*	*1.02*
Carbon tetrachloride	....	0.92	1.00	1.01	*1.02*	*1.03*
Chlorobenzene	*1.06*	1.03	1.00	0.97	0.95	*0.94*
Chloroform	*0.88*	0.94	1.00	*1.05*	*1.09*	*1.13*
Ethyl acetate	*1.12*	1.06	1.00	0.94	*0.87*	*0.81*
Ethyl alcohol (100%)	*0.74*	0.87	1.00	1.13	1.25	*1.36*
Ethyl alcohol (40%)	*0.47*	0.72	1.00	*1.28*	*1.55*	*1.83*
Ethyl bromide	*0.93*	0.97	1.00	1.02	*1.03*	*1.04*
Ethylene glycol (50%)	*0.52*	0.74	1.00	1.28	1.58	*1.92*
Ethyl ether	*0.91*	0.96	1.00	1.05	*1.09*	*1.13*
Ethyl iodide	*0.78*	0.89	1.00	1.10	1.19	*1.27*
Glycerol (50%)	*0.59*	0.77	1.00	*1.33*	*1.73*	
Heptane	*0.91*	0.96	1.00	1.03	*1.06*	*1.09*
Hexane	*0.91*	0.95	1.00	1.04	1.07	*1.10*
Methyl alcohol (100%)	*0.79*	0.91	1.00	*1.06*	*1.10*	*1.13*
Methyl alcohol (90%)	*0.76*	0.89	1.00	*1.08*	*1.14*	*1.20*
Methyl alcohol (40%)	*0.50*	0.76	1.00	1.20	*1.36*	*1.50*
Octane (n)	*0.91*	0.96	1.00	*1.05*	*1.09*	*1.13*
Pentane (n)	0.94	0.97	1.00	*1.02*	*1.03*	*1.04*
Propyl alcohol (iso)	*0.53*	0.77	1.00	1.21	1.38	*1.52*
Sulfur dioxide	1.03	1.01	1.00	*0.99*	*0.99*	*0.98*
Sulfuric acid (60%)	....	*0.80*	1.00	1.09	1.19	*1.28*
Toluene	0.90	0.95	1.00	1.04	1.07	*1.09*
Water	....	0.77	1.00	1.16	1.30	*1.41*

* Average temperature of the film = (temperature of the tube wall + temperature of the main body of the liquid) ÷ 2.

† Values in italics are based on extrapolated values of the physical properties of the fluids.

Diameter-correction Factor for Case 2

Outside Tube Dia., In.	F_d
0.250	1.84
0.375	1.54
0.500	1.36
0.625	1.23
0.750	1.14
1.00	1.00
1.25	0.91
1.50	0.84
2.00	0.74
2.50	0.67
3.00	0.62
3.50	0.58
4.00	0.54

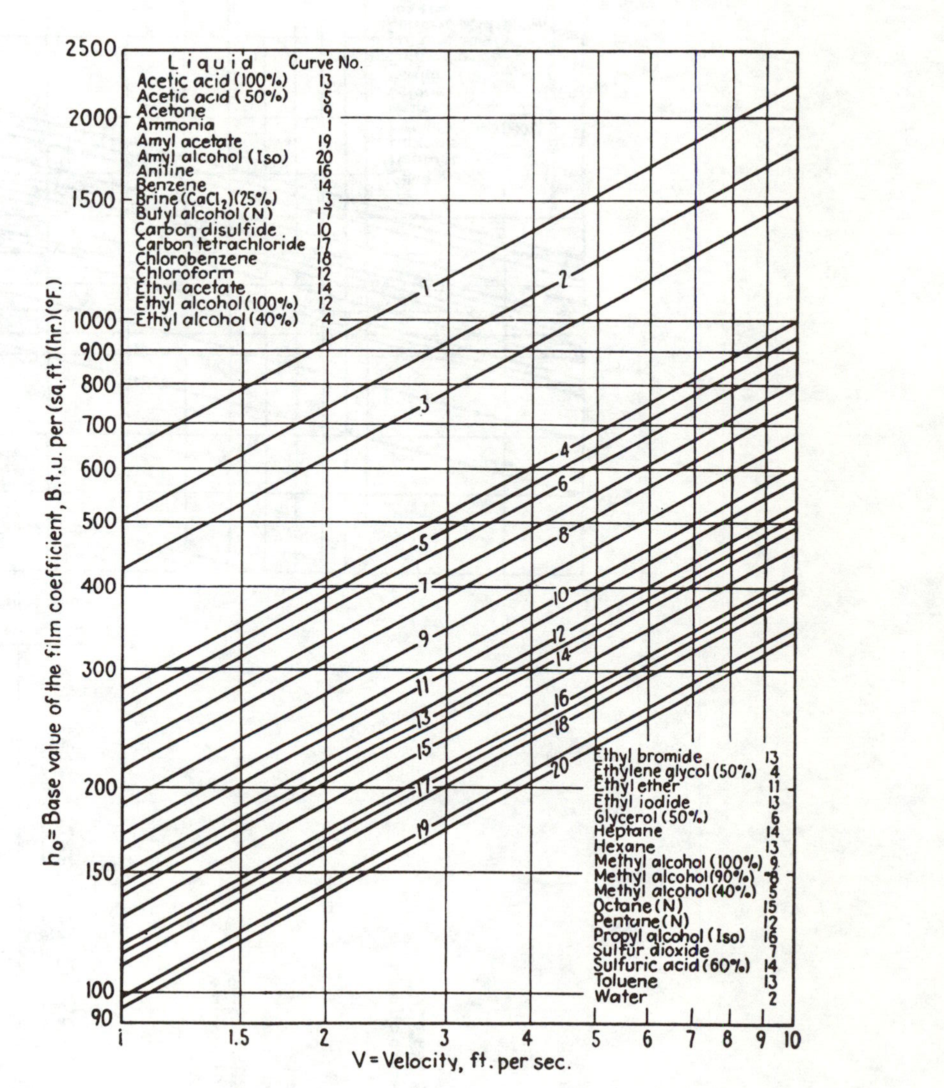

Table 6-8. Liquids Heated Outside Horizontal Tubes, Natural Convection

TEMPERATURE-CORRECTION FACTOR FOR CASE 3

Liquid	Ave. temp. of the film*					
	0°F.	50°F.	100°F.	150°F.	200°F.	250°F.
	F_t†					
Acetic acid (100%)	....	....	1.00	0.95	*0.90*	*0.87*
Acetone	....	0.97	1.00	*1.01*	*1.02*	*1.03*
Ammonia	0.79	0.89	1.00	1.11	*1.22*	*1.33*
Benzene	....	0.91	1.00	1.08	1.16	*1.23*
Carbon disulfide	0.97	0.98	1.00	1.01	*1.02*	*1.03*
Carbon tetrachloride	....	0.95	1.00	0.99	*0.98*	*0.96*
Chlorobenzene	*1.11*	1.05	1.00	0.95	0.91	*0.88*
Chloroform	*0.88*	0.95	1.00	*1.05*	*1.09*	*1.14*
Ethyl acetate	*1.17*	1.08	1.00	0.92	*0.85*	*0.78*
Ethyl alcohol (100%)	0.76	0.88	1.00	1.12	1.23	*1.33*
Ethyl alcohol (40%)	*0.49*	0.73	1.00	*1.29*	*1.58*	
Ethyl bromide	0.95	0.98	1.00	1.02	*1.03*	*1.04*
Ethyl ether	0.90	0.96	1.00	1.03	*1.06*	*1.08*
Ethyl iodide	*0.80*	0.90	1.00	1.08	1.16	*1.24*
Heptane	*0.95*	0.98	1.00	1.03	*1.05*	*1.07*
Hexane	*0.89*	0.95	1.00	1.04	1.07	*1.09*
Methyl alcohol (100%)	*0.83*	0.93	1.00	*1.06*	*1.10*	*1.14*
Methyl alcohol (90%)	*0.74*	0.89	1.00	*1.09*	*1.17*	*1.25*
Octane (n)	*0.92*	0.96	1.00	1.04	*1.08*	*1.11*
Pentane (n)	0.96	0.98	1.00	1.02	*1.05*	*1.07*
Sulfur dioxide	1.04	1.02	1.00	0.99	*0.98*	*0.97*
Sulfuric acid (98%)	....	0.76	1.00	*1.24*	*1.48*	
Sulfuric acid (60%)	....	0.80	1.00	1.17	1.30	*1.38*
Toluene	0.90	0.95	1.00	1.04	1.07	*1.08*
Water	....	0.77	1.00	1.15	1.26	*1.36*

* Average temperature of the film = (temperature of the tube wall + temperature of the main body of the liquid) ÷ 2.
† Values in italics are based on extrapolated values of the physical properties of the fluids.

DIAMETER-CORRECTION FACTOR FOR CASE 3

Outside Tube Dia., In.	F_d
0.250	1.41
0.375	1.28
0.500	1.19
0.625	1.13
0.750	1.08
1.00	1.00
1.25	0.93
1.50	0.90
2.00	0.84
2.50	0.80
3.00	0.76
3.50	0.73
4.00	0.71

Liquid	Curve No.
Acetic acid (100%)	11
Acetone	5
Ammonia	1
Benzene	11
Carbon disulfide	7
Carbon tetrachloride	13
Chlorobenzene	15
Chloroform	9
Ethyl acetate	11
Ethyl alcohol (100%)	10
Ethyl alcohol (40%)	3
Ethyl bromide	10
Ethyl ether	7
Ethyl iodide	11
Heptane	11
Hexane	10
Methyl alcohol (100%)	6
Methyl alcohol (90%)	4
Octane (N)	12
Pentane (N)	8
Sulfur dioxide	3
Sulfuric acid (98%)	14
Sulfuric acid (60%)	14
Toluene	11
Water	2

h_o = Base value of the film coefficient, B.t.u. per (sq.ft.)(hr.)(°F.)

Δt = Temperature difference between the retaining wall and the liquid, °F.

EXAMPLE

Determine the overall coefficient of heat transfer, U, for a shell and tube heat exchanger using the three methods discussed.

Given: Shell - 50% ethylene glycol, average temperature 130°F (55°C), flow velocity 2 fps (0.6 m/s). Tube water, average temperature 120°F (49°C), flow velocity 4 fps (1.2 m/s).

Method 1: Formulas. Find the shell outside film coefficient, h_o, assuming turbulent flow at 2 fps (0.6 m/s).

$$\frac{h_o D}{K} = 0.36\,(Re)^{0.55}(Pr)^{0.33}\,, \qquad (6\text{-}26)$$

Use physical properties for 50% ethylene glycol at 130°F (55°C) average temperature. Assume 3/4 in. OD copper tubing with equivalent diameter d_e = 0.55 in. (D_e = 0.046 ft) with tubes on 15/16-in. triangular pitch. Refer to Table 6-3 for fluid physical properties.

$$D_e = 0.046 \text{ ft } (0.014 \text{ m})$$
$$V = 2 \text{ fps} = (2)(3600) = 7200 \text{ ft/h}$$
$$\rho = 65.15 \text{ lb/ft}^3$$
$$\mu = 3.97 \text{ lb/h-ft}$$
$$K = 0.24 \text{ Btu/h-ft}^2\text{-°F/ft}$$
$$C_p = 0.835 \text{ Btu/lb-°F}$$
$$Re = \rho\frac{VD}{\mu} = \frac{(7200)(0.046)(65.15)}{3.97} = 5435$$
$$Pr = \frac{\mu C_p}{K} = \frac{(3.97)(0.835)}{0.24} = 13.81$$
$$h_o = \frac{K}{D}\, 0.36(Re)^{0.55}(Pr)^{0.33}$$
$$= \frac{0.24}{0.046}(0.36)(5435)^{0.55}(13.81)^{0.33}$$
$$= (1.88)(113.33)(2.38)$$
$$= 507 \text{ Btu/h-ft}^2\text{-°F } (2881 \text{ W/m}^2\text{-°C})$$

Find the tube side film coefficient, h_i, assuming turbulent flow at 4 fps (1.2 m/s).

$$\frac{h_i D}{K} = 0.023(Re)^{0.8}(Pr)^{0.4} \qquad (6\text{-}27)$$

Physical properties for water at 120°F (49°C) average temperature, with 3/4 in. copper tube. Refer to Table 6-2 for fluid physical properties.

$$D = 0.785 \text{ in.} = 0.0654 \text{ ft}$$
$$V = 4 \text{ fps} = (4)(3600) = 14{,}400 \text{ ft/h}$$
$$\rho = 61.7 \text{ lb/ft}^3$$
$$\mu = 1.367 \text{ lb/h-ft}$$
$$K = 0.367 \text{ Btu/h-ft}^2\text{-°F/ft}$$
$$Pr = 3.7$$
$$Re = \frac{VD\rho}{\mu} = \frac{(14{,}400)(0.0654)(61.7)}{1.367} = 42{,}725$$
$$Pr = \frac{\mu C_p}{K} = 3.7$$
$$h_i = \frac{0.367}{0.0654}(0.023)(42{,}725)^{0.8}(3.7)^{0.4}$$
$$= (0.129)(5065)(1.69)$$
$$= 1104 \text{ Btu/h-ft}^2\text{-°F } (6273 \text{ W/m}^2\text{-°C})$$

Example (continued on next page)

Example (concluded)

Find the overall U-factor - Method 1.

Assume fouling factor, ξ = 1000; then

$$U = \frac{1}{\frac{1}{h_i} + \frac{1}{h_o} + \frac{1}{\xi}} = \frac{1}{\frac{1}{1104} + \frac{1}{507} + \frac{1}{1000}} \qquad (6\text{-}28)$$

$$U = 256 \text{ Btu/h-ft}^2\text{-°F } (1455 \text{ W/m}^2\text{-°C})$$

Method 2: Charts. Find the outside film coefficient, h_o, assuming turbulent flow at 2 fps (0.6 m/s). Refer to Table 6-7, with 50% ethylene glycol at 2 fps (0.6 m/s) and 130°F (54°C) average temperature, and 3/4 in. type "L" copper tubing.

From Table 6-7,

$$h_o = 410 \text{ Btu/h-ft}^2\text{-°F } (2330 \text{ W/m}^2\text{-°C})$$

The correction factor, F_d, for tube OD at 0.785 in. is 1.07. The correction factor, F_t, for average temperature at 130°F is 1.168.

Therefore,

$$h_o \text{ corrected} = (410)(1.07)(1.168) = 512 \text{ Btu/h-ft}^2\text{-°F } (2909 \text{ W/m}^2\text{-°C})$$

Find the inside film coefficient, h_i, assuming turbulent flow at 4 fps (1.2 m/s).

Refer to Table 6-6, Case 1, with water at 4 fps and 120°F average temperature and 3/4 in. type L copper tubing.

From Table 6-6,

$$h_i = 980 \text{ Btu/h-ft}^2\text{-°F } (5568 \text{ W/m}^2\text{-°C})$$

The correction factor, F_d, for tube ID at 0.750 in. is 1.065. The correction factor, F_t, for 120°F average temperature is 1.088.

Therefore,

$$h_i \text{ corrected} = (980)(1.065)(1.088) = 1168 \text{ Btu/h-ft}^2\text{-°F } (6452 \text{ W/m}^2\text{-°C})$$

Find the overall U-factor - Method 2.

$$U = \frac{1}{\frac{1}{1136} + \frac{1}{512} + \frac{1}{1000}} = \frac{1}{0.00381} \qquad (6\text{-}29)$$

$$U = 261 \text{ Btu/h-ft}^2\text{-°F } (1483 \text{ W/m}^2\text{-°C})$$

Method 3: Assume overall U-factor for a glycol-to-water heat exchanger is 250 Btu/h-ft^2-°F (1420 W/m^2-°C).

Comparing the three methods for determining the overall coefficient of heat transfer for a shell and tube heat exchanger using glycol and water:

Method 1 - formula 256 Btu/h-ft^2-°F (1455 W/m^2-°C)

Method 2 - charts 262 Btu/h-ft^2-°F (1483 W/m^2-°C)

Method 3 - assumed 250 Btu/h-ft^2-°F (1420 W/m^2-°C)

General Design Guidelines

When specifying a heat exchanger for a system, the following items should be detailed: type, materials, fluids, exchange capacity, design temperature and pressure, effectiveness, pressure drop and flow rate requirements, insulation, protection method against high pressure and temperature, and single or double separation between potable water and nonpotable liquids. (See Fig. 6-8 for typical heat exchanger piping.)

Heat exchangers must be designed to withstand increased corrosion from glycol heat-transfer fluids at elevated temperatures. Two specific precautions should be taken when an antifreeze solution is used: (1) system piping must be cleaned and flushed before filling, and (2) solution reactions with inhibitors, sealants, and piping materials must be prevented.

Heat exchangers should be designed and constructed in accordance with the ASME Code for Unfired Pressure Vessels and be so stamped. They should be designed for a maximum 50 psig (345 kPa) on both primary and secondary sides and tested at 100 psig (690 kPa); they should be provided with ASME relief valves and drain valves. Solar heating systems should be designed so that leaky dampers and thermosiphoning will not cause freezing of heat exchangers or water piping.

Internal-Coil or Traced-Tank Heat Exchanger (for natural convection)

A problem with coils in tanks or with traced tanks involves the natural convection heat-transfer coefficient on the tank side. Forced convection heat-transfer coefficients are normally determined entirely by flow conditions. Natural convection coefficients, however, are determined by heating (or cooling) surface geometry, by the temperature difference between surface and fluid, and by fluid properties.

Conduction between the inside tank wall and tube fluid is analogous to that in a flat-plate collector with tubes bonded below the plate. Heat-transfer rate is given by the product of inside water film coefficient, h_t, inside tank heat-transfer area, A_t, and fluid-to-water temperature difference, Δt. The heat transfer correction factor F_t, is defined by

$$F_t = \frac{1}{\dfrac{Bh_t}{\pi D_o h_i} + \dfrac{Bh_t}{K_{bond}} + \dfrac{B}{D_o + (B - D_o)F}} \quad (6\text{-}30)$$

where

B = spacing between tubes (ft) (m)

D_o = outside diameter of coil tube (ft) (m)

h_i = heat-transfer coefficient on tube inside (Btu/h-ft^2-°F) (W/m^2-°C)

K_{bond} = conductance of tank to coil bond $\approx \dfrac{4T_{wall}K_{wall}}{D_o}$ (Btu/h-ft-°F) (W/m-°C)

T_{wall} = thickness of tank wall (ft) (m)

K_{wall} = conductivity of tank wall (Btu/h-ft-°F) (W/m-°C)

F = fin efficiency of tank wall between the tubes, for heat losses to the water

Figure 6-9 shows the relationship among the parameters of the traced tank.

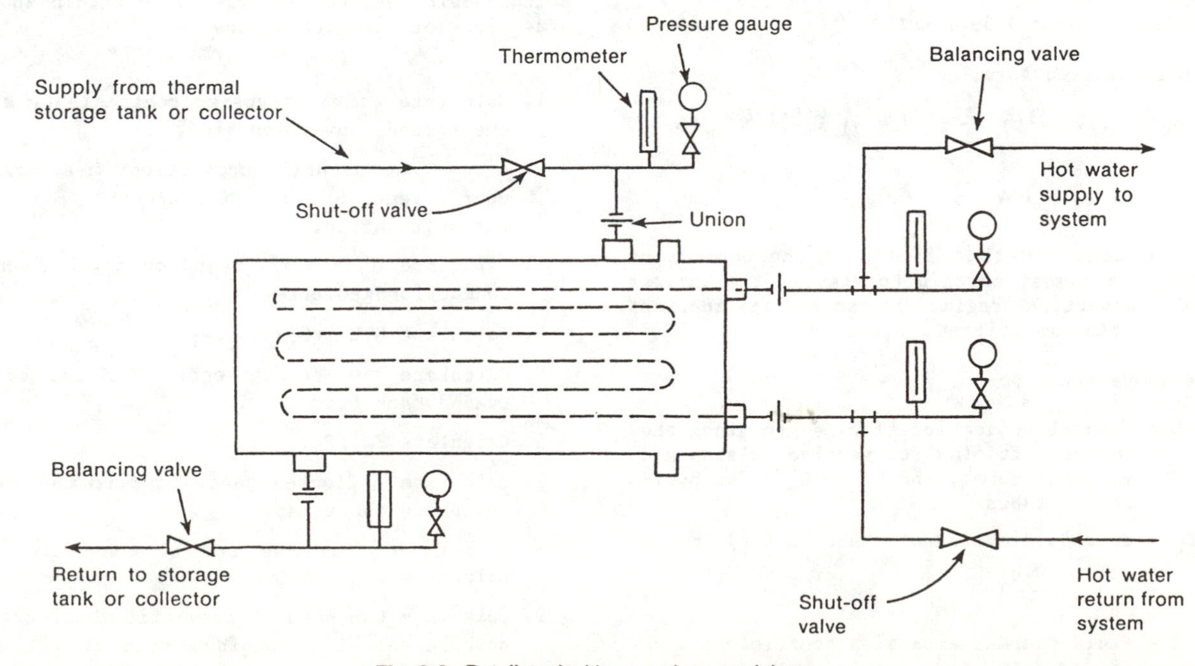

Fig. 6-8. Detail, typical heat exchanger piping

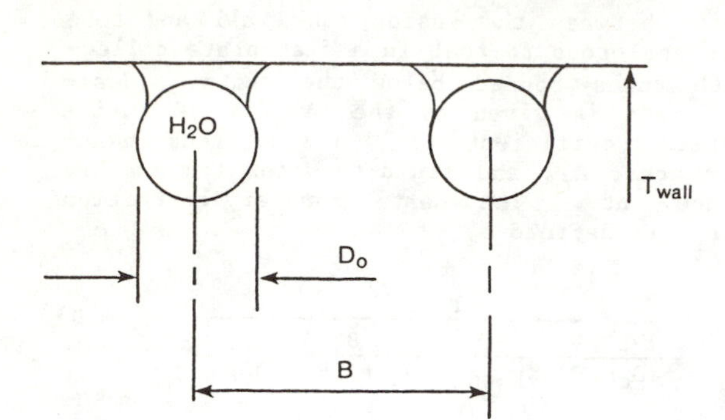

Fig. 6-9. Traced-tank heat exchanger dimensions (Cole et al. 1979)

In the turbulent regime, the heat-transfer coefficient for vertical plates is

$$\frac{h_t L}{K} = 0.13\ (kL^3\Delta t)^{1/3} \qquad (6\text{-}31)$$

where $kL^3\Delta t = Nu = PrGr$.

In simplified form,

$$h_t = 0.13\ Kk^{1/3}\Delta t^{1/3} = \gamma_t \Delta t^{1/3} \qquad (6\text{-}32)$$

where $\gamma_t = 0.13\ Kk^{1/3}$.

For the laminar regime, defined by

$$10^4 < (kL^3\Delta t) < 10^9 \qquad (6\text{-}33)$$

the heat-transfer coefficient for both vertical plates and horizontal tubes is given by

$$\frac{h_t L}{K} = 0.59\ (kL^3\Delta t)^{1/4} \qquad (6\text{-}34)$$

or, in simplified form,

$$h_t = 0.59\ Kk^{1/4}\ (\frac{\Delta t}{L})^{1/4} = \gamma_1\ (\frac{\Delta t}{L})^{1/4} \qquad (6\text{-}35)$$

where $\gamma_1 = 0.59\ Kk^{1/4}$.

It should be noted that flow over the outside of the tubes is almost certain to stay in the laminar natural convection regime in solar applications unless the tank is stirred.

In the above equations:

L = natural convection flow length along the surface (ft) (m) for vertical plates and vertical tubes, and $L = \pi\ D_o/2$ for horizontal tubes

D_o = outside diameter of the tube (ft) (m)

$k = \frac{\rho^2 g}{\mu^2}\ \beta\ (\frac{C_p \mu}{K})$

β = fluid thermal expansion coefficient (ft^3/ft^3-°F) (m^3/m^3-°C)

Δt = temperature difference between wall and fluid (°F) (°C)

h_t = natural convection heat-transfer coefficient (Btu/h-ft^2-°F) (W/m^2-°C)

ρ = fluid density (lb/ft^3) (kg/m^3)

g = acceleration of gravity = 4.18×10^8 ft/h^2 (1.28×10^8 m/h^2)

C_p = heat capacity (Btu/lb-°F) (kJ/kg-°C)

μ = fluid viscosity (lb/h-ft) (kg/h-m)

K = thermal conductivity of the fluid (Btu/h-ft-°F) (W/m-°C)

γ_t and γ_1 are defined by Eq. (6-32) and (6-35).

Values of K, k, γ_t, and γ_1 for water are given in Table 6-9 as a function of temperature t in °F.

Table 6-9. Convection Factors for Water

t (°F)	k	K	γ_t	γ_1
60	0.337×10^9	0.338	30.58	27.02
80	0.557×10^9	0.351	37.54	31.81
100	0.959×10^9	0.363	46.54	37.69
120	1.453×10^9	0.372	54.77	42.85
140	2.189×10^9	0.379	63.97	48.37
160	2.785×10^9	0.385	70.42	52.18
180	3.660×10^9	0.390	78.13	56.60

Since the natural convection heat-transfer coefficient is a function of the temperature difference, it is necessary to iterate to determine a final heat-transfer coefficient. The recommended scheme below should lead to convergence to within about 1% after four or five iterations.

1. Calculate a heat-transfer coefficient, h_i, on the forced convection side.
2. Assume a natural convection heat-transfer coefficient, h_t, of 100 Btu/h-ft^2-°F to start the calculation.
3. Calculate $U_x = h_t F_t$, based on h_i, h_t, and the conduction geometry.
4. Calculate $NTU = U_x q_t / C_{coll}$.
5. Calculate the effectiveness for the coil or traced tank from $\varepsilon = 1 - e^{-NTU}$.
6. Calculate F_R'/F_R.
7. Calculate collected heat, Q, from the collector performance map.
8. With Q, h_t, and the natural convection area, calculate $\Delta t_{avg} = Q/h_t A$.
9. Calculate the natural convection heat-transfer coefficient, h_t, obtained with this temperature difference using Eq. (6-32).

10. Return to Step 3 and repeat the calculations until the numbers in successive iterations no longer change appreciably.

This calculation applies to two system types:

- an external heat exchanger between collectors and storage
- a collector loop for heating a traced storage tank or a storage tank with an immersed coil; this involves water heated by natural convection.

If the load water is heated by flow through a heat exchanger immersed in the tank, a higher effectiveness heat exchanger is needed than is indicated by the above analysis since the load water is heated only when load is drawn.

Load Heat Exchanger

There are several types of heat exchangers that can be used to transfer heat from storage to space heating load. Solar heated water may circulate through radiant panels, fan-coil units, baseboard heating strips, or duct coils of central air heating systems. If supply water temperature to a radiant wall, floor, or ceiling panel is 100° to 120°F (38°-49°C) and a large panel area is used, room heating will be adequate. For fan-coil units and duct heating coils, water temperatures of 120° to 150°F (49°-65°C) are needed.

We will discuss load heat exchangers of the water-to-air crossflow type. As with liquid-to-liquid heat exchangers, water-to-air heat exchangers can be sized to deliver heat to the space at any desired rate. For practical reasons, the load heat exchanger should be sized such that water at about 145°F (63°C) can meet the design heating load. If heat delivery rate equals heat loss rate from the space,

$$\frac{\varepsilon C_{min}}{(UA)_L} = \frac{t_R - t_a}{t_S - t_R} \qquad (6\text{-}36)$$

where

C_{min} = the lower of two capacitance flow rates in the heat exchanger

ε = load heat exchanger effectiveness

t_R = room temperature, °F (°C)

t_a = design outdoor temperature, °F (°C)

t_S = storage water temperature, °F (°C)

$(UA)_L$ = load heat-loss coefficient

For a location where t_a = 0°F (-17°C), t_R = 70°F (21°C), and t_S = 140°F (60°C),

$$\frac{t_R - t_a}{t_S - t_R} = \frac{70-0}{140-70} = 1$$

If the heat exchanger has been designed to meet this situation, so that $\varepsilon C_{min}/(UA)_L = 1$, then when extremely cold temperatures are experienced outdoors, $t_R - t_a$ would be larger than 70°F (21°C) and water supply temperature must be higher than 140°F (60°C) to provide adequate heat to the space. However, if a larger heat exchanger is used (large εC_{min}), water at lower temperature can supply the heating rate to meet the load. Let us assume that $\varepsilon C_{min}/(UA)_L = 3$. In the example used above,

$$3(t_S - t_R) = t_R - t_a$$
$$3(t_S - 70) = 70 - 0$$
$$t_S = 93°\text{F } (34°\text{C})$$

Storage water at a temperature of 93°F (34°C) through the heat exchanger would deliver heat at a rate sufficient to heat the space air. Load heat exchanger effectiveness and size can be determined from Fig. 6-10.

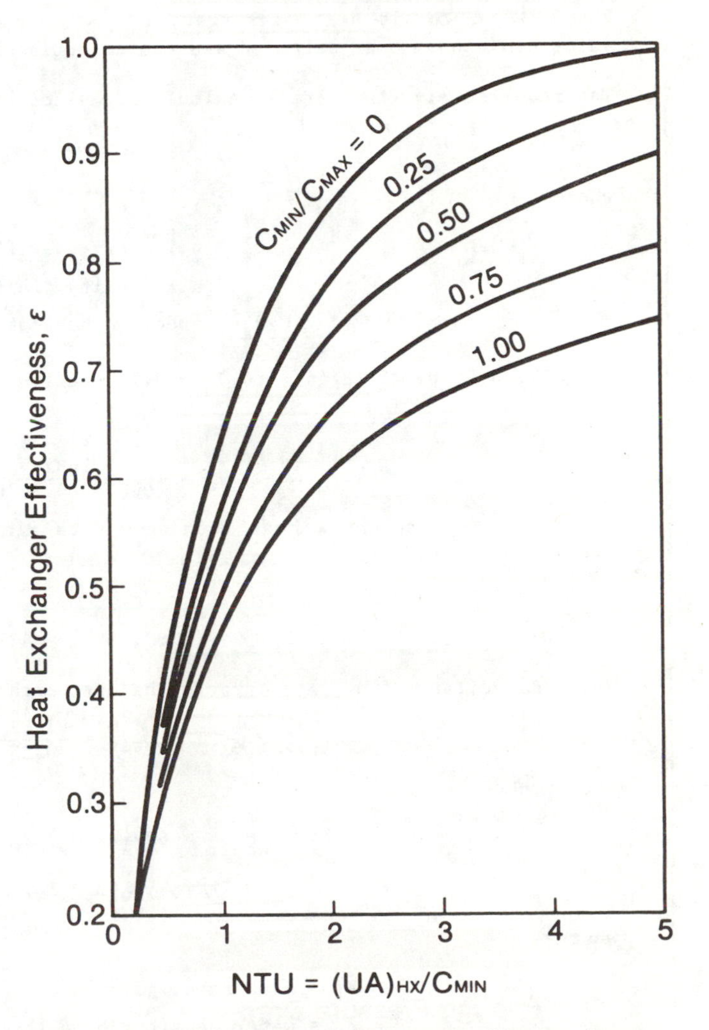

Fig. 6-10. Effectiveness of crossflow water-to-air heat exchanger

An example will illustrate the method. A common design point uses $\varepsilon C_{min}/UA$ equal to 2 (see Ch. 7). Higher values typically result in little improvement in performance.

EXAMPLE

Select a load heat exchanger such that when the storage water temperature is 100°F (38°C), sufficient heat will transfer to meet a design load. A heat loss calculation has been made, and $(UA)_L$ is 715 Btu/h-°F (379 W/°C) with a design outdoor temperature of 0°F (-17°C). Determine design air flow and water flow rates through the heat exchanger.

Solution - To determine $\varepsilon C_{min}/(UA)_L$, use Eq. (6-36).

$$\frac{\varepsilon C_{min}}{(UA)_L} = \frac{70-10}{100-70} = \frac{60}{30} = 2.0$$

Select a heat exchanger effectiveness of 0.85. Then

$$C_{min} = \frac{2.0 \times 715}{0.85} = 1682 \ \frac{Btu}{h\text{-}^oF} \ (891 \ W/^oC)$$

For a water-to-air heat exchanger, C_{min} is usually the heat capacitance rate for air. A C_{min} value could have been selected and the effectiveness, which must be less than one, determined.

The required air flow rate is calculated as follows:

$$C_{air} = 60\dot{V}_a\rho_a(C_p)_a \ , \qquad (6\text{-}37)$$

where

$\dot{V}_a$ = volumetric air flow rate, ft^3/min

ρ_a = air density, lb/ft^3

$(C_p)_a$ = specific heat of air, Btu/lb-°F

At 70°F, $\rho_a = 0.075 \ lb/ft^3$, $(C_p)_a = 0.24$ Btu/lb-°F.

Thus,

$$\dot{V}_a = \frac{1682}{(0.075)(0.24)(60)} = 1557 \ cfm \ . \qquad (6\text{-}38)$$

From Fig. 6-10 select a heat exchanger with minimum surface area. Using the curve C_{min}/C_{max} = 0.25, NTU = 2.6,

$$NTU = U_xA_x/C_{min}$$

$$U_xA_x = (2.6)(1682) = 4373 \ Btu/h\text{-}^\circ F \ (2318 \ W/^\circ C) \qquad (6\text{-}39)$$

Water capacitance flow rate through the heat exchanger is calculated as:

$$C_{water} = \frac{1682}{0.25} = 6728 \ \frac{Btu}{h\text{-}^oF} \qquad (6\text{-}40)$$

and because

$$C_{water} = 60\dot{V}_l\rho_w(C_p)_wS_w, \text{ English units} \qquad (6\text{-}41)$$

$$\dot{V}_l\rho_w(C_p)_wS_w, \text{ SI units}$$

where

$\dot{V}_l$ = flow rate, gal/min

ρ_w = water density, 8.34 lb/gal

$(C_p)_w$ = specific heat of water at a given temperature, ~1 Btu/lb°F

S_w = specific gravity of water at a given temperature

Therefore, in this example (using a water temperature of 100°F)

$$\dot{V}_l = \frac{6728}{(8.34)(0.997) \times (0.993)(60)} = 13.5 \ gpm \ .$$

Example (continued on next page)

Example (concluded)
The required air flow rate is large for a building with a heat loss rate of 715 Btu/h-°F, and the heat exchanger size is large. If the design storage water temperature is 140°F (60°C), a smaller air flow rate and heat exchanger size would result. With a storage temperature below design temperature, the energy transfer rate from solar storage would be unable to meet the design load, and auxiliary heating would be required. The design load occurs infrequently, however. Overall system performance should be evaluated to determine the cost effectiveness of the design; design can be changed if desired.

DUCT DESIGN

Sizing Criteria

To design a system that supplies optimal flow rates to the collectors with minimal initial and operating expenses, several factors must be considered. Initial system cost can be reduced by using smaller ducts. However, as air velocity and friction losses increase, operating expenses will also increase due to higher fan power requirements. Minimum life cycle cost will be realized when the sum of operating energy costs and installed system costs is lowest (see Fig. 6-11). Optimum design parameters are achieved by satisfying both economic criteria and system constraints.

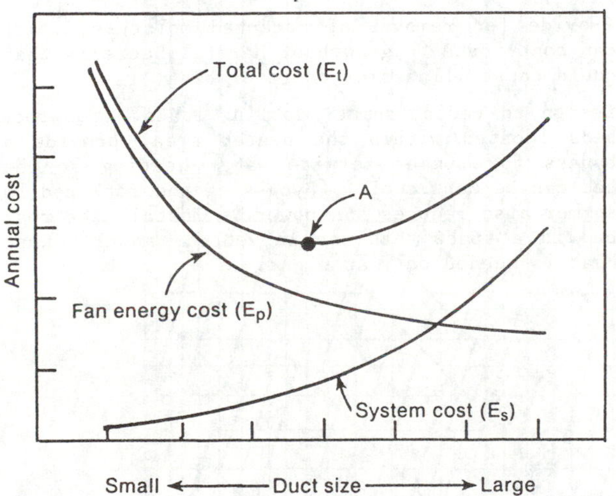

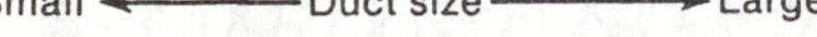

Fig. 6-11. Duct system economic analysis

For duct design consider the following:

- Minimize all duct lengths.
- Keep elbows and transitions (e.g., diffusers) to a minimum.
- Use turning vanes in all elbows.
- Try to minimize connections used in supply and return loops so that air leakage is reduced.
- Carefully tape and insulate ductwork. Air flow and thermal losses can severely reduce system performance.

Duct Design Method

The most common air duct system design methods are (1) equal friction; (2) velocity reduction; (3) static regain, including such variations as total pressure; and (4) constant velocity. Of the four, the equal friction method is by far the most commonly used for active solar air heating systems. Reasons for this include its simplicity in calculation, availability of diagrams for rough calculations, and its applicability to the low flow rates commonly found in solar heating systems.

In the equal friction method, system ductwork (including manifolds and collectors) is sized for a constant pressure loss per running foot of duct. Reducers, fittings, and elbows must be represented in terms of equivalent lengths of straight duct. However, total equivalent length in the critical flow path must be estimated from the total pressure-loss coefficients, given for most fittings, reducers, and elbows in the ASHRAE Handbook of Fundamentals (ASHRAE 1985).

It may be desirable initially to refer to a graph, such as Fig. 6-12, and try to size a duct that lies within the design band. In some cases, especially where higher air flow rates are anticipated, possible noise generation problems must be analyzed and appropriate sound attenuators placed in the system.

After a system is sized initially, total pressure drops are calculated for the main manifold, heat exchanger (if any), and system storage (if any) sections. With flow rates suggested by the collector manufacturer, a total pressure gradient line (see Fig. 6-13) can be established. Note that we will want total pressure (p_T) to change from positive to negative through the collector. This minimizes the difference between average pressure in the collector and ambient pressure and thus helps reduce collector air leakage (both infiltration and exfiltration), generally the greatest system air loss. Appropriate sections and fittings are then redesigned to balance the system approximately without use of dampers. After installation, it is usual to test, adjust, and balance the inlet air flow among arrays fed from a central trunkline.

When the system is sized, approximately balanced, and adjusted for elevation, temperature, and maximum velocity, the appropriate fan size to supply the required air flow at the calculated loop friction loss can be determined. Details of fan selection will be covered later.

Details of the fluid dynamics of air flows in ducts are discussed in the ASHRAE Handbook of Fundamentals (ASHRAE 1985, Ch. 33). Readers unfamiliar with these standard terms and calculations should consult this reference.

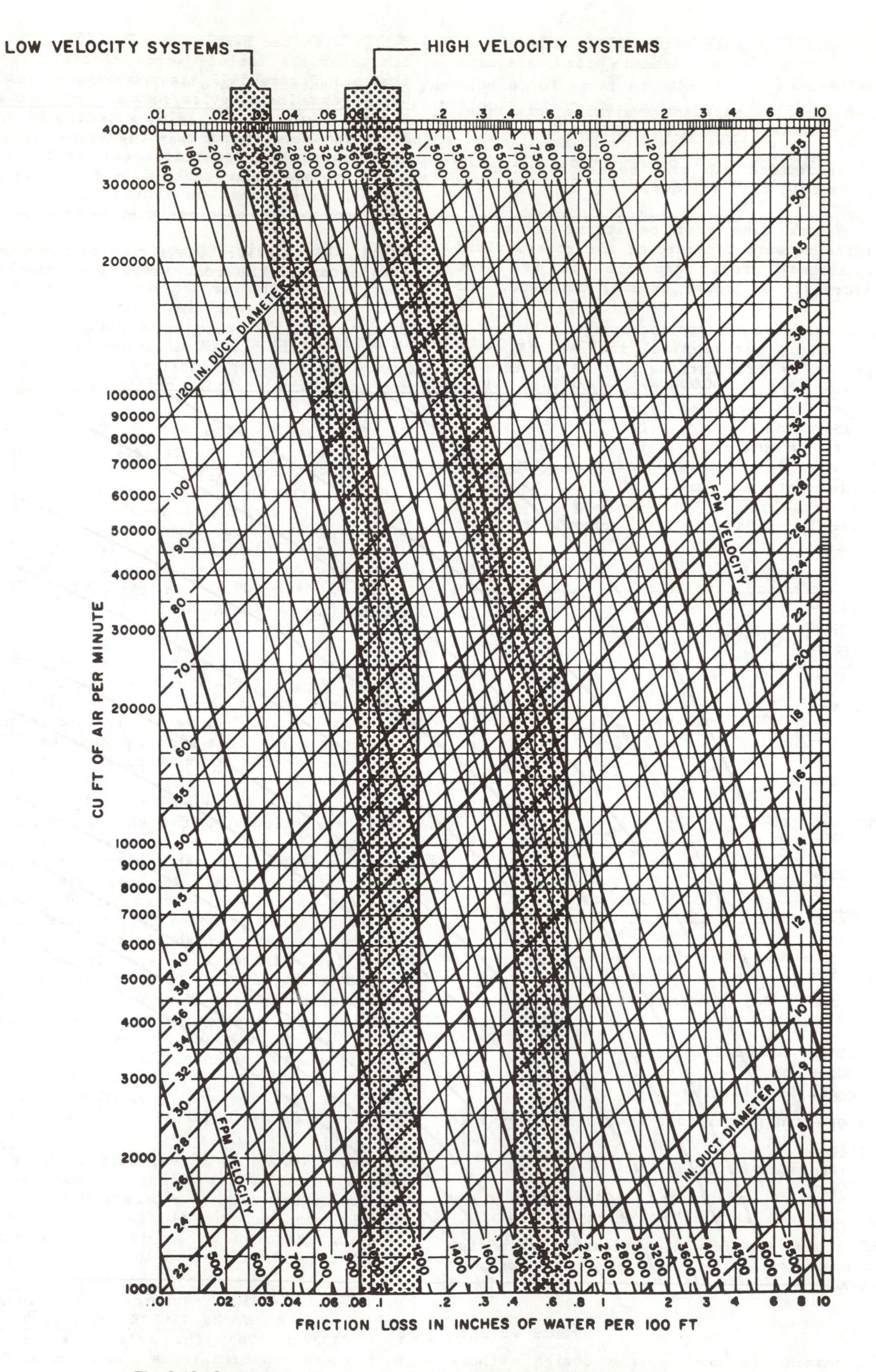

Fig. 6-12. Suggested velocity and friction rate design limits (ASHRAE 1981a)

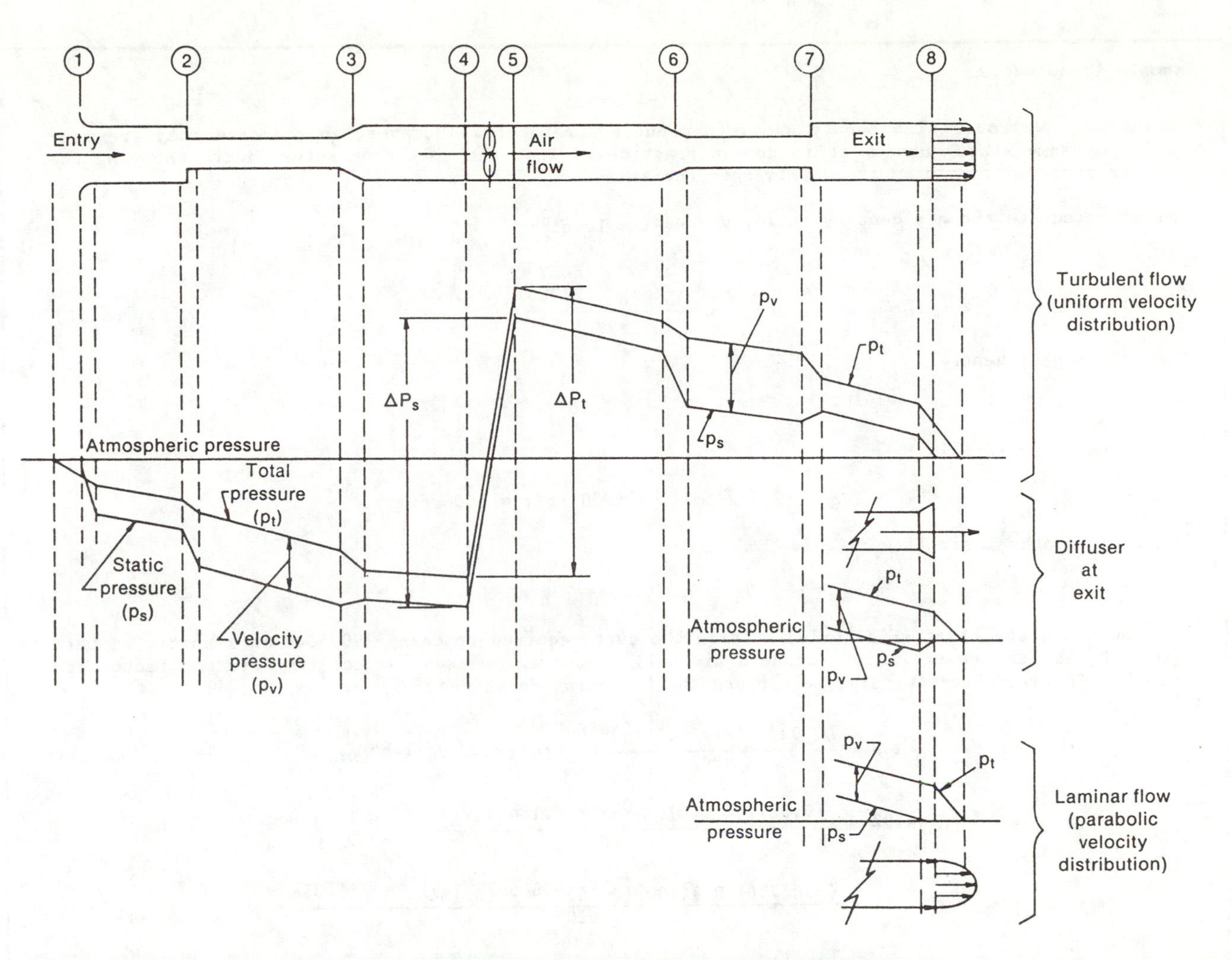

Fig. 6-13. Pressure changes during flow in ducts (ASHRAE 1981a)

The following information is needed for solar system duct design.

- Site elevation
- Design air temperature
  - Typically 70°F for return air from building or supply air to collectors
  - Typically 120°F for heated air to building or return air from collectors
- Insulation friction scaling factor (accounts for added friction due to duct liners)
  - $f_i = 1$ if no insulation used inside ducts
  - Obtain from insulation manufacturer; normally $f_i = 1.05$ to 1.40
- Maximum duct friction factor, f_{max}. Normally, $f_{max} = 0.08$ in. H_2O/100 ft of duct.
- Maximum duct air velocity, V_{max}. Normally, $V_{max} = 1000$ ft/min.

These inputs and basic air engineering concepts are sufficient for proper duct sizing. Inputs are based on extensive experience with residential solar and air distribution systems; they will provide quiet and cost-effective systems.

EXAMPLE

Size supply and return ducts for a solar collector array of 300 ft^2; assume 2.0 scfm/ft^2 of collector area. Use a rectangular duct at $f_i = 1.05$. The system is to be located near Ft. Collins, Colorado (elevation = 5000 ft).

Example (continued on next page)

Example (concluded)

Solution. We could size supply and return ducts separately. However, in practice they are usually the same size. Hence, it is common practice to size the collector return duct (carrying the hotter air) and then make the supply duct the same size.

First calculate the air density ratio, $\hat{\rho} = \rho/\rho_{sea\ level}$.

$$\hat{\rho} = \frac{530\, e^{-\frac{ELEV}{27000}}}{(460 + t)} \qquad (6\text{-}42)$$

use t = 120°F; hence

$$\hat{\rho} = 0.76$$

Collector flow rate is

$$\dot{V}_C = (2.0\ \text{scfm/ft}^2)\ (300\ \text{ft}^2) = 600\ \text{scfm}$$

Hence, the actual air flow rate is

$$\dot{V}_A = 600/0.76 = 790\ \text{ACFM}$$

We could use the chart (Fig. 6-14) to size the duct required to carry 790 ACFM at a specified head loss, H, but the chart is for standard air only. However, we can correct the friction factor and use Fig. 6-14. Alternatively, we can use the following equation:

$$H = \frac{(0.027)\ (f_i)\ (L)\ (\hat{\rho})^{0.845}}{D_e^{4.92}} (0.184 \times \dot{V}_A)^{1.85} \text{ or} \qquad (6\text{-}43)$$

$$(D_e)^{4.92} = \frac{(0.027)(f_i)(L)(\hat{\rho})^{0.845}(0.184 \times \dot{V}_A)^{1.85}}{H}$$

$$= \frac{(0.027)(1.05)(100)(0.76)^{0.845}\ (0.184 \times 790)^{1.85}}{0.08}$$

$$= 281{,}367$$

where

H = head loss, in. water column (W.C.) in inches H_2O

L = equivalent duct length, in.

$\hat{\rho}$ = air density ratio

D_e = equivalent round duct diameter, in.

$\dot{V}_A$ = actual air flow rate, ft^3/min

f_i = insulation friction scaling factor

Therefore

$$D_e = (281367)^{1/4.92} = 12.8\ \text{in.}$$

Select a 13 in. diameter duct.

Alternative Method. Rather than calculating the diameter using the formula, use Fig. 6-14 after correcting for f_{max}. First calculate a modified design friction factor from

$$f_{mod} = \frac{\text{desired friction rate, } f_{max}}{f_{DF} \times f_i} \qquad (6\text{-}44)$$

where duct friction correction factor (f_{DF}) = 0.182 + 10.911ρ and insulation friction scaling factor (f_i) = 1.0 to 1.40; ($\rho = 0.07\ \hat{\rho}\ lb/ft^3$). Using Fig. 6-14 with an air flow rate of 790 ft^3/min and f_{mod} = 0.1, a duct size of approximately 13 in. results.

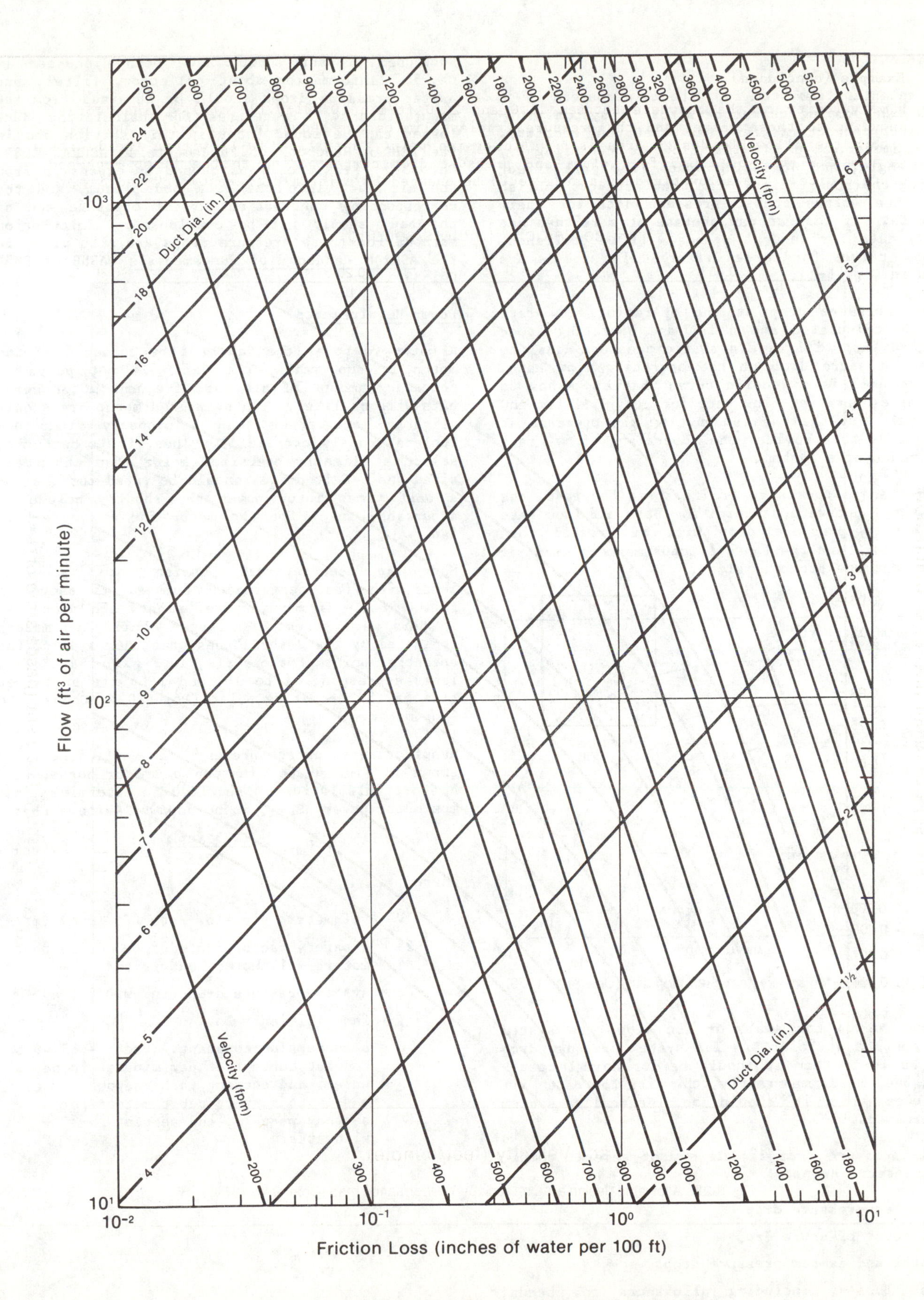

Fig. 6-14. Friction loss in a round straight duct (ASHRAE 1981a)

FAN SELECTION

When selecting a system fan, the main concern is to match fan pumping characteristics to system pressure drop characteristics (see subsequent section). Fan diameter, type (axial flow or centrifugal), blade angle, and operating speed (rpm) all affect pumping characteristics. Fan manufacturers publish flow rate versus static pressure data for their products. If the fan can operate at more than one speed, data for several speeds will be published. Typical curves for three different fan speeds are shown in Fig. 6-15.

System pressure drop at the operating flow rate must be calculated and a fan and operating speed selected that will give a static pressure equal to system pressure drop at the operating flow rate. (Figure 6-15 is a generic curve that shows how to determine operating flow rate for a typical fan and system.) For the data shown, the fan operates at 1100 rpm to provide the operating flow rate required by the system.

If the fan is too large or its speed too fast, the system will operate at Point A. Both air flow rate and system pressure drop will be greater than planned for, and the fan will consume more electric power than one properly sized.

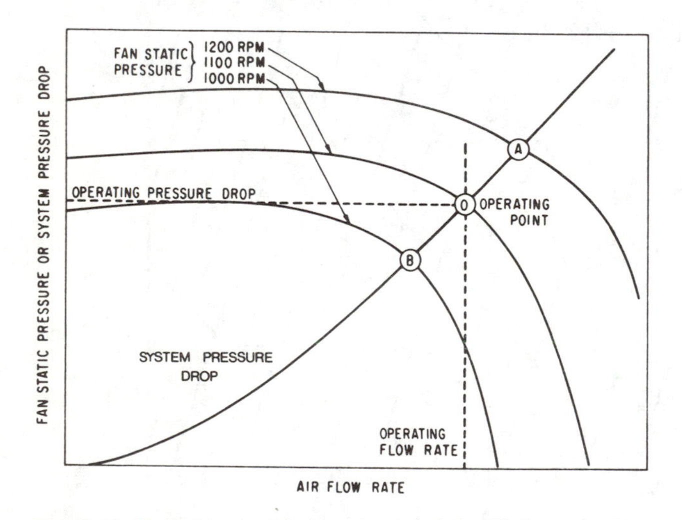

Fig. 6-15. Typical fan and system characteristics (Cole et al. 1979)

If the fan is too small or too slow, the system will operate at Point B. The system pressure drop will be lower than it should be for normal operation; but, more importantly, the air flow rate will be lower than it should be, degrading system performance.

System pressure drop is the sum of several component pressure drops:

- Rock bed pressure drop
- Collector pressure drop
- Filter and damper pressure drops
- Duct losses, including allowances for bends, branch ducts, and expansions or contractions.

Rock bed pressure drops have been discussed in Ch. 5. Information about collector, filter, and damper pressure drops should be obtained from the manufacturers. Procedures for calculating duct losses can be found later in this section and in the ASHRAE Handbook of Fundamentals (ASHRAE 1985, Ch. 33). Pressure drops caused by expansion from the air duct into the rock bed plenum, and the corresponding contraction at the opposite end of the bed, should not be overlooked. Calculation methods for these pressure drops are also given in the ASHRAE Handbook of Fundamentals (ASHRAE 1985, Ch. 33).

Power Requirements

A motor to power the fan must be chosen. If the motor is too small, the fan will not pump the necessary amount of air, and frequent motor burnouts will be likely. An oversized motor draws only slightly more power than a properly sized one (unless grossly oversized). Thus, it is better to select a slightly oversized motor than an undersized one. Belt drives should be rated for one and a half times motor power and should include an adjustable sheave on the motor for a single-belt fan.

Many fan manufacturers publish motor requirements with fan performance curves, as shown in Fig. 6-16a. If manufacturer's data are presented in this way, select the larger of the two motors indicated by the dashed lines on either side of the operating point. For example, in Fig. 6-16a, dashed lines corresponding to 3/4 and 1 hp lie on either side of the operating point (Point O). Choose the 1 hp motor.

Manufacturers often present fan efficiency, as shown in Fig. 6-16b, instead of motor horsepower. A short calculation is required to determine minimum motor power, P_{min}, in horsepower (kilowatts):

$$P_{min} = \frac{1.25\ \dot{V}\Delta p}{\alpha \eta_f} \qquad (6\text{-}45)$$

where

$\dot{V}$ = volumetric air flow rate (ft^3/min) (m^3/s)

1.25 = a safety factor included to ensure that motors will not be undersized

Δp = system pressure drop (in. W.C.) (Pa)

η_f = fan efficiency %

α = a conversion constant. Use $\alpha = 63.46$ to convert cubic feet per minute, inches of water, and percent to horsepower; use $\alpha = 10$ to convert cubic meters per second, pascals, and percent to kilowatts.

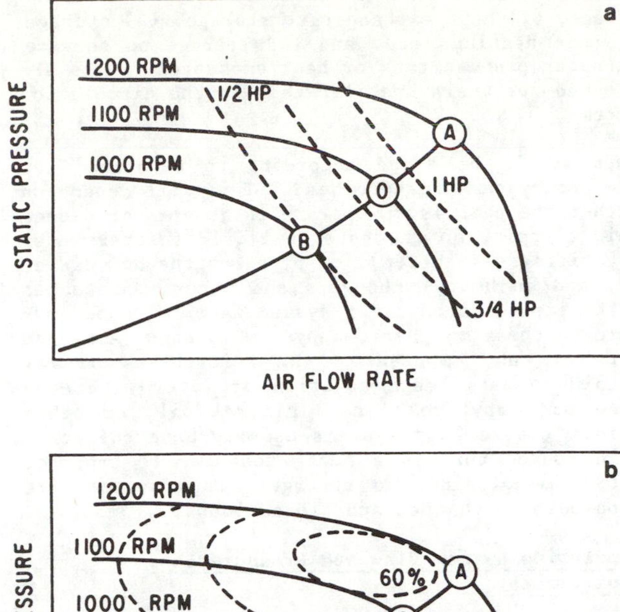

Fig. 6-16. Typical fan performance curves (Cole et al. 1979)

Fan Installation

The air temperature a fan must handle in a solar system can sometimes present a problem not often encountered in conventional heating systems. If the fan will encounter a maximum air temperature greater than 100 F° (38°C), it must meet the following specifications:

- Since fan bearings must be able to operate continuously at maximum air temperature, special bearings may be required. Alternatively, bearings can be located outside the hot air stream, and shaft seals can be specified to minimize leakage.
- Motor and drive belts must be outside the hot air stream, or a Type B motor must be connected directly to the fan.
- Fan selection should be based on a modified operating point (Point M in Fig. 6-17) instead of on the previous operating point (Point O). The modified operating point is found by multiplying both air flow rate and fan static pressure at the operating point by a factor, ϕ

$$\phi = \frac{t + t_o}{t_R + t_o} \qquad (6\text{-}46)$$

where

t = air temperature in the duct (°F) (°C)

t_o = reference temperature (freezing point of water) expressed in terms of absolute temperature. Use t_o = 491°R or 273 K

t_R = room temperature. Use t_R = 70°F or 20°C.

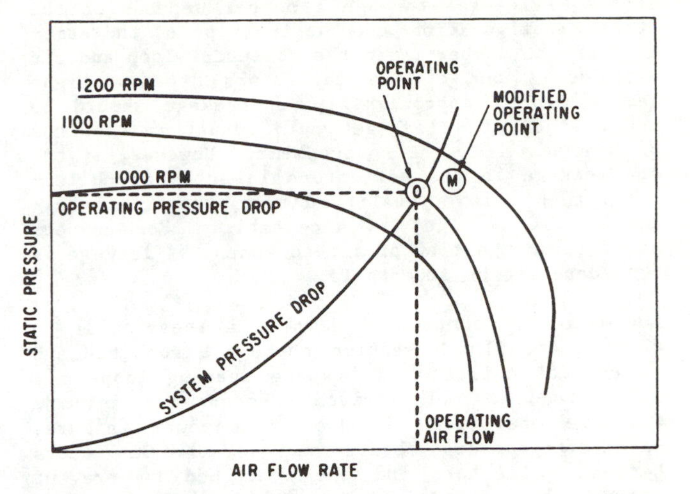

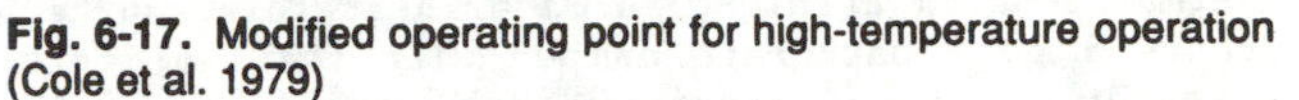

Fig. 6-17. Modified operating point for high-temperature operation (Cole et al. 1979)

The modified operating point applies only to fan selection and should not be used for other calculations.

Since the major operating expense of an air-based system is the cost of electricity to power the fan, it is important to install the fan so that it will operate at its highest efficiency. We recommend connecting the fan inlet to the ductwork with a straight duct section at least five duct diameters long. The duct work should match fan inlet diameter so there will be no sudden contraction or expansion as air enters the fan. If a transition from a rectangular duct to a round fan inlet must be made, its slope should not exceed 4 in 12 in. (18°). It is especially important to avoid use of bends or elbows near the fan inlet, because the turbulence they cause reduces fan efficiency. Use similar care in designing outlet ductwork.

Other Air System Components

Filters

Filters for air systems should be located at both the rock bin inlet and outlet where they are easily accessible for service or replacement. Filter mounts should minimize air bypassing the filter and leakage escaping the duct.

Filter face velocity (air flow rate divided by filter area) should not exceed 300 ft/min (1.5 m/s). If the filter is larger than the duct cross-section, a transition to full filter size must be made, with a slope not exceeding 4 in 12 in. (18°).

Install a filter-replacement indicator gauge at each filter. The gauge can be self-indicating or remote-indicating; in either case, the indicator must be located where it is easy to see.

Dampers

Since dampers have proven to be the least reliable component in existing air systems, it is worthwhile to invest in high-quality units. The amount of allowable leakage through a damper depends on the system configuration and the location of the damper. If the damper is in the collector loop and air leakage is out of the collectors, then a tight damper having approximately 2% leakage should be used, since this leakage would result in the loss of heated air to the environment. However, if the air leakage in the collectors is into the collectors, then a lower quality damper, having approximately 10% leakage, is acceptable. For dampers controlling the flow of air to zones, 5% leakage is considered to be acceptable.

Automatically controlled dampers are essential to control air flow direction through a rock bed and to control collector and space heating loops. A spring-loaded, motor-driven damper can provide failsafe operation in case of a power failure. Backdraft dampers should be installed in ducts between collectors and the rock bed to prevent thermosiphoning at night--a major heat loss source. Choose a good backdraft damper with seals of felt or other resilient material such as silicone rubber. Backdraft dampers must close by either gravity or springs and must remain tightly closed until reopened.

Air Handlers

Air handlers, including a fan and up to four motorized dampers in one package, are available. Air handling units require less installation labor than separate components and can be purchased specifically designed for solar applications. It is recommended that an air handler be used in preference to individual components if one is available that meets system flow rate and control requirements.

Temperature Sensors

Use temperature sensors recommended by the controller manufacturer. Two temperature sensors should be placed in the rock bed, one 6 in. (15 cm) below the top of the rocks and the other 6 in. (15 cm) above the bottom of the rocks. To avoid damaging temperature sensors and to make replacement simple, we recommend that sensors be placed inside pipes that extend from outside the bin's inner wall to the center of the bed. Sensor leads are run from the pipe through the insulation to an electrical box for connection to a controller. Low-temperature readings by both bin sensors indicate that little heat remains in the rock bed; the auxiliary heater must supply heat. High-temperature readings by both sensors indicate the rock bed is fully charged. A high reading at the top and a low reading at the bottom indicate partial charging.

PUMPS

Liquid systems other than thermosiphon systems require a pump in each separate circuit to circulate heat-transfer fluid. Pumps are typically required for each of the following circuits: (1) collector to heat exchanger or directly to storage, (2) heat exchanger to storage, (3) storage to space heating load, and (4) storage to service hot water preheat tank or heat exchanger. Pumps are selected for their specific task in the circulation system.

Pumps should be sized according to required flow rate and system pressure head. This will depend on whether the pump is to operate in an open or closed fluid circuit, on the type of liquid (either water or antifreeze solution), on pipe lengths and diameter, and on heat exchangers and other resistances in the piping circuit. If system water is open anywhere to the atmosphere, a system is open. A closed system is not exposed to the atmosphere and may contain an antifreeze solution of water with ethylene or propylene glycol, mineral oil, or other fluid. Closed loop systems usually have collector fluid routed through a heat exchanger to transfer heat from collector to storage. Many systems are composed of both open and closed loops.

Circulating System Flow and Pressure Loss Requirements

Different methods are used to calculate flow requirements and pressure loss resistance in open and closed systems. Flow requirements are based on energy delivered and attainable temperature rise through the collectors, which vary with available radiation. Pressure losses result from pipe friction, flow resistance of fittings and heat exchangers, and collector flow resistance. Similarly, pressure losses in open systems occur from the same sources and, in addition, from net vertical fluid rise in the system. For instance, a drain-back system for freeze protection is normally an open system; it can, however, also be closed. A closed drain-back system requires substantially more pump pressure to recharge than sustain and thus may require an additional pump. A pump must deliver sufficient head to fill the system, but, if a siphon return is established in the downcomer, then for most of the operating time, the pump has considerably smaller head requirements. If the downcomer is not filled with water, then the pump head during operation is the same as that during fill. The effects of this fill resistance must be evaluated during normal operation.

Once maximum flow and head requirements are established, the proper pump can be selected by examining pump performance curves. These usually list flow on the horizontal axis and head (pressure) on the vertical axis. If the point at which head and flow requirements meet is on or below the performance curve of a pump, the pump should be adequate.

Hypothetical system requirements of flow and head result in a point A below the performance curve of the pump in Fig. 6-18. The system will operate on the system curve; therefore, adjustments using speed selector switches available on some pumps may be necessary. The installer must accurately adjust pump performance to system requirements (i.e., B to A).

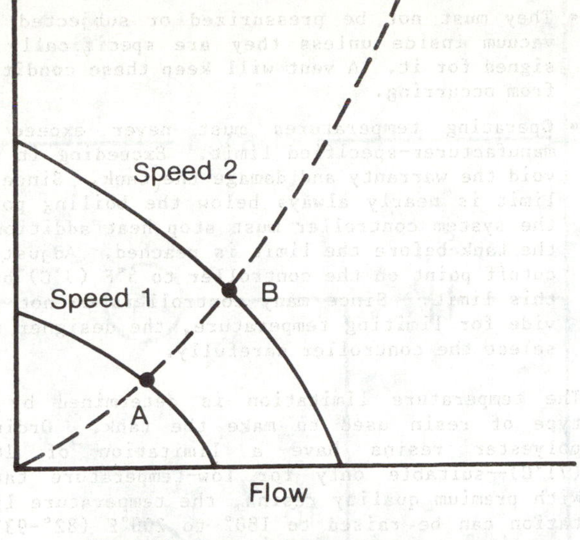

Fig. 6-18. Pump performance curve

In a well-designed system, pump circulation power should be no more than 1% to 3.5% of solar energy collected. Nevertheless, energy consumed by pumps and controllers must always be taken into consideration when realistically evaluating total system efficiency.

Pump efficiency can be determined as follows:

$$\eta_p \text{ (efficiency)} = \frac{\text{Output}}{\text{Input}} \qquad (6\text{-}47)$$

where output is work performed by the pump on the fluid, and input is electrical energy or work supplied. In a closed system with cold water at 8.33 lb/gal, the formula for pump horsepower reduces to hp = (gpm × head in ft)/3960. Pump efficiency, (reflecting motor and drive line losses), can be determined by

$$\eta_p = \frac{\text{Pump hp}}{\text{Motor hp}} \qquad (6\text{-}48)$$

Pumps for Higher Head Requirements

Traditionally, single-impeller hot water circulation pumps have been designed for relatively low head and large flow rate. Typically, a circulating pump is most efficient at half of maximum flow. Staging two or more smaller circulators in series is sometimes done to efficiently supply the necessary head. For example, a pump providing a maximum head of 14 ft (4.3 m), when staged in series with another of the same capacity will result in a total head of 28 ft (8.5 m) and can become more efficient than a single pump designed to operate at the required head/flow rate combination (Fig. 6-19).

Two or more single-speed pumps in parallel also represent a viable configuration for systems in which variable pump speed is desired. A large pump in parallel with a smaller pump is analogous to a single pump with three speeds; each pump can be operated alone, or the two can be operated together. Such flexibility can significantly reduce parasitic power. When they are working together, both

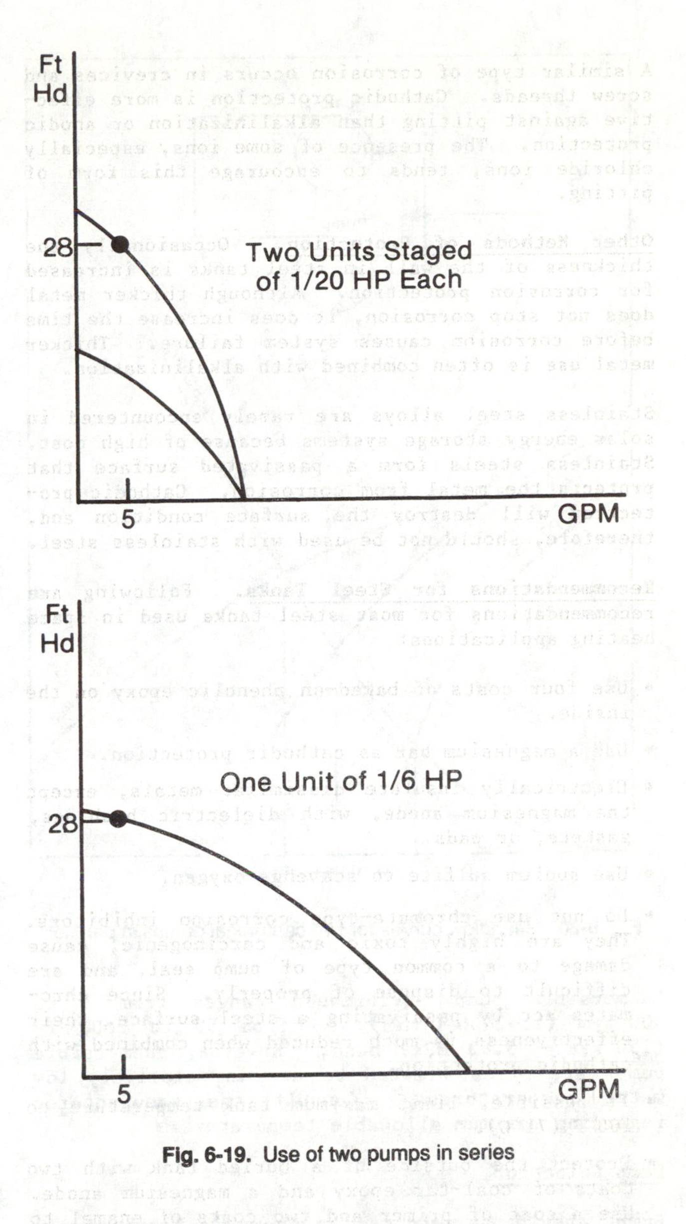

Fig. 6-19. Use of two pumps in series

pumps operate against the same head but contribute only a portion of the total flow rate. An example H-Q curve for parallel centrifugal pumps is shown in Fig. 6-20. However, the size difference between pumps must not be so great that the smaller one cannot operate under the head produced by the larger.

Two circulators may be necessary in a system--two to boost the pressure and charge the system initially and one or the other to operate the system. A time relay wiring scheme can be used (Fig. 6-21).

Submersible pumps offer another way to meet high head requirements in solar energy systems. They have the advantage that waste heat generated by pump inefficiency goes into the water instead of being lost to the environment.

They can be installed directly in a storage tank and used to either recharge the system only or circulate water only, depending on system design.

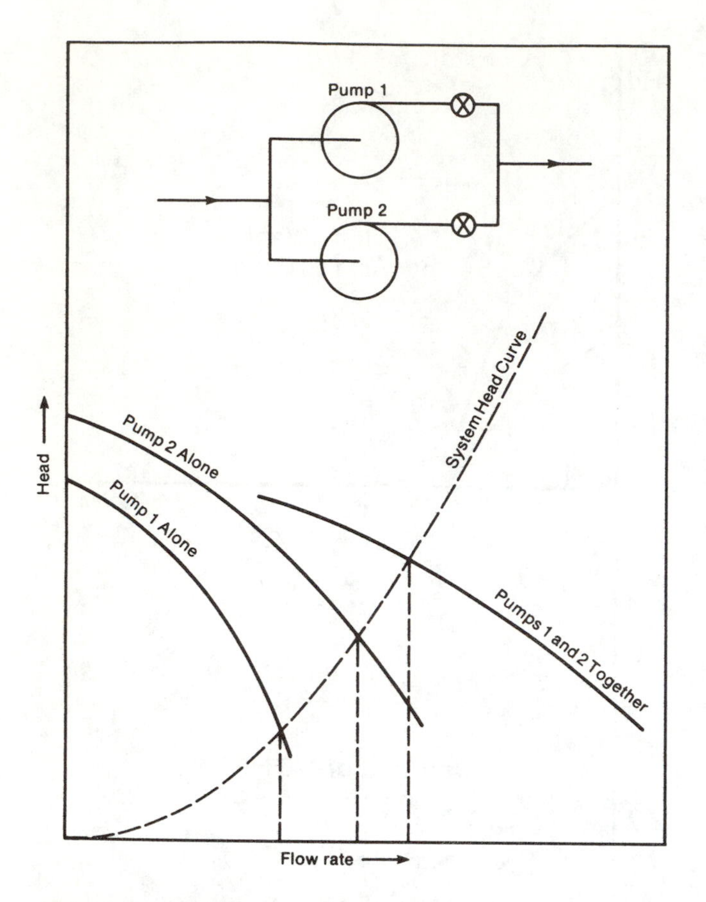

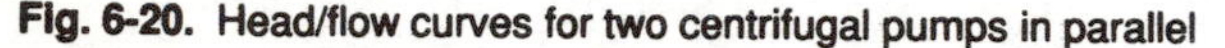

Fig. 6-20. Head/flow curves for two centrifugal pumps in parallel

Submersible pump performance ranges from 3 to 900 gpm (11-3400 L/min) and from 50 to 2000 ft (150 kPa - 5.63 MPa) head. However, submersible pumps are often limited to use in relatively low water temperatures. Consult the manufacturer regarding maximum allowable temperature.

Pump Selection

Pumps should be selected so there is little pressure rise when water flow is throttled. In systems with considerable throttle, a pump should be selected to operate on the flat portion of the "head-versus-flow" curve.

Brake horsepower (Bhp) should not exceed motor horsepower at any point on the pump curve (non-overloading), and maximum shut-off head should not exceed 125% of design head. Overall pump efficiency is the product of motor efficiency and hydraulic efficiency. It can be as low as 10% for small pumps.

$$\text{Bhp} = \frac{\text{lb/min} \times \text{ft head}}{33{,}000\ \text{ft-lb/min-hp} \times \text{pump efficiency}} \qquad (6\text{-}49)$$

For commercial and residential solar applications, centrifugal pumps should be used. (Positive displacement pumps can damage a system if they continue to operate while a system is valved off.) Centrifugal pumps operate only slightly above rated pressure if the fluid loop becomes blocked and provide increased flow rate as fluid temperature

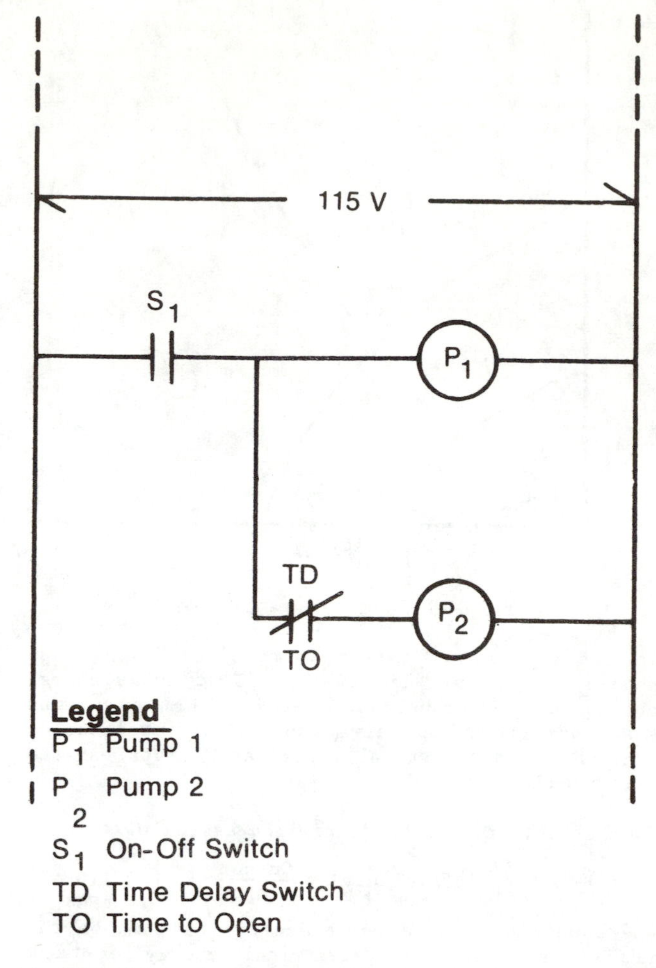

Fig. 6-21. Schematic wiring diagram for a two-pump (fill/run) system

increases. Centrifugal pumps, with built-in overload impedance protection removed, are available for pumping fluids with temperatures as high as 290°F (150°C).

A centrifugal pump always should be selected for calculated pump head, without addition of safety factors, because newly installed pipe has less than design friction. If a pump is selected for the calculated head plus safety factors, it must handle a larger than design fluid quantity, using more energy. When this occurs without provision to throttle or bypass excess flow, the pump motor may become overloaded.

Multiple pumps may be interconnected to the same header, with piping connections as illustrated in Fig. 6-22. This allows each (identical) pump to handle the same fluid quantity. Under part load conditions and at reduced fluid flow or when a pump is off line, the pumps still handle equal fluid quantities.

Since pump head is exactly balanced by system head losses, head losses in piping loops, across collectors, and through heat exchangers must be estimated before selecting pumps. A chart for estimating friction losses in copper tubing is shown in

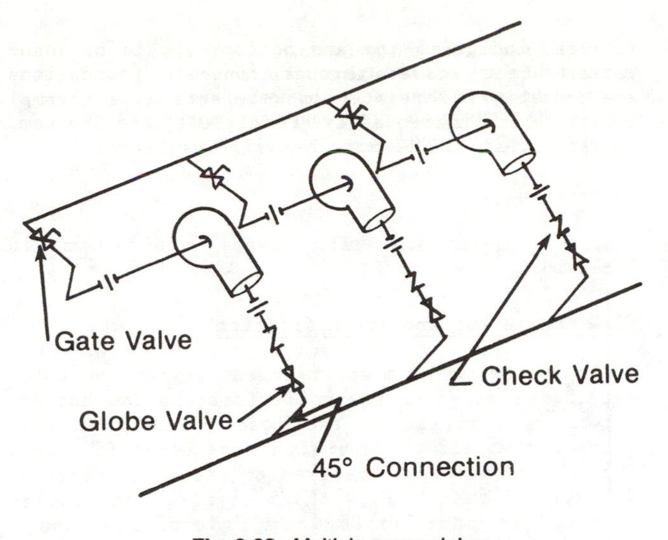

Fig. 6-22. Multiple pump piping

Fig. 6-23. Similar charts are available from suppliers for other piping. A nomograph for estimating head losses across valves and fittings is shown in Fig. 6-24. Collector and heat exchanger head losses are given in manufacturers' catalogs and brochures.

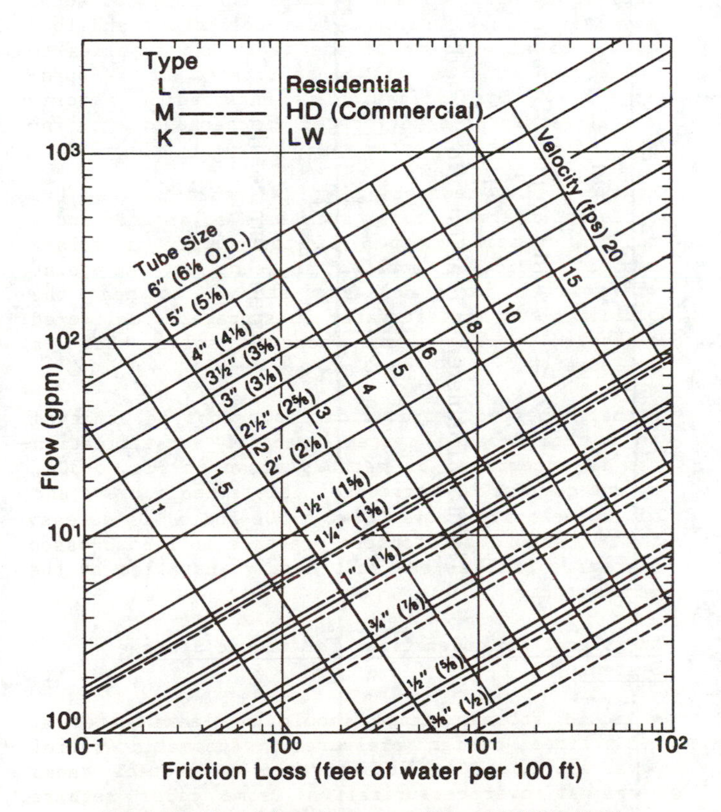

Fig. 6-23. Friction loss in copper tubing (ASHRAE 1985)

To estimate head losses in a loop, a piping layout (schematic) should be made with valves, fittings, filters, and other loop components shown. Pipe sizes should be selected so flow velocities will not exceed about 7 fps (2.1 m/s). With discharges selected and pressure losses calculated for each circulation loop, pump selections can be made from manufacturers' catalogs.

Some additional considerations in pump selection are as follows:

- Closed loops may use a circulator pump with an iron housing; open loop systems must use a pump in which all parts in contact with water are manufactured of stainless steel or bronze. Use of stainless steel circulators has eliminated many rust corrosion problems of bronze-lined pumps used in open and potable systems. Stainless steel pumps now compete with bronze-lined pumps in cost.
- Close-coupled pumps should be avoided to prevent heat transfer from the fluid through the shaft to the motor, which could cause motor overheating.
- Modular pumps for one-third to two-thirds of the load should be chosen when the system permits reduction in fluid volume. Larger pumps may have variable volumes for complete modulation over the entire load range.
- Some pumps are not self-priming and require a minimum 3 ft (0.91 m) of static head on the inlet side at all times. This ensures adequate fluid for bearing lubrication. Failure to provide this static head will result in bearing seizure and pump failure.
- The following should be assured:
  - pumps are constructed of vertically split casing with screwed connections in connector sizes 2 in. (5 cm) or smaller, flanged in sizes 2-1/2 in. (6.4 cm) or larger
  - pressure gauge tappings are in all flanges
  - internal parts can be serviced without disturbing piping
  - shaft is constructed of stainless steel
  - enclosed centrifugal impeller is lubricated
  - mechanical seals are suitable for 225°F (107°C) or maximum operating temperature, whichever is higher
  - coupling is spring type and flexible
  - pump is hydrostatically tested at the factory to 150% of design pressure.

PIPE, FITTINGS, AND INSULATION

A primary source of parasitic loss in a solar energy system is the energy-transport system (i.e., piping, fittings, and pipe insulation). Therefore, great care must be taken to ensure that thermal energy is not lost because piping was incorrectly circuited or insufficiently insulated, that shutdowns do not occur because piping accessories were not carefully selected or properly installed, or because durable materials were not selected. Pipe sizing and design must conform to established standards. Physical characteristics of the piping system designed to ensure against thermal and pumping power losses and to ensure system durability and operating continuity will be discussed here.

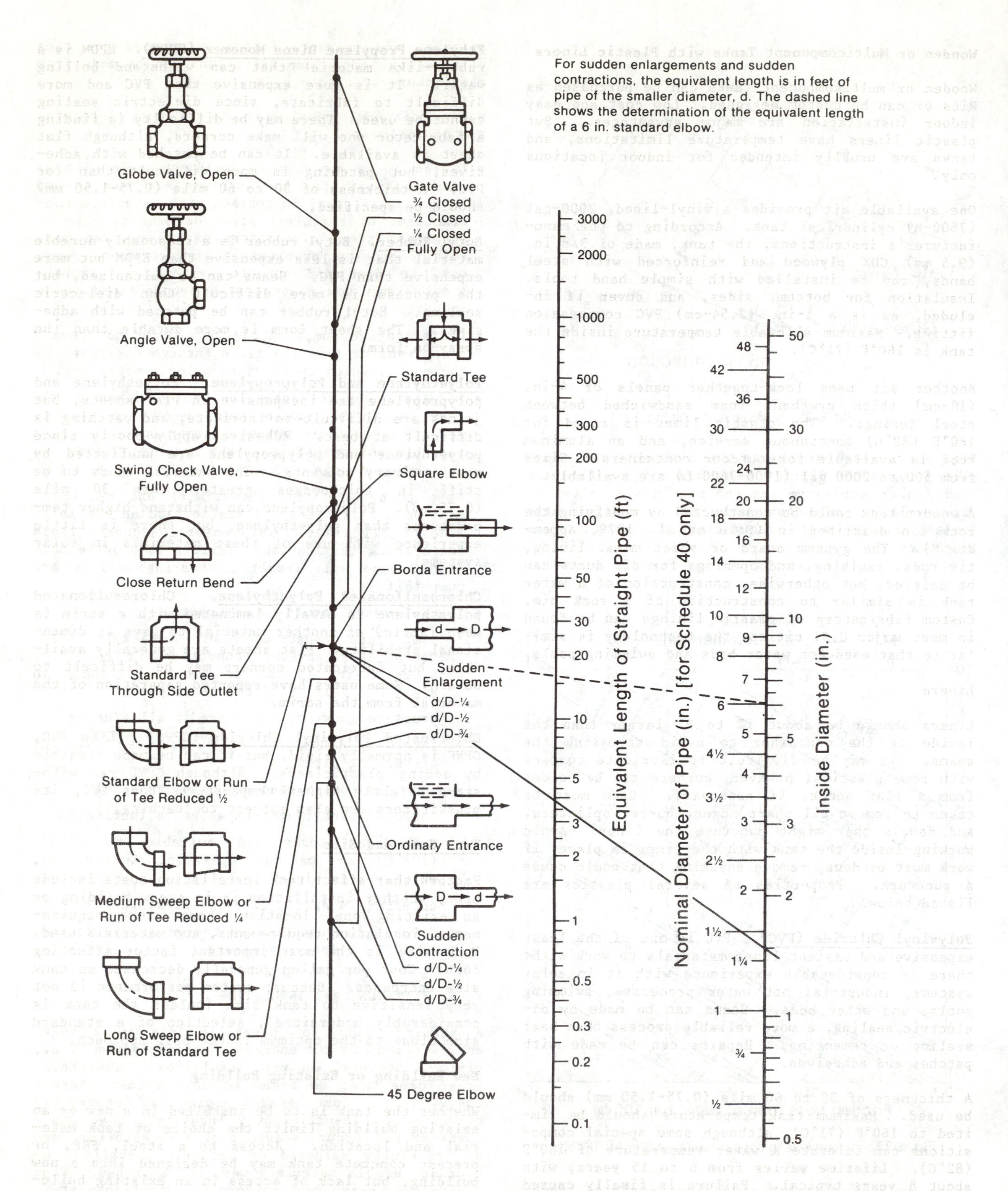

Fig. 6-24. Pressure loss in various elements (DOC 1980)

Piping Layout

Piping and manifolding at solar collectors are important considerations in solar system design. Design should minimize installed cost, leakage and maintenance problems, and heat loss; it should provide uniform flow to all collectors to maximize collector efficiency, minimize pump power requirements, and provide for air purging.

Two basic categories of collector manifolds are external and internal. External manifolds consist of insulated piping, external to the collector, that interconnects all collectors. Internal manifolds use piping internal to the collector as the major interconnecting manifold. Internal manifolds can reduce piping and insulation costs but require careful design to maintain proper flow to all collectors.

Collector piping is also classified as to circuit arrangement. Basic arrangements are shown in Fig. 6-25. These use series or parallel flow and external or internal headers. Arrangement should be carefully considered in terms of system size, type (e.g, drain-down, filled, etc.), and application (heating, cooling, or hot water). Reverse-return arrangements are generally preferred because they tend to promote more uniform flow to all collectors (provided low, uniform resistance is maintained in the manifolds). However, they generally require more piping and, therefore, additional cost. Satisfactory flow distribution will usually be achieved if head loss through each collector is about 70% of total head loss from A to B (Fig. 6-25). Balancing of large collector arrays dictates use of balancing valves. Direct-return is not generally recommended (especially for internal manifold collectors), except for small systems with large, low-resistance manifolds and balancing valves. Series flow can be used with only a limited number (2 or 3) of collectors. Efficiencies for series-connected systems can be equivalent to those for parallel systems, provided flow rates are adequate to provide the same overall Δt. Series systems usually have greater resistance, however, and require larger pumps, with greater parasitic losses. Series combinations can be advantageous for collectors with relatively low flow resistance.

Piping Details

Supply and return manifolding for the collector array should be sized to provide uniform flow through the collectors.

Expansion and contraction of manifolding due to temperature changes has led to many problems in solar installations. The linear expansion of copper tubing is shown in Table 6-10.

Pipe Sizing and Flow Rates

Flow rate through each system loop (collector-storage, storage-distribution, and other subsystems) depends upon thermal load, temperature rise, and fluid used. The piping system must be sized and designed to handle the flow rate required for the collectors. The advantages of smaller pipe, with corresponding cost savings in pipe,

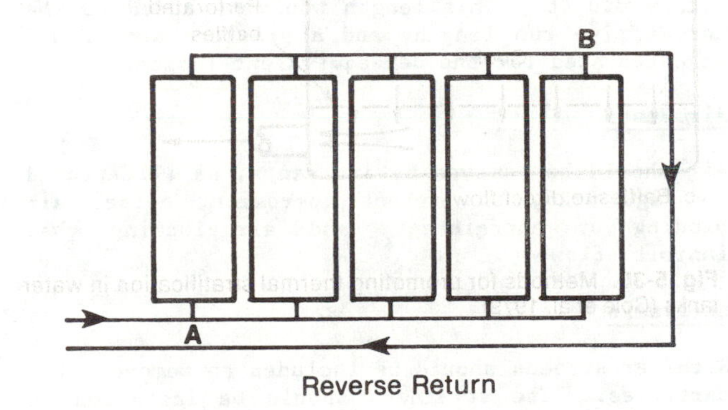

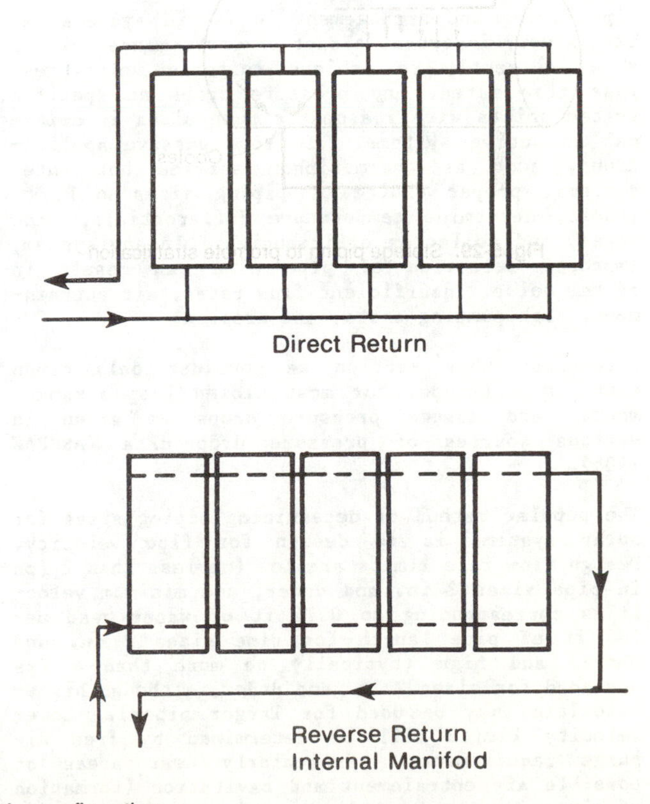

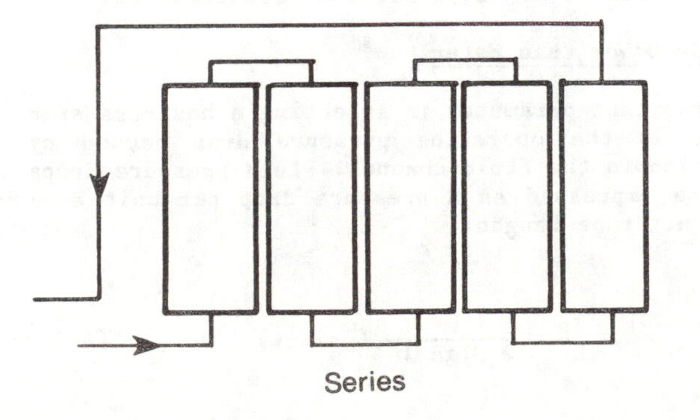

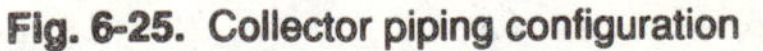
Fig. 6-25. Collector piping configuration

Table 6-10. Linear Expansion of Copper Tubing

Temperature Change (°F)	Expansion (in./100 ft)
0	0
50	0.56
100	1.12
150*	1.69
200	2.27

*Typical value

insulation, and fittings, versus the disadvantages of higher frictional resistance, larger pumps, and greater pump energy consumption must be evaluated on a life-cycle cost basis.

Pipe Friction Loss

Pipe friction loss in a system depends on fluid velocity, pipe diameter, interior surface roughness, and pipe lengths. Varying any one of these factors influences total friction loss in the pipe. Velocities recommended for water piping depend on three factors: (1) pipe use, (2) effects of erosion, and (3) noise resulting from excessive flow velocities.

If the system pressure drop is great enough, fluid pressure may drop to the boiling point and cause unstable water circulation and possible pump cavitation. If system pressure drops below atmospheric, air drawn in at vents and pump seals can collect in pockets and stop water circulation.

Pipe sizing and arrangement in liquid systems can have a very important effect on performance. Piping sized correctly to achieve suitable velocities, mass flow rates, and pressure drops at specific system points with reasonable pump sizes is critical in active systems. In some passive applications, such as thermosiphon service hot water systems, proper choice of piping sizes will, by itself, determine temperature differentials, flow rates, and collector efficiencies. In any system, improper selection of pipe size can result in system noise, insufficient flow rates, air entrainment, high pumping costs, and erosion.

Throughout this section we consider only clean metal pipe (proper for most closed loop arrangements) and assume pressure drops as given in various sources of pressure drop data (ASHRAE 1985).

The popular method of determining piping sizes for solar systems is to design for flow velocity. Design flow rate limits are low (no less than 2 fps in pipe sizes 2 in. and under, and minimum velocities corresponding to 0.75 ft of water head per 100 ft of pipe length for pipe sizes 2 in. and above) and high (typically no more than 4 fps designed for sizes 2 in. and under, although higher velocities may be used for larger pipes). Lower velocity limits will be determined by free air purge requirements, particularly near areas of possible air entrainment and cavitation (formation of air pockets and bubbles) and lower pressures (at system high points and in downcoming return lines). Upper limits are imposed to reduce erosion-corrosion and possible noise problems. Velocity noise in most systems is caused by free air, not water, in areas of sharp pressure drops and low static pressures. Where hydronic systems are carefully designed to maintain high static pressures and to remove entrapped air, flow velocities may be increased.

When flow velocities and mass transfer rates have been determined, piping friction losses, needed for selection of appropriate pump size and capacity, can be calculated.

EXAMPLE

Assume 16 gpm flowing in the collector loop and Type L (Residential) pipe.

Tube Size (in.)	Δp (ft H_2O/100 ft)	V (fps)
1.0	17	6.5
1.5	2.8	3
2.0	0.66	1.7

It is also important to calculate head loss in fittings and piping; data for this purpose are shown in Fig. 6-23.

Pressure drop in fittings is calculated from the equivalent length of the item. For example, refer to Fig. 6-24. For a 1-1/2 in. diameter pipe elbow, the equivalent length of pipe would be approximately 3.9 ft. This length would be added to the normal pipe run length, and a pressure drop could be calculated for the net equivalent length.

Air Vents

Air should be removed by air vents, as illustrated in Fig. 6-26, to prevent corrosion, noise, air binding or entrainment, and air locking that inhibits flow.

Strainers

Strainer screens should be included to remove solid particles. The strainers should be installed at pump inlets; a "Y" type strainer is preferred.

Fluids Other than Water

An important parameter in selecting a heat-transfer fluid is the operating pressure drop caused by friction in the fluid channel. Fluid pressure drops can be expressed as a pressure drop per unit area per unit tube length:

$$\frac{\Delta p}{L} = \frac{f\,\dot{m}^2}{2\,D_i g \rho\,144} \text{(psi/ft)} \qquad (6\text{-}50)$$

where

f = friction factor

= 16/Re, for Re < 2500 (laminar flow)

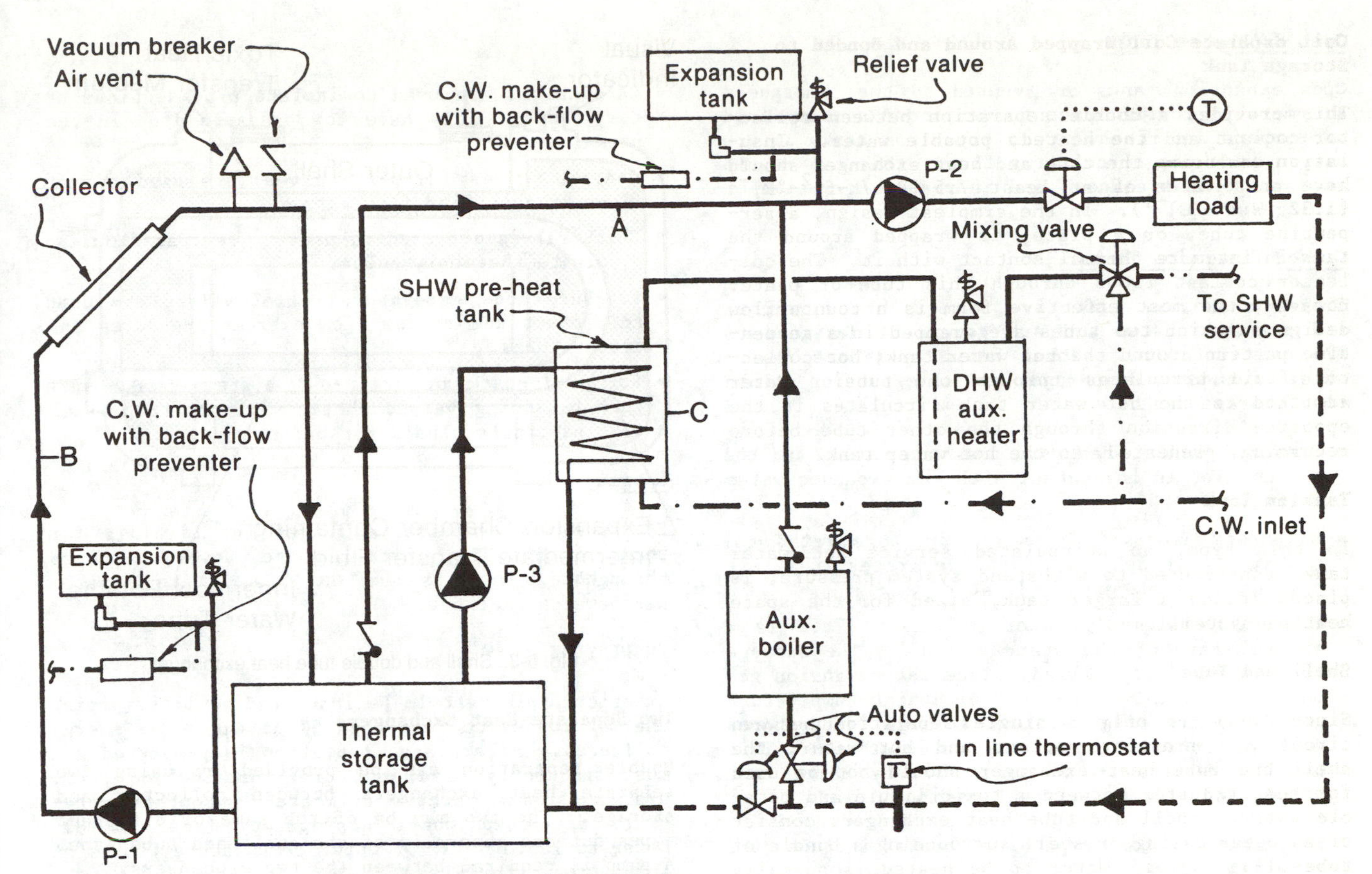

Fig. 6-26. Solar system flow schematic

$= 0.0014 = 0.125/Re^{0.32}$, for $Re > 2500$ (turbulent flow in smooth-walled tubes)

Re = Reynolds number (see previous discussion on heat exchangers)

$= \dot{m}D_i/\mu$

μ = fluid viscosity (lb/ft-h)

$\dot{m}$ = mass velocity through tube (lb/ft^2-h)

D_i = tube inside diameter (ft)

g = acceleration of gravity = 4.18×10^8 ft/h^2

ρ = density of fluid (lb/ft^3)

Equation (6-50) is applicable to collector, heat exchanger, and traced tank tubes but it neglects entrance and exit effects. It can be reduced to the Darcy equation in the form

$$\frac{\Delta p}{L} = \frac{0.216f\ \dot{V}_1{}^2}{D_i{}^5}$$

where

$\dot{V}_1$ = flow rate (gal/min)

D_i = tube inside diameter (in.)

Tube size greatly affects pressure drop within the tube. Some fluids require larger tube sizes than water does because they have higher pressure drops.

EXPANSION TANKS

Although expansion (or compression) tank sizing is considered a relatively straightforward procedure, solar system design has brought a demand for minimized cost, reduced maintenance and energy use, and systems of great complexity. Solar systems require extra consideration because they generally have large storage volumes that can result in large, expensive expansion tanks. Also, system operating temperatures vary widely, including high stagnation temperatures, and require careful attention to system pressure control.

An expansion tank limits all system pressures to the allowable working pressure. It maintains minimum pressure at all operating temperatures to provide positive air venting and to prevent flashing and pump cavitation.

Open Expansion Tanks

Open expansion tanks are vented to the atmosphere and merely provide space for expanded system fluid as temperature rises. Open expansion tanks absorb oxygen into the circulating fluid; this can result in antifreeze breakdown and increased corrosion of piping systems.

Closed Expansion Tanks

Closed expansion tanks are watertight, airtight, and pressure rated and are generally designed to 125 psi (860 kPa). Three basic types are available: standard, diaphragm, and an expansion volume in a storage tank.

- Standard Tank. These are pressure rated, and the gas cushion is in contact with the expanded water volume.
- Diaphragm Tanks. These are pressure rated and contain a diaphragm that separates the air cushion from the expanded water.
- Expansion Volume in Storage Tank. This is a special case of the standard tank. The storage tank is partially filled. Since the expansion gas volume is in direct contact with high-temperature water, increased gas volume at high temperatures must be included in sizing calculations. This usually results in an unnecessarily large tank volume and is generally not recommended.

Considerations for expansion tanks include the following:

- Provide for at least 3 to 5 psia (21-34 kPa) at the top of the system for positive venting at initial fill.
- Allow a safety margin of 15°F (8°C) above maximum operating temperature for flash protection.
- Calculate maximum operating pressures at system components with pump in operation (i.e., add or subtract pump pressures from maximum allowable pressures).
- Connect expansion tank to suction side of pump.
- Connect standard tanks to points of highest temperature and lowest pressure.
- Do not install shutoff valves in connection lines.
- Do not insulate.
- To reduce tank size:
  - Increase maximum allowable pressure
  - Reduce initial pressures (i.e., elevate the tank)
  - Mechanically pressurize standard tanks at initial fill
  - Be certain expansion tank is sized to allow for thermal expansion of the fluid used over the widest temperature range expected during system operation.
- Raise initial fill temperature before sealing.
- Verify pump suction pressure at maximum operating temperature.

Valves

- Gate valves are used to isolate pipe sections or components. They have low pressure drop in the normal mode.
- Control valves (circuit valves) are used to set flows at specified flow rates.
- Check valves are used to prevent reverse flow and eliminate thermosiphoning.
- Pressure and temperature relief valves are used to protect the system from extreme pressures and temperatures.
- For drain-back or drain-out systems, make sure that balancing valves chosen will not trap water when partially closed for balancing.

Gauges

Pressure and temperature gauges (or fittings for them) should be installed at several points throughout the system to facilitate system performance checkout.

INSULATION

Exterior piping should be insulated to limit piping thermal losses to less than 5% of the solar energy collected. Fiberglass insulation is preferred and most frequently used, but urea-formaldehyde foam, urethane foam, calcium silicate, and other materials are used where appropriate (see Ch. 8). Refer to Table 6-11 for upper temperature limits of insulation materials.

Based on fiberglass, all pipe insulation should have a K value (thermal conductivity) below 0.23 Btu/h-ft^2-°F/in. at a mean temperature of 75°F, in accordance with the following schedule for thickness (Table 6-12).

Composite insulation, jacket, and adhesive should bear the Underwriter's label, have flame spread of 25 or less, and smoke-developed rating of 50 or less, conforming to ASTM-E-84 and NFPA 225.

All exterior glycol solution supply and return piping (other than buried) should be wrapped, after insulation, with two layers of 30 lb roofing felt and finished with weatherproof plastic reinforced with glass membrane.

Fittings, valves, strainers, etc., should be insulated with a flexible fiberglass blanket (2-4 lb density) to a thickness equal to adjoining pipe insulation and finished with a smoothing coat of insulating and finishing cement.

Underground piping should have sufficiently sealed closed-cell insulation to render it permanently waterproof. It should be protected against undue compaction. Backfilling with sand and keeping pipes below the frost line and away from rocks and construction debris are standard practice.

CONTROLS

An automatic temperature control system is the heart of any system, solar or conventional. All

Table 6-11. Upper Temperature Limits for Insulation Materials for Collectors, Tanks, Pipe, and Fittings

Material	Density (lb/ft^3)	Thermal Conductivity at 200°F ($Btu/h-ft^2-°F/in.$)	Temperature Limits (°F)
Fiberglass with organic binder	0.6	0.41	350
	1.0	0.35	350
	1.5	0.31	350
	3.0	0.30	350
Fiberglass with low binder	1.5	0.31	850
Ceramic fiber blanket	3.0	0.4 at 400°F	2300
Mineral fiber blanket	10.0	0.31	1200
Calcium silicate	13.0	0.38	1200
Urea-formaldehyde foam	0.7	0.20 at 75°F	210
Urethane foam	2-4	0.20	250-400

Table 6-12. Recommended Pipe Insulation Thickness for 4 lb/ft^3 Foam Insulation and Pipe Fluid at 175°F

Nominal Pipe Size Diameter (in.)	Insulation Thickness (in.)
0.5	0.5
0.75	0.5
1.25	0.5
1.50	1.0
2.0	0.5
2.5	0.5
3.0	1.0

functions--"run-stop," "off-on," "open-close," "modulate"--depend on the type of controls used and the control sequences adopted. For large buildings, a central computerized energy management and control system (EMCS) can be integrated with the solar control system.

Control and system operation logic must be determined first. The objective may be any one or all of the following: (1) to collect and use as much solar energy as possible; (2) to reduce operating time and wear and tear on active components and auxiliary energy supply equipment; (3) to reduce electric peak demand; (4) to minimize auxiliary energy use in space heating and/or cooling and/or service water heating; (5) to provide predictable and continuously satisfactory internal environmental temperature and humidity conditions; (6) to assure continuity of operation, minimizing unscheduled shutdown; (7) to increase equipment lifetime; and (8) to minimize initial installation costs. All objectives may not be, and usually are not, mutually compatible.

__Control Strategies__

Select desired operation modes in a logical order of occurrence, including the following:

1. Heat from collector to storage
2. Heat from collector directly to load (heating and cooling)
3. Heat from storage to load
4. Heat from storage (or direct) to evaporator in heat pump operation
5. Preheat service hot water with solar collectors directly, or from solar-heated storage tank
6. Heat from auxiliary for service hot water and/or space heating, and/or absorption or Rankine cycle cooling
7. A combination of modes 1 through 6, inclusive.

With different operating modes, the controls switch the system from one mode to another. The selection of operating mode is based on temperature differences between collector and storage and between storage and load, and on whether solar heat can be used or auxiliary heat is necessary.

An automatic control system consists of some or all of the following devices:

- Controller. This measures a controlled variable, such as temperature or humidity, with a sensing element and compares the value with an input signal to produce a suitable action or impulse for transmission to the controlled devices. Thermostats, humidistats, and pressure controllers are examples.
  - Space-heating thermostat. This is designed to control the introduction of heat to occupied spaces, and it often has two stages. The first operates the system blower or pump to circulate hot fluid from storage to load; the second operates the auxiliary energy source.
  - A differential controller should be used to control operation of the heat-transfer fluid pump or blower. The collector pump should operate only when useful collected energy exceeds a minimum value. In practice, this is done by comparing fluid temperatures in the collector exit header and in the collector exit port of the storage tank. (See Fig. 6-27.) A pump or blower operates only

when this temperature difference exceeds a set value, usually 10°F (6°C) in liquid systems and 15° to 20°F (8°-11°C) in air systems; a shutdown difference of about 3° to 5°F (2°-3°C) is commonly used.

If a collector is drained when not in use, the control should allow slight collector overheating before startup and some thermal response lag in the collector sensor to avoid premature shutdown when the first cool water enters the collector.

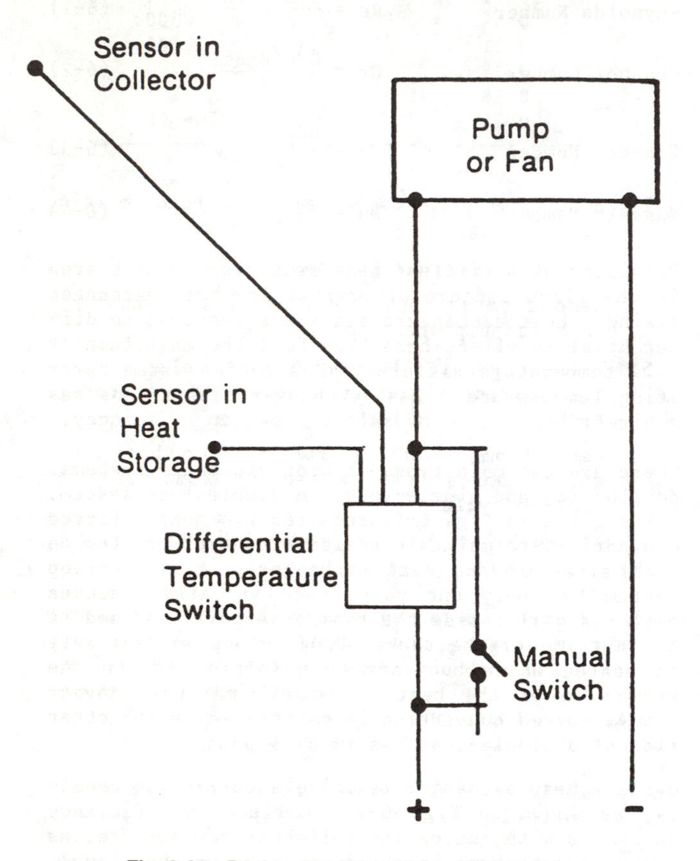

Fig. 6-27. Basic solar differential control system

- A Controlled Device. This has two components, an operator and a final control element.
- Safety Device. In addition to a controller and a controlled device, every control system should contain one or more safety devices that stop action, or initiate a new action, whenever a dangerous condition develops.
- Controlled Variable. If temperature, humidity, or pressure is being controlled, it is the controlled variable.

Controls for Solar Space Heating

A solar heating system consisting of collectors, storage system, and auxiliary heater can be controlled by some or all of the following:

- Space Thermostat. The space thermostat may have two or more stages for heating and an additional two or more for cooling (if applicable). Two-stage action in a heating mode operates as follows: when the first stage demands heat, the solar system is called upon to satisfy demand. If building heat loss is greater than the solar system can replace, and if space temperature continues to drop, the second stage will activate the auxiliary heating system. The auxiliary system should meet the entire space heating requirement by itself, or with help from the solar system, until space temperature rises to the lower temperature limit of the first stage, where auxiliary energy is cut off.
- Temperature Sensors measure air or liquid temperature at the collector discharge side, thermal storage temperature, and service hot water storage tank temperature. Sensing elements commonly used include thermocouples, platinum resistance temperature detectors (RTD), and thermistors. Only sensors provided by the controller manufacturer should be used.

A central control panel should be provided to consolidate circuits, fuses, and relays that provide control functions.

Fail-Safe Controls

A control subsystem should be designed so temperatures and/or pressures developed in the solar system because of a power or system component failure will not damage system components or the building, or endanger occupants. For example, high-temperature limiting devices should be installed to protect the system from damage from rupture, decompression, etc.; a low-temperature limiting device should be installed to protect the system against damage from freezing. If flammable or combustible fluids are involved, an alarm should be installed to indicate failure of the fluid transfer system or the fact that relief valves have been activated. Where a light sensor (i.e., solar cell) is used to initiate circulation of antifreeze to a heat exchanger, low ambient temperature operation can cause heat exchanger freezing. Where light sensors are used, a low limit cutout should be installed in series to prevent circulation of below -35°F (2°C) antifreeze through heat exchangers containing water. If snow can cover the light sensor, backup temperature sensors are needed in the return piping.

Control Sequence Example for Solar Space Heating

- If building temperature is satisfactory, and collector temperature is at least 10°F (6°C) above the storage tank temperature, the loop pump is activated, and control valves are positioned to deliver heat from collector to storage.
- If room temperature drops below 68°F (20°C) when occupied, or 58°F (15°C) when unoccupied, and storage temperature is at least several degrees above room temperature--the differential is determined by system configuration--valves are positioned and the load circulator is activated to deliver heat from storage to load.
- If the space requires heat but the solar tank temperature is below the useful temperature,

valves are positioned to shut off the supply pipe from storage to load, and auxiliary heat is delivered to the space. In some cases, storage tank temperature may be too low to supply useful heat, but the collector can supply heat directly by proper valving.

- For cooling, heat is never delivered directly from the collector to the absorption generator or Rankine cycle engine because the surge due to collection temperature variations is undesirable; solar heat is added through the storage tank.

Controller Selection

A controller primarily controls the action of fluid movers (pumps or fans) in a solar heating or cooling system. In a collector loop the pump or fan should be turned on to circulate fluid through the collector whenever it is advantageous to collect energy. Fluid movers in the distribution loop should be turned on whenever the building thermostat calls for heating or cooling. Kent, Winn, and Huston (1981) discuss controls for service hot water systems.

A typical controller has three component subsystems: a sensor subsystem, comparators, and output devices. Sensors sense temperatures at various system points, such as in the thermal storage unit, in the building, and at the collector outlet. A comparator subsystem compares temperature differences between various points with set-point temperatures to determine whether a fluid mover should be turned on. An output subsystem sends a signal to a fluid mover to control it.

A differential controller controls fluid movers and valves in almost all active solar systems. This device is popular because of its energy management capability and functional simplicity. It produces a typical daily system temperature history, as illustrated in Fig. 6-28. A fluid mover (pump or blower) is controlled by comparing temperature-dependent signals from collector and storage sensors to initiate flow whenever collector temperature is high enough to economically transfer energy to storage, i.e., it switches power "on" when $t_{coll} - t_{storage} > \Delta t_{ON}$. A "dead band," or hysteresis, is required to minimize on-off cycling. Such cycling results from thermal transients that exceed a system's thermal time constant, e.g., rapid collector cool-down when flow begins with low-temperature storage water.

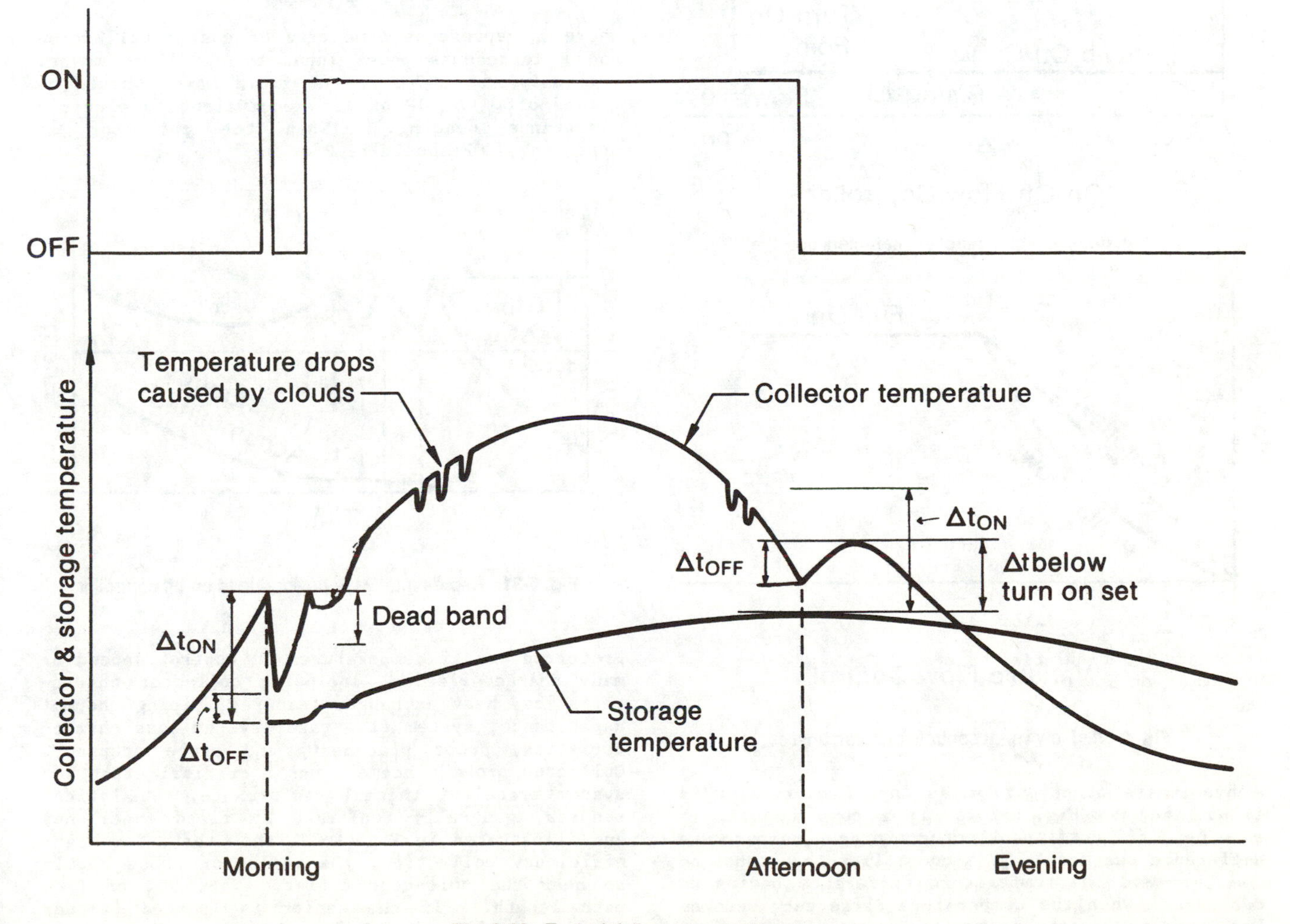

Fig. 6-28. Typical daily system temperature

Several firms offer differential thermostats for sale to the solar industry. Although there are variations in design, all are functionally similar and have collector and storage sensors, low-voltage, solid-state control logic, line voltage input, and logically switched line voltage outputs.

Flow Control Techniques

The solar industry uses two types of differential controller outputs. The first provides make-or-break power switching to operate a collector pump either full on or off. The second modulates power supplied to a pump during "on" time to vary pump flow rate as a function of collector-to-storage temperature difference. Flow characteristics of these operations are illustrated in Figs. 6-29 and 6-30.

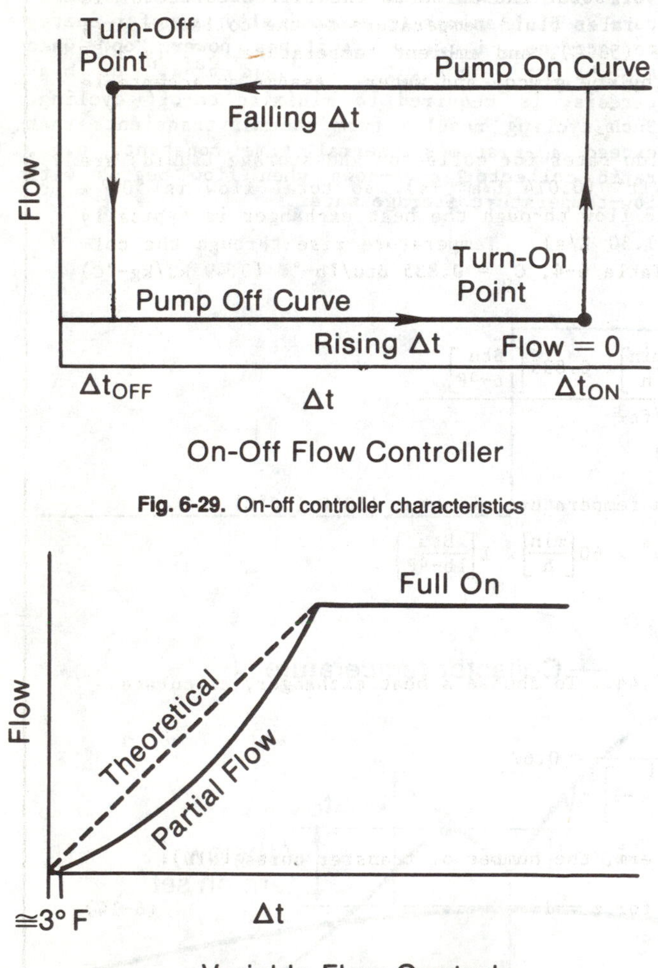

Fig. 6-29. On-off controller characteristics

Fig. 6-30. Flow vs. Δt controller characteristics

A hysteresis function for an on-off controller is illustrated in Fig. 6-29. If a pump is off, it remains off until collector-storage temperature difference reaches Δt_{ON}. A controller then turns on the pump and the temperature difference begins to decrease. When the temperature difference becomes less than Δt_{OFF}, the controller turns off the pump.

Flow characteristics of a variable flow rate pump are illustrated in Fig. 6-30. The dashed line indicates a linear theoretical flow rate as a function of temperature difference, Δt. However, actual flow rate is a nonlinear function of Δt, as illustrated by the solid line.

On-Off Flow

The most widely used pump and control configuration is on-off operation of a fixed flow rate collector pump. Make or break power switching implemented within a controller is familiar to installers, electricians, etc. On-off control is used to switch any motor with appropriate voltage and current ratings. Manufacturers use both electromechanical and solid-state relays in these controllers.

As shown in Fig. 6-31, choice of on and off temperature difference is critical if excessive pump cycling is to be avoided without significant reduction in system collection duty cycle and coefficient of performance (COP). COP is defined as the ratio of energy collected per unit time by collector fluid to external power delivered to the pump or fan motor. That is,

$$COP = \frac{\dot{Q}_u \text{ (kW)}}{P \text{(kW)}} \tag{6-51}$$

where $\dot{Q}_u$ represents time rate of energy collection and P represents power input to the fluid mover. Obviously, if COP ≤ 1, the fluid mover should be turned off. A COP of 1 is equivalent to electric resistance heating. When the collector is operating, COP should be 2 or 3.

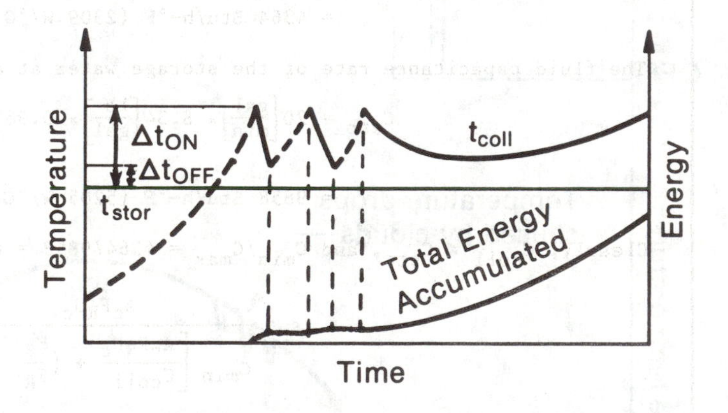

Fig. 6-31. Representative behavior due to on-off controller

Preferred on-off temperatures in control depend on many characteristics, including collector characteristics, heat exchanger features, piping thermal capacitance, system flow rate, system loss characteristics, probe placement, and probe accuracy. Collector probe placement has a critical effect on system transient thermal performance. Analytical results, generally confirmed by field experience and illustrated in Fig. 6-32, show that, for high-efficiency collection, the collector probe should be near the collector outlet, i.e., 90% of flow path length. If the sensor is located farther downstream, it turns the fluid mover on later. If the sensor is located too far upstream, effects of either boundary layer heat losses and/or natural convection are exaggerated and excessive cycling occurs. A location farther downstream may be preferable for a single collector probe system, if located in a high solar radiation area with much radiation to the night sky.

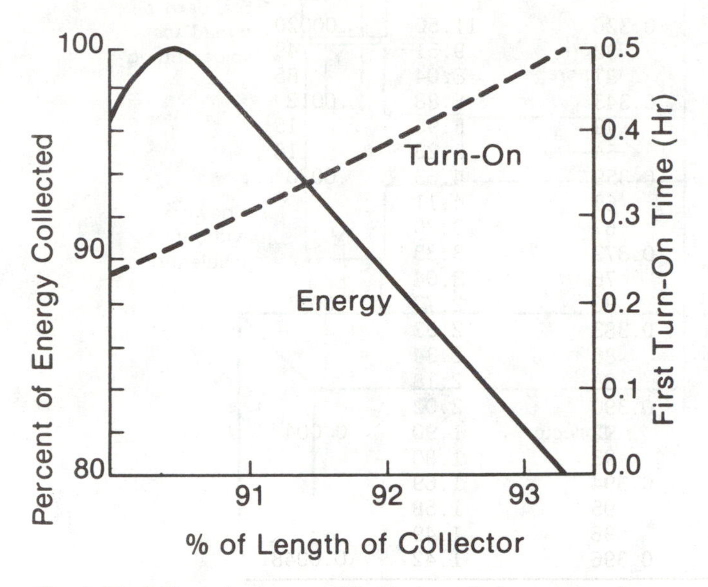

Fig. 6-32. Collected energy and first turn-on time vs. sensor location

The dead band between "off" and "on" temperature settings minimizes pump cycling during morning start-up and afternoon shut-down or intermittent cloud conditions. An "off" setting must be high enough to accommodate normal sensor inaccuracies but low enough to allow collection of as much solar energy as possible. It must ensure that collection stops when pump parasitic power exceeds solar energy gain. If an on-to-off temperature ratio of from 4:1 to 6:1 is maintained, with Δt_{OFF} between 2° to 4°F, most flat-plate liquid transport systems operate satisfactorily. Higher Δt_{ON}-to-Δt_{OFF} ratios apply when Δt_{OFF} is low. That is, if Δt_{OFF} is near 2°F, Δt_{ON} should be approximately 12°F. Similarly, lower ratio values apply when Δt_{OFF} is high. Thus, hysteresis in liquid systems is normally 12° to 16°F, with some systems as high as 20°F.

Air transport systems normally require a Δt_{OFF} of 10° to 20°F with a 15° to 25°F hysteresis. This slightly wider hysteresis range is needed if a COP > 1 is to be maintained to compensate for effects of high parasitic energy costs in air systems, relatively low specific heat of air, and relatively low blower efficiency (as compared to pumps). Δt_{OFF}, for COP = 1, is proportional to collector loop pressure drop and inversely proportional to fluid mover (fan or pump) efficiency and transport fluid specific heat. Because of air's low specific heat, an "on" temperature twice the "off" temperature is usually adequate to reduce cycling.

Poorly built controllers typically have been a major cause of system failure. Studies at SERI have indicated that, in many cases this can be caused by sensor (thermistor) drift. Because the Δt_{OFF} is only 2° to 4°F, sensor drift can actually lead to a negative value for Δt_{OFF} (i.e., the pump will not shut off until the collector outlet temperature falls below storage temperature). Since the pump supplies warm storage water to the collector, this can result in the pump's running all night.

Location of Control Sensors

The collector temperature sensor should be factory installed and on the absorber plate near the flow path exit. Plate mounting is preferable to mounting the sensor on the pipe close to the collector exit. The reason is that the latter will not directly record collector fluid temperature until after the fluid has begun to circulate. (There will be an indirect pickup because of radiation and conduction.) However, if the collector is not instrumented, the sensor preferably should be inserted through the pipe at the collector exit and into the heat-transfer fluid. Sensor probes should not protrude into collector piping so as to measurably impede heat-transer fluid flow. If the sensor has no "T" joint or similar pressure fitting to insert into the piping system, it may be attached directly to the absorber plate or to the first 3 in. of collector outlet pipe at the top of the collector (see Fig. 6-33); it should be well insulated from outside ambient temperature changes. In closed loop systems using high-viscosity oil, probe threads should be sealed with Teflon® tape.

A sensor can be attached to the absorber plate with a clip or bolt and thermal cement but should not touch collector box walls. Do not pierce the flow channels on the absorber plate. A sensor can be attached to the collector outlet pipe with a hose clamp or pressure clip if it cannot be installed in any other position. During installation, do not bend sensors, and do not solder them to the absorber plate or storage tank, since extremely high temperatures affect their performance.

Be sure good thermal contact exists between sensor and underlying metal. Thermal cement that can withstand 350°F (160°C) can be applied between sensor and underlying metal to ensure good heat transfer, but thermal cement alone should not be relied on to maintain a good mechanical attachment. Protection of the sensor from corrosion, which could result from condensation in a cold collector, should be considered. Sensors installed on external pipes should be adequately insulated so they are not affected by ambient air temperature.

Mount a storage tank temperature sensor as near the tank bottom (coldest point) as possible but not on the bottom. In baffled, horizontal tanks, locate the sensor in the cold end. The storage sensor should not be located too near the point where solar-heated liquid enters the tank. If it is, premature pump shutoff and cycling can occur. Some sensors can be mounted with thermal epoxy on an outside metal surface. Do not drill into a lined

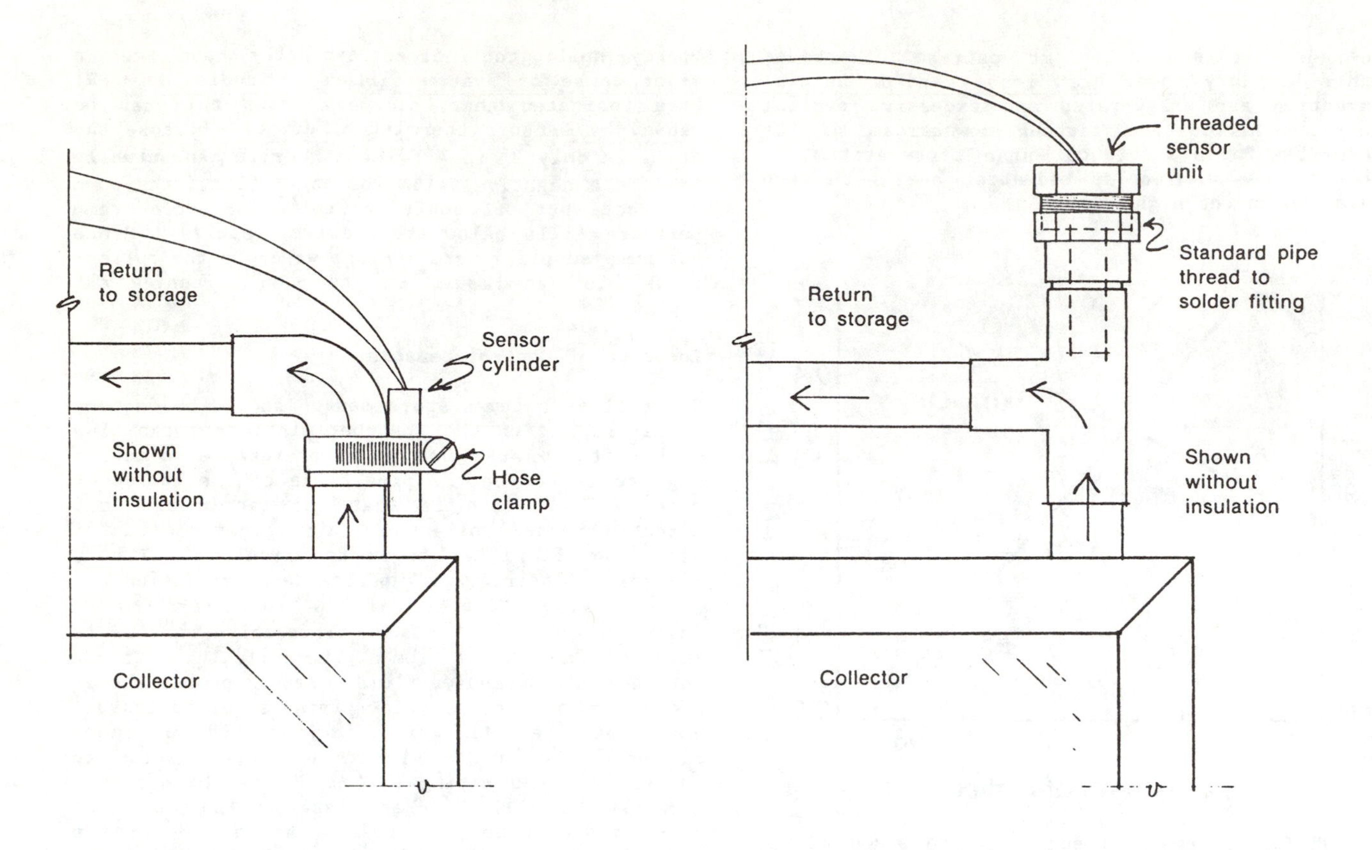

Fig. 6-33. Method of installing two different types of sensors at the collector outlet

storage tank wall since the tank must remain watertight. Sensors of almost any length desired are available in threaded, pre-assembled, sealed units. Even the best seals show periodic condensation, so try to mount the probe at a slight angle. If the probe is mounted vertically, it is advantageous to fill (if possible) the probe housing with a mineral or similar heavy oil so that condensed water cannot reach the sensor.

Storage sensors can be installed directly on or into the piping containing heat-transfer fluid between storage and collector (see Fig. 6-34). It is best to place a sensor very near, or into, the storage tank. If a collector temperature sensor is immersed in the heat-transfer fluid, the storage tank sensor should be immersed as well. Conversely, if a collector sensor is "clamped on," then the storage sensor should also be clamped on to keep the temperature bias of the sensors consistent. The bias is built into the controller to implement the Δt_{ON} and Δt_{OFF} set points.

System Alarms

High- and low-level alarm switches should be provided on expansion tanks. They should be of float-actuated, packless construction with siphon bellows. Electrode, two-probe type, liquid-level alarm switches should be provided for water storage tanks.

Flow switches with electrical contacts should be provided to actuate alarms for all pumped circuits. They should signal alarms only when the pump is electrically energized but no flow results.

Guidelines

For final design, verify that

- Control strategy matches desired modes of operation
- Collector loop controls recognize solar input, collector temperature, and storage temperature
- Controls allow collector loop and load loop to operate independently
- Control sequences are reversible and will always revert to the most economical mode
- All controls are "fail safe."

Accessories

The following additional design considerations should be implemented, when appropriate:

- Solar collector connections suitable for specific antifreeze coolant.
- Vacuum breakers of float or disc type with shutoff valve. Collectors, pipes, tanks, and heat exchangers should be designed with vacuum relief valves for protection against possible collapse.

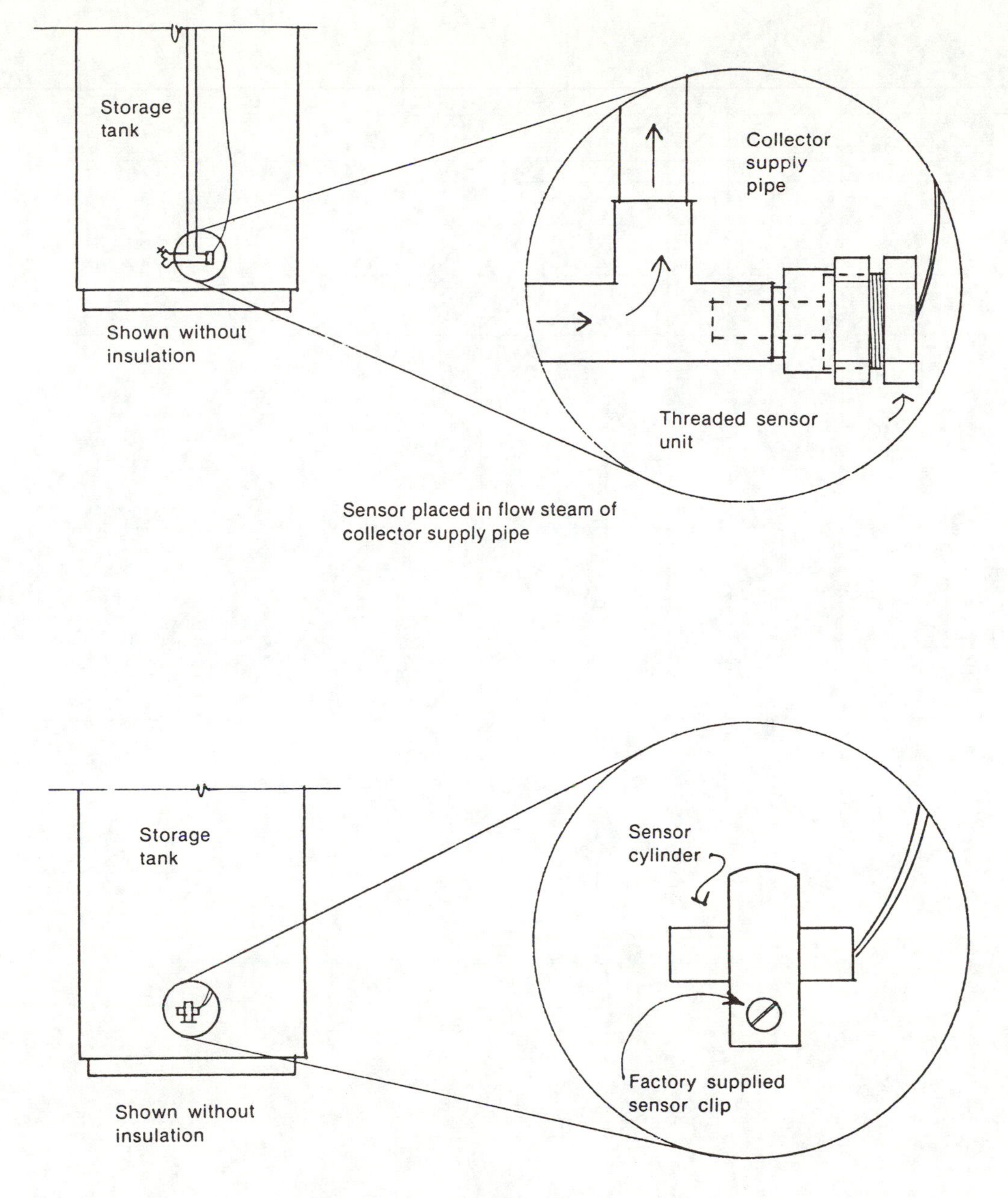

Fig. 6-34. Methods of installing two different types of sensor units for sensing the storage tank temperature

- Automatic pressure relief devices, adequately sized and installed in any energy transport subsystem containing pressurized fluids. Each system section capable of being heated should have a pressure relief device so no section can be valved off or otherwise isolated from a relief device.

- Antifreeze loss indicator in antifreeze loops. Can be pressure gauges or sight tubes in conjunction with colored antifreeze.

- Line strainers at pump inlets, control valves, etc. Strainers for pump protection should be bronze and no less than 40 mesh. A control valve needs greater protection than a pump, requiring a finer mesh strainer.

- Air vents at high points of all mains and risers. Vents in solar and heating coil piping should be suitable for antifreeze when used.

- Thermometers and gauges to monitor water temperature as it enters and leaves collectors, storage, and heat exchangers are usually considered to be essential; monitoring pump suction and discharge pressures is also essential.

Chapter 7

Active System Design and Sizing Methods

INTRODUCTION

Analysis and design methods for active solar heating, cooling, service hot water, and industrial process heat systems will be described. Included are manual methods, hand-held calculator and microcomputer methods, and detailed computer simulations. Methods that have been tested against actual installed solar systems are emphasized. In using any of the energy prediction methods, it is important to keep systems analysis in a proper perspective. Because collector area has a broad optimum and because auxiliary is always sized to supply 100% of the load, a precise prediction of energy delivery is not necessary. In fact, variables such as the weather make such predictions very difficult. No amount of analysis can take the place of proper system design and installation.

Several purposes are served by active solar system analysis and design methods. First, appropriate collector array size can be determined, based on the life-cycle cost of the solar system as compared to fuel cost savings during the system's lifetime. These fuel cost savings are reflected in system performance in terms of the energy fraction met by solar (the solar contribution), which is determined by analysis using a design method. Furthermore, a space conditioning system supplies heat (or cooling) energy while conserving the heat (or cooling) energy supplied. Since an economical system design will be based on a balance of these two aspects, design methods must quantitatively balance energy supply costs and conservation costs. This will lead to lower life-cycle costs than would result from supply systems designed to meet a fixed load (Barley 1979).

Second, once collector size is determined, other system components (e.g., storage, heat exchangers) and the system configuration and control strategy can be selected through performance or cost-optimization studies. Such studies are done by analyzing dynamic system performance, usually using computer programs. Integrated energy quantities, such as useful solar energy supplied to load and auxiliary energy needed, can be computed over long time periods (usually a year). One can also determine when these energy flows occur. Results of such evaluations can be generalized (correlated) for generic system types through parametric studies that examine system performance sensitivities to a set of design parameters (e.g., collector tilt, storage size, control strategy). For new or unusual systems, or for unusual applications, detailed simulation studies should be carried out during the design process. However, for most applications, generalized correlations of previous analyses, contained in graphs or tables, can be used to make most component sizing decisions.

Third, analysis and design methods are used to determine the economic feasibility of a proposed solar design using life-cycle cost analysis. Results of such studies cannot readily be generalized because they are specific to local weather, system design, and time-dependent system and fuel costs.

Subcomponents such as pumps, blowers, heat exchangers, valves, and controls are usually selected after the overall system is sized by using accepted rules of thumb that comply with existing codes and good engineering practice. Many of these rules are given and illustrated in this chapter.

Solar and conventional energy systems differ in three key ways that relate to system performance (Hittle, Holshouser, and Walton 1976).

- Solar energy is not available continuously; it may not be available at all if peak demand comes at night or during cloudy weather. Since solar energy cannot meet all demands economically, auxiliary energy is required. Solar system components are therefore sized based on overall system performance, rather than on peak load calculations; the auxiliary system is sized for peak loads.
- Both energy demand and thermal storage mass influence collector operating temperatures and therefore influence the amount of incident solar energy a collector array can collect. Consequently, seasonal building heating and cooling load variations strongly influence solar system annual performance.
- Solar energy is typically collected and delivered over a large temperature range. This influences the performance of heating coils and absorption chillers, so conventional peak-load design methods are not acceptable.

SOLAR SYSTEM SIMULATION

Virtually all design methods used today were developed using simulation models, whether they are exercised manually, by hand-held calculator, or by computer.

Solar heating or cooling system performance depends on solar radiation available to the collector, outside ambient air temperature and wind conditions, collector design, inlet fluid temperature from storage, and system thermal load (See Fig. 7-1). System simulation is required to understand how these factors affect system performance in various locations. Solar systems always operate in transient modes and are driven by constantly varying weather conditions. Relationships among system components are complex, and systems respond nonlinearly to solar radiation; thus prediction of system performance is difficult. Therefore, systems cannot be analyzed on the basis of their response to average weather conditions.

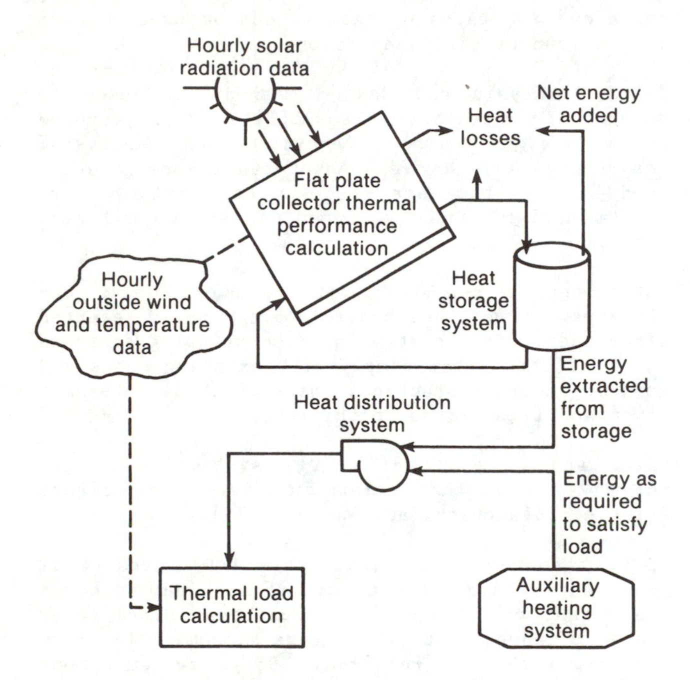

Fig. 7-1. Typical simulation model

Short-term performance information can be helpful in selecting components, but long-term performance is required to size the system properly. Long-term performance can be predicted fairly accurately with a computer program that simulates operation hour by hour. Parametric runs can be made in series to estimate design parameter sensitivity. Sound design tools will result if simulations are backed by laboratory and field verifications.

Simulation programs are developed by first formulating mathematical models for components. These component models may be based on first principles or may be empirical in nature; they may be simple or highly detailed. Next, means are added to solve these models simultaneously, using time-dependent weather data as forcing functions. Such programs may be general, versatile, and applicable to a range of processes, or they may be special-purpose programs for simulation of specific processes. Costs and the intended purposes of the simulation determine appropriate levels of detail (Duffie 1978; Jordan and Liu 1977).

Results of many simulation studies can be used to develop generalized performance data that correlate a system's performance with its design parameters and the weather. Furthermore, these results can be used to develop design methods that will not require a computer.

CLASSIFICATION AND USE OF DESIGN METHODS

Correlation methods and computer simulation methods are the two basic types. The distinction is not always clear, but most designers find it useful. Although correlation methods have resulted from computer analysis, most are executed manually on hand-held calculators or by use of graphs, tables, and manually solved equations. However, some are executed as microcomputer programs. Computer simulation methods are executed on microcomputers or main-frame computers. They generally use hour-by-hour computation for a full year, but this is not always the case. By and large, correlation methods are referred to as simplified methods.

Whether correlation or a computer simulation, the method selected should be appropriate to the task. In early design phases, when little is known about the building and the proposed solar system and when quick performance estimates are needed, simplified correlation methods are appropriate. However, in advanced stages of design optimization for large buildings or in the design of unusual systems, computer simulation methods are usually more appropriate. Comprehensive compendia (Feldman and Merriam 1979; Merriam 1981; SERI 1980, 1985) of both types of solar system design and analysis methods provide descriptions of available public- and private-domain methods.

Major advantages of simplified correlation methods are simplicity in use, good computational speed and low cost of repetitive use, rapid turnaround (especially important in iterative phases of design), and the ability to be used by persons with little technical experience (Feldman and Merriam 1979). Disadvantages include limited ability for design optimization, lack of control over assumptions made, and limited selection of systems that can be simulated. Simplified correlation methods have been developed mainly for direct solar space heating and service hot water applications; very few have been developed for heat pumps or cooling. [For a design method covering industrial process heat, see Kutscher et al. (1982)]. Furthermore, they are limited to standard system configurations and load characteristics. Combined solar and heat pump systems, for example, are currently covered by simplified design methods for parallel systems only (Anderson, Mitchell, and Beckman 1980). Thus, if systems under design are significantly nonstandard in application, configuration, or load characteristics, computer simulation methods are required to achieve accurate and detailed results.

Computer simulation methods can be used for design and optimization studies and, in some cases, for HVAC system tradeoff analyses and peak load and energy use (demand) analyses (Feldman and Merriam 1979). These methods accurately simulate the details of complex systems, incorporate a large variety of system types, and can be used in optimization studies. They also are complex, have long running times and high costs, need access to computing facilities, and require the user to learn a complex program to perform accurate simulations.

CORRELATION METHODS

Many correlation design methods have been developed for active solar heating and cooling systems (SERI 1980, 1985).*

Solar system performance, whether heating or cooling, depends strongly on design temperature of the load (the minimum temperature to which storage may fall before auxiliary energy is required) and, to a lesser extent, on the load profile. These must be considered in selecting a design method. Most simplified methods have been developed for residential buildings, where major loads occur at night, and do not account explicitly for a variation in load temperature. This temperature is significant for commercial heating and cooling systems and for industrial process heat (IPH) systems. For residential buildings, a design temperature of 70°F (22°C) is generally used, but design temperatures for commercial building loads can be much higher. Design methods have been developed for domestic hot water systems (Winn 1980), but most methods developed for residential space heating can be applied to domestic hot water. The ϕ-F-Chart method (Klein and Beckman 1979) and the methods developed by Lunde (1979; 1980) are available for predicting solar performance for cooling and IPH systems over a range of design temperatures.

Three of the many available correlation design methods, the ones most used for nonresidential building applications, are described below:

- F-Chart (and its derivatives)
- Solar Load Ratio (SLR)
- U.S. Army Construction Engineering Research Laboratory (CERL) methods.

The F-Chart Method

One of the most comprehensive and widely used simplified methods, F-Chart (Beckman, Klein, and Duffie 1977), was developed at the University of Wisconsin using the TRNSYS computer program (U. of Wis. 1983, 1985; Klein, Beckman, and Duffie 1976). (TRNSYS is discussed later in this chapter). F-Chart was developed using TRNSYS to simulate representative solar heating and service hot water systems in several geographical locations, with either air or water as the transport fluid. Correlations were determined between monthly solar fractions and two dimensionless parameters specific to the system being analyzed. One parameter is a measure of monthly solar radiation absorbed by the collector; the other is a measure of collector thermal energy losses. The correlations result in a graph (or chart) that shows the solar fraction, f, as a function of the two dimensionless quantities (Figs. 7-2 and 7-3 for liquid and air systems, respectively). The curves in the figures may be represented analytically. The solar fraction for a liquid system, f_1, as a function of dimensionless quantities, X and Y, is given by

$$f_1 = 1.029Y - 0.065X - 0.245Y^2 + 0.0018X^2 + 0.025Y^3. \qquad (7\text{-}1)$$

The solar fraction for an air system, f_a, is given by

$$f_a = 1.04Y - 0.065X - 0.159Y^2 + 0.00187X^2 - 0.0095Y^3. \qquad (7\text{-}2)$$

For service hot water systems, Eq. (7-1) may be used, where X is modified as follows:

$$\frac{X}{X_o} = \frac{-66.2 + 1.18t_W + 3.86t_M - 2.32\bar{t}_A}{212 - \bar{t}_A}, \text{ English}$$

$$= \frac{11.6 + 1.18t_W + 3.86t_M - 2.32\bar{t}_A}{100 - \bar{t}_A}, \text{ SI}$$

where

X_o = collector loss (X) for hot water heating system

t_W = water set temperature

t_M = supply water temperature

$\bar{t}_A$ = monthly average ambient temperature

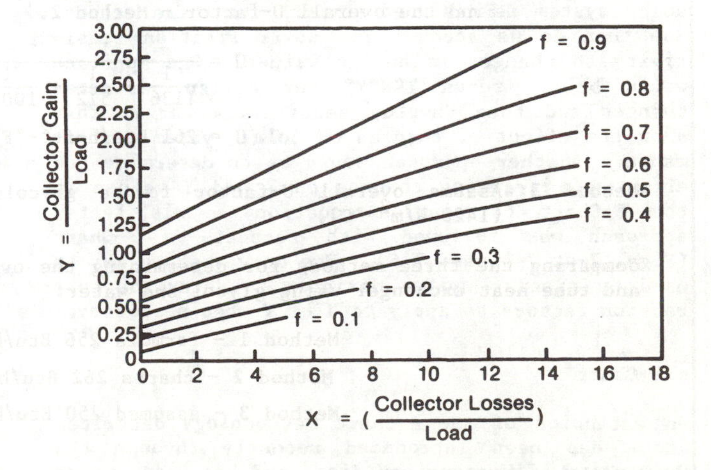

Fig. 7-2. F-Chart for a liquid system

*See also Beckman, Klein, and Duffie (1977); Hunn (1980); Ward (1976); Strickford (1976); Lunde (1978; 1979); Barley and Winn (1978); Hittle et al. (1977); Schurr, Hunn, and Williamson (1981); Swanson and Boehm (1977); and Klein and Beckman (1979).

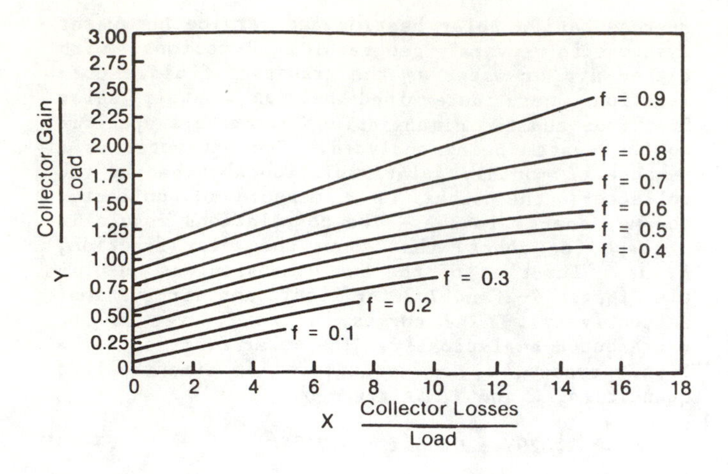

Fig. 7-3. F-Chart for a solar air-heating system

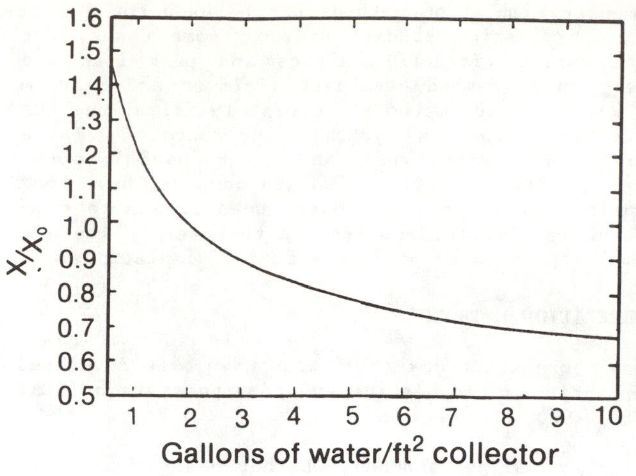

Fig. 7-4. Storage size correction factor for F-Chart

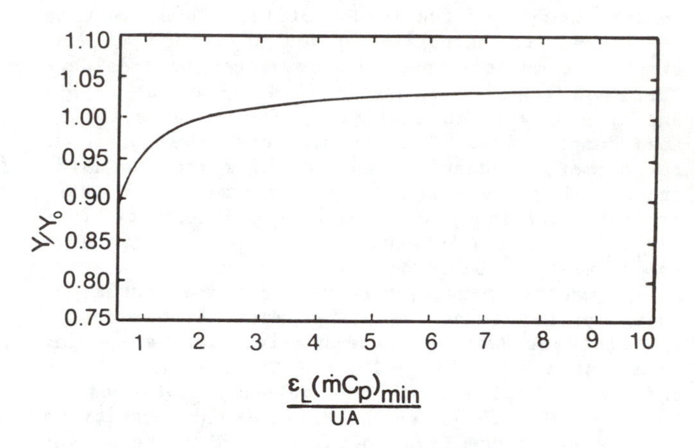

Fig. 7-5. Load heat exchanger correction factor for F-Chart

Solar fractions for each month are determined by calculating X and Y values and then either using one of the above equations or referring to the appropriate F-Chart and reading monthly solar fractions directly.

Effect of Parameter Changes

Correlation curves and equations developed for F-Chart are based on certain design parameter values. For example, storage was assumed to be 0.75 ft^3 of rocks per square foot of collector (0.25 m^3/m^2_c) for air systems and 2 gal of water per square foot of collector (80 L/m^2_c) for liquid systems. In addition, the load heat exchanger was assumed to have a value of 2 for the quantity $\varepsilon_L(\dot{m}\ C_p)_{min}/UA$, where ε_L is heat exchanger effectiveness, $(\dot{m}\ C_p)_{min}$ is minimum fluid capacitance rate, and UA is the area conductance product of the building, which represents the load. Also, the collector flow rate for air systems was assumed to be 2 $ACFM/ft^2_c$ (0.010 m^3/m^2_c). These values were selected because they are considered to be nominal design values for solar systems, but not every solar system is designed strictly to these values. How then do we account for solar fraction sensitivity to changes in design values? One approach would be to rerun TRNSYS for design parameter changes and then develop sensitivity curves that show the effect of changes on solar system performance. Another approach would be to determine the effect of parameter changes on X and Y values in the F-Chart correlation equation. This latter approach was followed with respect to F-Chart (Figs. 7-4, 7-5, and 7-6). The curves shown may be used by a solar system designer to determine correction factors to apply to X or Y when necessary.

$\bar{\phi}$-F-Chart

An extension of the F-Chart methodology described above has been introduced recently through the University of Wisconsin-Madison and has been incorporated in the latest F-Chart computer program, Version 4.2 (U. of Wis. 1985). This formulation, called $\bar{\phi}$-F-Chart, offers much more versatility than previous versions and uses the concept of utilizability, which represents the fraction of radiation above a minimum level that can be collected and stored (Klein and Beckman 1979; Duffie and Beckman 1980). Analysis of various types of systems is allowed, e.g.:

- Space Heating
  - Air- or liquid-based active solar
  - Ambient sourced or solar-assisted heat pumps
- Domestic Water Heating
  - Water heating only systems
  - Combined space and water heating
- Process Heating
  - Closed-loop heat delivery through heat exchanger or heat pump

The latest F-Chart version also offers many analysis options, including:

- Collectors
  - Flat-plate
  - Compound parabolic concentrators (liquid-based systems only)

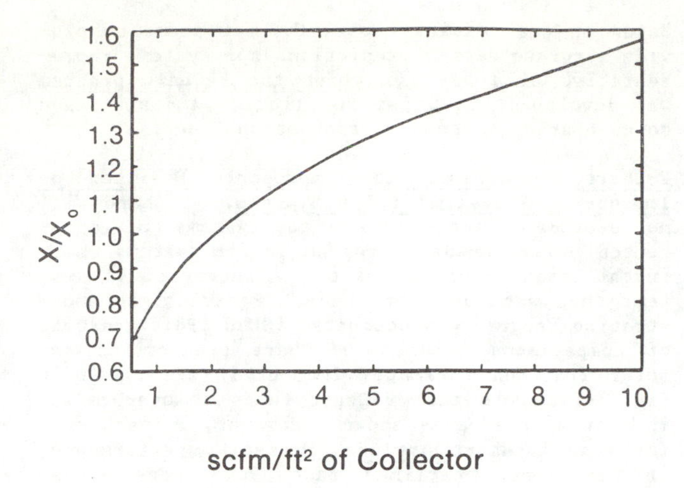

a. air flow rate

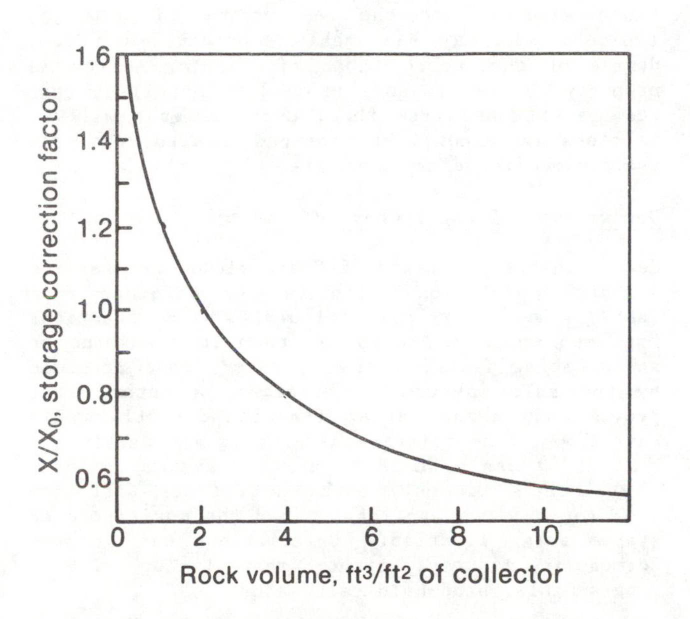

b. storage

Fig. 7-6. Air flow rate F-Chart correction factors for air systems

- Imaging concentrators (1 or 2-axis tracking; liquid system only)

- Collector-Storage Loop
  - Pipe or duct heat losses
  - Air duct leaks
  - Heat exchanger

- Storage Devices
  - Liquid tank
  - Rock bed
  - Phase-change material (air-based space heating)

- Storage-Load Heat Delivery
  - Heat exchanger
  - Heat pump (liquid-based systems)

- Load Calculation Methods
  - Space heating
    - User-supplied monthly loads
    - UA-degree day
  - Water heating
    - Based upon usage, set temperature, and mains temperature (includes auxiliary tank losses)
  - Process heating
    - User-supplied monthly totals

- Auxiliary (solar system back-up)
  - Conventional fuel (gas, electricity, oil)
  - Ambient-source heat pump (space heating)

- Economics
  - Life-cycle purchase and operating costs
  - Comparisons between alternatives
  - Optimization of any system parameter

The $\bar{\phi}$-F-Chart methodology offers a design tool for assessing many commercially available system types that previously have not been directly evaluated except by detailed simulations. The previous F-Chart version was for specific system types, with subcomponent size limitations. $\bar{\phi}$-F-Chart was formulated to relieve various constraints, but some constraints are still imposed. For example, $\bar{\phi}$-F-Chart requires some knowledge of collector inlet temperature, and the load must be in a closed loop with storage, with a conversion efficiency independent of temperature. Consequently, mechanical-energy conversion systems are precluded. Storage temperature (or load temperature) is not constrained to a minimum level except by the user, so the methodology can be used to assess energy inputs for many solar system types, including those with absorption chillers.

A disadvantage of $\bar{\phi}$-F-Chart is that solar collectors must be oriented due south. Slight variations from due south do not typically make much difference; i.e., performance predictions will not vary significantly from those obtained assuming collectors are oriented due south. The F-Chart 4.0 computer program is interactive and is now available for personal computers. A user's manual is available from the Solar Energy Laboratory at the University of Wisconsin-Madison.

F-Chart calculations can be performed manually, with the aid of a programmable hand-held calculator, with a desktop computer, or with a main-frame computer. Performing calculations manually is tedious and time consuming and is not recommended except on a one-time basis to gain an increased understanding of the method. An example of such a calculation is given in Cheremisinoff and Dickinson (1980b, Ch. 43). Main-frame computer use is the most efficient in terms of time, particularly when several different systems are to be analyzed. However, a main-frame computer is not always available, and overhead costs tend to be rather high. The program is available on several commercial computer service bureaus and time-sharing networks. Also, several hand-held calculator and microcomputer versions are available (SERI 1981). For most users, the PC version of F-Chart is most appropriate.

Validation Results

Agreement between F-Chart results and TRNSYS results is very good, but this is to be expected since F-Chart is based on a correlation to results obtained from TRNSYS. The question of interest is "How well do results obtained from F-Chart agree with results observed in practice?" F-Chart has been used to model several existing solar systems, and the results obtained from the program were compared with results obtained from the actual systems (Lantz and Winn 1978; Gary 1978). These systems included the Colorado State University Solar House II, the Eco-Era #2 House, the SEECO House, and several HUD demonstration houses; results are presented below.

CSU Solar House II. Agreement between F-Chart results and CSU Solar House II observed results is shown in Table 7-1. These results are for an air system operated during 1977. Based on this rather limited sample, the mean error in monthly fractions of the load supplied by solar is 2%, and the standard deviation is 7% between F-Chart results and measured results. This agreement is considered to be excellent.

Eco-Era #2 House. A single-family residence in Fort Collins, Colorado, was constructed as part of Cycle 1 of the HUD Residential Demonstration Program. The house has been occupied since early 1977, and auxiliary energy use has been monitored since that time. Observed values of energy use for two years have been used to calculate the solar fraction provided by the air solar system (Table 7-2). The mean and standard deviation between F-Chart and observed monthly results are 2% and 11%, respectively. This agreement is considered to be very good.

SEECO House. A liquid-based solar system was also installed in 1976 in Fort Collins, Colorado, as part of Cycle 1 of the HUD Residential Demonstration Program. Auxiliary energy use has been monitored since the house was occupied, and solar fractions have been determined. The comparison between monthly F-Chart studies and observations on the liquid system is shown in Table 7-3. The mean error is 0% and the standard deviation is 12%, again representing very good agreement.

Based on the results above, F-Chart seems to provide accurate design predictions for systems representative of those for which the F-Chart program was developed, that is, for liquid- and air-based solar heating systems of conventional design.

F-Chart Comparisons for Improperly Designed or Improperly Installed Solar Systems. F-Chart will not provide accurate results for systems not represented by the F-Chart program or for systems where installation problems exist. A survey of 31 systems that were part of the HUD Residential Demonstration Program was conducted (SEEC 1981); results of comparisons between F-Chart predictions and solar fraction estimates (measured) are shown in Fig. 7-7. Note that F-Chart tends to overpredict in most of the cases shown. However, in each case for which F-Chart overpredicts solar performance, it has been determined that there were either design problems, installation problems, or both. In many cases, hot water loads were different from those assumed, and the weather varied from the typical. We may reasonably conclude, with some degree of confidence, that if a solar system is properly designed and properly installed, the results obtained from the F-Chart program will be in close agreement with measured results, provided the assumed loads are accurate.

Derivatives of the F-Chart Method

Recall that the original F-Chart method is based on monthly correlations with results obtained from TRNSYS. An F-Chart user calculates solar fractions for each month and combines them to determine an annual solar fraction of the energy load provided by the solar system. Therefore, a method that provides an annual solar fraction directly would save time. The relative-areas program (Barley and Winn 1978) can be used by a solar system designer to calculate directly both the optimal collector area for a given application and the corresponding system solar fraction. Calculations can be performed in minutes, either manually or with a programmable, hand-held calculator.

The relationship between collector area and annual solar load fraction obtained from F-Chart calculations is similar for most collectors and locations. Differences observed among collector types and

Table 7-1. CSU Solar House II (Karaki, Armstrong, and Bechtel 1977)

Month	Solar Radiation (Btu/ft^2-Day)	Degree Days	Avg. Ambient Temp. (°F)	Hot Water Load 10^6Btu	Solar Fraction f_{meas}	Solar Fraction f_{Fchart}
November	1259	690	42	1.06	0.65	0.73
December	1409	899	36	1.64	0.69	0.69
January	1398	1240	25	2.12	0.58	0.52
February	1478	756	38	2.21	0.79	0.76
March	1793	806	39	2.21	0.82	0.85
April	1096	510	48	1.74	0.88	0.77

F_R = 0.57; F_RU_L = 1.30 Btu/h-ft^2-°F; Area = 722 ft^2 (690 ft^2 effective)

Table 7-2. Comparison Between F-Chart Calculations and Measured Results for Two Years of Observations: Air System, Fort Collins, Colorado

Month	f^*_{meas}	f_{Fchart}
October	0.81	0.91
November	0.68	0.55
December	0.60	0.42
January	0.44	0.46
February	0.46	0.53
March	0.66	0.59
April	0.69	0.74

*Includes pilot lights

Table 7-3. Comparison Between F-Chart Calculations and Measured Results for Two Years of Observations: Liquid System, Fort Collins, Colorado

Month	f^*_{meas}	f_{Fchart}
October	0.75	0.99
November	0.66	0.68
December	0.65	0.54
January	0.67	0.58
February	0.65	0.65
March	0.82	0.72
April	0.82	0.86

*Includes pilot lights

locations can be practically eliminated by plotting the solar fraction as a function of relative collector area, defined as $A/A_{0.5}$, where $A_{0.5}$ is the collector area corresponding to an annual solar load fraction of 0.5. With the relative collector area as the abscissa, solar fraction curves as a function of relative collector area are forced to intersect at the point $A/A_{0.5} = 1$, $f = 0.5$. Since the curves have similar shapes, they are closely grouped for all practical values of relative area and may be represented analytically by the equation

$$f = C_1 + C_2 \ln(A/A_{0.5}), \tag{7-3}$$

for $f > 0.15$,

which gives the solar fraction as a function of relative collector area and two location-dependent parameters, C_1 and C_2. (C_1 varies somewhat from its "forced" value of 0.5(=f) in order to obtain best fit curves in different regions.) C_1 and C_2 have been calculated for the 172 locations included in F-Chart, Version 2.0 and are tabulated (Barley and Winn 1978) for liquid- and air-based solar space heating and service hot water systems. By reading C_1 and C_2 values from the tables, a system designer can easily determine the annual solar fraction as a function of relative collector area. If $A_{0.5}$ is known, the solar fraction as a function of collector area can also be determined very quickly. An equation for $A_{0.5}$ has been derived and is given as

$$A_{0.5} = \frac{A_s(UA)}{F'_R\overline{\tau\alpha} - F'U_RZ_L}, \tag{7-4}$$

where A_s and Z are also location-dependent parameters that have been tabulated (Barley and Winn 1978), UA represents the UA factor for the building load, and $F'_R\overline{\tau\alpha}$ and F'_RU_L are collection parameters. The standard deviation in $A_{0.5}$ from the value given by F-Chart is about 1%.

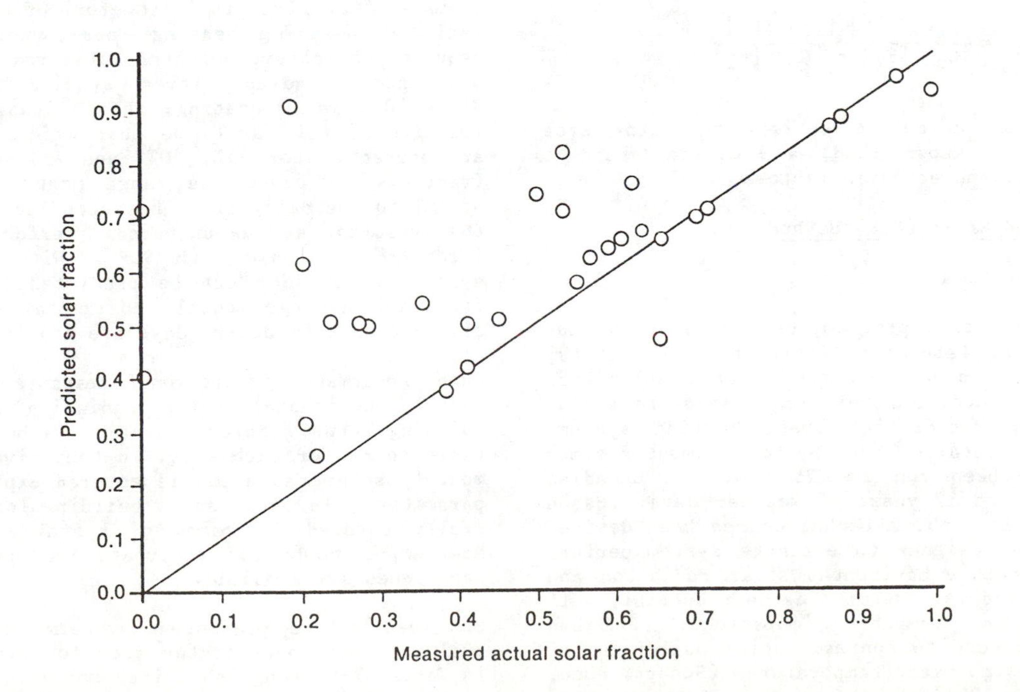

Fig. 7-7. Estimated (measured) solar fraction vs. predicted fraction from F-Chart for 31 systems, some of which had design and/or installation problems

Equations (7-3) and (7-4) provide a very simple method for calculating annual solar fractions as functions of collector area. Optimal collector area, i.e., the collector area that results in minimizing system life-cycle cost, can be calculated in an equally simple manner. The total cost of purchasing and operating a solar heating system over its lifetime can be expressed analytically as

$$C_T = (C_b + AC_a)E_1 + AC_oE_2 + C_mE_3 + (1 - f)LC_fE_4, \tag{7-5}$$

where

C_b = solar heating system base cost, \$

C_a = collector area dependent first cost, \$/ft^2

C_o = collector area dependent operation cost, \$/ft^2-yr

C_m = maintenance cost, \$/yr

C_f = fuel price, \$/million Btu delivered by furnace

L = total (space + SHW) annual heat load, million Btu/yr,

E = factors for converting cash flows to present worth.

For minimum total cost,

$$\frac{\partial C_T}{\partial A} = 0 = C_aE_1 + C_oE_2 - LC_fE_4 \frac{\partial f}{\partial A} . \tag{7-6}$$

From Eq. (7-3),

$$\partial f/\partial A = C_2/A. \tag{7-7}$$

Therefore, optimal collector area may be expressed as

$$A_{opt} = \frac{(C_2)\,(L)\,(C_f)\,(E_4)}{(C_a)\,(E_2) + (C_o)\,(E_1)} . \tag{7-8}$$

Hence, one can calculate optimal collector area directly for a known load, a location-dependent parameter, C_2, and economic factors.

The Solar-Load Ratio (SLR) Method

Residential Systems

An hourly simulation program, developed at the Los Alamos National Laboratory (Balcomb et al. 1976), has been used to develop the monthly Solar-Load Ratio (SLR) method, a simplified design and sizing procedure for residential space heating systems (Hunn 1980). Detailed hour-by-hour computer simulations have been run for 25 U.S. and Canadian cities for up to 12 years of weather data. Based on these simulations, a technique has been devised that allows a designer to estimate system performance from monthly horizontal solar radiation and heating degree-days. Results agree reasonably well with hour-by-hour computer simulations. SLR has also been extended to nonresidential buildings for a range of design water temperatures (Schurr, Hunn, and Williamson 1981).

The basic difference in approach between F-Chart and SLR is that while F-Chart treats basic solar collector characteristics as input variables, SLR presents performance for reference liquid and air systems where the collector characteristics are fixed at nominal values. Liquid reference system characteristics (for residential systems) are given in Table 7-4. A system designer can use parametric study results (Hunn 1980; ERDA 1976), where each system and collector characteristic is varied singly, to determine performance of systems that differ from the reference systems.

Tables of sizing results have been developed only for the reference liquid system. Although some researchers (Oonk et al. 1976) have found that air and liquid system performances can differ substantially, depending on climate, system characteristics, and load fraction met by solar, the Los Alamos simulations (Hunn 1980) indicate very little difference between the two reference system types. Thus, the simplified method presented here can be used for either system type, but air system performance should be approximated by adjusting liquid system results with a small correction factor (ERDA 1976).

SLR gives comparable results, on an annual basis, when compared to F-Chart (DOE 1977), the design method in most widespread use. Furthermore, in a comparison of several solar system computer simulation programs, the parent programs of F-Chart and SLR were also shown to be in substantial agreement, although program formulations are somewhat different (Freeman, Maybaum, and Chandra 1978).

Use of the SLR Method for Residential Systems. Collector array sizing information will be presented for the reference liquid system in two formats. The first is in the form of tables of the ratio of building heating load-to-collector area required to achieve selected solar fractions in the U.S. and Canadian cities studied (Hunn 1980). Since 100% solar heating, though feasible in some locations, will rarely be cost effective, results are presented for 25%, 50%, and 75% solar heating fractions to cover the range most likely to be useful to the designer. The second format presents the results as a universal performance curve (Fig. 7-8) for use with SLR. With this curve, system performance can be predicted for any city for which average monthly horizontal solar radiation and heating degree-days are available.

Both performance prediction formats require average monthly horizontal solar radiation and monthly building loads. Solar radiation is built into the table entry for each city; in the universal curve, monthly solar radiation is entered explicitly as a parameter. In both cases, building loads are generally entered in terms of monthly degree-days; however, a number of alternate load determination techniques are available.

SLR results are presented in terms of ratios of building load-to-collector area for liquid systems in Tables 7-5a (English units) and 7-5b (SI units). These tables present ratios in terms of the solar

Table 7-4. The Standard Liquid System (Residential Applications)

Values of parameters used for the "standard" solar heating system that includes liquid-heating solar collectors, a heat exchanger, and water tank thermal storage. The values are normalized to 1 ft^2 of collector (ft_c^2).

Parameter	Nominal Value
Solar Collector	
1. Orientation	Due south
2. Tilt (from horizontal)	Lat +10°
3. Number of glazings	1
4. Glass transmissivity (at normal incidence)	0.86 (6% absorption* 8% reflections)
5. Surface absorptance (solar), α	0.98
6. Surface emittance (IR)	0.89
7. Coolant flow rate ($\dot{m}C_p/ft^2{}_c$)	0.04 gpm water/$ft^2{}_c$ (20 Btu/h-°F-$ft^2{}_c$)
8. Heat transfer coefficient to coolant (UA/$ft^2{}_c$)	30 Btu/h-°F-$ft^2{}_c$
9. Back insulation U-value (UA/$ft^2{}_c$)	0.083 Btu/h-°F-$ft^2{}_c$
10. Coolant heat capacity rate ($\dot{m}C_p/ft^2{}_c$)	1 Btu/h-°F-$ft^2{}_c$
Collector Piping	
11. Heat loss coefficient, to ambient (UA/$ft^2{}_c$)	0.04 Btu/h-°F-$ft^2{}_c$ (U = 0.3 Btu/h-°F-ft^2 pipe or 1 in. insulation of k = 0.025 Btu/h-ft-°F)
Heat Exchanger	
12. Heat transfer coefficient (UA/$ft^2{}_c$)	10 Btu/h-°F-$ft^2{}_c$
Thermal Storage	
13. Heat capacity	1.8 gal water/$ft^2{}_c$ (15 Btu/°F-$ft^2{}_c$)
14. Heat loss coefficient (UA/$ft^2{}_c$), assuming all heat loss is to heated space	0 Btu/hr-°F-$ft^2{}_c$
Controls	
Building maintained at 70°F	
Collectors on when advantageous	

*These values apply for normal incidence on ordinary double-strength glass (1/8 in.). For other angles of incidence, the Fresnel equation is used.

heating percentage of the total space heating requirement (solar fraction). The Load-Collector Ratio (LC) is defined as

$$LC = \frac{\text{Building Load (Btu/}^{o}\text{F-day)}}{\text{Collector Area (ft}^2\text{)}} \quad (7\text{-}9)$$

or

$$LC = \frac{\text{Building load (kJ/}^{o}\text{C day)}}{\text{Collector Area (m}^2\text{)}}$$

To use the building load/collector area (LC) ratio for determining collector size for one of the cities listed (or an equivalent location), the building thermal load must be determined and the percentage of heating the solar system is to provide must be decided.

__Changes in Design Parameters__. Changing any design parameter for the reference liquid or air systems will have some effect on system performance (Hunn 1980; ERDA 1976). However, it should be noted that these data reflect changes that occur when only one parameter is varied at a time. Because of nonlinear effects, a change in more than one parameter may be quite different from the algebraic sum of the results for each individual parameter change. Applications of parametric results to systems much

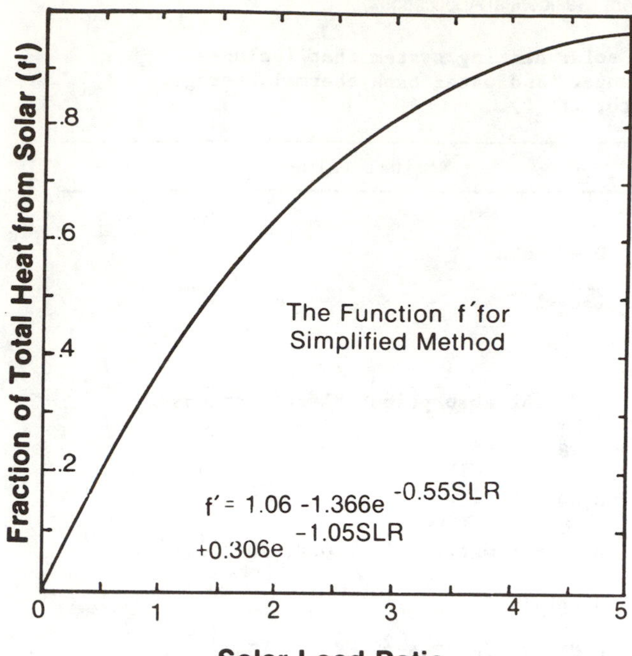

Fig. 7-8. Universal performance curve for the monthly SLR method

different from the reference systems should be made with caution.

Service Hot Water Add-On. Addition of a service hot water preheat capability is an almost universal feature of a space heating system. This is a good marriage because it allows solar collectors to serve a large winter load, thus increasing their effectiveness, and to be used in the summer when they would otherwise be idle.

The LC values given in Table 7-5 do not include effects of this additional load--reference liquid or air systems studies are for space heating only. It is no longer possible to characterize the performance in terms of a single parameter, like LC, when a hot water capability is added because different buildings will have different thermal load balances. A set of different tables could be generated for different proportions of space and water heating loads, but this complication would defeat the intent of the simple approach taken.

The designer of a combined system may choose to estimate collector area requirements, based on the space heating load only, using LC values given here or to resort to the somewhat more complex Monthly Solar Load Ratio method to obtain a more accurate estimate.

Year-to-Year Variations at a Fixed Location. Year-to-year variations in solar heating fractions of ±5% can be expected to result from year-to-year variations in climate.

The Monthly Solar-Load Ratio (SLR) Method. For locations not covered by the cities given in Table 7-5, the monthly SLR method can be used. It is based on correlations developed for many hour-by-hour computer simulations from a variety of locations.

The following five-step process can be carried out with a hand calculator with exponential functions.

1. Obtain the following data for each month of the year for the system site:

   a. heating degree-days, per month
   b. total solar radiation incident on a horizontal surface, per month.

2. Correct the incident solar radiation for tilt angle of latitude plus 10°, using the following approximate formula:

$$H_{M\Sigma} = \left[\begin{array}{l}\text{Total monthly radiation}\\ \text{on tilted surface, Btu/ft}^2\end{array}\right] \simeq 1.025d - 8200,$$

where

$$d = \frac{\text{Total monthly radiation on horizontal surface, } (H_{MH}),\ \text{Btu/ft}^2}{\cos\ (\text{latitude-solar declination at midmonth})}$$

   The solar declination at midmonth is approximated as 23.45° [cos (30mo-187)]; where

   mo = month number (January = 1, December = 12).

   The resulting values are the monthly solar radiation on a plane surface tilted at an angle equal to the latitude plus 10° and oriented due south.

3. Determine building thermal load, in units of Btu/°F-day or kJ/°C-day. This is total building heat required per day for a 1° difference between inside and outside temperatures.

4. Determine the solar load ratio (SLR) for each month from the following formula:

$$\text{SLR} = \frac{\text{Solar Collector Area} \times \text{Total Radiation Flux on Tilted Surface}}{\text{Building Thermal Load} \times \text{Heating Degree-Days per month}}.$$

   SLR is the dimensionless ratio of total solar energy incident on the collectors to total energy required to heat the building.

5. The annual heating fraction, f, is estimated from the following formula:

$$f = \frac{\sum_{mo=1}^{12} (\text{Degree-Days})\ (f')}{\sum_{mo=1}^{12} (\text{Degree-Days})},$$

where

$$f' = 1.06 - 1.366\ e^{-0.55\ \text{SLR}} + 0.306\ e^{-1.05\ \text{SLR}},$$
(for SLR ≤ 5.66);

$f' = 1$ (for SLR > 5.66).

The function f', plotted in Fig. 7-8, is not a very good estimate of the monthly solar heating fraction because it is designed to compensate (on an annual basis) for the fact that the heating load based on

Table 7-5a. Values of LC for Reference Liquid System for Selected Cities (English units)

Location	Latitude °N	Elevation ft	Degree Days	LC where solar provides 25%, 50%, 75% of heating load: 25% Btu/DD ft^2_c	50% Btu/DD ft^2_c	75% Btu/DD ft^2_c
UNITED STATES						
ARIZONA						
Page	37	4280	6632	128	48	23
Phoenix	33	1139	1765	300	118	59
Tucson	32	2440	1800	301	118	59
ARKANSAS						
Little Rock	35	276	3219	126	48	24
CALIFORNIA						
Davis	39	50	2502	198	72	33
El Centro	33	12	1458	547	206	97
Fresno	37	336	2492	195	70	32
Los Angeles	34	540	2061	416	157	75
Riverside	34	1050	1803	391	152	74
Santa Maria	35	289	2967	353	142	67
COLORADO						
Grand Junction	39	4832	5641	119	46	22
FLORIDA						
Apalachicola	30	46	1308	324	129	65
Tallahassee	30	64	1485	283	113	57
GEORGIA						
Atlanta	34	1018	2961	154	59	29
Griffin	33	1001	2136	217	84	42
IDAHO						
Boise	44	2895	5809	108	39	17
ILLINOIS						
ANL, Lemont	42	750	6155	79	30	14
INDIANA						
Indianapolis	40	819	5699	86	32	15
KANSAS						
Dodge City	38	2625	4986	126	49	24
LOUISIANA						
Lake Charles	30	60	1459	244	96	48
Shreveport	32	220	2184	179	70	35
MAINE						
Caribou	47	640	9767	68	26	12
MARYLAND						
Silver Hill	39	292	4224	111	43	21
MASSACHUSETTS						
Blue Hill	42	670	6368	82	31	15
Boston	42	157	5624	86	33	16
East Wareham	42	50	5891	97	37	18
MICHIGAN						
East Lansing	43	878	6909	76	28	13
Sault Ste. Marie	46	724	9048	74	27	12
MINNESOTA						
Saint Cloud	46	1062	8879	71	27	13
MISSOURI						
Columbia	39	814	5046	102	38	18
MONTANA						
Glasgow	48	2109	8996	105	41	20
Great Falls	47	3692	7750	93	35	16
NEBRASKA						
Lincoln	41	1316	5864	104	39	19
North Omaha	41	1323	6612	89	34	16
NEVADA						
Ely	39	6279	7733	119	47	23
Las Vegas	36	2188	2709	218	84	42
Reno	39	4400	6632	125	47	22
NEW JERSEY						
Seabrook	39	110	4812	97	37	18
NEW MEXICO						
Albuquerque	35	5327	4348	161	64	31

Table 7-5a. Values of LC for Reference Liquid System for Selected Cities (English units) (concluded)

Location	Latitude °N	Elevation ft	Degree Days	LC where solar provides 25%, 50%, 75% of heating load: 25% Btu/DD ft^2_c	50% Btu/DD ft^2_c	75% Btu/DD ft^2_c
Los Alamos	36	7200	6600	107	41	21
NEW YORK						
Ithaca	42	951	6914	68	24	11
New York	41	187	4871	88	34	16
Sayville	41	56	4811	98	38	18
Schenectady	43	490	6650	63	24	11
NORTH CAROLINA						
Greensboro	36	914	3805	128	50	24
Hatteras	35	27	2612	204	79	39
Raleigh	36	440	3393	133	52	25
NORTH DAKOTA						
Bismarck	47	1677	8851	78	29	14
OHIO						
Cleveland	41	871	6351	71	26	12
Columbus	40	760	5211	77	29	13
Put-In-Bay	42	580	5796	68	24	11
OKLAHOMA						
Stillwater	36	910	3725	132	52	25
Oklahoma City	35	1317	3725	134	53	26
OREGON						
Astoria	46	22	5186	127	45	19
Corvallis	45	236	4726	120	42	18
Medford	42	1321	5008	107	38	16
PENNSYLVANIA						
State College	41	1230	5934	78	29	14
RHODE ISLAND						
Newport	41	50	5804	97	37	18
SOUTH CAROLINA						
Charleston	33	69	2033	210	82	41
SOUTH DAKOTA						
Rapid City	44	3180	7345	97	37	18
TENNESSEE						
Nashville	36	614	3578	117	44	21
Oak Ridge	36	940	3817	111	42	20
TEXAS						
Brownsville	26	48	600	517	218	110
El Paso	32	3954	2700	228	88	44
Fort Worth	33	574	2405	186	73	37
Midland	32	2885	2591	202	79	39
San Antonio	30	818	1546	262	103	52
UTAH						
Salt Lake City	41	4238	6052	107	40	19
Flaming Gorge	41	6273	6929	111	43	21
VERMONT						
Burlington	44	385	8269	63	24	11
VIRGINIA						
Sterling	39	276	4224	111	43	21
WASHINGTON						
Prosser	46	840	4805	117	41	18
Pullman	47	2583	5542	100	36	16
Richland	47	731	5941	100	35	15
Seattle	48	110	4785	94	33	13
Spokane	48	2356	6655	90	31	14
WISCONSIN						
Madison	43	889	7863	76	28	13
WYOMING						
Lander	43	5574	7870	108	42	21
Laramie	41	7240	7381	106	42	21
CANADA						
Toronto	44	443	6827	72	27	13
Winnipeg	50	820	10629	63	23	11

Table 7-5b. Values of LC for Reference Liquid System for Selected Cities (SI units)

Location	Latitude °N	Elevation m	Degree Days	LC where solar provides 25%, 50%, 75% of heating load: 25% kJ/°C-day-m^2_c	50% kJ/°C-day-m^2_c	75% kJ/°C-day-m^2_c
UNITED STATES						
ARIZONA						
Page	37	1304	3684	2615	981	470
Phoenix	33	347	981	6129	2411	1205
Tucson	32	683	1000	6149	2411	1205
ARKANSAS						
Little Rock	35	84	1788	2574	981	490
CALIFORNIA						
Davis	39	15	1390	4045	1471	674
El Centro	33	4	810	11175	4209	1982
Fresno	37	102	1384	3984	1430	654
Los Angeles	34	165	1145	8499	3208	1532
Riverside	34	320	1002	7988	3105	1512
Santa Maria	35	88	1648	7212	2901	1369
COLORADO						
Grand Junction	39	1473	3134	2431	940	449
FLORIDA						
Apalachicola	30	14	727	6619	2635	1328
Tallahassee	30	20	825	5782	2309	1165
GEORGIA						
Atlanta	34	310	1645	3146	1205	592
Griffin	33	305	1187	4433	1716	858
IDAHO						
Boise	44	882	3227	2206	797	347
ILLINOIS						
ANL, Lemont	42	229	3419	1614	613	286
INDIANA						
Indianapolis	40	250	3166	1757	654	306
KANSAS						
Dodge City	38	800	2770	2574	1001	490
LOUISIANA						
Lake Charles	30	18	811	4985	1961	981
Shreveport	32	67	1213	3657	1430	715
MAINE						
Caribou	47	195	5426	1389	531	245
MARYLAND						
Silver Hill	39	89	2347	2268	878	429
MASSACHUSETTS						
Blue Hill	42	204	3538	1675	633	306
Boston	42	48	3124	1757	674	327
East Wareham	42	15	3273	1982	756	368
MICHIGAN						
East Lansing	43	268	3838	1553	572	266
Sault Ste. Marie	46	221	5027	1512	552	245
MINNESOTA						
Saint Cloud	46	324	4933	1451	552	266
MISSOURI						
Columbia	39	248	2803	2084	776	368
MONTANA						
Glasgow	48	643	4998	2145	838	409
Great Falls	47	1125	4306	1900	715	327
NEBRASKA						
Lincoln	41	401	3258	2125	797	388
North Omaha	41	403	3673	1818	695	327
NEVADA						
Ely	39	1914	4296	2431	960	470
Las Vegas	36	667	1505	4454	1716	858
Reno	39	1341	3684	2554	960	449
NEW JERSEY						
Seabrook	39	34	2673	1982	756	368
NEW MEXICO						
Albuquerque	35	1624	2416	3289	1308	633

Table 7-5b. Values of LC for Reference Liquid System for Selected Cities (SI units) (concluded)

Location	Latitude °N	Elevation m	Degree Days	LC where solar provides 25%, 50%, 75% of heating load 25% kJ/°C-day-m^2_c	50% kJ/°C-day-m^2_c	75% kJ/°C-day-m^2_c
Los Alamos	36	2194	3667	2186	838	429
NEW YORK						
Ithaca	42	290	3841	1389	490	225
New York	41	57	2706	1798	695	327
Sayville	41	17	2673	2002	776	368
Schenectady	43	149	3694	1287	490	225
NORTH CAROLINA						
Greensboro	36	279	2114	2615	1022	490
Hatteras	35	8	1451	4168	1614	797
Raleigh	36	134	1885	2717	1062	511
NORTH DAKOTA						
Bismarck	47	511	4917	1594	592	286
OHIO						
Cleveland	41	265	3528	1451	531	245
Columbus	40	232	2895	1573	592	266
Put-In-Bay	42	177	3220	1389	490	225
OKLAHOMA						
Stillwater	36	277	2069	2697	1062	511
Oklahoma City	35	401	2069	2738	1083	531
OREGON						
Astoria	46	7	2881	2595	919	388
Corvallis	45	72	2626	2452	858	368
Medford	42	403	2782	2186	776	327
PENNSYLVANIA						
State College	41	375	3297	1594	592	286
RHODE ISLAND						
Newport	41	15	3224	1982	756	368
SOUTH CAROLINA						
Charleston	33	21	1129	4290	1675	838
SOUTH DAKOTA						
Rapid City	44	969	4081	1982	756	368
TENNESSEE						
Nashville	36	187	1988	2390	899	429
Oak Ridge	36	286	2121	2268	858	409
TEXAS						
Brownsville	26	15	333	10562	4454	2247
El Paso	32	1205	1500	4658	1798	899
Fort Worth	33	175	1336	3800	1491	756
Midland	32	879	1439	4127	1614	797
San Antonio	30	249	859	5353	2104	1062
UTAH						
Salt Lake City	41	1292	3362	2186	817	388
Flaming Gorge	41	1912	3849	2268	878	429
VERMONT						
Burlington	44	117	4594	1287	490	225
VIRGINIA						
Sterling	39	84	2347	2268	878	429
WASHINGTON						
Prosser	46	256	2669	2390	838	368
Pullman	47	787	3079	2043	735	327
Richland	47	223	3301	2043	715	306
Seattle	48	34	2658	1920	674	266
Spokane	48	718	3697	1430	633	286
WISCONSIN						
Madison	43	271	4368	1553	572	266
WYOMING						
Lander	43	1699	4372	2206	858	1287
Laramie	41	2207	4101	2166	858	1287
CANADA						
Toronto	44	135	5905	1471	552	266
Winnipeg	50	250	3793	1287	470	225

degree-days per month is lower than the heating load based on hour-by-hour calculations.

The key to the accuracy of the method is the determination of f'. This function has been carefully determined so that the resulting error in predicting the solar heating fraction (compared to the simulation result) will be minimized without using a different function for each locality.

Determination of f' is based on hour-by-hour computer simulations of a "test year" for 25 locations for five different collector sizes in each location. The standard deviation of the prediction error is 4.4% of the solar heating. Values for f' from Fig. 7-8 versus simulated results for the 125 cases are given in (ERDA 1976).

The SLR method, it should be emphasized, is intended for design purposes and not for precise determination of system performance.

Nonresidential Buildings

Using hour-by-hour simulations with the DOE-2 computer program (York, Tucker, and Cappiello 1982), which included the component based solar simulator, the SLR method described above has been extended to nonresidential buildings (Schurr, Hunn, and Williamson 1981). Generic single- and double-glazed collectors and evacuated-tube collectors, characterized by typical collector efficiency curves, were used in a liquid system to simulate an office building service hot water load. Also, a combined space and service hot water load for an office building was simulated using both a liquid solar system and an air solar system. Simulations were carried out for seven U.S. cities that represented a variety of weather patterns. From hundreds of simulations, a series of solar fraction versus SLR correlation curves was developed. Annual correlations were used for the service hot water system, and monthly correlations were used for the combined space and service hot water system.

A standard liquid system similar to that defined in Table 7-4, but specific to commercial buildings, was defined by specifying values for those parameters that do not significantly affect the results. In addition, parameters such as collector tilt, collector flow rate, and storage tank volumes are assigned values (Table 7-6) that were found to be optimum in previous parametric studies (Hunn 1980; ERDA 1976). Design parameters that were varied included hot-water supply temperature, collector area, and collector type.

Collector types are designated by efficiency curves, specified by

$$\eta_n = A + Bx + Cx^2 , \qquad (7\text{-}10)$$

where

$$x = (t_f - t_a)/(IK_{\alpha\tau}) , \qquad (7\text{-}11)$$

and η_n is the efficiency for normal incidence.

Collector coefficients, A, B, and C, depend on the specific collector and whether one uses an average or an inlet fluid temperature for t_f. t_a and I correspond to ambient temperature and total solar radiation incident on the collector. $K_{\alpha\tau}$ (see Eq. 5-15) is an incident-angle modifier defined by $K_{\alpha\tau} = 1 + b_o[(1/\cos\theta)-1]$, where θ is the incident angle, and the incident-angle modifier coefficient, b_o, has values of -0.1 for single-glazed collectors and -0.18 for double-glazed collectors. The collector coefficients used were intended to correspond to typical collectors of four generic types and were based on a study of performance data supplied by manufacturers of many different collectors. Coefficients selected for the collector types used are given in Table 7-7.

Hourly simulations were run for various design parameters combinations using the DOE-2 program. The range of collector areas used resulted in a corresponding range of SLR values. The solar fraction is the fraction of the load (total energy required for water heating or combined water and building coil load) supplied by solar. SLR is as defined above.

Service Hot Water System. The first system studied was that shown in Fig. 7-9. A heat exchanger separates the collector coolant from storage water, and the solar-heated storage tank acts as a preheater for a conventional hot-water tank. The system is therefore classified as a double-tank, indirect system. A control scheme provides auxiliary heat, as needed, to maintain delivery water temperature at a specified level. The SHW demand schedule (Fig. 7-10) has a typical office building hot-water profile.

Typical results showing climate effects on service hot water system performance are shown in Fig. 7-11. Because load characteristics are uniform throughout the year, system performance is best correlated using annual load parameters. A set of design curves, recommended for various ranges of heating degree-days, is given in Fig. 7-11 for a hot water delivery temperature of 130°F (54°C) and single-glazed, selective-surface collector.

The next parameter investigated was hot-water delivery temperature. Several runs were made for a single-glazed, selective-surface collector, using water as the coolant (Fig. 7-12). As expected, lower delivery temperatures result in lower collector operating temperatures and higher efficiencies. This has a significant effect on overall system performance as represented by the solar fraction.

The effect of cold-water supply temperature was also investigated. SLR curves were produced for fixed supply temperatures that ranged from 40° to 60°F (4.4°-15.6°C). Curve differences were less than 4%, so the results presented would be applicable to most geographic locations without a correction for the cold-water supply temperature.

The effect of collector type on system performance is shown in Fig. 7-13. Note that the performance difference for the three generic collectors considered is relatively minor.

Table 7-6. The Standard Liquid System (Commercial Building Applications)

Parameter	Nominal Value
SOLAR COLLECTOR	
Orientation	Due South
Tilt (from horizontal)	Latitude + 10°
Coolant flow rate[1]	0.05 gpm/ft^2_c (113 kg/h-m^2_c)
HEAT EXCHANGER	
Effectiveness	0.70
Cold side flow rate	0.05 gpm/ft^2_c (113 kg/h-m^2_c)
STORAGE TANKS	
Capacity	1.8 gal of water/ft^2_c (73 kg/m^2_c)
Height to diameter ratio	3.0
Loss coefficient	0.05 Btu/h-ft^2-°F (0.28 W/m^2-K)
Environment temperature[2]	70°F (21°C)
Cold water supply temperature (to SHW system)	60°F (16°C)

[1]Water, water/glycol, or nonaqueous collector coolant could be used.

[2]Assumes that storage losses do not contribute to meeting the heating load.

Table 7-7. Efficiency Curve Coefficients Used to Characterize the General Types of Collectors

Type	Fluid	A	B (Btu/h-ft^2-°F)	B (W/m^2-K)	C (Btu/h-ft^2-°F)^2	C (W/m^2-K)^2
Single-glazed, selective	LIQ	0.705	-0.887	-5.04	0	0
Single-glazed, nonselective	LIQ	0.780	-1.32	-7.50	0	0
Double-glazed, nonselective	LIQ	0.643	-0.880	-5.00	0	0
Evacuated tube, single-glazed	LIQ	0.642	-0.0604	-0.343	-0.1076	-3.475
Single-glazed, selective	AIR	0.55	-0.86	-4.89	0	0
Single-glazed, nonselective	AIR	0.59	-1.1	-6.25	0	0
Double-glazed, nonselective	AIR	0.475	-0.73	-4.15	0	0

The fluid temperature used in Eq. (7-11) is the inlet temperature except for the evacuated tube case for which the average fluid temperature is used.

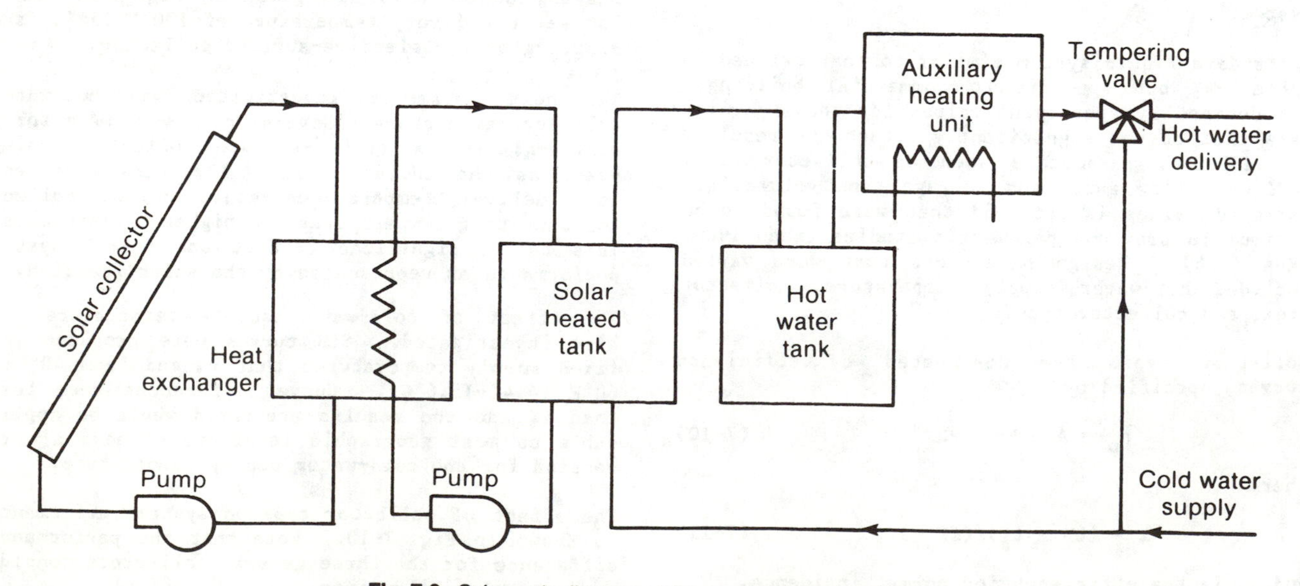

Fig. 7-9. Schematic diagram of the service hot water system

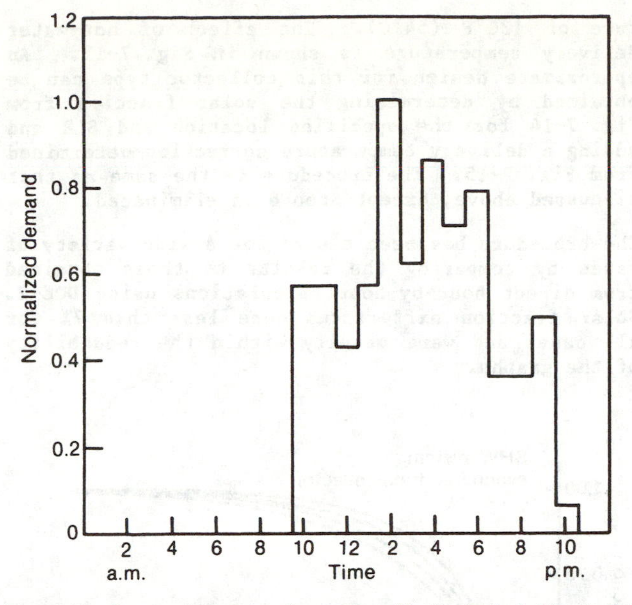

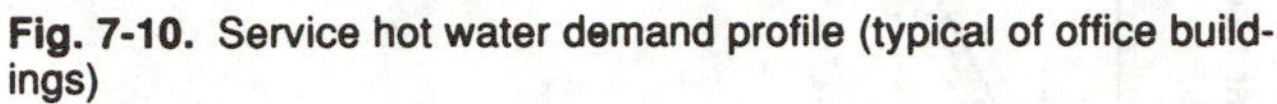

Fig. 7-10. Service hot water demand profile (typical of office buildings)

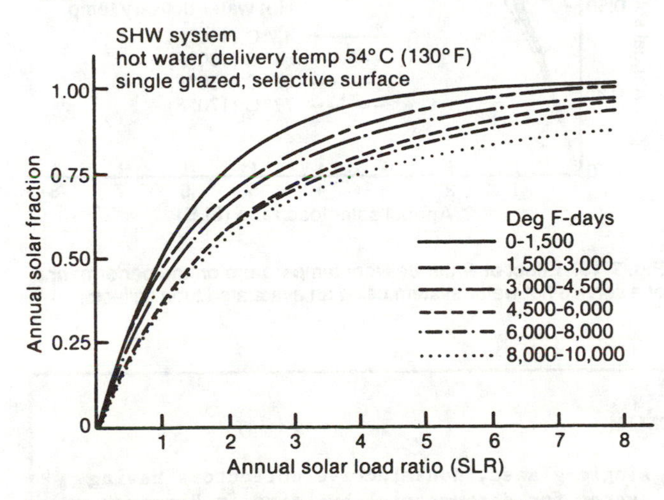

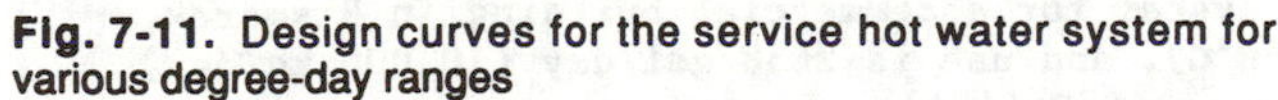

Fig. 7-11. Design curves for the service hot water system for various degree-day ranges

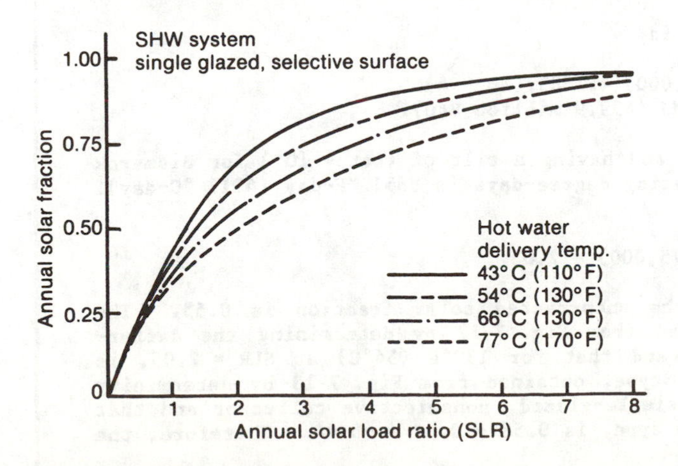

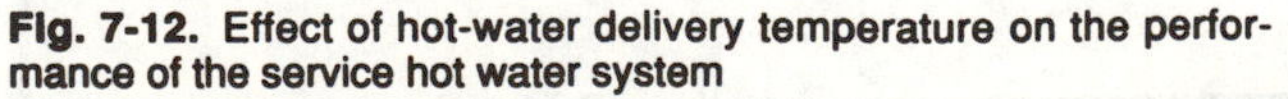

Fig. 7-12. Effect of hot-water delivery temperature on the performance of the service hot water system

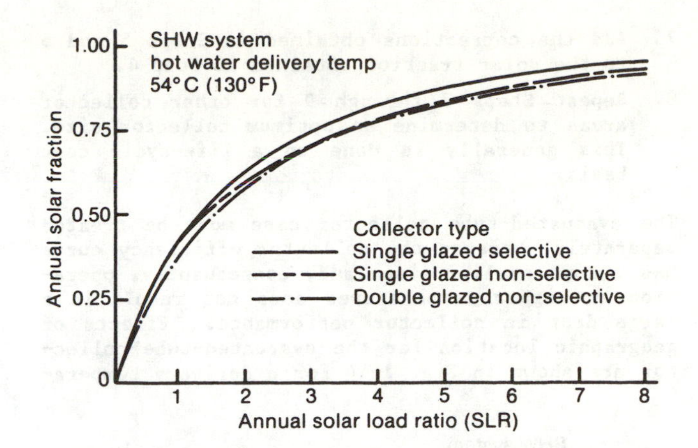

Fig. 7-13. Effect of collector type on the performance of the service hot water system

__Design Procedure for a Service Hot Water System.__ A flat-plate collector system can be sized with the following procedure. A sample problem using this recommended procedure will be given later.

1. Given a use schedule, cold-water temperature, and hot-water supply temperature, calculate the annual service hot water load.

2. Determine solar radiation incident on the collector at a given geographic location and at a specified tilt.

3. Select a collector area. Multiply the area by the result of Step 2 to determine annual solar radiation incident on the tilted collector. Divide the annual solar radiation by the load to obtain the annual SLR.

4. Use Fig. 7-11 to determine the solar fraction for the specified location (degree-day range) and the SLR from Step 3.

5. Correct for hot-water delivery temperatures other than 130°F (54°C) using Fig. 7-12. Measure the vertical distance (solar fraction correction) between the curve being used and the curve for 130°F (54°C) at the given SLR. The correction will be positive for temperatures less than 130°F (54°C) and negative for higher temperatures.

6. Correct for collector type using Fig. 7-13. An uncorrected value was obtained for a single-glazed, selective-surface collector whose efficiency curve has an intercept of 0.705 and a slope of -0.887 Btu/h-ft^2-°F (-5.04 W/m^2-K). If the collector efficiency curve is known, compare with the cases available in Fig. 7-13 and interpolate or extrapolate to make the correction.*

---

*It is recognized that this procedure is approximate and requires judgment on the part of the user. The corrections are usually small, so the effect of some error in this step will not greatly affect overall accuracy.

7. Add the corrections obtained in Steps 5 and 6 to the solar fraction obtained in Step 4.

8. Repeat Steps 3 through 9 for other collector areas to determine an optimum collector size. This generally is done on a life-cycle cost basis.

The evacuated-tube collector case must be treated separately, because the collector efficiency curve has a much smaller slope and, consequently, operation at higher temperatures does not result in a large drop in collector performance. Effects of geographic location for the evacuated-tube collector are shown in Fig. 7-14 for a delivery temperature of 130°F (54°C). The effect of hot-water delivery temperature is shown in Fig. 7-15. An approximate design for this collector type can be obtained by determining the solar fraction from Fig. 7-14 for the specified location and SLR and adding a delivery temperature correction determined from Fig. 7-15. The procedure is the same as that discussed above, except Step 6 is eliminated.

The procedure has been tested for a wide variety of cases by comparing the results to those obtained from direct hour-by-hour calculations using DOE-2. Solar fraction differences were less than 7% for all cases and were usually within the readability of the graphs.

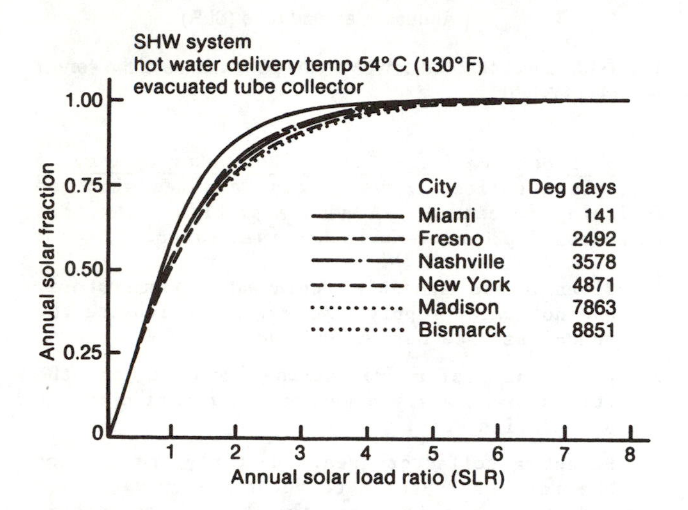

Fig. 7-14. Effect of location on the performance of a service hot water system using an evacuated tube collector

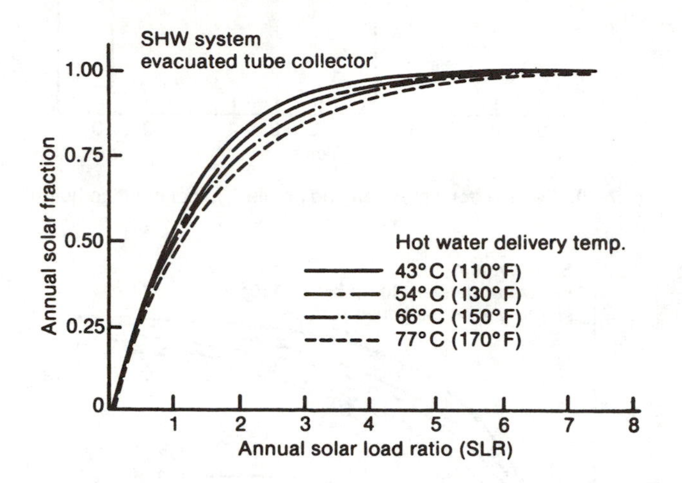

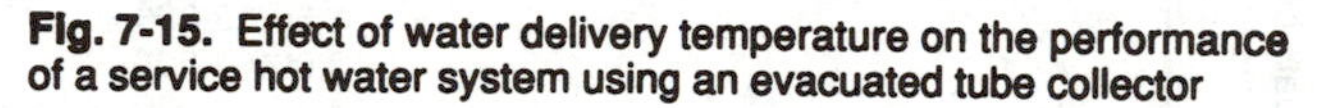

Fig. 7-15. Effect of water delivery temperature on the performance of a service hot water system using an evacuated tube collector

EXAMPLE PROBLEM

Determine the solar fraction supplied by an array of single-glazed, nonselective collectors having an area of 2153 ft^2 (200 m^2) for a service hot water system for a commercial building in Bismarck, North Dakota. Water delivery temperature is 150°F (66°C), and use is 2638 gal/day (10,000 kg/day) for week days. City water temperature is estimated to be 40°F (4°C).

Solution. The annual load, assuming 260 week days, is

$$Q = \dot{m}\, C_p\, \Delta t = (260)(10{,}000)(4.1889)(66 - 4)$$
$$= 675{,}000 \text{ MJ } (639.9 \text{ million Btu}).$$

Annual solar radiation on a collector facing south and having a tilt of (Lat + 10°) for Bismarck is 602,000 Btu/ft^2 (6840 MJ/m^2). The number of heating degree-days is 8851 °F-day (4917 °C-day). SLR is

$$SLR = (200)(6840)/(675{,}000) = 2.03 .$$

From Fig. 7-11 for SLR = 2.03 and 8851 °F-days, the uncorrected solar fraction is 0.55. The correction for water delivery temperature, obtained from Fig. 7-12 by determining the difference between the solar fraction for 150°F (66°C) and that for 130°F (54°C) at SLR = 2.05, is 0.55 - 0.62 = -0.07. The correction for collector type, obtained from Fig. 7-13 by determining the difference between the solar fraction for the single-glazed, nonselective collector and that of the reference case, the single-glazed, selective type, is 0.58 - 0.62 = -0.04. Therefore, the corrected solar fraction is 0.55 - 0.07 - 0.04 = 0.44.

Space Heating Liquid System (SHLS). A space heating liquid system provides both space heating and service hot water for a commercial building using solar collectors with a liquid coolant. A schematic diagram of the system studied is shown in Fig. 7-16.

The building selected for analysis is a medium-sized, three-story, double-loaded corridor office building having 16,600 ft^2 (1542 m^2) of floor space. It consists of offices, a computer room, a small unfinished space, restrooms, hallways, and stairwells. Internal loads are primarily from people, lighting, and computer equipment. The heating, ventilating, and air conditioning (HVAC) system is multizone, with a proportional thermostat having a 4°F (2°C) throttling range and setpoints of 68°F (20°C) in winter and 78°F (26°C) in summer. Hot water for space heating is supplied to a duct heating coil at a temperature that varies with outside air temperature. The service hot water demand schedule was obtained by multiplying values in the normalized schedule of Fig. 7-10 by a use factor of 0.47 gpm (107 kg/h).

An additional parameter that must be specified is the minimum water supply temperature for the main heating coil. A series of preliminary runs, made to determine the effect of that temperature on the solar fraction, found the solar fraction to be relatively insensitive to coil temperature changes for values less than 100°F (38°C). However, the solar fraction began to drop sharply as the coil supply temperature was increased above that level. Because a 100°F (38°C) minimum coil temperature is typical, it was selected for all subsequent runs.

Monthly loads for the SHLS vary widely because little, if any, space heating is required in the summer. The collector array is oversized for that season, the solar fraction is unity, and excess heat must be dumped. Thus results are better correlated in terms of monthly, rather than annual, load parameters.

A series of runs was made to determine the effect of location. Results indicated that, because of the resulting reduced data scatter compared with the service hot water case, a single performance curve, for a given collector, adequately represents all cities (Schurr, Hunn, and Williamson 1981).

Collector type effects are shown in Fig. 7-17. The curve for each collector type represents the best fit for 216 data points representing 6 cities, 3 collector areas, and 12 months. The standard deviations between monthly points and the curves are in the range of 4% to 7%. Note that universal performance curves for each collector type apply to all locations.

Effects of service hot water delivery temperature and use, HVAC system type, and building type were investigated; but, in all cases, variations in results over normal ranges of these parameters were small. Therefore, it is recommended that the curves of Fig. 7-17 be used for preliminary design and sizing of liquid-based solar systems for all commercial buildings in all locations. However, the following precautions should be observed:

- Results are not expected to be accurate if the service hot water load is greater than about 20% of the total load.
- Results do not apply to buildings with passive solar energy features that meet a significant portion of the load.
- Heat losses in pipes and ducts have not been included in the analysis. If they are significant, suitable adjustments should be made to the results presented here.

Space Heating Air System (SHAS). A system using air collectors was also studied. A schematic diagram of an SHAS used to provide both space heating and service hot water for a commercial building is shown in Fig. 7-18. The reference building used is

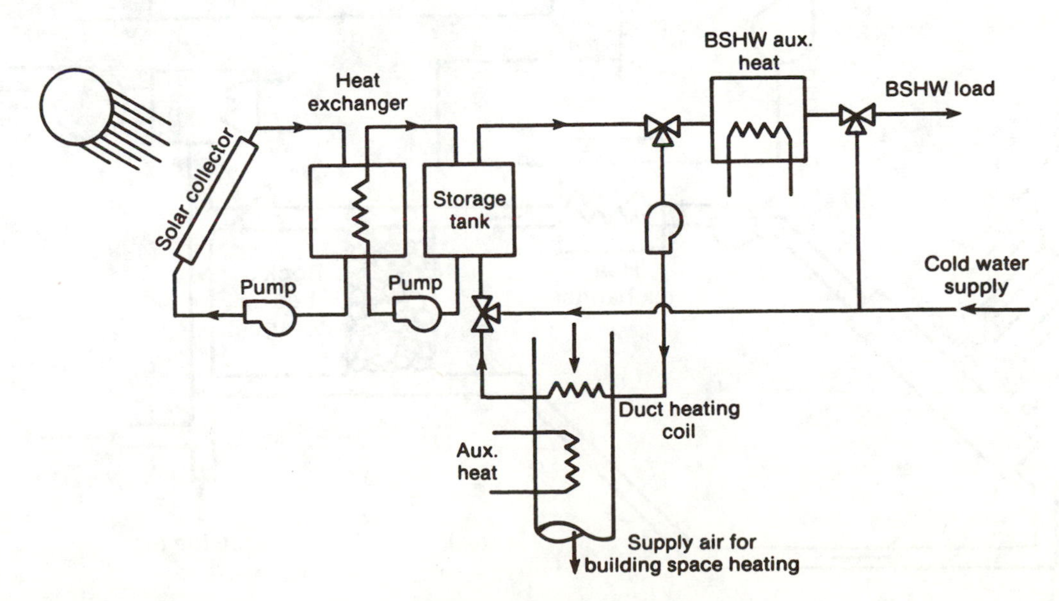

Fig. 7-16. Schematic diagram of the SHLS

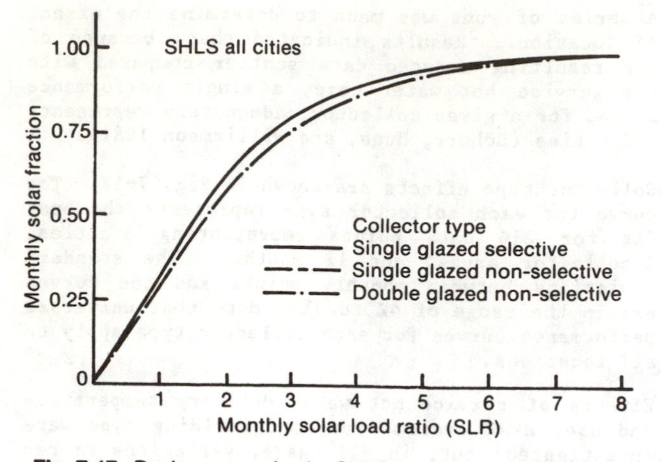

Fig. 7-17. Design curves for the SHLS for various collector types

the same as that used for the SHLS. A rock-bed volume of 0.72 ft^3/ft^2_c (0.22 m^3/m^2_c and a hot water storage mass of 0.75 lb of water/ft^2_c (3.67 kg of water/m^2_c are used. All other parameters are the same as for the SHLS.

Results for the three air collectors types listed in Table 7-7 are shown in Fig. 7-19. These may be considered monthly universal design curves and may be used for preliminary design of air heating systems for commercial buildings.

For the SHAS, effects of building service hot water demand and supply temperature on solar system performance were found to be minor.

Service Hot Water--Design Parameter Variations. Using the standard liquid collector system for residential applications (parameters listed in Table 7-4) as a reference case, a parameter variation of a residential domestic hot water load profile in Fresno, California, has been performed using both one- and two-tank systems. The two-tank system performed best thermally because lower operating temperatures are possible for its solar-heated portion. However, economic considerations often justify use of a single-tank system. Significant results are as follows:

- **Collector Tilt.** Because the load is constant over the year, collector tilt effects are quite different from those in combined space and water heating. Optimal tilt is about latitude + 5° for a two-tank system and equal to the latitude for a one-tank system. One-tank system performance drops rapidly if the tilt is more than 20° from optimum; two-tank system performance is less sensitive.

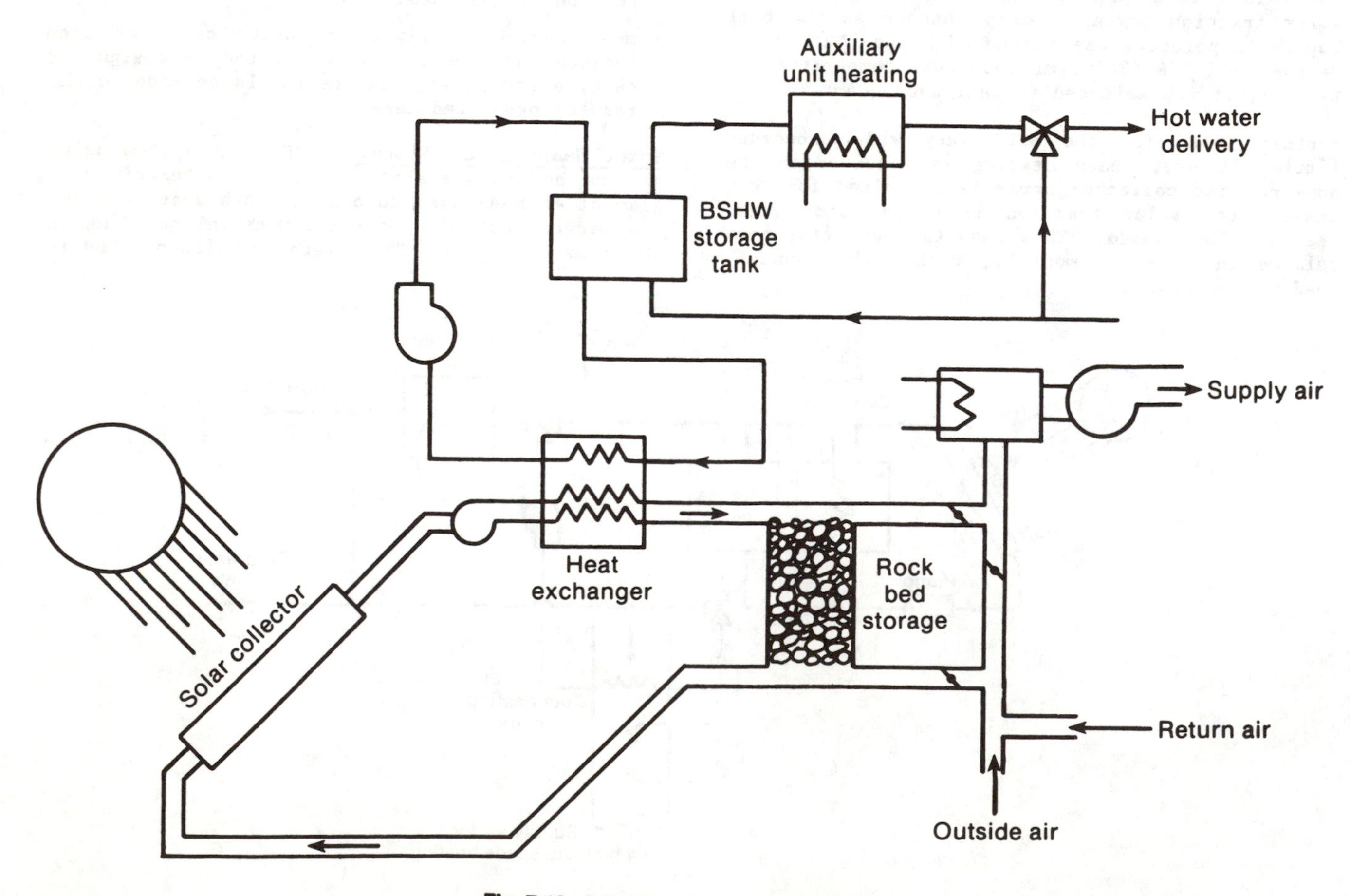

Fig. 7-18. Schematic diagram of the SHAS

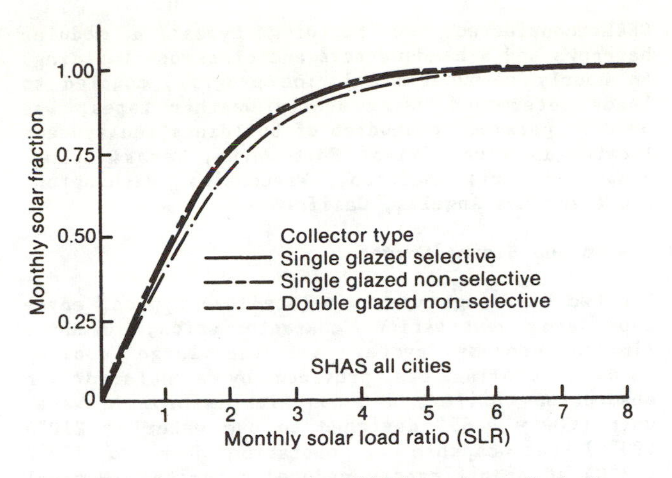

Fig. 7-19. Design curves for the SHAS for various collector types

- **Storage Mass.** Performance drops rapidly at storage parameter values below 10 Btu/°F-ft^2_c or 1.2 gal water/ft^2_c (48 kg/m_c^2) for both one- and two-tank systems. However, there is little to be gained by storage mass increases beyond 15 Btu/°F-ft^2_c or 1.8 gal water/ft^2_c (72 kg/m^2_c). This is the recommended design value, corresponding to about 1 to 2 days' storage.

The effect of other system parameters is essentially the same as in combined space and water heating, below.

Combined Space and Service Hot Water Heating--Design Parameter Variations. Studies conducted at Los Alamos (Hunn 1980; ERDA 1976), simulating residential building performance in several cities, indicate the effects on performance when parameters are varied from their nominal values. Major results are as follows:

Liquid Space Heating/Hot Water System

- **Collector Orientation.** Due-south collector orientation is optimum. Variations of up to 30° east or west reduce performance only by about 3% in all except the coldest climates, where variations should be less than 20°.
- **Collector Tilt.** Collector tilt angle is important. The oft-quoted latitude + 15° rule of thumb applies in some locations but not in all. Optimum tilt also is influenced by whether the heating load is concentrated in a short period of two or three months or is spread out over half the year. Relatively major deviations from optimum tilt affect performance only slightly.

  Many other considerations, such as ease of assembly and repair, shedding of snow and rain, architectural integration, and avoidance of potential overheating, may be more important than maximizing performance. A vertical collector may well be best in some situations.
- **Heat Exchanger Heat Transfer Coefficient.** A heat exchanger overall heat transfer coefficient of at least 10 Btu/h-°F ft^2_c (60 W/°C-m^2_c) should be achieved. For example, if total energy being collected at a particular time is 100 Btu/h-ft^2_c (300 W/m^2_c), then a temperature difference between one side of the heat exchanger and the other of 5° to 10°F (3°-6°C) is practical, and the exchanger should be sized to provide this temperature difference (see Ch. 6).

  Some collector liquids that have poor heat-transfer characteristics, such as paraffinic oils, can make achieving this kind of heat exchanger effectiveness difficult, but water is an excellent heat-transfer fluid.
- **Thermal Storage Heat Capacity.** A water tank for thermal storage is considered part of a standard liquid system. Nominal storage capacity is 15 Btu/°F-ft^2_c or 1.8 gal of water/ft^2_c (72 kg/m^2_c). For a 50-50 ethylene glycol and water solution, the corresponding thermal storage capacity is 2.0 gal/ft^2_c (82 kg/m^2_c).

  Simulation analysis indicates that having enough thermal storage to counteract more than one day of poor weather does not improve yearly performance very much. However, fairly severe performance losses are predicted if storage capacity is less than 10 Btu/ft^2_c or 1.2 gal water/ft^2_c (48 kg/m^2_c).
- **Number of Glazings.** Double glazing usually does not reduce heat losses enough to justify its added cost. Single glazing with a selective absorber surface is preferred. However, for low-temperature applications [< 120°F (49°C) collector temperature] or in mild climates, as in solar-assisted heat pump systems, a nonselective, single-glazed collector may be appropriate.
- **Heat Losses from Storage.** The best location for a heat storage tank is within or beneath the space to be heated. Then, heat lost from storage tends to contribute to the heated space. The tank should be insulated for times when the living space does not require heat and when tank heat losses would contribute to overheating. A U-value of less than 0.2 Btu/h°F-ft^2 (1.1 W/°C-m^2) of tank is recommended.
- **Collector Glazing Transmittance.** A 1.0% reduction of glass transmittance reduces performance by only 0.25%.
- **Collector Surface Absorptance.** Surface absorption should not be less than 90%.
- **Collector Surface Emittance.** A highly selective surface can increase performance of single-glazed collectors significantly; but it offers less advantage to double-glazed collectors, and durability can be a problem.
- **Collector Coolant Flow Rate.** The flow rate should be designed to keep peak coolant temperature rise to about 20°F (11°C). This generally corresponds to a flow rate of 10 lb/h-ft^2_c or 0.02 gpm water/ft^2_c (49 kg/h-m^2_c).
- **Collector Heat-Transfer Coefficient.** A heat-transfer coefficient (UA/ft^2_c) greater than 10 Btu/h°F-ft^2_c (57 W/°C-m^2_c) will give near maximum performance.

- **Collector Back Insulation.** UA/ft^2_c values of 0.1 $Btu/h°F\text{-}ft^2_c$ (0.6 $W/°C\text{-}m^2_c$) should be integrated into the collector design. This corresponds to 3 in. (76.2 mm) of insulation that has a thermal conductivity (k) of 0.025 Btu/h-ft-°F (0.043 W/°C-m).
- **Collector Heat Capacity.** Collector design should minimize metal mass and fluid inventory in collectors.
- **Distribution Pipe Insulation.** Collector-to-storage distribution pipe heat losses can be kept relatively low by insulating to a U-value of 0.15 $Btu/h\text{-}°F\text{-}ft^2$ (0.9 $W/°C\text{-}m^2$) of pipe or 2 in. (50.8 mm) of insulation that has a k of 0.025 Btu/h-ft-°F (0.043 W/°C-m).

Air Space Heating/Hot Water System

The above remarks about the standard liquid system generally apply to the standard air system. An air collector, coupled with a rock bed, was simulated using the same locations and weather data. The design parameters, and the effects of varying them, are essentially the same for both the liquid and air systems, except for those listed below.

- **Collector Heat-Transfer Effectiveness** (product of heat transfer coefficient times heat transfer area divided by collector area). This should be approximately 4 $Btu/h\text{-}°F\text{-}ft^2_c$ (23 $W°C\text{-}m^2_c$).
- **Collector Air Flow Rate.** Performance drops severely at air flow rates below 1 cfm/ft^2_c (5 $L/s\text{-}m^2_c$); 2 cfm/ft^2_c (10 $L/s\text{-}m^2_c$) is recommended.
- **Thermal Storage Heat Capacity.** The capacity of the thermal storage does not rise rapidly beyond a value of about 10 $Btu/°F\text{-}ft^2_c$ (200 $kJ/°C\text{-}m^2_c$), corresponding to 50 lb of rock/ft^2_c (230 kg/m^2_c). Actually, 15 $Btu/°F\text{-}ft^2_c$ is a recommended value.
- **Rock-Bed Temperature Distribution.** A short air flow path in the rock bed is preferable to a long one to sustain minimum pressure drop, but it should be longer than about 12 rock diameters to take advantage of spatial temperature distribution benefits.

Recommended Solar System Design Parameters. The above parametric study results can be summarized by identifying recommended solar system design parameters. These nominal design parameters (Table 7-8) are recommended for the standard system configurations studied. They are not, for example, applicable to systems that combine solar-heat pumps or solar cooling systems.

Solar Cooling Systems

The U.S. Army Construction Engineering Research Laboratory (CERL) has developed performance curves for liquid solar systems that provide both space heating and cooling (Hittle et al. 1977). The method is recommended for assessment of the feasibility of such systems, but hourly computer simulation should be used for system design.

CERL considered two building types: a modular barracks and a headquarters and classroom building. An hourly computer simulation program, coupled to loads determined using hourly weather tapes, was used in parametric studies of buildings and systems located in five cities: Fort Worth, Texas; Columbia, Missouri; Madison, Wisconsin; Washington, D.C.; and Los Angeles, California.

Design and Sizing Curves

The two building types simulated had typical envelope and ventilation characteristics, used a limited economy cycle, and had large cooling loads. Cooling was provided by a solar-driven absorption chiller, a 3-ton lithium bromide water unit (COP = 0.65) designed to use water at 210°F (99°C) but capable of operation down to 170°F (77°C) at significantly reduced capacity. Nominal collector operating temperature was about 200°F (93°C). Heating was provided by a coil in the air supply duct that was heated by water from solar storage and boosted by an auxiliary heater if the supply temperature fell below a specified level.

The resulting universal performance curves (Fig. 7-20) can be applied to almost any building, at almost any location, that has a dominant cooling load and known annual heating and cooling loads and known annual solar radiation. Data are correlated on an annual, rather than a monthly, basis because the combined heating and cooling load is fairly uniform on an annual basis. However, the performance curves should not be applied to lightweight or uninsulated buildings or those with large glass areas because their thermal properties may differ greatly from those studied. Simulations used to develop these curves were limited, so care must be taken in extrapolating results to other applications. Several assumptions were made, including use of a particular outside air schedule and a single storage tank, with auxiliary boost and a fixed heating coil design temperature.

Figure 7-20 shows solar system performance for storage capacities of 16, 31, and 62 $Btu/°F\text{-}ft^2_c$ (325, 630, and 1260 $kJ/°C\text{-}m^2_c$). Each curve is described by the general equation

$$f = B(r - 0.082\ r^2), \qquad (7\text{-}12)$$

where f is the annual load fraction supplied by solar energy and B is a parameter shown in Fig. 7-20. r is a modified SLR that corrects for different collector types:

$$r = \frac{H_\Sigma A_C}{Q_L M} \qquad (7\text{-}13)$$

H_Σ, A_C, and Q_L are the annual incident solar radiation incident on a tilted surface, the collector area, and the annual load respectively, and M is the multiplying factor for various collectors given in Table 7-9. Note that proper operation of the absorption chiller requires a solar collector that achieves a temperature of 180°F (82°C) or greater. Thus, solar heating and cooling systems must include a high-performance collector, and a collector multiplying factor for a selective surface should be used in design evaluations.

Table 7-8. Recommended Solar System Design Parameters

Parameter	Recommended Design Value
1. Collector area	Determined by life-cycle cost analysis
2. Collector tilt	Latitude +5° (water heating only) or latitude +15° (space and water heating)
3. Collector orientation	Due south
4. Storage mass	15 lb water/ft^2_c = 1.8 gal/ft^2_c (liquid system) 15 Btu/°F-ft^2_c = 75 lb rocks/ft^2_c (air system)
5. Heat exchanger heat transfer coefficient	10 Btu/h-°F-ft^2_c
6. Collector heat transfer coefficient	10 Btu/h-°F-ft^2_c (liquid system) 4 Btu/h-°F-ft^2_c (air system)
7. Number of glazings	1
8. Collector surface absorptance	0.9 with selective surface
9. Collector coolant flow rate	10 lb/h-ft^2_c = 0.02 gpm water/ft^2_c (liquid system) 2 cfm/ft^2_c (air system)
10. Collector back insulation	U = 0.1 Btu/h-°F-ft^2_c
11. Distribution pipe insulation	U = 0.15 Btu/h-°F-ft^2_c of pipe

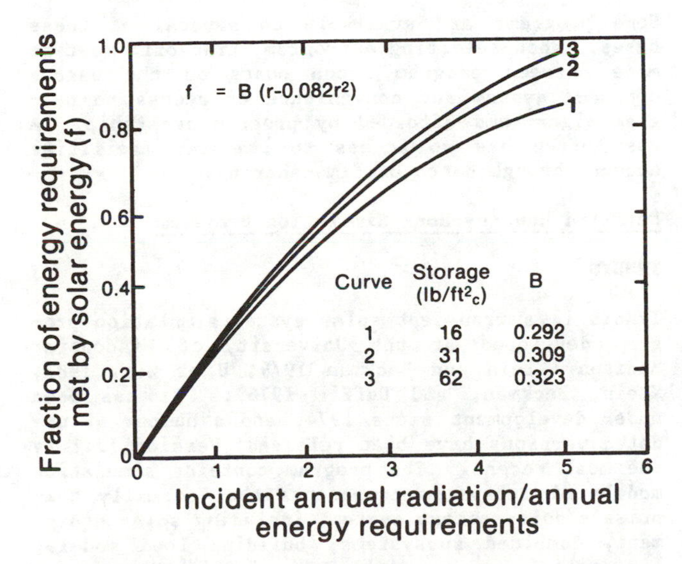

Fig. 7-20. Approximate universal performance curves for three storage capacities in a combined heating and cooling system

Design Parameter Variations

Selected system parameter variations were studied through hourly simulations using weather data for the five cities that CERL studied. For any system size, solar energy use varied greatly at the different locations (Hittle et al. 1977), indicating the critical impact of building load on overall solar system performance.

Comparison of load fraction met by solar energy as a function of thermal storage capacity shows that

Table 7-9. Collector Area Multiplying Factor, M, for Different Collector Characteristics

α = Absorptance
ε = Emittance
N = Number of glazings

Absorptance/Emittance α/ε	Multiplying Factor, M N = 1	N = 2
0.96/0.96	1.55	1.09
0.94/0.30	1.09	0.97
0.90/0.10	1.00	0.93

performance is improved greatly by increasing storage from 8 to 15 lb of water/ft^2_c (40-80 kg/m^2_c), but it is improved considerably less by further increases.

The effect of collector tilt on performance of a combined solar heating and cooling system shows that optimum tilt is approximately latitude - 10°. This is quite different from service hot water only or space heating only systems, which optimize at higher tilt angles, because dominant cooling loads occur during the summer when lower tilt angles are more effective.

Other Applicable Methods

Solar design methods (Klein and Beckman 1979; Lunde 1979, 1980) that include effects of different design water temperatures can be applied to solar cooling systems if a minimum load temperature is set that is appropriate to absorption cooling equipment.

Industrial Process Heat (IPH) Systems

Most IPH systems are characterized by high-temperature output needed to meet load requirements. Such systems, therefore, often use concentrating collectors with highly nonlinear performance curves. To account for higher temperatures when flat-plate collectors are used, methods from Klein and Beckman (1979) or Lunde (1979, 1980) can be used; computer analysis methods are recommended when concentrating collectors are used. Kutscher et al. (1982) give performance estimation and system sizing methods for IPH systems.

COMPUTER SIMULATION METHODS

Two types of main-frame computer simulation programs are applicable to solar system design. Some building energy analysis computer programs have solar simulation capabilities; examples are DOE-2, BLAST, ACCESS, and TRACE. There are also detailed programs that have been developed specifically for solar system simulation but that generally have some capability to simulate load and HVAC systems; examples are TRNSYS, SOLCOST, SIMSHAC, EMPSS, and WATSUN. Many in this second program set offer specific features comparable to those in large energy analysis programs but are usually only appropriate for envelope-dominated buildings, where loads are readily computed using degree-days. Building energy analysis computer programs are most appropriate for commercial buildings, which are interior-load dominated and require calculations of coupled heating and cooling loads. Compendia of both types of solar simulation computer programs are available (Merriam 1981; SERI 1980). Winn (1980) presents a comprehensive discussion of such programs.

Two approaches used in computer simulation of active solar heating and cooling systems are the fixed-schematic approach and the component-based approach (Feldman and Merriam 1979), with the first (with options) most prevalent. A set of generic system schematics is fixed, with a user selecting design options that are usually limited to equipment configuration and performance, control strategies, and auxiliary fuel types. Usually, not all program capabilities are exercised in a given run. When modeling a system, a user defines only those components required; other components are deactivated. A fixed-schematic approach is less expensive to develop and execute than component-based methods. However, the method is less flexible, and distinct system schematics must be programmed separately; as additional systems are added, program size increases.

The component-based approach is organized around components rather than systems. User inputs include a definition of a system schematic that identifies the library of components to be used and how they are to be interconnected. Additional components may be added to the library. This approach provides flexibility in simulating a variety of system types, but input data preparation is more complex and run times are increased.

Solar heating and cooling simulation programs are available by purchase, lease, time-share, or batch processing. In addition, some programs are proprietary; a user has limited access to program documentation and assumptions, and limited opportunity to make program changes to meet special needs. Source code (and documentation) purchase is only useful if a user has access to computer facilities. Under a lease agreement, the program leasor installs the program on the leasee's computer and provides instruction and documentation for a limited time, but this also requires in-house computer facilities.

Time-share programs are available on national computer service bureau time-sharing networks. Costs are generally proportional to computer time used. Problems of initial implementation and maintenance are avoided, but a user forfeits the flexibility to modify the program. When programs are operated in batch mode, input data are submitted to the program owner for execution, with charges usually based on computer time used. Turnaround may vary from a few hours to a few days. Programs (e.g., DOE-2) are increasingly becoming available for use on the IBM PC, and this is an alternative unless many annual simulations are needed.

Some programs are available on several of these bases. When selecting a program, tradeoffs must be made between program setup costs on the user's computer system and convenience of access to program algorithms afforded by program ownership. A user often has no access to computer facilities except through batch or time-sharing.

Detailed Hour-by-Hour Simulation Programs

TRNSYS

TRNSYS is a transient solar system simulation program developed at the University of Wisconsin-Madison (Klein and Beckman 1979; U of Wis. 1983; Klein, Beckman, and Duffie 1976). It has been under development since 1974, and a number of updated versions have been released; Version 12.1 is the most recent. The program contains simulation models of subsystem components that normally comprise a solar energy system, including solar equipment, combined subsystems, building load models, and utility routines (see Table 7-10). A user can "hook up" components representative of a particular solar system to be analyzed and then simulate that system's performance. The level of detail can be user selected.

TRNSYS was developed to fill the need for a flexible, component-based, public domain computer program for the analysis and design of a wide variety of solar heating and cooling systems. An additional use of such a program is to develop design methods, such as F-Chart, and design handbooks for use in design of representative solar system types. Once correlation methods have been developed from simulations, detailed computer simulations are no

Table 7-10. TRNSYS Library Components

Utility Components	Equipment	Hydronics
Data Reader	On/Off Auxiliary Heater	Pump/Fan
Time-Dependent Forcing Function	Absorption Air Conditioner	Flow Diverter/Mixing Valve/Tee Piece
Algebraic Operator	Dual-Source Heat Pump	Pressure Relief Valve
Radiation Processor	Conditioning Equipment	Pipe
Quantity Integrator	Part-Load Performance	
Psychrometrics	Cooling Coil	**Controllers**
Load Profile Sequencer		Differential Controller with Hysterisis
Collector Array Shading	**Heat Exchangers**	Three-Stage Room Thermostat
Convergence Promoter	Heat Exchanger	Microprocessor Controller
	Waste Heat Recovery	
Solar Collectors		**Output**
Linear Thermal Efficiency Data	**Combined Subsystems**	Printer
Detailed Performance Map	Liquid Collector-Storage System	Plotter
Single or Bi-Axial Incident Angle Modifier	Air Collector-Storage System	Histogram Plotter
Theoretical Flat-Plate	Domestic Hot Water	Simulation Summarizer
Theoretical CPC	Thermosiphon Solar Water Heater	Economics
Thermal Storage	**Building Loads and Structures**	**User-Contributed Components**
Stratified Liquid Storage (finite-difference)	Energy/(Degree-Hour) House	PV/Thermal Collector
Algebraic Tank (plug-flow)	Detailed Zone (transfer function)	Storage Battery
Variable Volume Tank	Roof and Attic	Regulator/Inverter
Rockbed	Overhand and Wingwall Shading	Electrical Subsystem
	Window	
	Thermal Storage Wall	
	Attached Sunspace	

longer necessary for design of the standard representative systems.

Program Description. An information flow diagram, which describes the system to be simulated, is first developed for a TRNSYS simulation. It defines the components to be used and how they are connected. There must be an information flow diagram, typically referred to as a component diagram, for each subsystem component used. A component diagram for a collector is shown in Fig. 7-21. Inputs are fluid inlet temperature, mass flow rate, ambient temperature, and radiation incident on the tilted collector surface. Outputs are fluid outlet temperature, mass flow rate, time rate of change of useful energy collected, a transmittance-absorptance product, and collector loss coefficient. Parameters used in modeling the component include collector area, an effectiveness factor, specific heat of the transport fluid, collector absorptance, collector loss coefficient, and collector cover transmittance.

Suppose we wish to simulate the performance of the service hot water system shown schematically in Fig. 7-22. Subsystem components include a collector, a pump, and a controller. Suppose the controller turns the pump on whenever the difference between fluid inlet temperature and fluid outlet temperature exceeds 0.9°F (0.5°C), and turns the pump off whenever this temperature difference is 0°F. Suppose, furthermore, that supply water is at a constant temperature of 68°F (20°C), and that the mass flow rate is constant at 0.9 gpm (200 L/h) whenever the pump is on. An information flow diagram for this system is shown in Fig. 7-23. The manner in which component outputs are connected to inputs of other components is illustrated in this diagram. For example, output t_o from the collector becomes input t_1 to the controller.

Now the TRNSYS program requires solar and weather input to drive the system. Hence, we must read in the necessary data. These are provided by a data reader, which may be modeled as simply another subsystem component. The data reader for TRNSYS has no inputs but has up to 20 outputs. These outputs include hourly radiation on the tilted surface, I_t, and ambient temperature, t_A. Input from the data reader (see the information flow diagram, Fig. 7-23, connecting I_t and t_A to the inputs of the collector component diagram), is needed to make the system operate.

Finally, we must provide for system output. TRNSYS output components include integrators, printers, and plotters. Suppose we want to determine a collector's average daily efficiency. This would be provided by

$$\eta = \frac{Q_u}{\Sigma I_t} \tag{7-14}$$

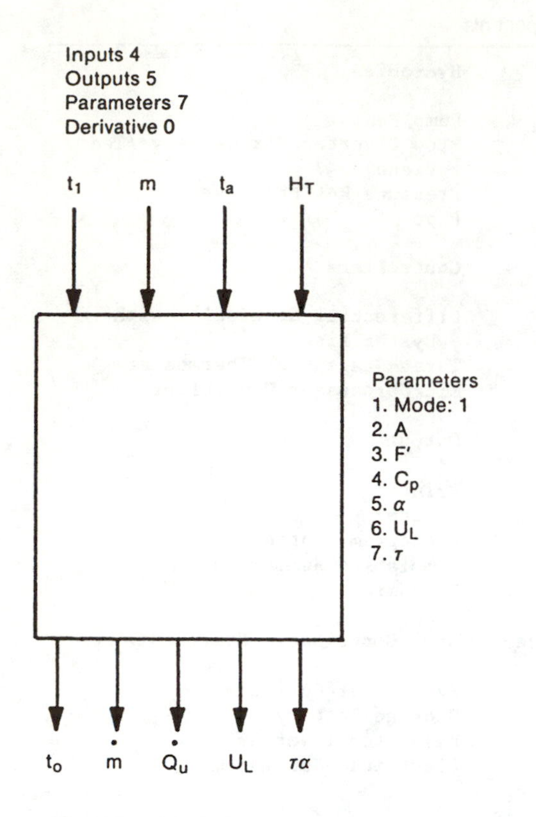

Fig. 7-21. TRNSYS component diagram for a collector

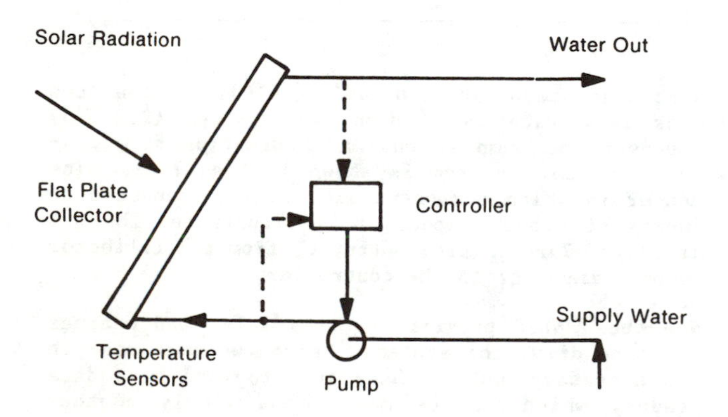

Fig. 7-22. Schematic diagram of a simple solar water heating system

Useful energy, Q_u, can be determined by integrating instantaneous values over 24 hours. These integrations are performed in TRNSYS by using a quantity integrator. Finally, tabular results can be obtained by using a printer subsystem, and graphical results can be obtained by using a plotter subsystem. Component diagrams for these subsystems can be added to the information flow diagram to obtain a complete information flow diagram for this example (Fig. 7-24).

Constructing information flow diagrams for reasonably simple solar systems is straightforward but can be rather involved. However, proficient use of TRNSYS for complex systems typically requires the talents of an experienced systems analyst.

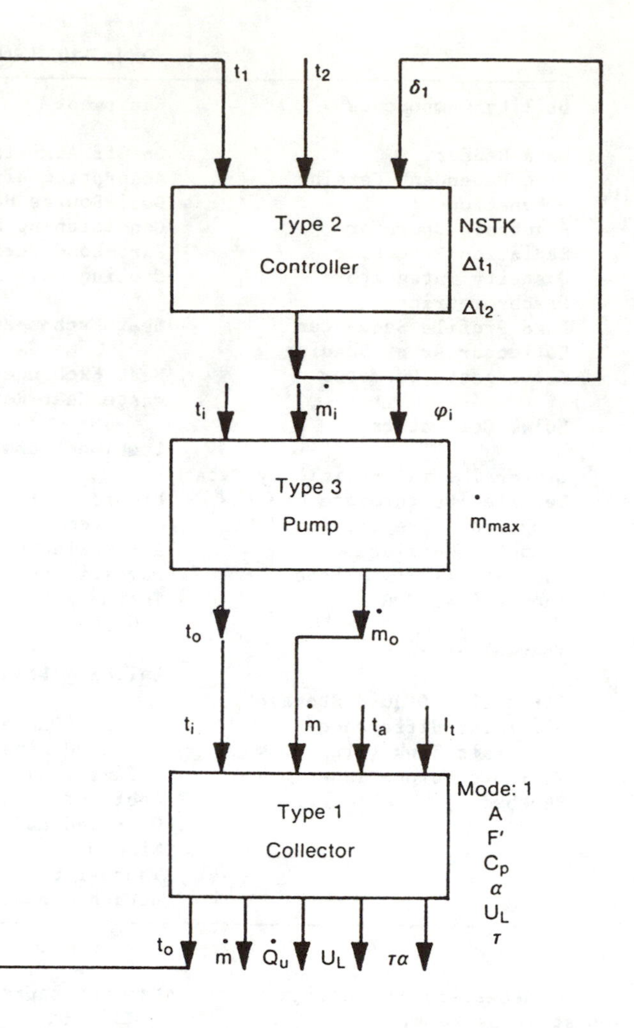

Fig. 7-23. Information flow diagram for a solar water heating system

A user wishing to model a subsystem component not included in the TRNSYS component library must write a subroutine and develop an information flow diagram for that particular component. This is usually straightforward, but at times it can be a significant problem. Components included in the TRNSYS library are listed in Table 7-10.

Validation Results. If a simulation program's results are shown to be accurate, the program can be used to study the performance of different systems under varying climatological conditions, or it can be used to develop simplified tools. TRNSYS validation studies have been conducted using data obtained from Colorado State University Solar House I (Winn and Duong 1976; Lantz and Winn 1978). Measured values of component outputs, including solar radiation, ambient temperature, storage tank temperature, enclosure temperature, collector inlet and outlet temperatures, collector flow rates, useful energy collected, and auxiliary energy supplied, were compared with simulated values obtained from TRNSYS. Differences between observed and simulated results were analyzed statistically. Results obtained from a simulation of Colorado State University Solar House I during the cooling season are shown in Table 7-11. TRNSYS can be seen

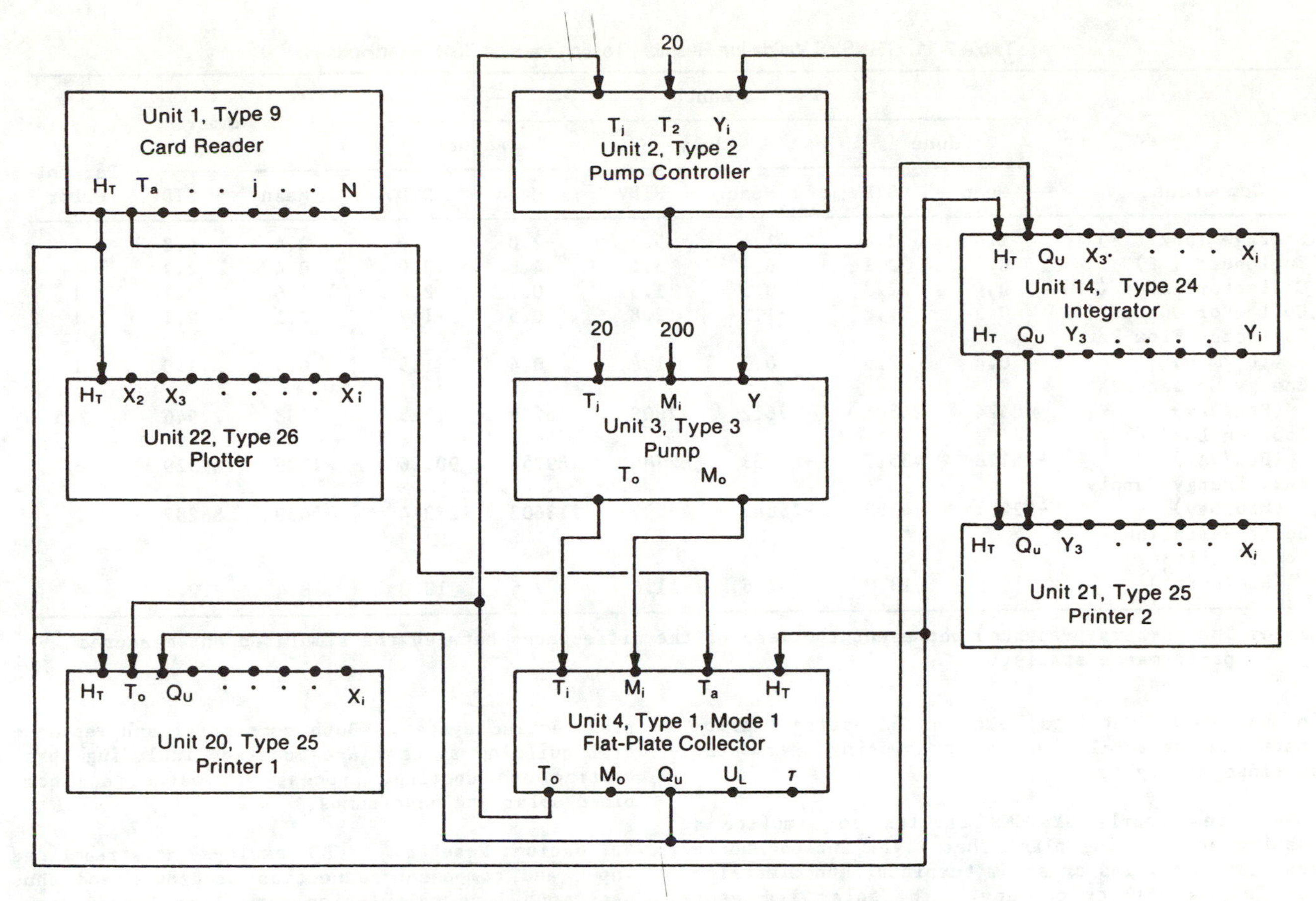

Fig. 7-24. Complete information flow diagram for a solar water heating system

to provide a very accurate tool for analyzing such systems.

Additional comparisons of measured and simulated performance for CSU Solar House I (Mitchell, Beckman, and Pawelski 1978) also show quite good agreement. This further suggests that TRNSYS simulations can adequately model this physical system. Therefore, one can proceed with confidence to analyze similar type systems in different geographic locations. In fact, if validated TRNSYS subsystem components are used, accurate results can be obtained from TRNSYS for any system type under any climatological conditions. TRNSYS has been used to model many system types, including space heating and cooling systems in both residential and commercial buildings, industrial process heat applications, and grain drying applications.

DOE-2/CBS

DOE-2 is an hour-by-hour building energy analysis program that also has extensive active solar systems simulation capabilities in its component based simulator (CBS). The main DOE-2 program was developed by Lawrence Berkeley Laboratory (LBL) with assistance from Los Alamos National Laboratory. Los Alamos developed the solar system simulator CBS based on the component based approach used in TRNSYS. The latest version of the program is DOE-2.1C (York, Tucker, and Cappiello 1982). It is available on several computer service networks and at the Solar Energy Research Institute (SERI). (SERI access is available only to governmental agencies or their contractors.) It also may be purchased from the National Technical Information Service (NTIS). A version of DOE-2 (without CBS) is now available for the IBM PC.

Program Description. Four programs, LOADS, SYSTEMS, PLANT, and ECONOMICS, form the heart of DOE-2. LOADS uses hourly weather data to perform hourly and design, or peak, energy-use calculations of heating and cooling loads on a building's HVAC system. It takes into account the type of building construction; the occupancy; building equipment; hours of the day and days of the week of occupancy and equipment use; solar radiation conducted through walls and transmitted through windows; and thermal masses of interior walls, floors, and furniture in which heat is stored. It also takes into account effects of building shading by trees, attached balconies, overhangs, fins, or nearby buildings.

SYSTEMS takes hourly LOADS results and simulates one or more of 21 HVAC systems, such as a variable-volume air system, floor-panel or baseboard heating, or a constant-volume air system. It also includes in its calculations effects of variations

Table 7-11. TRNSYS Validation Results - Testing Periods CSU Solar House I

Component	June Mean	June STDV	July Mean	July STDV	August Mean	August STDV	Summer Average Mean	Summer Average STDV	Percent Error
Storage Tank (°F)	1.4	2.8	1.6	5.7	-2.0	5.9	0.4	4.8	1
Enclosure (°F)	-0.1	2.1	6.1	3.1	-4.8	3.0	0.4	2.7	1
Collector Inlet (°F)	0.6	2.3	0.5	1.6	0.7	2.4	0.6	2.1	1
Collector Outlet (°F)	0.5	1.8	-1.2	2.8	0.5	1.7	-0.1	2.1	1
Collector Flow Rate (lb_m/min)	0.4	2.5	0.5	1.8	0.4	1.5	0.4	1.9	1
Energy Collected (Btu/Day)	8524	24506	7812	22609	6279	21725	7538	22946	2.5
Cooling Load (Btu/Day)	-25178	43517	-68253	62785	89753	90286	1226	65529	1
Aux. Energy Supply (Btu/Day)	-32271	44952	-75843	86597	113603	127314	5489	86287	2
Solar Radiation (45° tilt) (Btu/h-ft^2)	9.9	13.9	7.6	11.6	7.6	10.8	8.4	12.1	4

Note: The numbers presented represent the mean of the differences between the simulated and measured performance statistics.

in thermostat settings, such as adjusting thermostats for seasonal changes or lowering thermostat settings at night.

PLANT uses hourly SYSTEMS results to simulate a heating and cooling plant that might include boilers, chillers, gas or steam turbines, and electricity from a utility company. The solar simulator package, CBS, a set of subprograms that simulates elements of an active solar heating or cooling system, is an element of the PLANT program.

Finally, ECONOMICS uses PLANT results to calculate and compare costs of installation, maintenance, and fuel use of alternative systems over a building's lifetime. DOE-2's thermal energy results have been compared with recorded utility consumption data for many different types of systems (Diamond, Hunn, and Cappiello 1981, 1986).

The DOE-2 solar simulator, CBS (Roschke, Hunn, and Diamond 1978), is a multilevel analysis and design tool for both liquid (Fig. 7-25) and air (Fig. 7-26) active solar heating and/or cooling systems. It contains many component models from which a user can assemble a system to suit his or her own special requirements. Alternatively, the user can select a preassembled, standard liquid or air system from among several stored in the DOE-2 CBS library. To save time, most users will use a preassembled system. Both commercial and residential building systems are modeled, including space heating and cooling, process hot water, and combined solar and heat pumps.

<u>**Validation Results.**</u> CBS employs a streamlined input and component-connection procedure and thus uses much less calculation time than TRNSYS does. Several comparative runs (Freeman, Maybaum, and Chandra 1978; Eden and Morgan 1980; Roschke, Hunn, and Diamond 1978) have been made with TRNSYS for space and domestic water heating systems to verify the relative accuracy of CBS. Both hourly and monthly results were examined; solar fractions agreed to within 3.2% on a monthly basis and to within 0.8% on an annual basis.

Other Computer Simulation Programs

There are many other special-purpose computer programs, both proprietary and public domain, for solar system simulation. These programs, which include SOLCOST, BLAST, ACCESS, HISPER, RSVP, SIMSHAC, EMPSS, and TRACE, among others, can be quite useful in the design of certain solar systems. However, most have not been thoroughly validated. For descriptions of these and other programs, as well as access information, see Feldman and Merriam (1979), Merriam (1981), and SERI (1980, 1985).

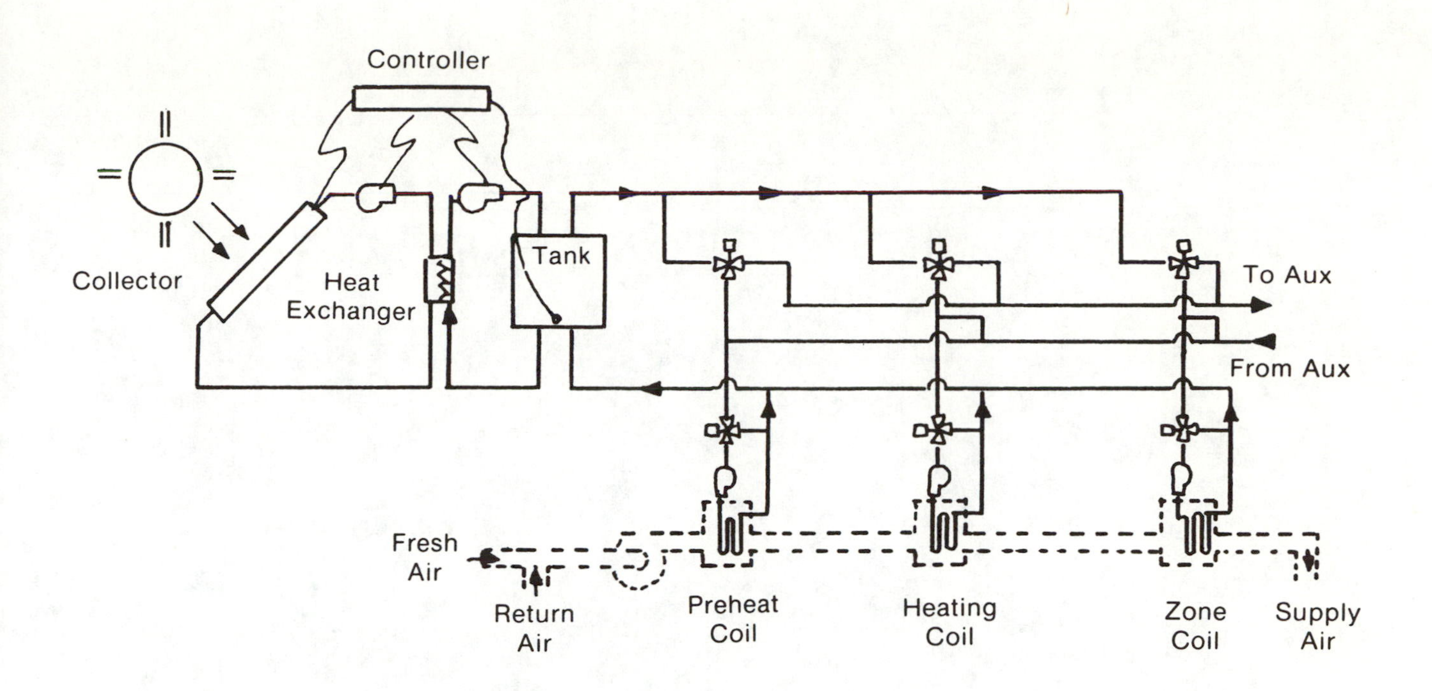

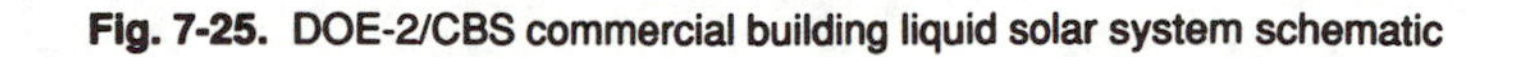
Dashed equipment items are not simulated in CBS.

Fig. 7-25. DOE-2/CBS commercial building liquid solar system schematic

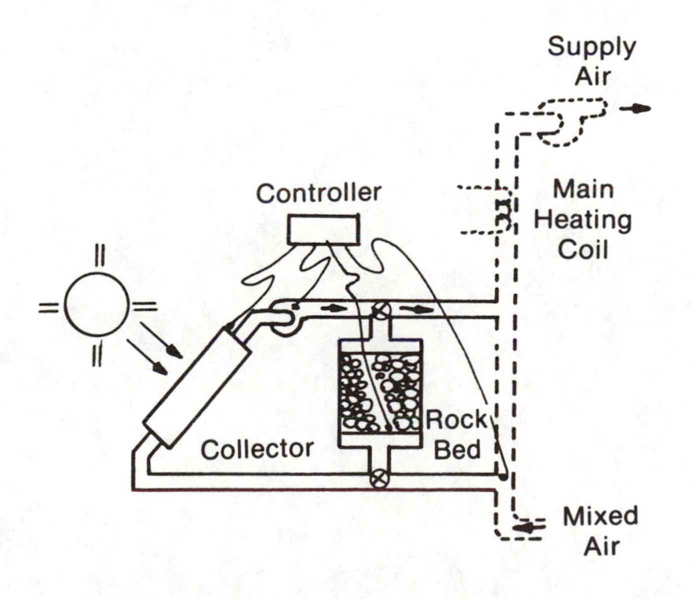

Dashed equipment items are not simulated in CBS.

Fig. 7-26. DOE-2/CBS commercial building air solar system schematic

Part III
Active Solar System Installation, Construction, and Operation

Chapter 8

Detailed Design, Construction, and Installation Considerations for Active Systems

INTRODUCTION

Experience gained from more than 100 DOE design reviews shows that construction problems, cost cutting, and inexperience contribute to most installation problems associated with active solar systems (DOE 1979a). The building trades are not experienced in solving problems associated with active systems. Whereas a typical residential or commercial building roofscape has minimal exposed construction, especially piping, a building with an active solar system has considerably more exposed construction. It is therefore important for a designer to isolate and solve construction and installation problems and not leave solutions to the contractor. In addition, a designer's desire to extract the maximum useful energy from a system may lead to installation of devices that are not cost effective, increasing the payback period and reducing system reliability.

This chapter concentrates on design of construction details for installing active solar systems rather than on engineering design issues, which are generally addressed during the design development phase (see Chs. 4-6). Emphasis is on reliability, durability, and cost effectiveness. In many cases, construction problems are just described; no solutions are offered since they will vary with location and with changing technology. Once a problem is understood, a designer can select the "best" approach based on specific conditions.

COLLECTOR INSTALLATION

General Considerations

Collectors must be properly handled during delivery and installation. When collectors are delivered, packages should be opened immediately and inspected for shipping damage. A damage claim should be filed promptly.

To prevent cover plate breakage, some collectors come with a protective adhesive paper. In extreme temperatures, the glue can freeze or melt and can be difficult to remove. Therefore, this paper should be stripped as soon as possible.

Where collectors are being mounted near high-traffic areas, reflective glare can cause public annoyance and opposition to installation. Collectors with black anodized aluminum and stippled finish on the glass are the least reflective. Lightning protection for the collectors and system should also be provided, where appropriate.

Collectors with large areas usually cost less per unit area to install than small collectors because installation involves approximately the same amount of work. Taller collectors require fewer vertical supports and, therefore, fewer roof penetrations for roof-mounted installations and fewer footings for field-mounted installations. On a pitched roof, wide collectors tend to require less wood blocking and consequently are more cost effective than tall collectors.

Support Structures and Collector Attachments

Designers tend to select tubular aluminum or steel space frames when long-span triangular-shaped structures are required. Although aesthetically pleasing, this approach can be extremely costly and labor intensive, and fabricators of space frames are hard to find. In general, simple hot-rolled steel, wide flange beams, and steel angle structural supports are less expensive and easier to install. It is important to minimize roof penetrations, based on economical structural spans, and to design for wind and thermal expansion forces.

For small installations, many collector manufacturers have developed simple predesigned support systems for their product. Simple racks are also available, constructed of either pipe sections or "Unistrut" sections or angles, and can be cost effective. However, when used for larger projects, they often lead to extensive roof penetrations that add to roofing cost and increase the probability of leaks.

Collector attachment details also can affect cost. Some collectors require special structural connections, such as pipe or Unistrut. If not integral parts of the structural support, they will cost extra and will detract from the cost effectiveness of the collector. Therefore, it is important to review collector connection requirements before designing the structure.

In general, modules with exterior manifolds tend to cost more to install than those with internal manifolds. Regardless of the type of collector and support used, different metals must be kept from coming in contact with one another and causing galvanic corrosion. This corrosion can occur between steel storage tanks and copper piping, between pumps and piping, between collectors and piping, and between different types of piping. Nonmetallic material should be placed between different metals. Bituminous materials, neoprene, and Teflon® are examples of materials that have been used successfully.

Designers should consider aesthetic implications of solar designs relative to scale, support design, and piping arrangement. Elevation drawings are particularly important to illustrate these implications.

Mechanical support systems for tracking collectors (Fig. 8-1) include motors and other moving parts. The two most common mechanical problems encountered with tracking collectors are failure of motors that drive the tracking mechanism and failure of rotating pipe connections. It may be advantageous to interview previous users of the model and, if possible, inspect an existing installation before designing or specifying a tracking concentrating collector. The reflector design also should be scrutinized. For example, the materials used for the mirror should be reviewed with respect to their responses to air pollutants, which might cause deterioration.

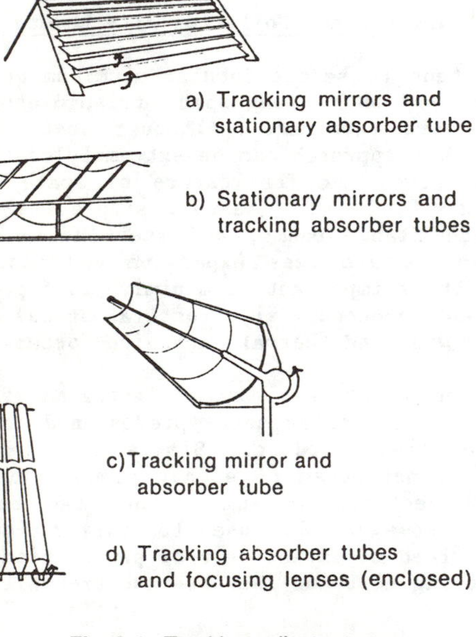

Fig. 8-1. Tracking collector types

Structural Loads

Structural loads imposed by solar installations are fairly well known. There is a dead load, a wind load (with horizontal and vertical components), and a snow load, each varying with design. Because of the extent of exterior construction, thermal expansion and contraction induce an additional stress (Fig. 8-2).

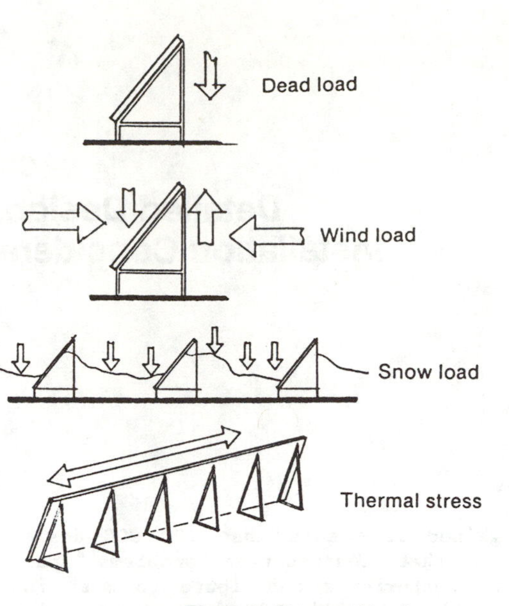

Fig. 8-2. Structural loads

Collector weight is seldom a dominant factor because the dead load of the collectors is relatively small compared with other load components. Most flat-plate collectors weigh about the same per unit area (Fig. 8-3).

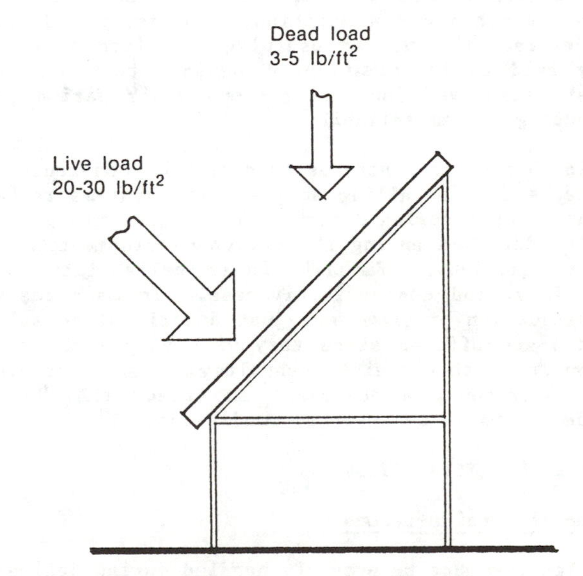

Fig. 8-3. Dead load vs. live load

The design of a collector structure must consider wind loads, although a collector enclosure provides some resistance to wind stresses. For small installations, where structural supports are provided by the manufacturer, wind stresses are usually accounted for in support design; however, it is advisable to check with the collector sales representative. In larger systems, cross-bracing is required in both horizontal and vertical planes.

Wind bracing must be designed on a system-by-system basis, based on structural requirements.

All structural design should be based on generally accepted engineering practice and should be in compliance with ANSI A58.1.

Wind loads are usually based on hurricane velocities (150 mph). Design wind loads should be calculated according to the procedures in ANSI A58.1. Flat-plate collectors mounted flush with the roof surface should use design loads that would have been imposed on the roof area they cover. Collectors mounted at an angle to or parallel to the roof surface on open racks can be lifted by wind striking their undersides. This wind load is in addition to the equivalent roof-area wind loads, and it should be determined according to accepted engineering procedures. Although wind load may not lift an array off the structure, it might promote vibration that can lead to leaks through pitch boxes and other mounting points. Negative pressures induced by wind can and often do exceed pressure loadings. Cover-plate retainers must be adequately designed to prevent separation from the collector frame and to prevent glazings from popping out of collectors.

Solar system components are subject to seismic forces generated by their mass. The design of all connections between components and the structure should allow for anticipated movements of the structure.

Minimum support member size should be 2 in. (50 mm) diameter or 2-in. × 2-in. × 1/4 in. (50 × 50 × 6 mm) structural angle. These dimensions are adequate for loads of up to 300 lb/ft^2 (1465 kg/m^2).

Horizontal Mounting

Flat roofs are a common collector location. Hoisting of collectors does not present a major problem, but roof surfaces must be protected and collectors should be placed so roof structures are not overstressed. The following factors should be considered to ensure that such mountings are reliable and cost effective.

Spacing, Height, and Arrangement of Arrays

A good rule of thumb for determining the distance between collectors is to allow enough space between rows so the shade angle does not greatly exceed the solar altitude angle on December 21 at noon. The highest continuous obstruction on the southernmost row should not cast a shadow on the glazed area of the next row. Thus, if exterior manifolds are used, the manifold top must be considered, not the collector top (Fig. 8-4). Decisions on whether to arrange collectors in rows one, two, or more collectors high should be based on spans, structural loads, roof area, obstructions, aesthetics, pipe configurations, and performance.

The distance between rows is usually less than 10 ft (3 m) when an array is only one collector

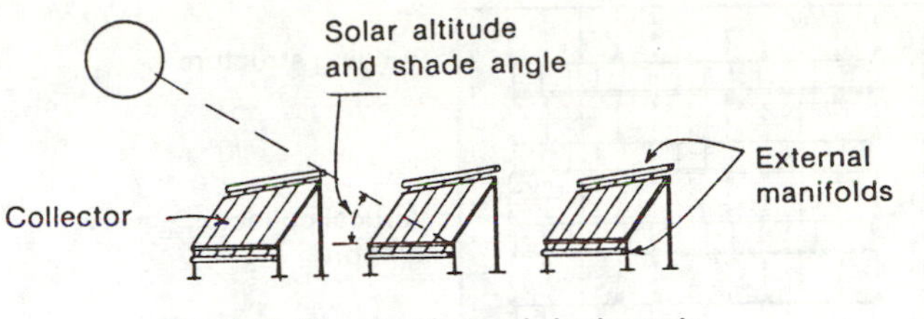

Fig. 8-4. Row spacing and shade angle

high, allowing piping to be supported on collector supports. If collectors are spaced farther apart, piping will require special supports, which must be detailed. Even so, it is usually more economical to install collectors in rows two collectors high, especially if the collectors are not tall. This approach minimizes piping and structural requirements (Fig. 8-5). For collectors with internal manifolds, most collector manufacturers suggest that a maximum of 6 to 10 collectors be installed in a row. This limits expansion problems and provides for more even flow distribution.

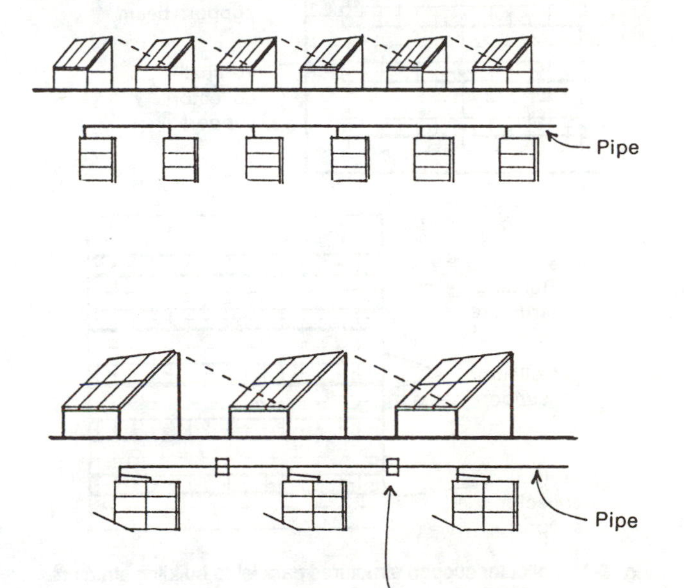

Fig. 8-5. Alternative array layouts

The orientation of collectors to the building structure, and the type of building structure--concrete, steel, or wood--will affect collector support. If collectors run at right angles to the building's secondary steel structure (joists), the legs of the collector support structure can be located at regular intervals (Fig. 8-6).

However, if collectors run parallel to the secondary steel members, vertical collector supports must be located directly over single secondary steel members. Members must support the total load, or an additional structural frame must be provided above the roof to carry the collector supports. A designer may elect to have the collector support span the entire roof (Fig. 8-7).

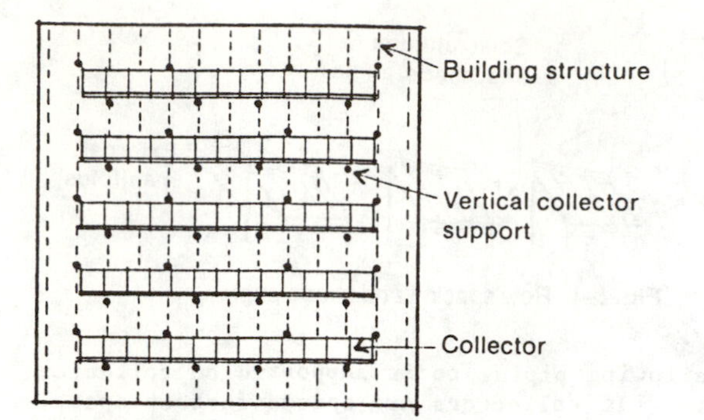

Fig. 8-6. Collectors at right angles to building structure

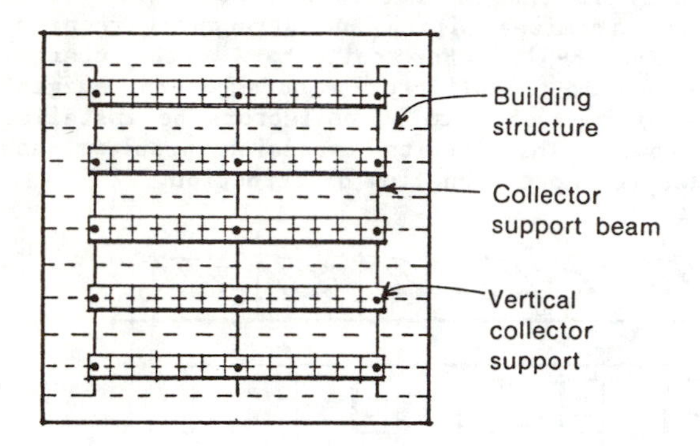

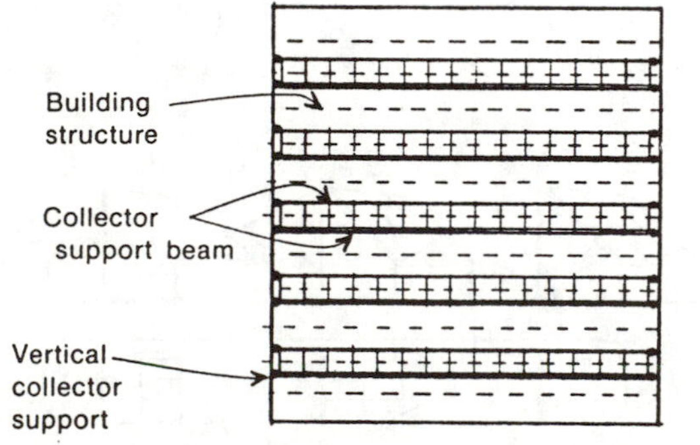

Fig. 8-7. Collector support structures parallel to building structure

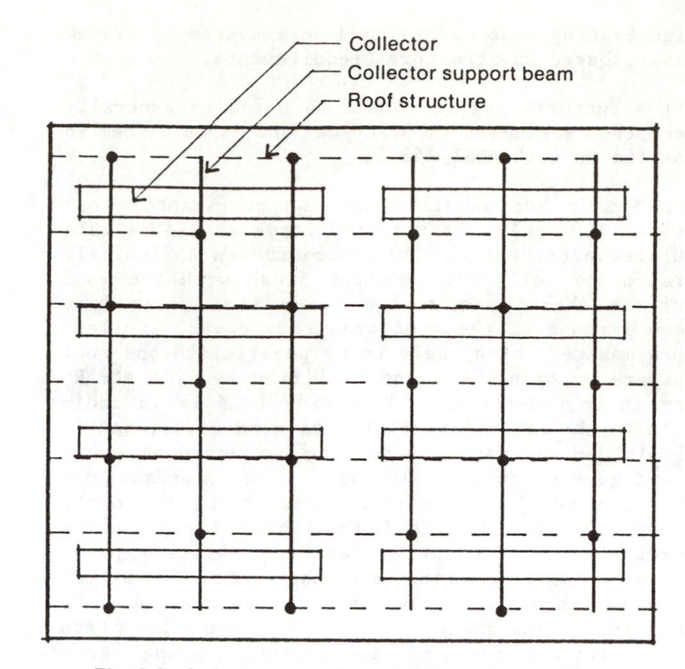

Fig. 8-8. Collector support run at right angles to collectors

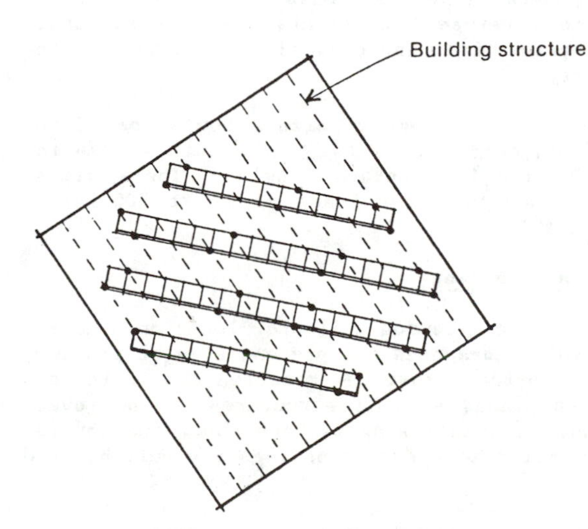

Fig. 8-9. Collectors set at an angle to the building structure

Another approach to supporting collectors installed parallel to the secondary steel members is to run the main supporting members at right angles to the collectors rather than parallel to them. This allows the collector load to be spread evenly over the roof (Fig. 8-8).

If collectors are set at an angle to the building, support legs may have to be located in an irregular pattern to conform to the roof structure (Fig. 8-9).

Generally, the distance between vertical members of the support structure attached to the roof varies according to the capacity of the building's structure to accommodate point loads. In one design, the building structure could not accept point loads, so a collector support structure was built with vertical legs 4 ft (1.2 m) on center, requiring 700 pitch pockets. In another design, the collector support structure spanned more than 50 ft (15.2 m) across the entire roof. However, the most economical span seems to be approximately 10 to 15 ft (3-4.5 m).

A structure may experience a differential temperature of 160°F (90°C) between daytime summer heat and a cold winter night, and piping could experience a 200°F (110°C) differential (Fig. 8-10). Thermal expansion cannot be overcome by using stronger members or additional bracing. It must be accounted for by proper design and the use of expansion joints. Adequate room for expansion should be provided between adjacent collectors; 1/8 to 1/4 in. (3 to 6 mm) minimum is required. Manufacturers should be consulted for recommendations.

Whenever possible, collectors should "float," compensating for expansion across the entire array.

Steel L = 100 ft (30 m) ΔL = 1.5 in. (3.8 cm)

Aluminum L = 100 ft (30 m) ΔL = 3.0 in. (7.6 cm)

Copper L = 100 ft (30 m) ΔL = 2.25 in. (5.7 cm)

Fig. 8-10. Thermal expansion for a temperature differential of 200°F (110°C)

Anchorage

There are many ways to anchor a collector support structure to a building. For any anchorage approach, connections must be designed to support all loads imposed on the collector, including dead loads, horizontal thermal movement, and snow and wind loads. Failure of anchorage design will result in roof leaks.

The anchorage approach selected will depend on the roof construction material. For steel supports and metal deck roofs, it is important to anchor steel supports directly to the roof, not to the deck; bolting or welding will work for final fastening. Expansion bolts work well on concrete roofs. Wood roofs present the most complex anchorage problems. It is not sufficient to anchor to the sheathing alone; anchorage to sheathing may resist downward loads but will not resist uplift. Heavy blocking, secured to rafters at each side, can be used to accept lag screws, or a piece of blocking set under and across the rafters can be used to accept long threaded bolts (Fig. 8-11). Once an attachment is designed, a waterproofing approach must be selected.

A pitch pocket (Fig. 8-12) is one of the oldest and most common anchorage approaches. One must specify them carefully and install them correctly. If pitch pockets are used to anchor piping, not more than one pipe should be installed in each pocket or differential movement may cause problems. The main problem with pitch pockets is that they require perpetual maintenance.

Curbs (Fig. 8-13) are another common anchorage method. If curbs are fabricated in the field, construction and roof anchorage should be carefully drawn and specified to withstand both horizontal and vertical loads. Manufactured curbs must also be selected and fastened to the roof to withstand expected future loads. If curbs are incorrectly installed, any movement that ensues could cause a roofing failure and leaks.

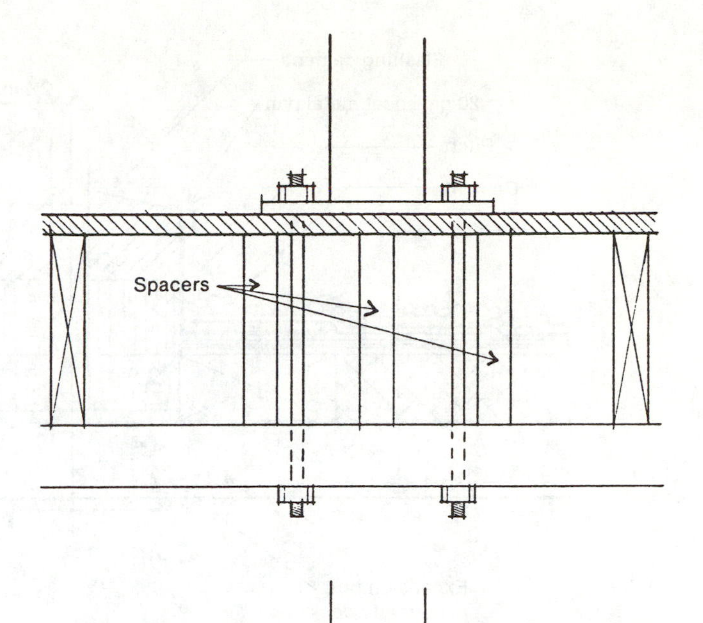

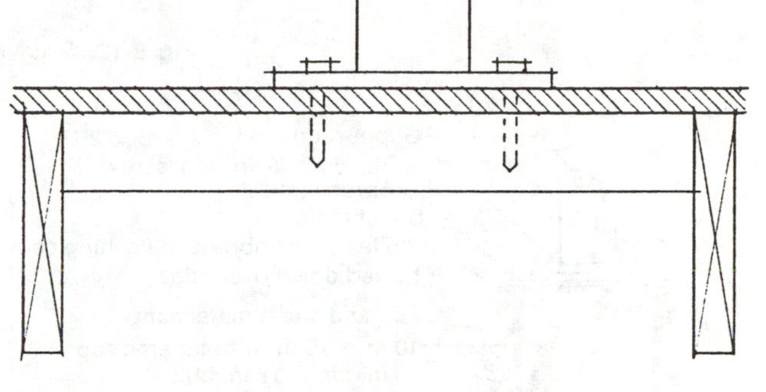

Fig. 8-11. Anchoring details

If a post base plate is fastened to a wood curb, for example, and only metal fasteners hold the metal base plate through metal flashings, problems will result (Fig. 8-14). In this approach, metal-to-metal contact will undoubtedly allow water to enter through holes made by the fasteners. Back-and-forth motions caused by wind and temperature differentials can cause fasteners to pull out slightly and lead to leakage. If the wood is not treated, it can deteriorate and cause further problems. Also, as the post rocks, the base plate will tend to cause fatigue in cap flashings at its edges.

In one case, a prefabricated curb was bolted to the structure, with a beam over the curb. Back-and-forth movement of the beam, from wind load or from expansion and contraction or from both, caused movement of the curb where it joined the roofing. This could result in a roofing failure (Fig. 8-15).

An easy way to avoid these problems is to specify that curb edges be secured to the roof structure. Also, when detailing wood curbs for construction in the field, sufficient lateral strength should be indicated. There is a tendency to put curbs close to the roof edge to use as much of the roof surface

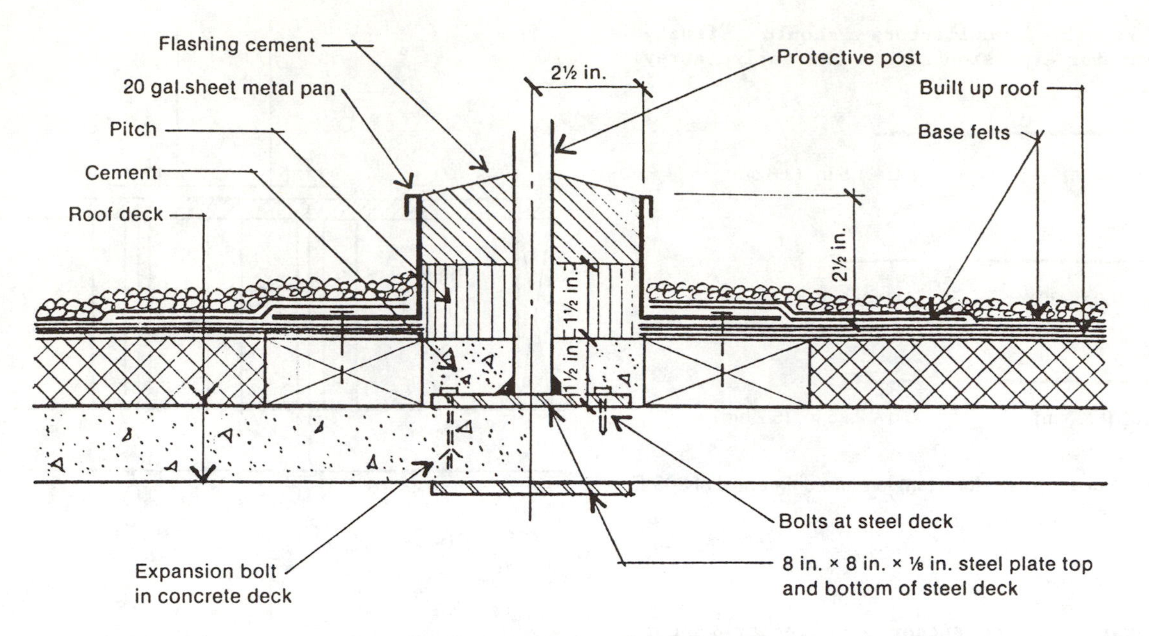

Fig. 8-12. Suggested pitch pocket detail

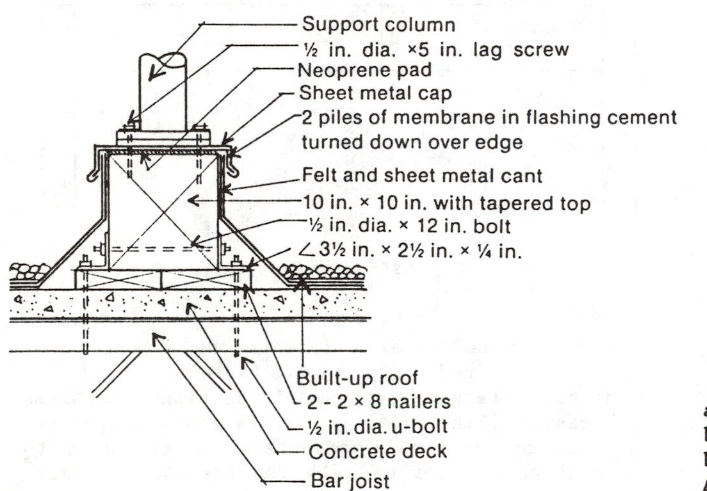

Fig. 8-13. Suggested curb detail

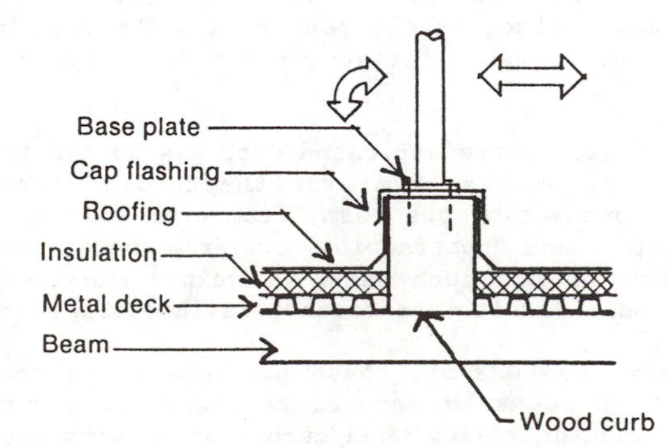

Fig. 8-14. Movement of post connected to wood curb

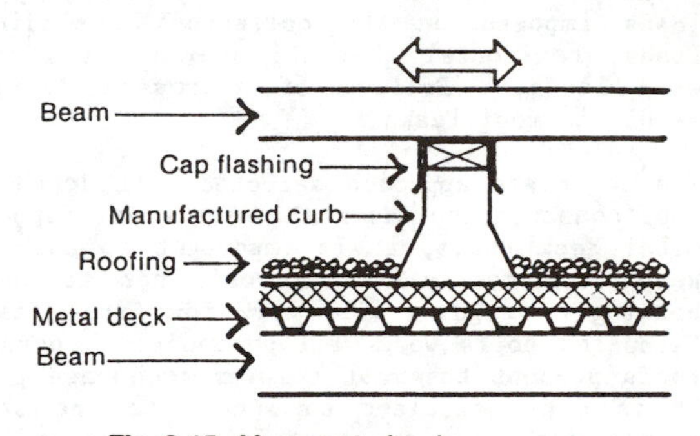

Fig. 8-15. Movement of curb

as possible. This will result in interference between the two sets of flashing. Flashings could be combined to correct this problem. The National Association of Roofing Contractors suggests providing at least 1 ft (0.3 m) of flat roof surface between flashings.

If a vertical collector support has a common shape (round or square), manufactured neoprene sleeve curbs are available to anchor support legs. However, one needs access to the structure below the roofing (Fig. 8-16). Neoprene collar flanges do not require sealant where a pipe passes through them.

Sleepers bolted directly to flat roofs are a simple attachment approach that has often been used, especially in retrofits. However, sleepers laid over the roofing are the cause of a number of roofing failures and are not considered a reliable installation approach for flat roofs. If all joints and openings are sealed, there is a better chance of success (Fig. 8-17), but the approach is not wholly reliable.

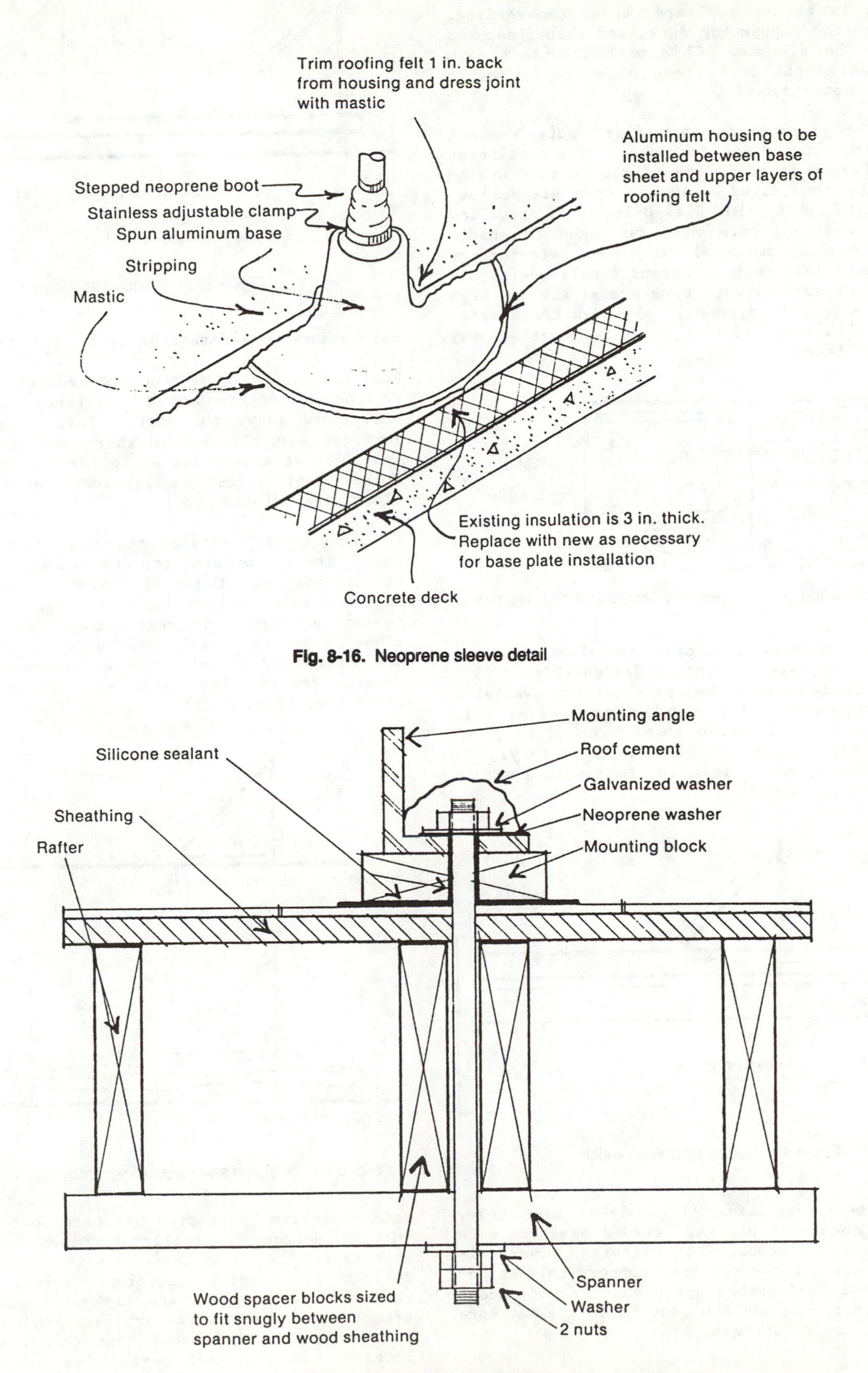

Fig. 8-16. Neoprene sleeve detail

Fig. 8-17. Suggested sleeper detail

A more durable and reliable approach, although more expensive, is to set a sleeper under the roofing, directly on the supporting deck, and then flash the roof over the sleeper. This will provide a firm platform above the roof plane on which to secure collector supports.

It should be remembered that flat roofs are not flat but are pitched for drainage. If a continuous curb or sleeper is used, the effect on roof drainage must be considered. An improper sleeper arrangement is shown in Fig. 8-18. If puddling cannot be avoided, care must be taken to ensure that the roofing material will not deteriorate. The only materials that withstand puddling with any degree of reliability over time are single-ply EPDM or neoprene roofing systems. Although these materials can withstand puddling, bolt penetrations may still cause leaks.

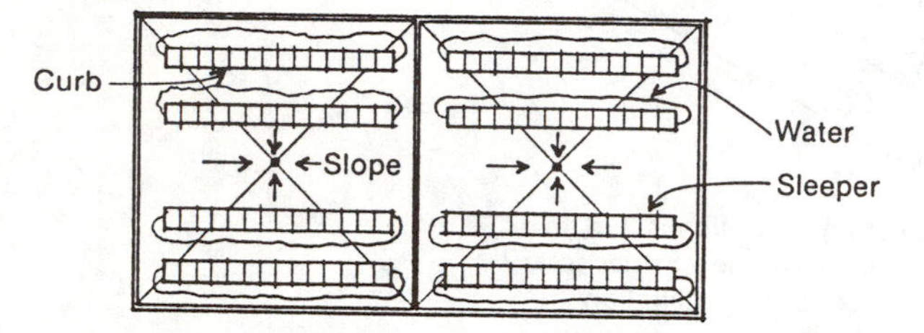

Fig. 8-18. Puddling due to improperly placed curbs or sleepers

Some designers have attempted to eliminate roof penetration completely. In one design (Fig. 8-19), structural members span the space between available vertical supports. Although in most cases this approach is more expensive than using pitch pockets, it should not be discounted completely. If spanning between walls is used, temperature expansion must be allowed for.

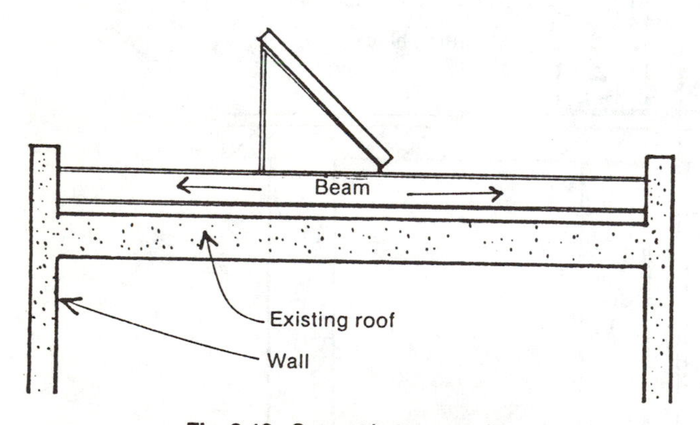

Fig. 8-19. Support between walls

Another interesting idea (Fig. 8-20) uses heavy dead-load concrete blocks, tied together with collector support members, to withstand wind horizontal forces. If the roof can support this structurally, it is a feasible approach. If not, additional costs to upgrade the roof structure may more than offset the advantages.

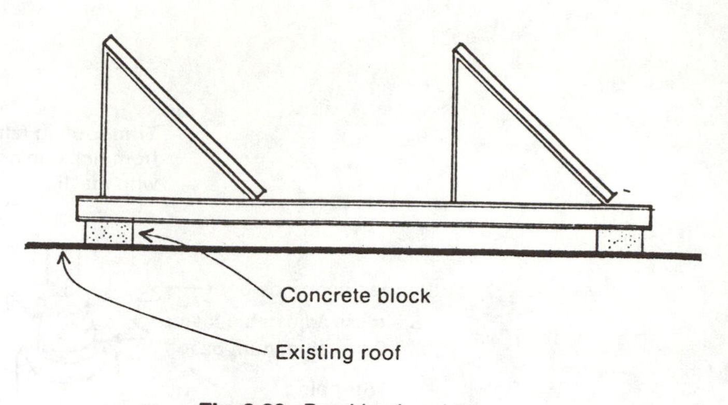

Fig. 8-20. Dead-load anchorage

Maintenance Considerations

Roof and collector maintenance requires easy access to the collectors and the ability to raise the collectors above the roof. This is particularly important in the South, where collectors may be installed at a very low angle for air conditioning. Unless roof access is provided, roof repair is almost impossible.

In snowy areas, raising the low point of collectors, their support structure, and all piping sufficiently off the roof (based on the amount of expected snow) allows the wind, rather than maintenance personnel, to remove snow. Collectors too close to the roof will tend to act as a snow fence and cause additional snow to accumulate. This creates considerable loads and reduces collector efficiency (Fig. 8-21).

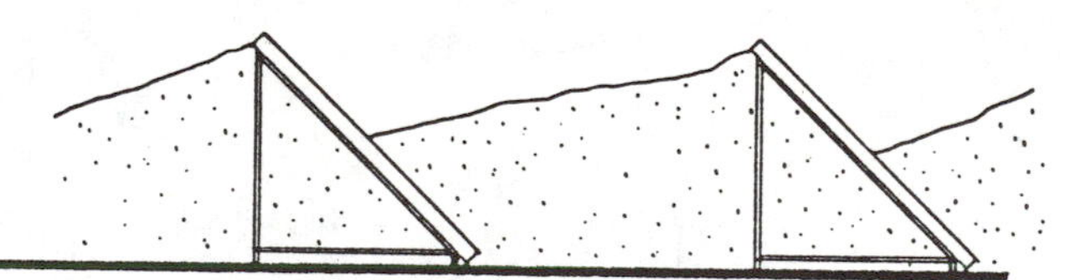

a) Collector too close to roof

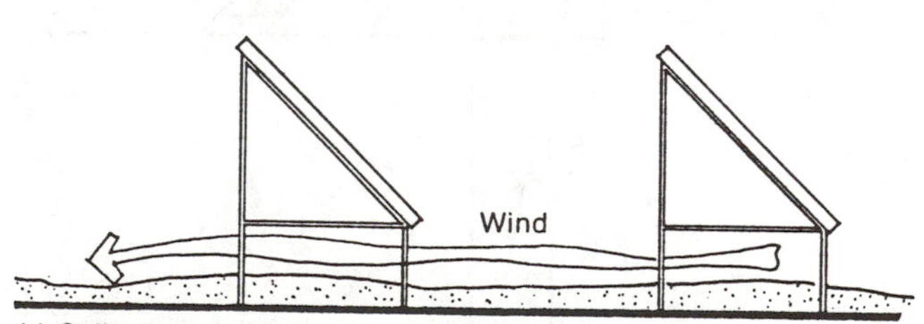

b) Collector raised above roof

Fig. 8-21. Avoiding snow accumulation by raising collectors

Much expensive overdesign has been done to provide instant access to collector pipes and valves, especially in conjunction with evacuated-tube collectors. Working platforms provided as part of the collector support are expensive and not cost effective. Access walks are probably unnecessary; no emergencies will occur that cannot wait for a ladder. If the roof surface is protected from point loading created by ladders, the need to repair the roof is minimized.

Because flat roofs are not actually flat, it is necessary to adjust the level of the collector structure. Roof slopes vary, and some leveling approach, either shimming, grouting, and/or varying vertical leg heights, should be detailed and specified.

Pitched-Roof Mounting

Roof Pitch Less Than Collector Angle

If a roof pitch is less than a desired collector pitch, many of the same collector spacing considerations encountered with flat roofs will apply. Shade angles, structural loading, and snow problems must be considered (Fig. 8-22).

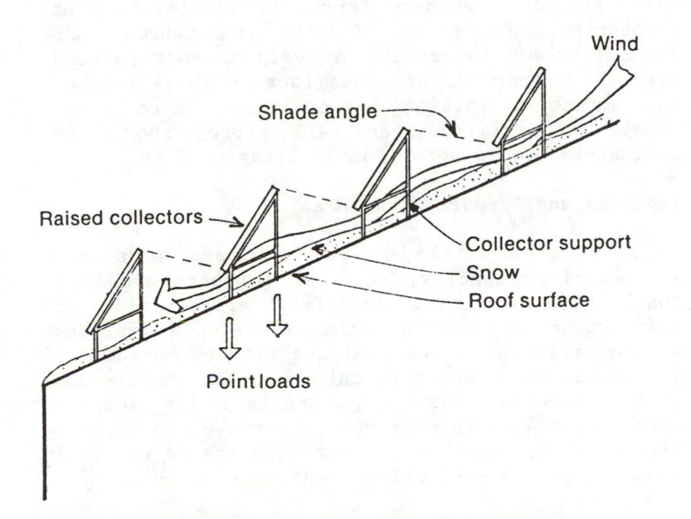

Fig. 8-22. Pitched roof collector considerations.

If collectors do not have sufficient roof clearance, snow will build up behind them and cause additional snow loads and roof leaks (Fig. 8-23). If only one row of collectors is used, these problems can be minimized if the row is set near the roof peak.

Large flashed arrays should have gutters or provisions for runoff. Asphalt roofing is designed for evenly dispersed water flow; reinforcement may be necessary if normal flow is locally increased.

Direct-Roof (Parallel) Mounting

If collectors are directly mounted parallel to a pitched roof, spacing requirements are all but eliminated. Collector weight increases the dead load applied to the roof but does not increase the live load. Snow load is not a problem, but avalanche protection must be provided for unwary pedestrians (Fig. 8-24). Building exits or walkways should be located away from areas where accumulated snow might slide from the roof, or

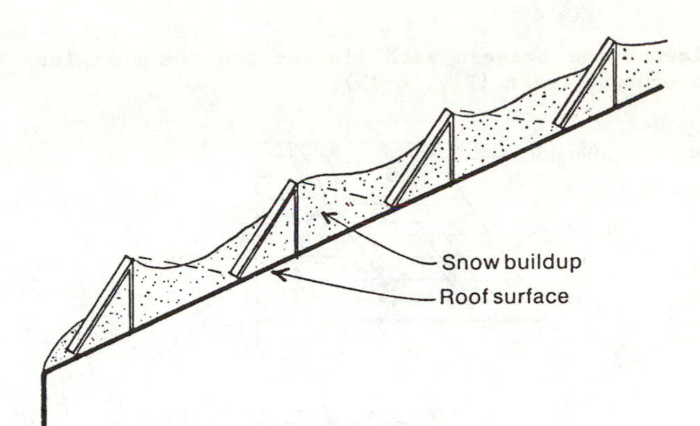

Fig. 8-23. Collectors and support structures contacting pitched roofs

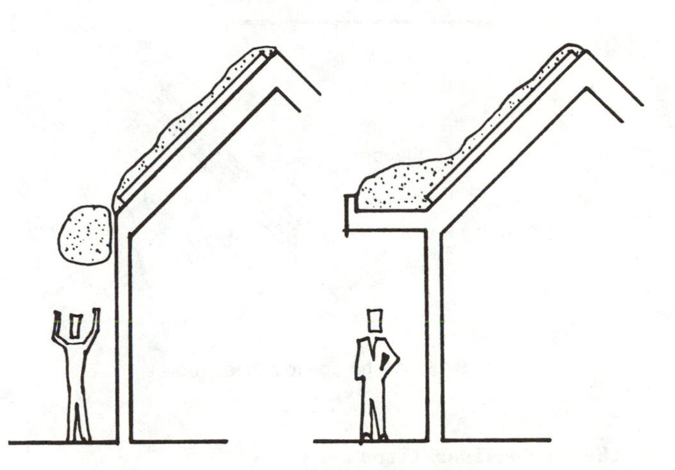

Fig. 8-24. Avalanche protection

overhangs should be provided to protect pedestrians.

Roofing under parallel-mounted collectors cannot be replaced without removing the collectors. All that can be done to waterproof the installation is to protect the connection between the collector support and the roofing from water running down the roof. Unless flush mounted, collectors should be mounted with a minimum of 1.5 in. (38 mm) spacing from the roof for air circulation to disperse moisture buildup and to prevent fungus growth.

Hoisting and installation of collectors onto pitched roofs must be carefully planned. If a crane or cherry picker is used, concern is minimal. If manpower is used, good working platforms must be provided on the roof and enough people used to ensure a safe installation.

Sleepers are very effective at providing collector anchorage support on pitched roofs. Sleepers fastened directly to the roof should always run down the roof slope, never across it. If they run across the roof, they will create a dam and cause leaks. During installation, roofing cement should be applied to each layer of shingles under each sleeper and between each sleeper and the shingles to prevent leaks (Fig. 8-25).

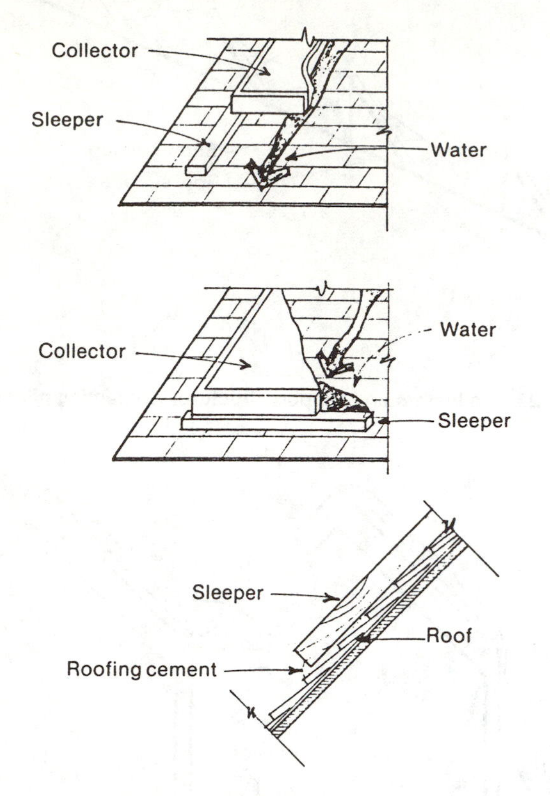

Fig. 8-25. Anchorage to pitched roofs

Aesthetic Considerations

A collector design for direct, parallel mounting to pitched roofs should consider aesthetic concerns such as collector slope (in drain-out systems), amount of exposed piping on the roof, and provision for pipe crossovers. If a drain-out system is used, collectors and pipes must be sloped to drain as indicated in Fig. 8-26. All exposed piping should be considered from both functional and aesthetic points of view. Pipe crossovers should be avoided or carefully detailed because there is little room between the roof surface and the top surface of the collectors. Pipe crossovers invariably protrude above the collector surface, making them both highly visible and unsightly.

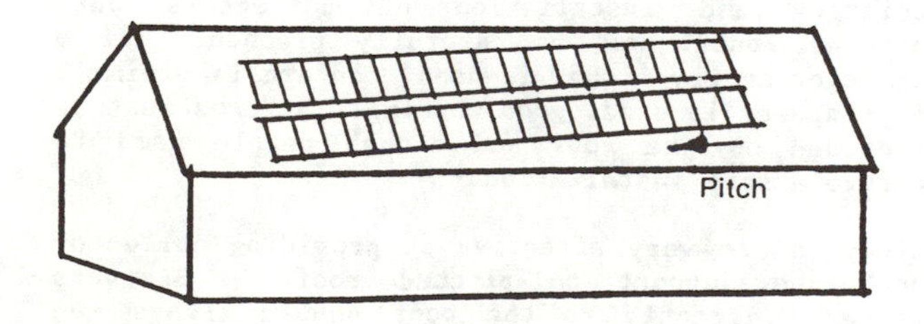

Fig. 8-26. Sloped collectors and pipes to allow drainage

When designing drain-out or drain-back piping layouts, positive pitch must be provided for drainage. To ensure good drainage, allow for a minimum of 1/8 in. of pitch per foot of run. The greater the pitch, the more reliable the system.

For site-built collectors, avoid rigid fastening of metal absorbers to roof rafters. Expansion can place damaging stresses on the structure, on ducting, or on pipe connections. In addition, a fire hazard may be created.

Ground Mounting

Piping

Ground, or site, mounting raises the issue of buried versus aboveground piping. In areas with a deep frost line or very wet soil, locating piping above ground is much preferred to facilitate pipe inspection and repair, despite the thermal and water problems incurred. A designer must contend with the appearance of pipes located above ground. The amount of piping buried under paved areas should be minimized, and all piping should be adequately tested before backfilling.

Footings and Support Structures

In areas with little or no frost and with good soil-bearing capacity, vertical supports are less costly than they are in deep-frost areas because of differences in footing requirements. A designer should arrive at a good balance between horizontal structural spans and vertical supports. Metals and pressure-treated wood are materials for support structures; wood is more cost effective. A prefabricated frame supported by site-fabricated vertical members adds to cost effectiveness (Fig. 8-27).

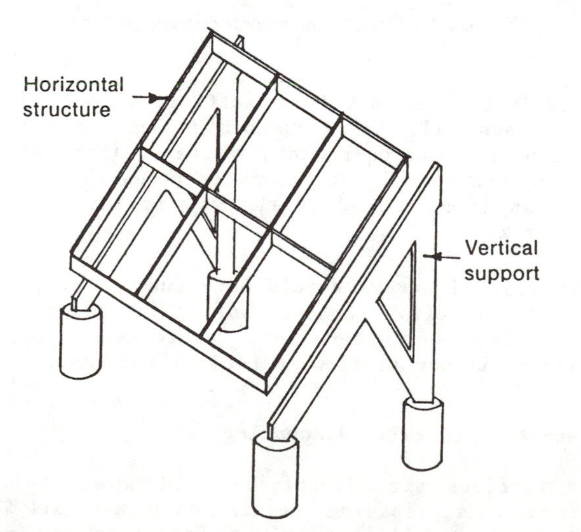

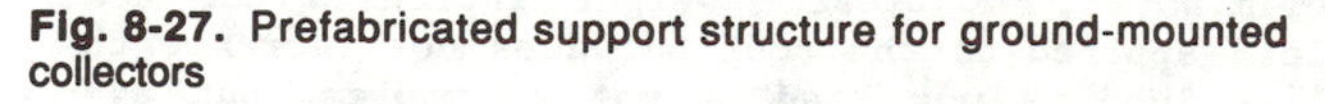

Fig. 8-27. Prefabricated support structure for ground-mounted collectors

Regional construction methods often provide cost effective designs. For example, pole-barn-type vertical supports constructed in upstate New York might be less expensive than standard footings.

Because of footing costs, it is less expensive to use arrays that are two collectors high, keeping collectors off the ground to allow for snow and undergrowth.

Vegetation

Vegetation presents a problem unique to ground-mounted collectors. The cost and inconvenience of mowing the area in which the collector field is located can be considerable. The area can be paved, or a chemical defoliant can be spread on the soil, which is then covered with a plastic sheet and gravel (Fig. 8-28); proper ground drainage must be included in the design.

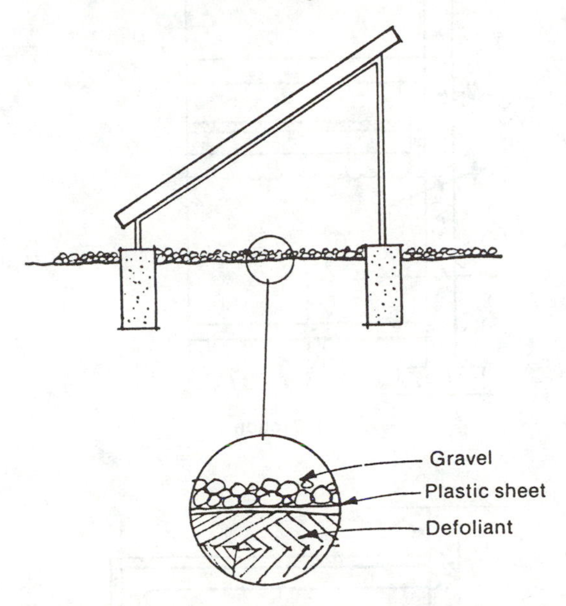

Fig. 8-28. Soil preparation near ground-mounted collectors

STORAGE INSTALLATION

Liquid Systems

General Considerations

The cost effectiveness of different tank types should be considered. If the tank is to be pressurized, an ASME-certified steel tank must be used. For an unpressurized tank, concrete, reinforced plastic, lined wood, and steel tanks are possible options. Each type offers cost or construction advantages and disadvantages, each of which should be weighed carefully. Reinforced plastic tanks, for example, are lightweight and easy to handle but have relatively low maximum temperature limitations.

Tank tapping is an important design consideration. Provisions must be made for tank drainage, temperature sensing, venting, a pressure/temperature relief valve, monitoring, and liquid makeup. If an existing steel tank is to be retrofitted, a number of openings might be combined to eliminate or minimize tappings. Some combinations are load return with liquid makeup, collector supply with tank drainage, and collector return with load supply and a pressure/temperature relief valve. Combining the collector return and the load supply is risky because the collector return is usually not the hottest liquid available.

Interior Storage

Storage inside a building is simple, if room can be found, but the space required may cost \$30 to \$60/ft^2 (\$320-\$650/m^2) for new construction or may take up valuable existing space. An interior storage installation is easy to insulate, does not require waterproofing, and tends to have short pipe runs. The storage tank must be insulated correctly, or energy will be lost from storage, thereby increasing summer cooling loads. Interior storage locations generally require less insulation than do exterior or buried locations.

In interior storage installations, attention should be given to the type of insulation used. When ignited, urethane foam (an isocyanurate compound) emits toxic fumes of cyanide gas. Urethane foams should be sealed or protected from direct ignition, as required by some building safety codes.

With steel tanks, supports--legs, saddles, or the tank bottom--are the most common sources of heat loss. With tank saddles or tanks placed directly on a slab, high-density insulation, such as foam glass, between tank and saddle or slab should be used. If saddles or legs are integral to the tank, insulate the supporting members, which should then be isolated from the slab with Neoprene® or cork pads, normally used as vibration isolators.

For nonpressurized vertical steel tanks, insulate between the tank bottom and the supporting slab with insulation capable of withstanding compressive loads. Foam glass or other high-density insulation is recommended.

Large steel tanks should not be installed in basements, crawl spaces, or other locations where building modifications would require moving or replacing the tank.

Nonpressurized storage containers constructed of wood and properly lined with a sheet material, such as reinforced EPDM, are cost effective and can solve many construction problems associated with minimal storage space (Fig. 8-29). However, liners tend to fail at the corners, and tensile failure may occur when a liner is poorly fitted.

On-site installation of fiberglass-reinforced plastic (FRP) tanks requires much more care than required for factory-insulated FRP tanks. Tanks must be protected from punctures and must be carefully supported to prevent rupture. Extra labor costs to insulate and install an on-site-insulated tank may make a factory-insulated tank less expensive. Some installations, however, benefit from the greater variety of sizes and shapes available in on-site-insulated tanks. One should not be installed, however, where major building modifications would be necessary for tank replacement.

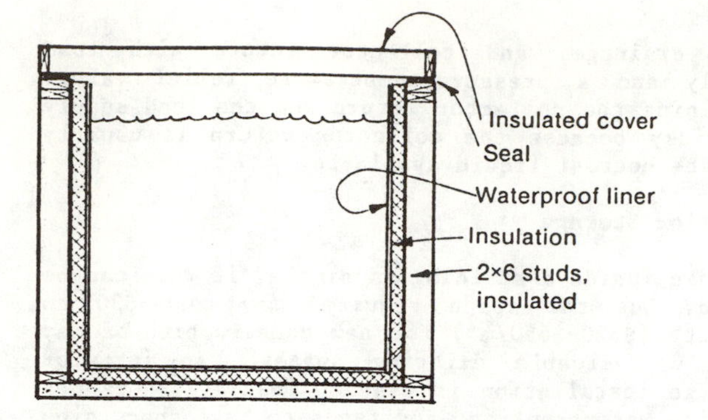

Fig. 8-29. Wood container for liquid storage

Before accepting delivery of an FRP tank, a designer should inspect it carefully and not accept a tank that has been dropped or shows any signs of physical damage. If the gel coat is cracked or if fibers are exposed, hot water can break the bond between the glass fibers and the resin, leading to premature tank failure.

Some concrete tank designs are either square or round with top, bottom, or midheight joints (Fig. 8-30). Construction joints and material porosity are major concerns. If a tank is supplied in two parts, the joint should be well sealed. The two parts should have grouted-in steel ties to prevent any movement that can cause leaks, and the coating should be able to span the minor cracks that often develop. Cement and epoxy waterproofings are less desirable because they cannot span cracks. Liquid-applied elastomeric (e.g., spray-on butyl rubber coating) or replaceable plastic liners can span minor cracks and should be considered if the temperature and liquid additives are compatible. Wall and bottom insulation is important.

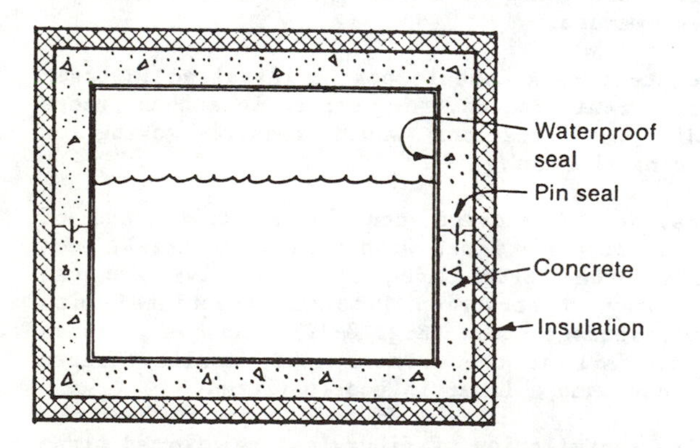

Fig. 8-30. Concrete container for liquid storage

Concrete tanks should not be installed where major building modifications would be needed for tank replacement. Cast-in-place tanks can be installed in some locations that do not permit use of precast tanks.

Some designers have tried to integrate a concrete storage container into the building structure by using foundation walls and the slab as the sides and bottom of the storage tank; only the side of the tank toward the basement is insulated. This approach, though possibly less expensive, has drawbacks. Heat tends to dissipate through the concrete walls into the soil, and heat is conducted along the walls and then radiates to the living space, adding to the cooling load (Fig. 8-31). This problem can be overcome by insulating the inside of the tank and then using a waterproof liner. Expansion joints must be provided between the storage tank and adjoining concrete slabs and walls.

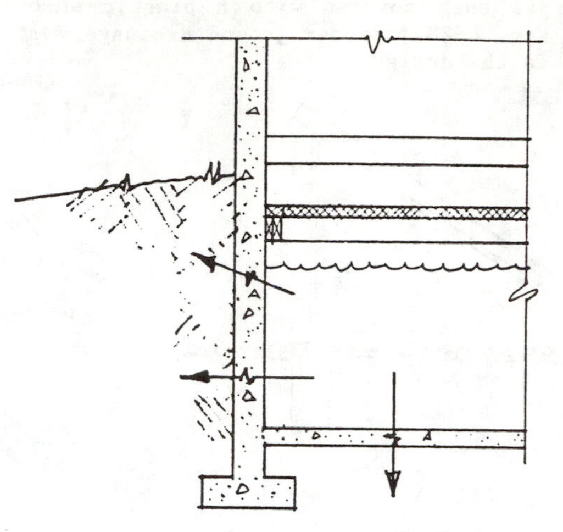

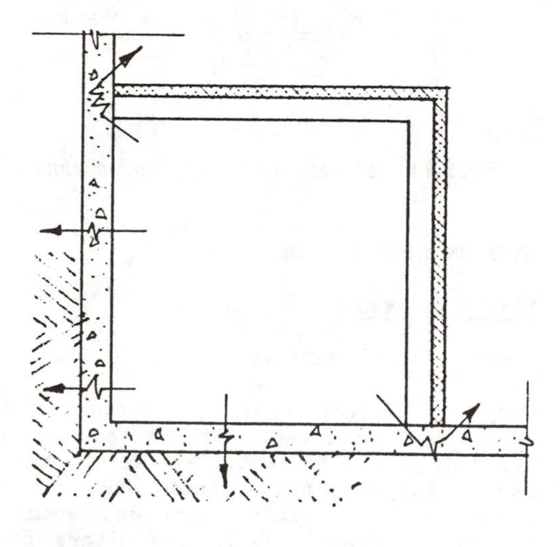

Fig. 8-31. Heat loss from storage

Exterior Storage

Storage placed outside a building may be less expensive than that placed inside, but problems and the chance of energy losses are considerably increased. Close attention must be paid to insulation and waterproofing. Insulation should be impervious to water in case a leak develops in the waterproofing. If the tank is supported on the ground by the insulation, the insulation must support the load of a filled tank without being crushed. If the tank sits on supports, the supports should be thermally isolated or insulated to minimize thermal losses. If waterproofing fails, insulation can be severely affected. All penetrations for valves, supports, piping, sensor wires, etc., must be waterproofed adequately to preserve the thermal integrity of the storage. Waterproofing requirements for exterior storage locations must be carefully detailed and specified.

Freezing can be a problem, not so much for a storage tank, which has a large thermal mass, but for pipes carrying fluids without antifreeze. Visual monitoring is always possible and is highly desirable for improved maintenance. Freezing, however, precludes normal use of tank sight glasses for monitoring water levels. Sight glasses should be drained or placed under an insulated panel to prevent freezing.

The aesthetics of exposed storage tanks are frequently ignored, since the tanks are often in the back of the building, and no effort to improve their appearance is felt to be necessary. Well-designed screens can be used to camouflage a tank, or a tank can be incorporated as a design element. In one case, storage tank waterproofing was white fiberglass-reinforced epoxy; it was attractive as a free-standing design element. Solar components should be designed into the building, not added as an afterthought.

Below Grade Storage

Water-table elevation is the first and main consideration before deciding to bury a storage tank. If the water table is above the height of the tank bottom, a special structure must be used to hold down the tank and waterproofing. This condition is best avoided. If there is absolutely no alternative, a support should be designed so the tank is held up when full and down when empty. Usually, a large concrete footing is used to hold a tank up, and steel straps are used to hold it down. Neither footing nor straps should break a tank's waterproof integrity. Insulation and waterproofing can be applied before installation. High-density insulation, such as foam glass, which is capable of withstanding point loads imposed by the support or by steel straps, should be used, and waterproofing in point-load areas must be strengthened (Fig. 8-32). Just as a support can crush insulation and tear waterproofing under compression, steel straps can do equal damage if a tank tends to float.

If there is no water-table problem, there is probably no need for a concrete support, especially if the soil has good bearing and drainage characteristics. A common installation approach is to prefoam a tank with urethane, wrap it with nylon fabric and a bituminous material, and set it in a sand bed. If done carefully, this approach works well (Fig. 8-33).

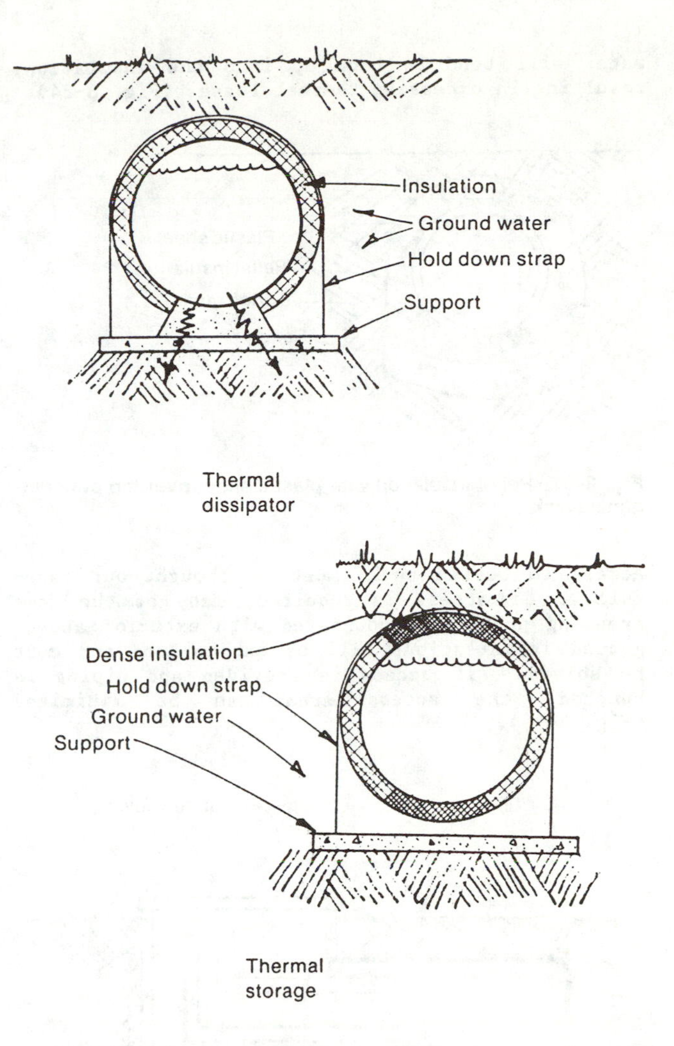

Fig. 8-32. Insulation and waterproofing in wet soils

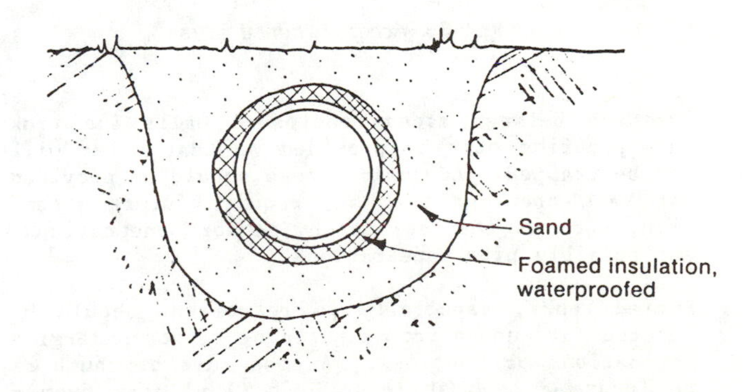

Fig. 8-33. Prefoamed underground tank with surrounding plastic sheet

Another approach, in dry climates, is to set the tank in granular insulation and place a plastic sheet over its top. If the surrounding ground is porous, this approach can work, but if it is not, water will tend to back up into the insulation, resulting in excessive thermal losses (Fig. 8-34).

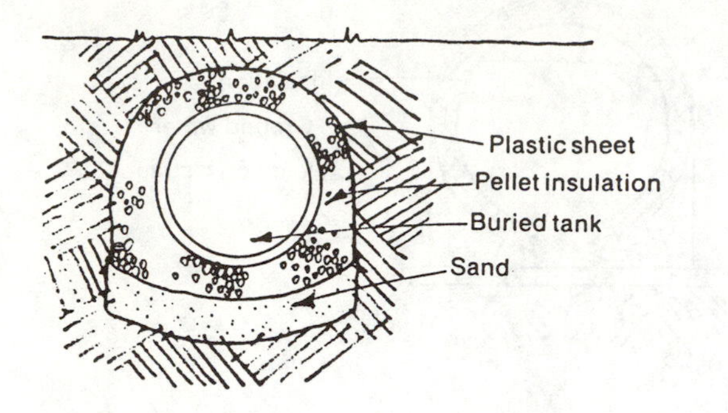

Fig. 8-34. Pellet insulation with plastic sheet over top of underground tank

Access to buried tanks must be thought out carefully. If access is required, many of the same freezing problems encountered with exterior above-ground installations will be encountered and must be solved. If access is provided and piping is bunched, the access area can be minimized (Fig. 8-35).

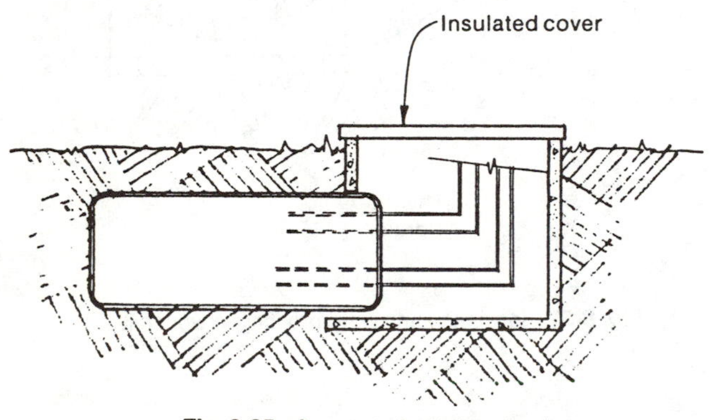

Fig. 8-35. Access to buried tanks

Drainage between access enclosure walls and tank waterproofing must be provided so that water will not be trapped. Adequate access should be provided for tank openings that may require future attention, such as the tank hatch, sensor penetrations, and possible pipe penetrations.

Buried tanks, especially in retrofits, should be located far enough from a building to not undermine foundations or footings. A good rule of thumb is to locate a tank at least 3 ft (1 m) away from a building perimeter for each foot (0.3 m) below the footing (Fig. 8-36).

If undermined, existing slabs or pavement over a tank will ultimately crack or break up. The slab should be cut back sufficiently to provide for tank installation and allow for proper backfill and repaving so that the surface will not be undermined. A structural engineer should check the site (Fig. 8-37).

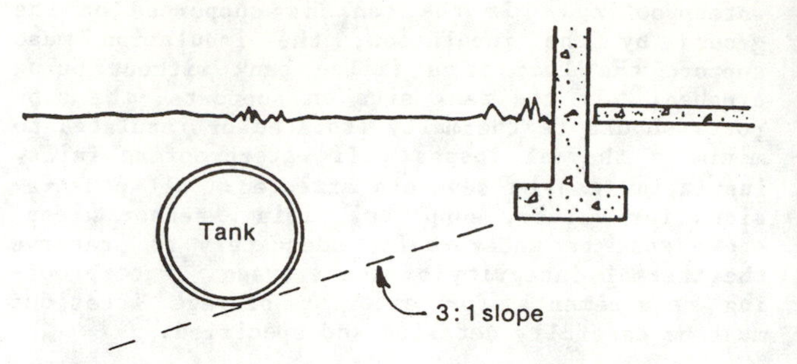

Fig. 8-36. Locating a tank near the building foundation

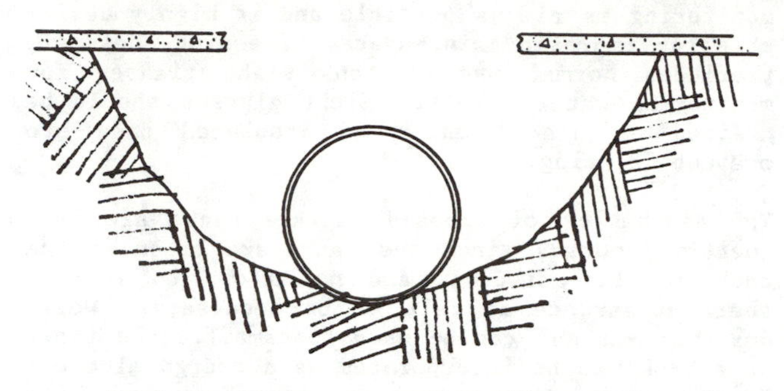

Fig. 8-37. Locating a tank under concrete slabs

Air Systems

Collectors

Air collectors can use ducting of various kinds; most use a simple flexible duct connector. Some collectors can be interconnected with internal manifolds so supply and return ducting does not have to be provided for each collector; but most require each collector to be manifolded separately, and some require a special boot connector to each. Costs associated with manifolding must be considered before a particular collector is selected.

Storage Containers

Materials suitable for constructing rock storage boxes for air systems are available. Enclosure floors are normally concrete slab-on-grade. Walls may be 8 in. (20 cm) of poured concrete, reinforced masonry, or wood stud and plywood. Enclosure tops are either poured-in-place concrete or constructed from wood and plywood.

Improper sealing of a wood cover to a storage box and of duct penetrations is common. This problem can be solved by considering the paths by which air can escape and by closing these with flashings, caulking, or glue (Fig. 8-38). The problem cannot be solved just by running a bead of caulking around duct penetrations, especially if one is caulking against 1/2-in. (12 mm) plywood.

Air leaks are not a concern when a poured-in-place concrete cover is used. However, construction should be planned to ensure that the formwork, if

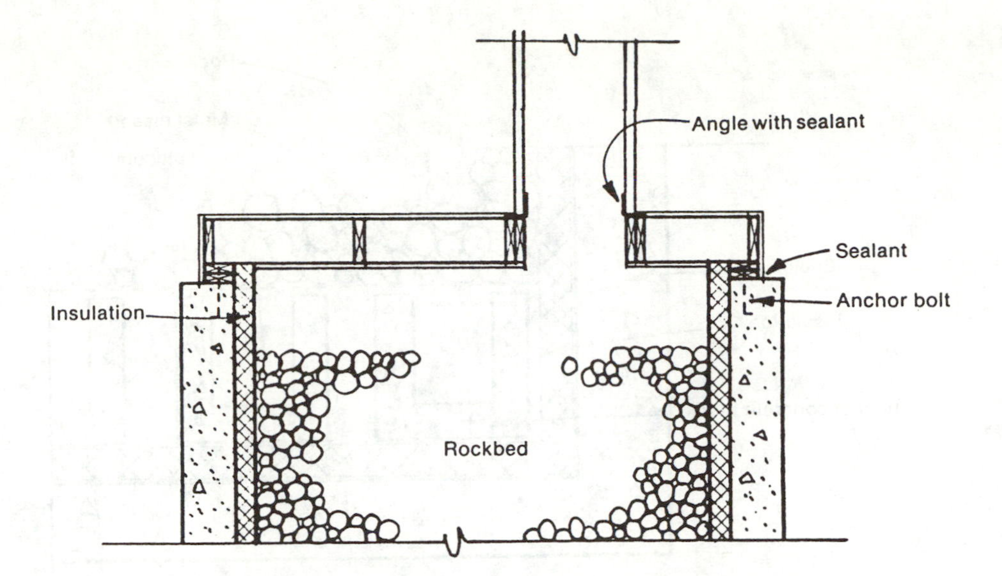

Fig. 8-38. Sealing a rock bed

any, can be removed, the top is adequately insulated, and the rocks are relatively level. An opening left in the top, that can be closed tightly after the rocks are installed, will accomplish the last.

Insulation is normally placed inside the box for continuity. Insulation used must conform to the following criteria: It should be noncombustible and not give off toxic gases, such as chlorine, when exposed to high temperatures. It must resist compressive loads imposed at the bottom and sides of the box. It should not rot or mildew and must be able to withstand the rock load. High-density fiberglass and plastic foam boards are available that meet these requirements.

Drainage is a common concern in rock boxes, especially with below-grade enclosures. Floor drains should not be used since they require maintenance to keep their water seals. If seals dry out, vermin or sewer gas will enter the storage box, and the system will have to be shut down. A water vacuum duct should be available in case water enters the box. A duct that extends to the box bottom, or a separate 3 in. (7.6 cm) pipe, can also be used for visual inspection of rock beds.

Flow Channeling

The most effective way to ensure adequate and even flow through storage is to supply air across the whole bottom area of storage, in a vertical-flow arrangement, and exhaust the air across the top. Air channels along the bottom and the plenum across the top must be sized for proper flow (Fig. 8-39). Flow-channeling in horizontal-flow beds is discussed in detail in Ch. 5.

Air Ducts and Dampers

Leaks are a serious concern in air collector systems. Performance is often degraded significantly because of leaks in duct seams, damper shafts, collectors, and rockbox joints. Duct seams should be caulked or taped with top-quality materials. Longitudinal seams should be caulked or taped prior to installation. Space should be allowed to seal cross seams after installation. Attention must be paid to tight, quality installation procedures to minimize air (and heat) losses.

Air systems have encountered significant problems when leaky dampers have allowed unregulated flow. Service hot water preheat coils can freeze when cold air from the collectors falls through a loose-fitting damper across a duct-mounted coil. Stored heat is lost when warm air rises from rock storage through a loose damper into cold collectors during cloudy winter days or winter nights. Tight, low-leakage dampers should be specified, and correct, high-quality installation procedures should be followed. Hot water coils should be located so they are protected by storage heat, yet heat losses should be prevented by an additional tight damper.

PIPE AND DUCT INSTALLATION

Where nonpotable fluids are used, fill connections should be labelled with a danger warning. The label should contain the fluid name, the date filled, the installer's name, and a warning not to connect to a potable water supply. A removable handle or lock wire is advisable if fill connections are located in public areas. Piping hookups must comply with local codes to prevent backflow and possible contamination of city water.

Galvanized piping is not recommended for solar installations. Sludge is formed when fluids containing glycol experience temperatures greater that 130°F (54°C) in such pipes.

Leakage is more likely with antifreeze, oil, or (especially) silicone transfer fluids than with water. Threaded fittings should be avoided; recommended pipe connectors and fittings are discussed below.

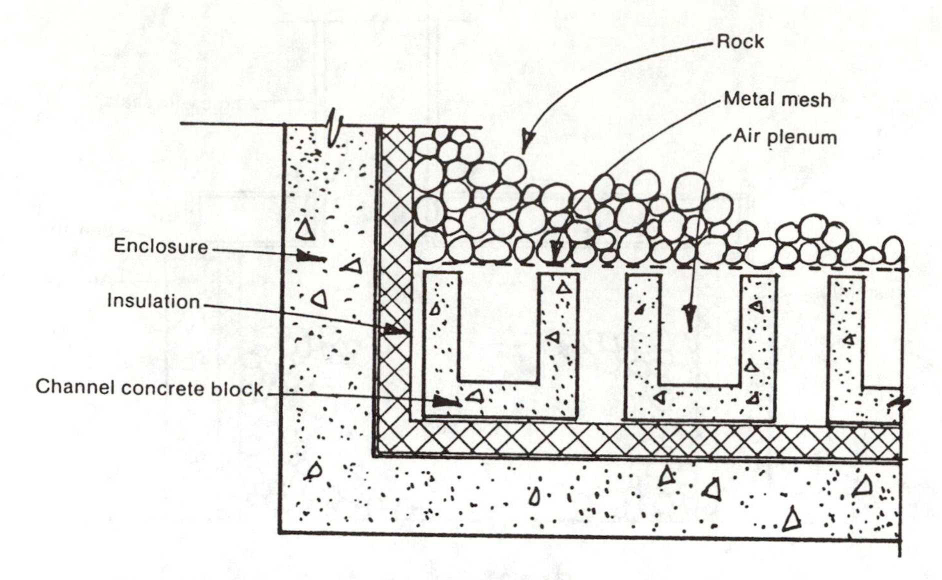

Fig. 8-39. Design to ensure adequate and even flow through storage

Pipe Expansion and Contraction

Expansion-Contraction Compensation

Temperature differentials can cause pipe movements of up to 2 in. (50 mm) or more. Movements are either in opposite directions or at right angles to each other. Allowance should be made for movement at ties or in long pipe lengths by using silicone bulbs, bellows, braided wire, expansion loops, or swings (Fig. 8-40). Flexible hose is often used to provide expansion compensation for collector connectors. Regardless of method, details should be designed and specified, not left to the contractor.

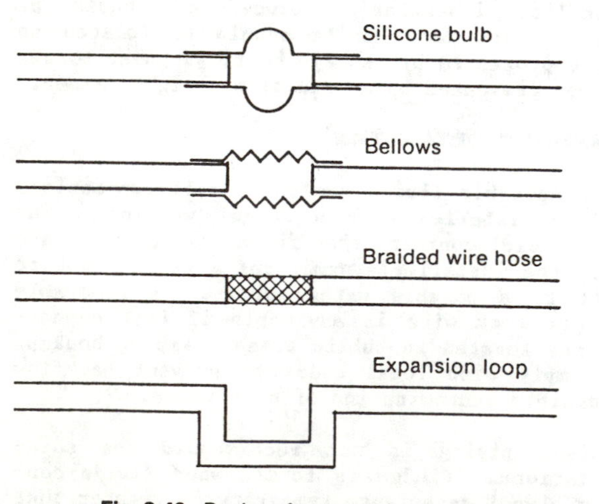

Fig. 8-40. Devices for expansion and contraction

Each compensation approach requires a different amount of restraint to function properly. Restraint must be provided by anchors attached to the structure. If the structure is not sufficiently rigid, an expansion compensator will be less effective. Hose clamps are often used, but overtightened screw clamps occasionally slice a hose; smooth liners are available for screw clamps to help avoid this. A recommended screw torque should be considered for the specific hose material used. Hoses may also take a "set" over time; a yearly maintenance procedure to check clamp tightness may be required.

Spring clamps (Fig. 8-41) successfully avoid hose damage and automatically compensate for hose "setting." Expansion cycling can gradually work a hose loose, since a spring clamp generally has less clamping force than a screw clamp does. This is of particular concern on straight hose connections between collectors with internal manifolds. A shoulder on the end of a pipe connection will help prevent this looseness. Crimp clamps are another option, but they cannot be adjusted and are not recommended.

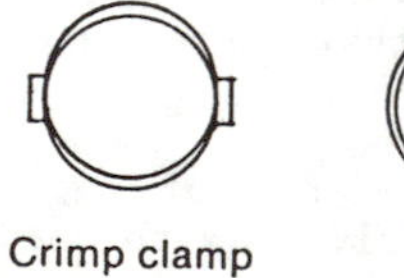
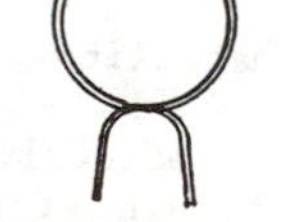

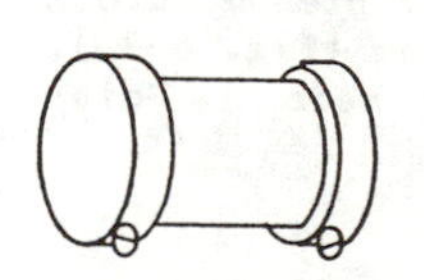

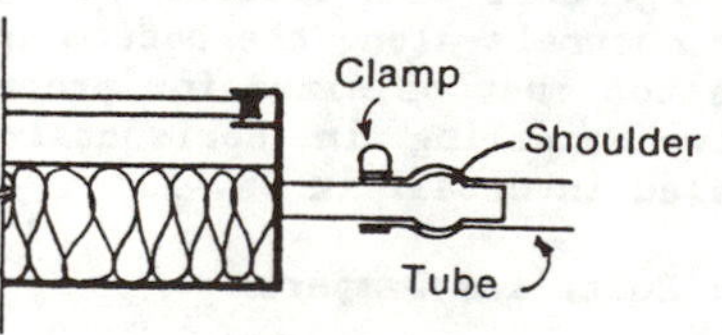

Fig. 8-41. Types of hose clamps

Special unions have been developed for solar installations. They are similar to a double screw clamp with an integral hose, but they are engineered specifically for interconnecting collectors and are reliable if installed properly.

Mating pipes must be deburred before hoses are installed. Burrs can cut small slits in the inside surface of flexible hoses and cause leaks. Proper alignment of pipes to be connected is extremely important for most hose-clamp arrangements.

Pipe-to-Collector Connections

Pipe connections to solar collectors are critical. On long collector arrays, pipe expansion and contraction can result in more than 1 in. (25 mm) of travel on supply and return headers, especially if expansion compensation is not adequate. If rigid pipe is used to connect collectors to manifolds, absorbers must be free to move. Absorber movement, however, can be limited by bind up or lack of room to move within the collector box. For collectors with exterior manifolds, a "pigtail" or offset in the rigid connector should be considered to help isolate collectors from pipe expansion (see Fig. 8-42).

Weld or sweat joints should be used on all collection loops and other piping where applicable, and tightly drawn, Teflon-taped joints should be made where needed. Some pipe dopes, butyl rubber expansion tank diaphragms, valve seats, discs, and sealants are not compatible with most glycols. Use 95/5 tin/lead solder on all collector-manifold sweat joints.

For collectors with interior manifolds, most collector manufacturers suggest that no more than 10 collectors be installed in a row. This limits expansion and provides for more even flow distribution.

Pressure Relief

Each collector bank in a solar array usually has shutoff valves in both inlet and outlet piping. Problems arise if both valves are closed while the collectors are full of liquid. If the sun heats the collectors, fluid expansion and steam generation can result; collectors may rupture and even explode. Pressure relief valves should therefore be installed on each collector bank between the inlet and outlet valves (Fig. 8-43). Discharges from each pressure relief valve should be piped to roof level or to a roof drain to avoid scalding someone who is next to a valve when it discharges hot fluid. A roof drain is preferred to avoid roof or structural damage, discoloration of building finishes, or plant damage.

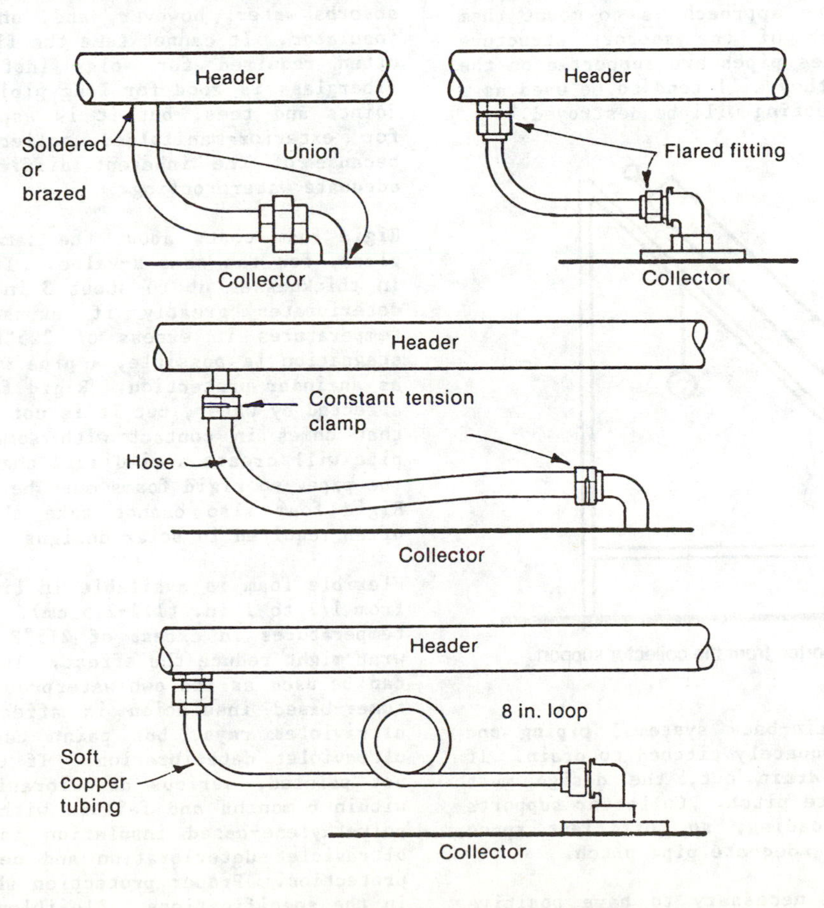

Fig. 8-42. Pipe-to-collector connections

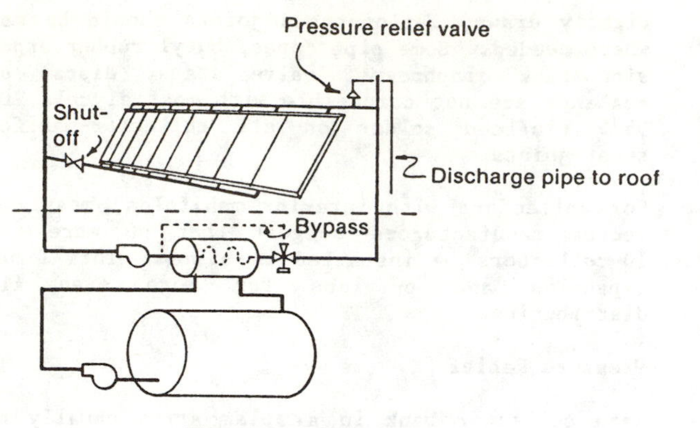

Fig. 8-43. Pressure relief

Pipe Support and Pipe Pitch

Pipe supports for collector arrays should be part of the design. Pipes can be supported by collectors, normally with soldered copper nipples. When soldering, an installer must avoid burning waterproof seals on the collectors. This is a constant problem because the solder flows at a high temperature.

If pipes are supported by the collector supports, space must be allowed for them in the support structure design. One approach is to mount them directly on the back of the support structure (Fig. 8-44). Sometimes pipes are supported on the roof surface; if so, they will tend to be used as a platform, and waterproofing will be destroyed.

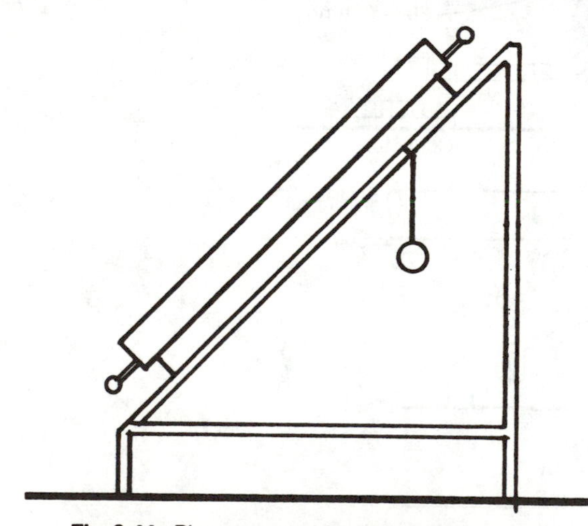

Fig. 8-44. Pipe supported from the collector support

For drain-out and drain-back systems, piping and collectors must be adequately pitched to drain. If the design requires drain out, the design must allow room for adequate pitch. Collector supports will deflect under loading, so sufficient space should be designed for adequate pipe pitch.

In theory it is only necessary to have positive pitch on collectors or pipes to allow for drainage in drain-out or drain-back systems. A minimal pitch approach might be acceptable where freezing is rare. Where frost is common, however, freezeups in systems with minimal pitch result from sagging piping or sagging collectors caused by poor workmanship or by workers' stepping on pipes. Water also collects and freezes in the bodies of improperly installed valves. For adequate pitch, between 1/8 to 1/4 in. (3-6 mm) drop per foot should be used to prevent freezeup due to sagging collector piping. Pitch should be taken in the true horizontal plane and not with respect to the slope of the structural support. Valves should be installed on the true vertical members, or valves should be selected that cannot hold water.

Pipe Insulation

Standard installation procedures should call for system pressure and water tests before pipes and storage tanks are insulated and waterproofed. Otherwise, insulation will have to be removed before leaks discovered at system startup can be repaired.

Types of Insulation

Rigid fiberglass is a commonly used insulator. It is relatively inexpensive, is available in thicknesses up to 3 in. (7.6 cm), and resists temperatures in excess of 400°F (200°C). It readily absorbs water, however, and, once wet, is a poor insulator. It cannot take the flexible shapes very often required for solar installations. Rigid fiberglass is good for long piping runs having few joints and tees, but it is especially unsuitable for exterior-manifolded collector installations because of the inherent difficulty of providing adequate waterproofing.

Rigid foam costs about the same as rigid fiberglass, for the same R-value. It too is available in thicknesses up to about 3 in. (7.6 cm), but it deteriorates greatly if exposed to stagnation temperatures in excess of 225°F (107°C). Where stagnation is possible, a pipe wrap should be used as an inner protection. Rigid foam is not readily affected by water, but it is not waterproof. Water that comes in contact with some rigid foams and pipe will create a mild acid that will deteriorate the pipe, so rigid foams must be well waterproofed. Rigid foam also cannot take the flexible shapes often required in solar designs.

Flexible foam is available in limited thicknesses, from 1/2 to 1 in. (1.2-2.5 cm). It is affected by temperatures in excess of 215°F (101°C), but pipe wrap might reduce the effect. It is waterproof and can be used as its own waterproofing. Older elastomer-based insulation is affected by the sun's ultraviolet rays, but paint can protect against ultraviolet deterioration. If this insulation is not painted, serious deterioration is often seen within 6 months and failure within a year. Newer polyethylene-based insulation is not affected by ultraviolet deterioration and needs no additional protection. Proper protection should be addressed in the specifications. Flexible foam can take any shape necessary and is well suited to exterior uses that include the many bends, short sections, or penetrations encountered with exterior manifolds or interconnecting collectors.

Piping with integral insulation and a protective waterproof cover is available for different types of pipe, insulation, and waterproofing materials. Copper pipe is most common, with urethane insulation and an ultraviolet-inhibited PVC jacket. Product makeup should be specified exactly. Preinsulated piping is most effective when there are a minimum number of bends and penetrations.

Waterproofing

Joints between collectors and piping, direction changes in pipes or pipe tees, and points where valves or monitoring equipment protrude require special attention for waterproofing. Waterproofing should have no open seams, and joint protection should be provided at all elbows and tees.

Pipes should not be bound to their supports. They should be free to move, and waterproofing and insulation should be designed assuming pipes will move.

Pipe supports should not puncture waterproofing, especially where they support pipes directly, to avoid creating a heat bridge that will result in thermal energy loss through conduction. Pipes should be supported loosely on the outside of the waterproofing (Fig. 8-45).

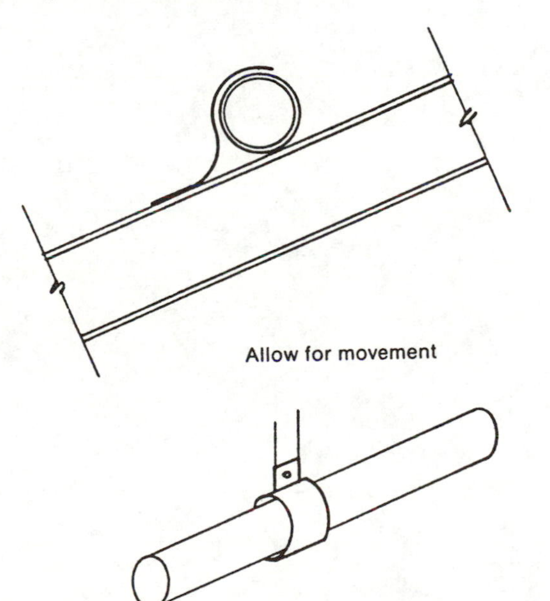

Fig. 8-45. Supporting pipes over insulation and waterproofing

Ends of sheet waterproofing should not be sealed to the pipe. Only seals between waterproofing layers tend to hold (Fig. 8-46).

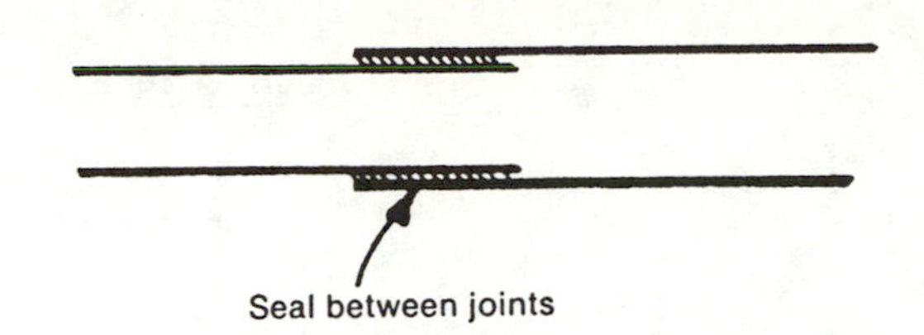

Fig. 8-46. Sealing joints of waterproofing

Contacts between different metals should be avoided if aluminum waterproofing is used. For steel structures or steel clamps, nonmetallic protection should be included to prevent galvanic action. Do not use aluminum waterproofing with a sheet of galvanized steel under it without galvanic protection between them. Flashing should be galvanically compatible with collector construction, including supports and fasteners. Proper dielectric insulation must be used where incompatibilities exist.

Chapter 9
Collectors: Specification and Prebidding

INTRODUCTION

Collector procurement is an important part of the purchase of a large-scale solar system. In this chapter, we consider possible approaches to collector specification and procurement, and the advantages and disadvantages of each approach. An in-depth discussion of the prebid method of procurement follows, with cost effectiveness and project control emphasized. Because prebidding with a performance specification may not be famliar, we provide instructions on developing such a specification.

COLLECTOR PROCUREMENT METHODS

We will discuss only flat-plate and evacuated-tube collector procurement since these are the collector types most commonly used. Details change for tracking concentrators, but the basic approach remains the same.

Three methods to procure collectors for a solar project are selection by the system designer (A/E), bidding as part of a total solar system, and prebidding.

The method selected should optimize on the following criteria:

- Collector should be most cost-effective in terms of energy delivered per installed cost.
- Collector must be reliable and durable.
- Collector must be easily integrated into the overall system.
- The A/E, the owner, or both must maintain control of the design and construction process.

Selection by the System Designer (A/E)

This simple procurement method is acceptable when the owner is in the private sector. When the owner is a public body, however, selection by an A/E or use of a proprietary specification is usually not legal. Furthermore, a more competitive cost can usually be obtained in an open-bid procurement.

Bidding Collectors as Part of a Total Solar System

This is the second easiest procurement method. A designer uses a "standard" collector as a part of the construction bid package and allows bidders to make substitutions. Unfortunately, there is no standard among collectors. Each is different in size, support requirements, piping arrangements, coolant pressure drop, performance, and materials. The specified collector also may not be the most cost-effective one.

With a substitution, many problems will arise. The bidder must find a collector that is more cost effective and prove to the A/E that its performance equals or exceeds that of the standard specified. The bidder is responsible for redesigning structural supports, piping, and array configuration. Few contractors are capable of specifying their own collectors, and the A/E may not accept the alternate design. A long bid period is required with this procurement method if bidders are to opt for substitutions and not just bid plans and specifications. The extra time required to properly bid a project that allows substitutions will partially offset the additional period required to prebid collectors.

The types and formats of information to be provided by a contractor making substitutions must be specified. There should be a procedure to prequalify a substitution prior to bidding to save a contractor the expense of preparing a bid on a collector system that may be rejected.

Prebidding Collectors

Prebidding collectors involves putting collectors out for bid early in the design stage and using the successful collectors for final design. Collectors are then provided to the contractor as owner-furnished equipment.

For large public projects, where everything must be bid, and for large private-sector projects, prebidding collectors is the best method. By prebidding, the most cost-effective collectors can be secured prior to completing system construction documents. Final design details can be based on the actual collectors to be used. A major drawback is the difficulty of writing bid documents to ensure that different collectors can be compared on an equal basis.

SYSTEM DESIGN USING THE PREBID APPROACH

The prebid approach allows bidders to base their bids on the cost to supply collectors that will satisfy a specified annual load, using "local" long-term weather data. The early solar system design process is therefore independent of the collector procurement method. Loads and solar availability are determined, and collector location is selected based on available area and structural capability. Storage and mechanical equipment locations are based on available space and pipe-run requirements. The required collector area is estimated by computer modeling, using a generic collector type, but the collector described at this point is only an approximation. Preliminary design may be based on a flat-plate collector, due to the load characteristics and the particular market, but the low bid may involve an evacuated-tube collector, or vice versa.

The major change in the design process occurs in scheduling. If collectors are either designer-selected or bid as a part of the total system package, system design and the development of construction documents are parts of a continuous process. However, when collectors are prebid, early design phases normally define collector requirements and develop prebid specifications. A collector array area is defined, and the total system is outlined and modeled; design is made final when collector bids are in and collectors selected.

A preliminary collector support design should be included in the collector prebid material. Final dimensions may vary, but a design is necessary for cost estimating and for evaluating attachment procedures. During the collector bid period, work can continue on below-the-roof system design. Storage, heat exchangers, pumps, and load interfaces can be designed and located since design of these items should not change significantly with collector selection. Only support structure sizing and spacing and above-roof piping layout will be affected.

THE PREBID SPECIFICATION PACKAGE

A prebid specification package, prepared by the owner, the designer, or a designated agent, contains four sections. The first includes typical boiler plate used to procure any piece of hardware and requires no further discussion here. The second contains the bid forms all bidders must complete. The third describes bid evaluation considerations and requirements, and the fourth contains collector specifications. The sections may be reversed or combined.

Bid Form Section

The bid form section is designed to elicit a cost proposal with supporting and clarifying descriptions from bidders. The information provided by the bidder includes (1) the base bid, (2) general collector information, (3) specific collector information, and (4) references.

Base Bid

This is the actual bid document on which collector purchase will be based. Since this bid will be submitted prior to bidding of a general construction contract, prices must be in effect until the remaining project costs are known, possibly up to 150 days. The following information should be required:

- Bidder's name, address, telephone number, and contact person.
- Collector manufacturer and model number.
- Number of collectors proposed to provide the minimum amount of energy required to satisfy the specified load. (This number is not contractually binding; the final number of collectors purchased may differ substantially. After verification by bid reviewers, this number is used to develop an "Adjusted Total Cost," discussed below.)
- Base-bid collector costs. These are the only contractually binding costs. Total cost for the final number of collectors purchased is based on the following cost information provided:
  - Collectors (cost per collector based on the approximate number noted above)
  - Spare glazings
  - Special hardware required for installation
  - Shipping
  - Warranties
  - Installation supervision and acceptance testing
  - Special sensors.
- List of all special hardware to be supplied, as priced in base bid cost.
- Number of working days to fabricate and deliver collectors to the site. This is required so that general construction can be scheduled. Liquidated damages may be tied to this figure.
- List of any exclusions or exceptions to bid-document requirements.
- A place for both parties to sign.

Shipping, warranties, and other guarantees referred to in the above should be referenced to the section on specific collector information, below.

General Collector Information

This section should list general material required to describe collectors, including

- Manufacturer's brochure describing the collector model proposed and all related special hardware
- Collector construction details
- Certified test reports covering NBSIR 78-1305A (Waksman et al. 1978) and ASHRAE Standard 93-77 (ASHRAE 1977) results (see "Applicable Publications," below)
- Manufacturer's warranties

- Mounting details, based on details provided in the appendix to the collector specification
- Manufacturer's installation instructions, including interconnecting piping details
- Bidder's F-Chart (U. of Wis. 1985) (see "Bid Evaluation Section," below) run output used to develop the base bid (optional).

Specific Collector Information

Material presented in this bid-package section is largely extracted from the general collector information described above. Information needed to confirm adjusted costs and F-Chart results is highlighted; it can also be used for detailing the final design. The information should include

- Collector physical description
  - Collector manufacturer name and model number
  - Overall height per collector, including protrusions
  - Overall width per collector, including protrusions
  - Aperture area
  - Gross area per collector, per ASHRAE Standard 93-77
  - Diameter of headers, or pressure drop at recommended coolant flow rate
  - Working pressure
  - Manufacturer's suggested coolant flow rate.
- Array size
  - Number of collectors proposed under base bid
  - Total area of collectors, computed as gross area per collector multiplied by number of collectors
  - Total length of collectors, computed as unit width plus 3/4 in. (for piping) multiplied by number of collectors
- Collector performance description (as required for F-Chart input), including
  - FR-UL PRODUCT (C1): the first-order term from a post-stagnation collector performance curve, based on gross collector area, in units of thermal conductance per unit area
  - FR-TAU-ALPHA (C3): the constant term from a post-stagnation first-order collector performance curve, based on gross collector area
  - Incident angle modifier constant: for evacuated tubes, the average modifier; for flat plates, the incident angle modifier. (Both should be derived from performance testing in accordance with ASHRAE Standard 93-77.)

References

From three to five references should be required, with some representing projects of a similar size to the one being bid. Each reference should include the name, address, telephone number, contact person, total array size, and date of startup of systems using the same collector as that being proposed. This information is particularly important if tracking or concentrating collectors are involved. Previous users can supply information on motors, leakage, and the reliability and durability of tracking and reflective components. At least two references should be to jobs of comparable size to the one being bid to ensure that the bidder and his suppliers have sufficient capacity to install a system in a timely manner.

Bid Evaluation Section

Determining Cost Effectiveness

Specifying a manufacturer and model number for a boiler, and permitting substitution of approved equals, will result in comparable bids. Solar collectors, however, are not sufficiently comparable to be bid as easily. No established standards exist for comparing collector performance. Definitions of collector efficiency and aperture area vary considerably in practice, particularly when flat-plate and evacuated-tube collectors are compared. The only reliable way to ensure prebid selection of the most cost-effective collectors is to base selection on energy delivered per dollar of installed collector cost for a given annual load and delivery temperature.

There are two ways to figure collector costs to evaluate cost effectiveness: (1) compare unit prices, as noted on the bid forms, times the total number of collectors, or (2) develop a total adjusted cost for the collector installed, including all piping, insulation, structure, etc. The second is better and strongly suggested. A small collector or a wide, short collector may have a low cost, but extra handling and greater piping and structural costs can result in an actual installed cost considerably greater than that for a large or tall, more expensive collector.

Annual Energy Requirements to be Met

To ensure a valid comparison, all bidders should use the same approach to arrive at a total number of collectors required to meet the specified load. The most reliable way is to require use of an easily accessed and widely accepted computer program that has a sufficiently large selection of weather data locations. F-Chart (U. of Wis. 1985) is an appropriate program for this purpose. To use F-Chart, the following inputs are provided to bidders:

- Solar system type
- Collector type
- Heat transfer fluid type
- Storage type
- City call number.

Inputs must be carefully defined to bidders to ensure that they all provide program inputs on the same basis (see Table 9-1). Input for heat transfer fluid parameters, storage parameters, delivery device parameters, load parameters, and auxiliary energy supply parameters should be given to bidders in the bid package. All other input parameters (except collector slope and azimuth and ground reflectance) should be furnished by the bidder, based on the performance of the collectors proposed.

Since bidders determine the number of proposed collectors, and their costs, required to satisfy the stated energy requirements, energy delivered per dollar cost can be computed from figures noted in the base bid. This does not take into account differing costs of installing different collectors. Prebidding is meant to ensure not only that the most cost-effective collectors are selected, but also that the selected collectors will still be the most cost effective when installation costs are included. Therefore, cost effectiveness should be computed using "total adjusted cost."

Total Adjusted Cost

Total adjusted cost is defined as

Base Bid Cost + Structural Support Cost
+ Collector Installation Cost
+ Special Hardware Installation Cost

Total adjusted cost equals collector costs plus costs as estimated for each of the nonbid items required to complete collector installation, as indicated above.

Base Bid Cost

This is the total cost of the collectors themselves. Since there can be significant differences between bidder and evaluator F-Chart results, the bid evaluator should not simply accept the number of collectors proposed by the bidder. To ensure compatibility, one bid evaluator should make all F-Chart model runs.

Table 9-1. F-Chart Program Input

Input	Description
C1. Collector Area	Gross area of collectors used in base bid. The gross area shall be in accordance with ASHRAE Standard 93-77.
C2. FR-UL Product	The value used is the first-order term from the post-stagnation collector performance curve.
C3. FR-TAU-ALPHA (Normal Incidence)	The value used is the constant term from the post-stagnation collector performance curve.
C4. Concentration Ratio C5. CPC Acceptance Half-Angle	For concentrating collectors based on proposed collector as tested.
C6. Number of Covers C7. Index Refraction C8. Extinction Coefficient Length (KL)	Should be omitted in favor of C9 test results.
C9. Incident Angle Modifier Constant	For evacuated tube, use average incident angle modifier derived from performance testing in accordance with ASHRAE Standard 93-77. For flat plate only, based on the incident angle modifier equation derived from performance testing in accordance with ASHRAE Standard 93-77.
C10. Collector Flow Rate-Specific Heat/Area (Btu/h-ft^2°F)	Based on the manufacturer's recommended coolant flow rate for a single panel multiplied by the specific heat of the coolant and divided by the collector area.
C11. Tracking Axis (1 = EW; 2 = NS; 3 = 2-Axis)	For concentrating collectors based on proposed collector as tested.
C12. Collector Slope	Provided to bidders by A/E.
C13. Collector Azimuth	Provided to bidders by A/E.
C14. Ground Reflectance	Provided to bidders by A/E.
C15. Incident Angle Modifiers (10, 20, 30, 40, 50, 60, 70, 80)	For concentrating collectors based on proposed collector as tested.

Structural Support Costs

Cost estimates by the bidder for structural supports should be based on the cost of a linear foot of support structure length. This will be multiplied by the total length of the collector array to arrive at an estimated structural-steel cost. Cost per linear foot of structure length should be estimated by first designing and pricing a typical support section, including painting and roofing. Difference in cost per linear foot will not vary significantly with collector height, especially in the 7 to 9 ft range. Structural members may vary somewhat in weight, but labor required to fabricate and install such members will be about the same. If additional support rows are required, support costs will increase significantly because additional parts, roof penetrations, and labor are required.

If collectors are to be mounted on a continuously sloping surface and parallel to it, a square-foot of roof cost approach will be most accurate. In this case, there would be no significant difference in structural cost if 5000 ft^2 of collectors were mounted on a 50 ft × 100 ft sloped surface in 6 rows or in 10 rows. If, however, access walks were provided every two rows, a combination of collector width and unit area approaches would be required.

Collector Installation Cost

Adjusted collector costs include costs for handling, uncrating, placing, securing, and piping or ducting the collectors. These costs can be estimated at a flat rate per collector.

Special Hardware Installation Cost

Costs to install special hardware (i.e., those items that must be attached to the collectors or to the structure for collector installation) should also be added to installation costs. Many collectors can be fastened directly to a flat support; others require special hardware. Extra handling and time required to install special hardware should be considered in determining an adjusted cost. Costs should be determined on a sliding scale, depending on the work involved. Since apportioning additional costs will be done in a somewhat subjective manner, the method used should be described either in the bid document or in a sealed memo filed prior to bid opening.

Once a total adjusted cost has been determined, a proper comparison between collectors can be made on the basis of annual energy delivered per dollar of adjusted cost.

Collector Specifications Section

This section of the bid is used by the bidder to amplify the information requested in the bid forms section. In addition, it further describes special collector requirements such as collector housing material and color, fluid passage and absorber materials, avoidance of trapped air in fluid passages as installed, internal or external manifolds, minimum or maximum pressure drops through collectors, etc. Such items as absorber finish, minimum or maximum collector efficiencies, number of glazings, type of glazing, etc., might also be specified, but they are best left to the economics of the bidding process. If a generic type of collector, such as a tracking collector, is not wanted, this should also be stated.

Since prebidding involves two separate contracts, one for the collectors and one for the system, full responsibility for proper collector installation cannot be placed under a single contractor. Therefore, the collector supplier's responsibilities must be carefully defined.

Scope of Work

The bidder should furnish a scope of work that includes information on how the following items or work, as required, will be provided: collectors, spare glazings, special hardware, shipping, delivery, warranties, supervision, installation of sensors provided by others, and special sensors supplied by the collector manufacturer.

Work of Other Contractors

Normally, someone other than the collector supplier will receive and install the collectors. If so, the duties of the installation contractor must be specified here.

Applicable Publications

Publications that deal with various collector components (e.g., cover glass, copper tubing, reflective material, insulation) should be provided to bidders in this section. General collector publications should also be referenced, especially those that define applicable testing procedures, such as those developed by the American Society of Heating, Refrigerating and Air Conditioning Engineers (ASHRAE) and the National Bureau of Standards (NBS). These include ASHRAE Standard 93-77 (ASHRAE 1977) (collector testing standard), NBSIR 78-1562 (DOC 1978) (solar system performance criteria), and NBSIR 78-1305A (Waksman et al. 1978) (collector testing procedures).

Testing

Collector testing, in accordance with ASHRAE Standard 93-77 and NBSIR 78-1305A, should be required. Testing should be conducted at an independent testing laboratory that will certify test results. Acceptable "independent testing laboratories" may be named, or an experience criterion (e.g., the laboratory must have performed at least "X" number of similar tests) should be stated. A bidder should include all data required for modeling with F-Chart or whatever analysis method is selected.

Submittals

It may be desirable to have a full-size collector submitted, prior to full shipment, for a general contractor prebid meeting. A statement from the collector supplier certifying that the collectors delivered are as proposed in the bid should be included in this section.

Special Hardware

Special hardware is defined as any separate pieces, not shipped as an integral part of a collector, that must be attached in the field so a collector can be installed on its support. Special hardware may include, but not be limited to, clip angles, pressed-steel channels, and special nonstandard bolts. Information on special support requirements must be provided to bidders so they can determine an installation approach and any need for special hardware.

Shipping

Special shipping requirements, such as use of trailers, should be noted in the bid. If the project requires a trailer load or more, it may be best to not unload and store collectors but to leave them stored on or in the trailer. Therefore, it might be cost effective to have the trailer remain at the site for a week or two, at the collector supplier's expense, with additional time charged to the contractor. If this were made an "add alternate," a similar "deduct alternate" could be put in the construction bid document. The deduct alternate would indicate what it was worth to the contractor not to have to unload the trailer but to use it for storage. Breakage responsibility questions can arise, but they may arise anyway. Will it be the fault of shipper, trucking company, or receiver if breakage occurs? This should be resolved in the bidding process.

Sensor Requirements

If thermal sensors are to be factory installed (as they should be), this should be so stated. Backup sensors should be installed on the control collector. If the proposed collectors rely on special sensors (e.g., pyranometers) or tracking devices, this requirement should be noted in the base bid. A bidder should be required to supply such special sensors and include the cost in the base bid.

Delivery

A time commitment for collector shipment must be included in the base bid, and all special hardware should be shipped with the collectors. Any collectors damaged during delivery should be replaced or repaired in a timely manner. Spare collectors and glazings (about 2% of the number installed) should be ordered as standard practice to cover damage potential.

Installation Supervision

A general contractor's price for collector installation is likely to be reduced if supervision of collector installation is provided as a part of the collector bid. A minimum of three days of installation supervision is suggested, with the supervisor certifying that the collectors were installed properly. (It is important to inform the general contractor in the specification that this service will be provided.)

Warranties

Responsibilities of each contractor for providing warranties must be clearly stated. These responsibilities should also be restated in the general contract specification. A suggested form would be:

- Collectors shall be fully warranted against defects in performance, design, materials, and workmanship for a period of three years from the date of installation completion. This warranty will cover for a period of three years the full cost of parts and labor required to remedy any manufacturing defect of a collector, including, if necessary, replacement at the site.
- The collector supplier will be responsible for collectors damaged in transit and for any manufacturing defects. The collector supplier will not be responsible for collectors damaged in installation once the collectors have been accepted by the general contractor. In cases of dispute as to responsibility, the (owner, A/E, government, other) will be the final arbitrator. The warranty will not be voided if the collector undergoes sustained stagnation.
- Upon notification of collector failure attributed to manufacturing defects, the supplier will provide an on-site inspection within 10 days, at no cost to the owner.

Solar Collectors

The remainder of the specification should deal with the specific collectors to be supplied. If more than one type is involved, each type should be covered separately.

Drawings

Drawings showing structural support details and a typical collector row configuration (with all connections detailed) should be included in this section. The support detail should be sufficiently large in scale that collector attachment can be seen in detail.

BID REVIEW AND CONTRACT AWARD

After all bids are received, they should be reviewed for completeness and accuracy. If there are any conflicts between the brochure information, the material noted on the forms, and the test data, the bidder should be contacted for clarification or additional information. Once reviewers are satisfied that all responses are complete and accurate, one reviewer should make all F-Chart model runs for all collectors bid to substantiate the number required to satisfy the specification requirement.

When collector numbers are agreed upon, adjusted costs should be figured by adding the following bidder-provided costs:

- Cost of collectors delivered (based on bid prices)

- Cost of structural support (estimated cost per linear foot times total collector array length)
- Handling and installation (estimated cost to install times total number of collectors)
- Additional cost for installing special hardware (estimated cost for work times total number of collectors).

This will provide an estimated total adjusted cost for the collectors. The successful bidder should have the lowest adjusted cost.

The contract is let solely on the unit prices noted in the bid form. All other numbers are developed purely for comparisons. The number of collectors should be based on the final design and may be more or less than originally bid. However, since the final price is derived from the unit price, there is no problem. No contract for collectors should be signed before the contract for the general construction is signed.

Chapter 10

Start-Up and Acceptance Testing of Active Solar Systems

INTRODUCTION

General Considerations

A documented start-up and acceptance-test procedure ensures that all components of a solar energy system operate as designed and specified. An acceptance test plan should clearly define expected system performance in terms of mechanical performance, thermal performance, or both. The plan should include all new and existing subsystems associated with system operation. Energy-collection (collectors, pressure relief valves, air vents, expansion tanks, etc.), energy-transfer (pumps, piping, heat exchangers, flow-measuring devices, control valves, etc.), energy-storage (storage tanks, pressure-relief valves, etc.), and energy-delivery (heat exchangers, heat pumps, chillers, etc.) devices all should be included in the test plan, along with all controls associated with their operation.

Most owners have neither the experience nor the knowledge to develop an appropriate testing procedure. Therefore, for commercial and industrial projects, the system architect/engineer should prepare a complete and comprehensive acceptance test plan. Once approved by the owner, the plan can be provided to the person or company responsible for system installation (hereafter referred to as the contractor) or to a contracted testing and balancing agency. Residential installations predominantly are of standard design. The homeowner mostly relies on the system's vendor to act as designer and contractor/installer. To be assured about the system and its installation, however, the homeowner should hire an independent consultant to check system design before it is purchased and to check system integrity before the system is accepted.

Scheduling and Duration of Acceptance Test

The contractor is responsible for scheduling and coordinating an acceptance test, and he should provide adequate notice to all involved parties. Acceptance test duration is a function of several variables, including system size and complexity. Up to three days may be required, provided unexpected problems are minimal. A system should be deemed acceptable only if it and all its components operate as designed.

Instrumentation

A minimum instrumentation level is required to perform an acceptance test. Data acquisition can be facilitated by taps in the system for receiving temperature and pressure measuring probes; "Pete's Plugs," or similar devices, can be very effective. A variety of portable devices are available for measuring temperatures, flows, and pressures.

All test instruments should be calibrated to an operating accuracy within a manufacturer's specified accuracy. All thermal indicators should be calibrated with a laboratory-standard thermometer. A thermal probe should be placed into a common-stirred thermal bath with the laboratory standard thermometer and calibrated in the design range of system operating temperatures. A detailed discussion of instrumentation appropriate to active solar systems is presented in Ch. 12.

Testing Agencies

System testing and balancing should be performed by personnel whose qualifications have been reviewed and approved by the architect/engineer. In general, the contractor is responsible for providing qualified personnel. For small, simple systems, the contractor often performs the entire acceptance test. For large or complex systems, an approved member of the Associated Air Balance Council (AABC) or the National Environmental Balancing Bureau (NEBB) usually performs testing and balancing. These agencies are regulated by industry and have no governmental ties.

The testing firm should be informed of any major system changes made during construction and should be given a complete set of "as-built" drawings. Contractor and test firm should schedule and coordinate final acceptance testing, and the testing firm should make site visits during system construction.

Owner, contractor, and architect/engineer representatives should be present during acceptance testing of a large system. The contractor is responsible for putting the system into full operation, maintaining its operation for the duration of the test, and providing all labor and materials needed to correct system problems without undue delay.

The testing firm should test, adjust, and balance flows in all equipment that will operate as an integral system part; specific testing and balancing procedures to be performed are discussed below. Once an acceptance test is completed, the testing firm should prepare a written report that includes diagrams, a description of test procedures, and all recorded test data. This report should be bound to provide a permanent record and should contain all certifications of proper system operation.

Acceptance Test Plan

Preparation of an organized and comprehensive Acceptance Test Plan (ATP) is extremely important for proper acceptance test performance and, consequently, for proper system operation. The architect/engineer should provide sufficient detail and scope in the ATP to assure the owner that system performance meets design requirements.

A typical ATP for a large or complex solar system can be separated into three stages: initial visual inspection of the system, preoperational system testing, and operational system testing. Table 10-1 presents an ATP outline prepared for a 2000 ft^2 (186 m^2) collector array for a liquid-based solar service hot water system located in Massachusetts (Mueller 1982).

VISUAL INSPECTION PROCEDURES

General Considerations

The first stage of the acceptance test is a visual inspection of the system to verify that proper components have been installed in proper physical configurations. To conduct this inspection, the architect/engineer reviews shop drawings submitted by the contractor and prepares a checklist, or "punch list," to check that all components are (1) of proper materials, (2) of proper size and capacity, (3) with no apparent physical damage, and (4) in their proper physical location. The check should also include an inspection of minimum straight-pipe runs both preceding and following all flow-measuring devices, as recommended by the device manufacturer.

In addition, all components should be inspected to verify that they are installed in accordance with manufacturer's specifications and instructions regarding mounting, alignment and leveling, charging with operating fluid, direction of rotation, lubrication, and other specified considerations.

The punch list must be in accordance with all architectural/engineering design and specification documents. These construction documents form the basis for resolution of any discrepancies found during a visual inspection; they must be prepared in a comprehensive manner.

If a solar system is to operate in conjunction with conventional HVAC distribution systems, these systems should be checked for proper operation before the acceptance test. If problems exist with these systems, they should be resolved prior to acceptance testing.

Visual Inspection Checklist

The extent and degree of detail in a punch list are functions of system complexity. A long and detailed punch list may be divided into subsystems, or categories, to be inspected. Table 10-2 presents a portion of a punch list used for the liquid-based system previously referenced (Mueller 1982). The list is divided into the eight subsystems listed under IIA in Table 10-1.

Table 10-3 presents a portion of a similar punch list prepared for an air-based system (DOC 1980).

PREOPERATIONAL TESTING PROCEDURES

The second stage of the acceptance test is called "preoperational testing." It consists of all tests performed after visual inspection, with the system not fully operational, including pressure and leak testing; flow balancing; and checking of components, sensors, and circuitry. Because of basic differences in test procedures, liquid- and air-based systems will be discussed separately.

Liquid-Based Systems

Data Requirements

Planning, data collection and review, instrumentation procurement, field checks, and actual testing should be carefully scheduled. Data required for testing, balancing, and adjusting a system consist of all applicable contract documents, including change orders, plus the following (Eads 1982):

- Pump curves and performance data
- Pressure, flow, and temperature characteristics and other performance data for all applicable equipment (boilers, chillers, coils, heat exchangers, etc.)
- Motor nameplate data
- Starter sizes and locations, and relay and thermal-overload-protection ratings
- Controller operating specifications
- Performance characteristics of balancing and flow-metering devices in all circuits
- Temperature-control diagrams and control-valve characteristics
- Special piping and control diagrams
- Ratings and settings of pressure-relief, temperature-relief, and pressure-reducing valves
- Operating and maintenance instructions for significant equipment
- Characteristics of automatic recorders for flow, pressure, temperature, etc.
- Availability of special handles required for valve adjustments
- Thermometer and other gauge data including ranges, graduations, and accuracy limitations.

Table 10-1. Acceptance Test Plan Outline for Liquid-Based Systems

I. INTRODUCTION
    A. Objective
    B. System Description
    C. System Schematic
    D. System Narrative
    E. Definition of Acceptance

II. SITE INSPECTION CHECKOUT
    A. Visual Inspection
        1. General
        2. Mechanical
        3. Collectors
        4. Instrumentation
        5. Electrical
        6. Structural and Civil
        7. Safety
        8. Major Component Identification
    B. Pre-Operational Testing and Procedure Verification
        1. General
        2. Expansion Tank Precharge Pressures
        3. System Power Circuitry
        4. Five-Minute Minimum Run Time/High Collector Temperature/Freeze Protection Circuitry
        5. Differential Thermostat Checkout
        6. Pump Current Measurement
        7. Pump Rotation Checkout
        8. Collector Loop Pump Failure Circuitry
        9. Intermediate Loop Pump Failure Circuitry
        10. Storage Loop Pump Failure Circuitry
        11. Collector Flow Balancing
        12. Intermediate Loop Cold Water Makeup Pressure Reducing Valve Setpoint
        13. Nighttime Pump Operation
        14. BTU Meter No. 1 "Solar Btu's" Test
        15. BTU Meter No. 2 "Service Hot Water Btu's" Test
        16. BTU Meter No. 3 "Recirc Btu's" Test
        17. Semi-Instantaneous Water Heater Pump Control
        18. Tempering Valve Operation
        19. Strainer Cleanliness (Collector Loop)/Glycol Hand Pump Checkout
        20. Strainer Cleanliness (Intermediate Loop)
        21. Strainer Cleanliness (Storage Loop)

III. OPERATIONAL TESTING
    A. Storing Solar Energy
        1. Mode Description
        2. Test Procedure
        3. Westover Job Corps Center Data Log Sheet
    B. Heating Service Hot Water
        1. Mode Description
        2. Test Procedure
    C. Recirculation Control
        1. Mode Description
        2. Test Procedure
    D. Performance Testing
        1. Performance Data Sheet
        2. Performance Data Sheet Notes

Appendix I - Equipment List

Detailed schematics should be available at test time for the system and all subsystems. These should include all system components and flows, pressures, pressure drops, and temperatures at major system points.

Flushing, Pressure Testing, and Filling

After visual inspection is completed, preoperational testing is initiated. All steps should be performed when solar radiation incident on the

Table 10-2. Portion of Visual Inspection Punch List for a Liquid-Based System

A. Visual Inspection

1. General

Item to be Verified	Initials	Date
Operation and Maintenance Manual is available on site.	______ /	______
As-Built Drawings are available on site.	______ /	______
Items requiring attention/comments: ______		

2. Mechanical

Item to be Verified	Initials	Date
Components are set, aligned, and bolted down in accordance with installation drawings.	______ /	______
Adequate space is provided around components for operation, maintenance, inspection, and servicing.	______ /	______
Tanks are provided with adequate access to nozzles, sight glasses, and instruments.	______ /	______
Piping runs are routed and sloped per drawings and specifications.	______ /	______
Piping and components can be adequately filled and drained.	______ /	______
Piping runs provide for thermal expansion.	______ /	______
Piping is adequately supported per drawings and specifications.	______ /	______
Piping connections are tight and no leakage is evident.	______ /	______
Piping and components are insulated and weather protected per specifications and drawings; insulation jacketing is lapped to shed water, and all joints are sealed.	______ /	______

collectors is very low or with the collectors covered. Testing begins by flushing the system of all metal filings, flux, solder, silt, and other contaminants that may have been introduced during installation. For systems that use silicone or hydrocarbon heat-transfer fluids, a small, high-head, self-priming pump connected to the fill valve can be used to flush the system using a full container of the specified heat-transfer fluid (HUD 1980). For water/glycol systems, the fill valve can be connected directly to the water supply. After flushing, strainer screens should be removed and cleaned. Before the system is filled, precharge air pressure in each expansion tank should be measured with a precision pressure gauge.

The next step is normally a system pressure or leakage test, conducted before any insulation is installed. Pressure testing should not be conducted when freezing is possible. The contractor usually performs the test and provides all required equipment and labor. Gauges, relief valves, and expansion tanks that might be damaged at test pressures should be removed before testing and replaced after testing. The system should be filled with water and pressurized for at least four hours. Should any leaks develop, the system must be drained, the leaks repaired, and the system refilled and retested. The system may be pressure tested in sections as approved by the architect/engineer.

Service water piping should generally be tested to 20 psig (138 kPa) above maximum expected makeup-water pressure (ANL 1981). All solar collector supply and return piping should be tested to at least 100 psig (690 kPa) unless the collector manufacturer's recommendations are different.

Drain pipes should be tested with a minimum pressure of 4 psig (28 kPa). Tests are normally required by code and, at a minimum, they should verify that all joints are watertight.

Table 10-3. Portion of Visual Inspection Punch List for an Air-Based System (including Rock-Bed Heat Storage Unit)*

<u>INSPECTION CHECK LIST - COLLECTORS</u>

1. Collector Array: Refer to plans.

2. Holding proper dimension from collector to collector - thus making airtight seal from port to port: Refer to specifications and/or plans.

3. Used specified material for port and end cap sealant: Refer to specifications and/or plans.

4. Relief tubes sealed and in place: Refer to specifications.

5. Confirm location and dimensions of 2 × 8 (1 1/2" × 7 1/8") frame.

6. Cap strips installed so proper and airtight seal is accomplished.

7. Perimeter insulation installed: Refer to specifications and/or plans.

8. Perimeter flashings installed properly: Refer to specifications and/or plans.

9. Connecting Collars: Refer to plans.

   a. Location
   b. Sealed properly

10. Heat Sensors: Refer to plans.

   a. Installed properly
   b. Correct Location

<u>INSPECTION CHECK LIST - ROCK-BED HEAT STORAGE UNIT</u>

1. Location: Refer to plans.

2. Location of Duct Openings: Refer to plans.

3. Dimensions of Unit: Refer to plans.

4. Dimensions of Duct Openings: Refer to plans.

5. General Construction: Refer to plans.

   a. If construction does not follow plans, make sure modifications are adequate to meet specifications.
   b. Check for exposed wood - all combustible surfaces must be covered with a non-combustible material.

6. Check all joints

   a. Sealed adequately
   b. Correct sealant used (i.e. suitable for temperatures around 180°F)

*Courtesy of the Solaron Corporation, used by permission.

Water cannot be used to test silicone or hydrocarbon systems since any water left in the system will separate from the heat-transfer fluid and could boil, freeze, or cause corrosion. Use of silicone or hydrocarbon fluid for a pressure test can be expensive and difficult, so the test is frequently performed with pressurized air from an air compressor. If system air gauge pressure drops more than 20% in 24 hours, the system leaks. All pipe joints and packings should be tested with soapy water; bubbles will appear at leaks. This test is not a replacement for the ASME pressure-vessel certification test (Cole et al. 1979).

Final system filling is performed after the pressure test. Fluid introduced at the bottom should be pumped through the system for about 20 minutes to purge air. A drain (return) valve should then be closed and the system allowed to pressurize. When the desired pressure is reached, the pump should be shut off and the fill valve capped and tagged with fluid type, fill date, and emergency procedures. A system can also be filled by pouring fluid into a fill valve installed at the system's high point. This is often done for nonpotable or toxic fluids since it minimizes initial entrapment of air and fluid handling (HUD 1980). At this point, the heat-transfer fluid should be tested to determine that its concentration, freeze point, pH, and corrosion inhibitors, as applicable, are proper.

Circuitry and Sensor Checkout

System circuitry, sensors, and instrumentation are tested next to verify proper installation and operation of system wiring, sensors, actuators, thermostats, meters, alarms, valves, etc. At a minimum, these tests should check the following components or subsystems:

- System power circuitry--to verify proper wiring from power source(s) to pumps, controllers, and indicators
- Time-delay relays--to provide minimum run times for pumps to avoid pump cycling, when applicable
- Collector sensors, relays, controllers, valves, and indicators--to detect and alleviate high collector temperatures
- Freeze-protection sensors, relays, controllers, valves, alarms, and indicators--to prevent freezing
- Storage-tank sensors, relays, controllers, valves, and alarms or other indicators--to detect and alleviate high storage-tank temperatures
- Differential thermostats--to ensure control of pump and valve operations
- Pump circuitry--to verify proper pump current, direction of rotation, and all elements of pump-failure circuitry, including flow switches
- Thermal instrumentation--to measure system thermal performance.

Most solar systems are designed so these circuitry tests can be conducted while the system is not operating. Temperature, flow, time, and other conditions required to conduct these tests can usually be simulated by disconnecting the required sensors, adjusting setpoints, immersing the sensors in ice or boiling water, etc. However, proper operation of sensors and circuitry should be checked later under actual operating conditions to verify sensor calibration.

Flow Balancing and General Equipment Checkout

The final stage of preoperational testing consists of flow balancing and verification of proper operation of certain components and subsystems that are checked when a system is in, or ready for, operation. Proper flow through collectors, heat exchangers, and piping must be established and maintained for efficient and reliable system operation. If flow through a collector or heat exchanger is too low, heat transfer will be reduced and system efficiency will suffer. If flow is too high, a pump motor may be overloaded and excessive erosion/corrosion may occur at heat-transfer surfaces.

Flow measuring and balancing are important for identifying blockages in fluid passages from solder or other debris and to identify collectors having unusually high pressure drops. If a pressure drop exceeds the manufacturer's specifications, some collector passages may be blocked, deformed, or leaking (ANL 1981).

Certain basic flow-balancing procedures are common to all liquid-based systems. These sequential procedures can be summarized as follows (Eads 1982):

1. Determine that all preliminary procedures and system equipment checks have been performed.
2. Determine that all manual valves are open or preset, as required, and that all automatic valves are in full-open position.
3. Measure system static pressure at the pump(s).
4. Place the system in operation, eliminate all air, and allow flow conditions to stabilize.
5. Measure pump speed, operating voltage, and amperage and compare them with nameplate ratings.
6. Determine shutoff discharge and suction pressures at the pump by slowly closing the shutoff valve in the pump discharge line. Using shutoff head, determine if the measured test point falls on the submitted pump curve. If it does, proceed to the next step; if it does not, plot a new pump curve parallel to the submitted curves, from zero to maximum.
7. Determine pump discharge pressure, suction pressure, and total head after slowly opening the balancing valve in the discharge line to its full-open position. From the measured head, determine flow using the corrected pump curve established in Step 6. Measured flow should be somewhat greater than design flow. If measured flow is less than design flow, a problem exists (e.g., blockage, leakage, undersized piping, defective motor) and will have to be corrected before testing continues. If measured flow is greater than designed, the balancing valve should be closed until flow is approximately 110% of design. Pump motor amperage should be measured and compared to the nameplate rating.
8. Determine pressure drops through system equipment (collectors, heat exchangers, etc.) and compare these readings with submitted data to determine flow. Make a preliminary adjustment in the balancing valves on all equipment with high flow, lowering flow to 10% above design flow.
9. Measure pump pressures and amperage again. If flow has fallen below design flow, open the balancing valve in the pump discharge line until the flow is from 105% to 110% of design.
10. Remeasure the pressure drop through system equipment and readjust balancing valves on all equipment with high flow, again lowering flow to about 10% above design flow.

11. Repeat Steps 9 and 10 until the actual flow through each piece of equipment is within ±10% of design.
12. After balancing is completed, mark or score all balancing valves at final setpoints or range of operation. Make a final check of pump motor amperage to determine that it is within nameplate rating.

At this stage, pressure-reducing valves, pressure-relief valves, temperature-relief valves, air vents, vacuum breakers, and tempering valves should be checked and adjusted to specified setpoints and tolerances.

Considerations for Specific System Types

In addition to general testing and balancing, certain components and subsystems specific to different types of liquid-based systems require particular attention and consideration. These components and subsystems prevent conditions that can cause serious and permanent system damage.

In drain-out or drain-back systems, where water is the heat-transfer fluid, freeze-protection operation must be carefully tested. Proper location, calibration, and setting of sensors and thermostats should be verified. Freeze conditions should be simulated at least once before operational testing by immersing the freeze sensor in ice water or by adjusting thermostat settings. The test should be performed when solar radiation is not being collected. For drain-out systems, proper operation of drain valves during freeze protection must be verified.

When a heat-transfer fluid other than water, such as a glycol or silicone, is used, overheat protection for collectors and storage tanks must be verified. Excessive temperatures may break down fluids and result in collector, piping, heat exchanger, and tank degradation. Glycol or silicone systems should be designed not to allow collector stagnation under overheat conditions. Overheat protection should be placed into operation at least once before operational testing by simulating high-temperature conditions by immersing the sensor in boiling water or by adjusting thermostat settings. Accurate calibration of sensors should be carefully checked.

Air-Based Systems

Data Requirements

Testing, balancing, and adjusting an air-based active solar system require the same careful planning and adherence to systematic procedures required for liquid-based systems. Similarly, data required to perform these functions consist of all applicable contract documents, including change orders, plus the following (Eads 1982):

- Fan curves and performance data
- Performance data for all heating and cooling coils
- Fan-motor nameplate data
- Starter sizes, locations, and thermal-overload protection ratings
- Belt-drive data including sheave sizes, pitch diameters, belt sizes, and adjustment limits
- Manufacturer's data and test recommendations on all applicable air-handling devices
- Normal air-pressure drops across louvers, filters, coils, and all other circuit devices
- Temperature-control diagrams and pertinent data sheets
- Controller specifications
- Manufacturers' catalogs and operating and maintenance instructions for all equipment.

As with liquid-based systems, detailed schematics should be prepared for air-based systems and subsystems, including all dampers, regulating devices, terminal units, and inlet and outlet ducts.

Duct Leakage Testing

The first step in preoperational testing for air-based systems is to test for air leakage from system ductwork. Some leakage will exist, regardless of precautions taken during fabrication and installation, but leakage should be minimized to avoid efficiency loss and other problems. The system designer should have specified a maximum leakage rate, usually expressed as a percentage of ductwork flow rate at design conditions. Leakage from ductwork is more difficult to detect than leakage from liquid piping, so air-leakage testing should be conducted with great care; excessive leakage drastically reduces system efficiency.

There are several ways to test for duct leakage. All are fairly similar and all involve controlled pressurization of the ductwork being tested. Standard procedures, as approved by the AABC, can be summarized as follows (AABC 1979):

1. Connect the pressurizing apparatus (either a rotary blower fan or a high-power tank-type vacuum cleaner) to a duct section with a flexible duct connection or hose.
2. Close the damper on the blower suction side to prevent excessive pressure buildup.
3. Start the blower and gradually open the damper on the blower's suction side.
4. Pressurize the duct section to at least 3 in. of water (750 Pa) pressure.
5. Maintain this pressure for 10 minutes and check ductwork for leaks.
6. Shut down the blower and repair identified leaks.
7. After repairs are completed, pressurize the duct section to design operating pressure and measure the pressure differential across a calibrated orifice plate between the blower and the duct section.
8. Leakage flow rate is determined from a calibrated chart for the particular orifice plate used.

9. If the resultant measured leakage is greater than that specified, Steps 2 through 8 should be repeated until leakage is satisfactory.

Two other techniques are sometimes used to locate ductwork leaks. The first introduces a nontoxic smoke from a smoke candle into the inlet of the pressurization apparatus during testing. Leaks can be identified visually. The second applies liquid soap to duct joints and connections (ANL 1981). Liquid is applied directly but gently to avoid bubble formation during application, and no more than 5 linear feet of duct joint should be tested at a time. Joints and connections are considered air-tight if there is no visible bubble formation within 20 seconds of applying the soap. Leaks are most likely to occur at seams of rock beds, at joints between duct sections, and at connections between ducts and other equipment. These locations should be checked very carefully.

Circuitry and Sensor Checkout

After ductwork is tested for leakage and repaired as necessary, system circuitry, sensors, and instrumentation tests are performed. As with liquid-based systems, these tests are conducted prior to flow balancing to verify proper installation and operation of system wiring, sensors, actuators, thermostats, meters, dampers, etc. At a minimum, these tests should check the following components or subsystems:

- System power circuitry to verify proper wiring from power sources to fans (and pumps if applicable), controllers, and indicators
- Time-delay relays to provide minimum run times for fans and pumps to avoid cycling
- Sensors, controllers, valves, dampers, pumps, and indicators that detect and prevent freezing, due to reverse thermosiphoning, of any system's air-to-water heat exchangers
- Differential thermostats used to control fan, damper, pump, and valve operation
- Fan and pump circuitry to verify proper current, direction of rotation, and all elements of fan- and pump-failure circuitry
- Thermometers, flow meters, and any instrumentation used to measure system thermal performance.

Depending on system design, all, or nearly all, of these circuitry tests can be performed while a system is not operating. Conditions required to conduct the tests can usually be simulated by disconnecting sensors, adjusting setpoints, immersing sensors in ice or boiling water, etc.

Flow Balancing and General Equipment Checkout

As with liquid-based systems, the final preoperational test stage for air-based systems consists of flow balancing and verification of proper operation of certain components and subsystems that are checked when a system is in, or ready for, operation. Proper flow through collectors, heat exchangers, ductwork, and piping is as important for air-based systems as for liquid-based systems for efficient and reliable operation. If air flow is too low through a collector or across a heat exchanger or rock bed, heat transfer, and, consequently, system efficiency, will be reduced. If flow is too high, a fan motor may be overloaded and air leakage may increase. Unusually high pressure drops during flow balancing can identify collectors and ductwork that may be blocked or poorly constructed.

Basic sequential flow-balancing procedures, common to all air-based systems, can be summarized as follows (Eads 1982):

1. Determine that all preliminary procedures and equipment and system checks have been performed.
2. Determine that all dampers are open or preset, as required.
3. Start system fans and adjust until design speed is attained.
4. Determine air volume being moved by fans at design speed (rpm) either by performing a pitot-tube traverse of the main supply duct or by using fan curves or fan performance charts. For the latter, measure fan amperages and voltages and calculate brake horsepower. Static pressure across fans also must be measured. With these quantities, fan manufacturer's data sheets can be used to determine air flow, in standard cubic feet per minute (scfm), as predicted by the manufacturer.
5. If airflow is not within 10% of design capacity at design rpm, review all system conditions, procedures, and recorded data. Measure pressure drops across filters, coils, eliminators, etc. to see if excessive loss is occurring. Duct and casing conditions at fan inlet and outlet should be carefully checked.
6. If measured scfm is still not within 10% of design scfm at design rpm, adjust fan drives to obtain the required scfm. Measure fan suction static pressure, fan discharge static pressure, and amperage. Verify that fan motors are not overloaded.
7. Make pitot-tube traverses on all major duct branches (e.g., ducts to separate collector banks) to determine air distribution. Investigate any branch that is very low in capacity to make sure that no blockage exists.
8. Adjust volume dampers in branch ducts to design scfm. Adjustments should be made sequentially from branches with the highest cfm to branches with the lowest cfm. This should be repeated until the system is in balance.
9. Recheck fan capacities and operating conditions (scfm, suction static pressure, discharge static pressure, and motor amperage) and verify that motors are not overloaded. Make final fan-drive adjustments, if necessary.

10. Verify that all fan shutdown controls and air flow safety controls are operating properly.

11. After balancing is completed, mark all volume dampers at their final position. Make a final check of fan motor amperages to determine that they are within nameplate rating.

Many air-based solar systems are designed to preheat service water. Most use a separate hydronic loop that includes an air-to-water heat exchanger and a preheat storage tank that interfaces with the conventional service water heating system. The same start-up, testing, and balancing procedures must be performed for this loop as for any other hydronic system. In particular, heat-exchanger freeze protection should be carefully checked.

OPERATIONAL TESTING PROCEDURES

The third and final acceptance test stage is designated as "operational testing." It consists of all testing performed when a system is fully operational and capable of providing energy to loads. Proper operation of each possible mode for the particular solar system is verified first. A "mode" is defined as a portion of system operation associated with a specific function, such as collecting and storing solar energy, preheating service water from storage, preheating service water from the collector array, space heating from storage, etc. These modes are first checked individually for normal operation and are then checked in combination.

Checking individual modes may require isolation or suppression of other operating modes. For instance, solar energy collection and storage for a liquid-based system is checked by putting the system in full, automatic operation (no operator intervention) and observing a normal morning start-up, daytime solar collection and transfer to storage, and normal shutdown at day's end. To determine thermal performance with the greatest possible accuracy, however, no service water would be drawn through the storage tank during checkout. With no service water load, energy delivered is simply the increase in storage-tank energy from start-up to shutdown.

Testing of individual modes involves checking various items and collecting data. For the example above, items to check would include normal start-up, no pump short-cycling, pump start-up sequence, no gurgling or air noises in fluid systems, normal shutdown, etc. In addition, performance data would be recorded at regular intervals (hour or half hour) during the test. These data would be used to calculate different measures of thermal performance and to verify proper system operation. Table 10-4 is an example of a data log sheet that might be used, and Table 10-5 is an example of performance data that can be measured or calculated using data from Table 10-4.

Once individual modes and the whole system have been checked and tested, the system should undergo a trial start-up and a setting into automatic operation. Trial operation is generally for a minimum of two weeks and can be as long as a month. The architect/engineer, being responsible for identifying any system operation anomalies, should set the trial operation duration. Any problems that occur must be corrected by the contractor, and the trial operation period should begin again.

Once the system has successfully completed this trial operation in all control modes, the architect/engineer can recommend system acceptance to the owner. When all work and all tests are completed, the contractor should furnish skilled labor and helpers to operate the system and equipment for a specified time (generally decided upon by the architect/engineer), during which the owner's representative should be taught full operation, adjustment, and maintenance of all system equipment. A bound manual should accompany this instruction.

Any system or component warranties should begin after the successful completion of the acceptance test. The contractor is generally responsible for correcting abnormalities in control setpoints and system operation during the warranty period. The architect/engineer should also monitor system operation from time to time, notify the owner of his findings, and recommend any appropriate corrective actions.

Table 10-4. Sample Data Log Sheet for a Liquid-Based System

Date	Time	Temperature (°F)															Pressure (psig)			Btu Registers (10^3)		
		C.W.	H.W.	Collectors		Int.	Stor. Loop		Storage		Htg. Loop		Water Heater			SHW	Coll	Int.	Stor	#1 Solar	#2 SHW	#3 Recirc
		t-10	t-11	t-12	t-13	t-14	t-15	t-16	t-17	t-18	t-19	t-20	t-21	t-22	t-23	t-24	PG-1	PG-2	PG-3			

Table 10-5. Sample Performance Data Sheet for a Liquid-Based System

Sunrise: __________

Solar Noon: __________

Sunset: __________

__________/__________
(Initials/Date)

Parameter Measured/ Calculated	Units	Expected Range	Measured/Calculated Values (at Specified Time of Day)							
Instantaneous Solar Irradiation	$\frac{Btu}{ft^2\text{-}h}$									
Integrated Solar Irradiation	$\frac{Btu}{ft^2}$									
Ambient Temperature	^oF									
Collector Array Inlet Temperature	^oF									
Collector Array Outlet Temperature	^oF									
Collector Loop Flow Rate	gpm									
Rate of Solar Energy Collection	$\frac{Btu}{h}$									
Instantaneous Collector Array Efficiency	%									
Collector Loop Pump Current	amp									
Collector Loop Pump Voltage	volt									
Collector Loop Pump Power	$\frac{Btu}{h}$									

Chapter 11
Developing an Operations and Maintenance Manual

INTRODUCTION

An operations and maintenance (O&M) manual will assist a building owner and his maintenance personnel in properly maintaining solar equipment, since the designer or installer may not be available when major repairs are necessary. It should provide a guide to systematic detection, diagnosis, and correction of breakdowns or minor defects that can occur. Where component replacements are required or further information is needed to diagnose a system deficiency, the manual should contain a parts list, including the name of the manufacturer's representative to be contacted regarding warranty and further maintenance and repair information. Manufacturers of standardized small-scale service hot water systems frequently provide their own O&M manual. Manufacturers of nonstandard systems do not, and development of an O&M manual for such systems is especially important.

This chapter outlines the general contents of an effective manual and then describes each aspect of the manual, including what information should be provided and why it is necessary, with accompanying examples of material and format. Blank forms and subject pages, with instructions that can be photocopied, are included at the end of the chapter.

Basic O&M manual preparation is the same for both air and liquid systems. A liquid system is selected as a model here because there is usually less accompanying O&M information from the manufacturer, except for small installations. For air systems, complete O&M information is often provided by the collector manufacturer.

A suggested manual format is a title sheet, four major parts, and appendices. All information needed to contact the designers, contractor, maintenance personnel, and O&M manual preparer should be listed on a title sheet. Part One of the manual should describe overall system installation, with accompanying schematic drawings. A comprehensive parts list should describe each component. Part Two should provide standard operating instructions for the specific system and should contain a complete description of start-up procedures. Part Three should describe maintenance procedures for individual components to ensure their effective and efficient operation. Part Four should describe corrective maintenance procedures to assist in problem diagnosis and offer solutions. Appendices should contain backup material, including all component catalog sheets, specifications, and other technical data for all components, warranties, available shop drawings and construction details, glossary of terms, photographs, and any other pertinent data. Appendices can be packaged separately.

TITLE SHEET

The title sheet should include the following information:

- Project name, address, contact person, telephone number
- Person responsible for maintenance, telephone number
- Designer, address, contact person, telephone number
- Contractor, address, contact person, telephone number
- Name, address, and affiliation of persons compiling the O&M manual.

Persons originally involved in the project may no longer be available, but contacts provided may aid in tracing the proper people. A blank title sheet is provided at the end of this chapter.

OVERALL SYSTEM DESCRIPTION

An effective approach to system description is to organize system information in the following categories:

- Systems
- Control sequence
- Parts list.

For each system's subsystem, locations of related components, such as collectors and piping runs, and mechanical layout are described in the manual. Schematics can be either new or taken from working drawings (Fig. 11-1).

An overall system description provides a basis for all subsequent information. Therefore, a schematic must illustrate a system as built and must identify all valves (manual, automatic, and check) and all major equipment and sensor locations so they can be cross referenced to other descriptive material. Valve tags should use the same identification

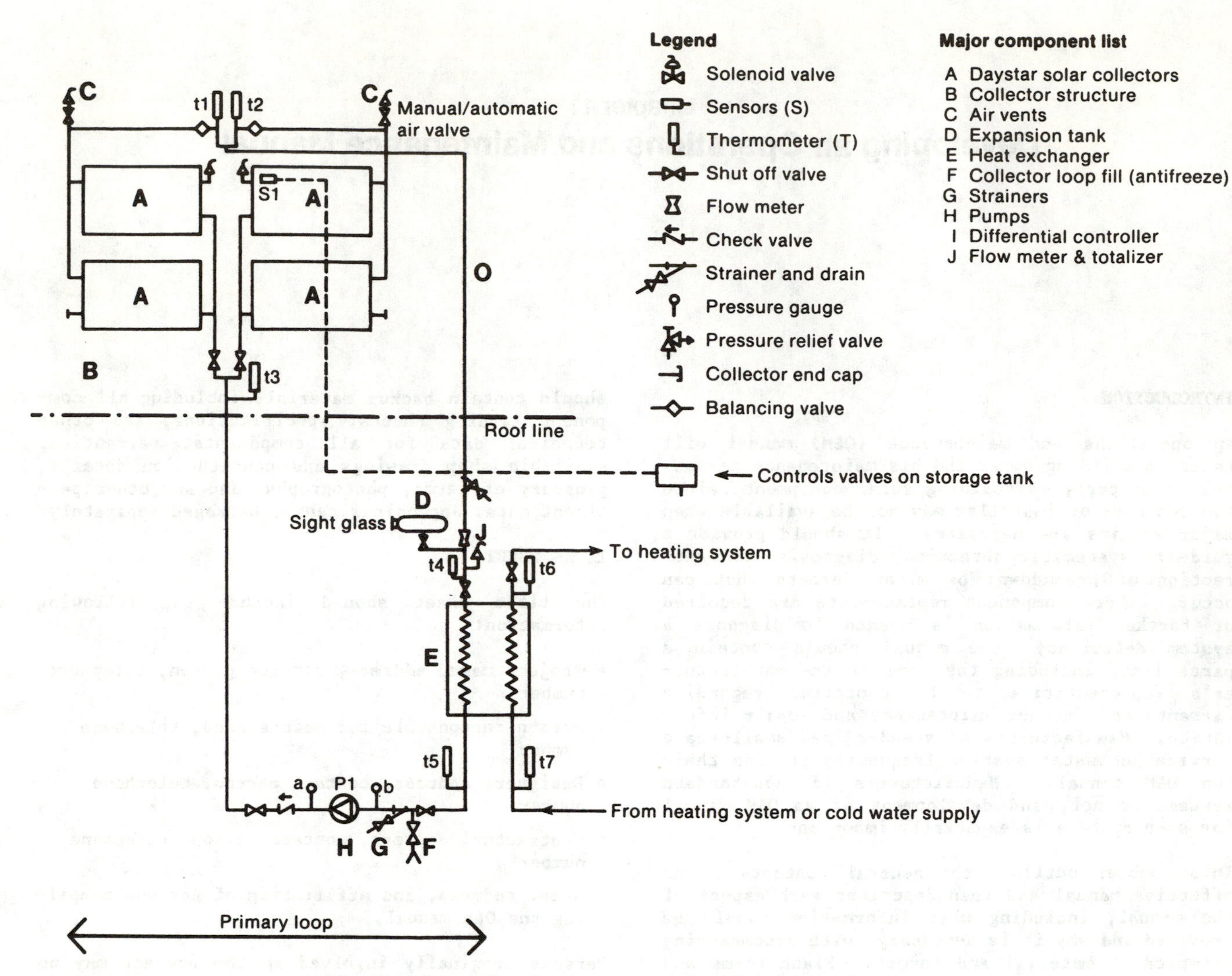

Fig. 11-1. Description of system — example partial system schematic

system. Photographs of key components in relation to their surroundings are recommended, with all elements labeled.

An easy-to-understand format divides systems into three subsystems:

- Collector subsystem (primary loop)
- Storage subsystem (secondary loop)
- Energy-load subsystem.

Pertinent items to be discussed in the three subsystems include

Collector Subsystem (Primary Loop)

- General description
- Collector
- Collector support
- Piping/ducts
  - between collectors
  - headers
  - insulation
  - valves/dampers
  - venting
- Pumps/fans
- Heat exchangers (HX)
- Expansion tank
- Air separators
- Sensors and controls

Storage Subsystem (Secondary Loop)

- General description
- Storage: tank/rock bed/phase-change storage
- Pumps/fans
- Piping/ducts
  valves/dampers
- Heat exchangers
- Expansion tanks
- Sensors and controls

Energy-Load Subsystem

- General description
  - load
  - storage to load connection

- Heating
  - heat exchanger/liquid/air
- Cooling
  - absorption/mechanical
- Service hot water
  - heat exchanger
  - preheat tank

A blank form, "Overall System Description," located at the end of the chapter, can be used for system description and schematic drawings.

Control System Description

To understand system operation, information on types of hardware used and how they function must be available. A schematic drawing (Fig. 11-1) and a parts list (Table 11-1) describe existing hardware. A control sequence should describe system operation. If only hardware and controls are described, a competent mechanic with reasonable knowledge of solar installations should be able to troubleshoot and perform most maintenance tasks. Therefore, sensor and control information must be complete and accurate.

A sensor locations and control diagram can either be included with the schematic (Fig. 11-1) or be a separate drawing (Fig. 11-2). Whichever approach is used, a complete written description of the control sequence should be provided to explain the diagram, including:

- Operation
  - turn on, turn off of each subsystem
- Freeze control
- Overheating control.

Information on all monitoring equipment, either visual or datalogging, should be supplied. If visual monitoring equipment consists of thermometers, pressure gauges, and flow meters only, their functions need not be described; indicating their positions on the schematic will suffice. An example control-sequence description follows, with references to the schematic in Fig. 11-2.

Table 11-1. Sample Parts List

* Part	Manufacturer	Model Number	Quantity	Warranty	Local Manufacturer's Representative
A. Solar Collectors	Daystar	21C	32	1 year; 5 year, Absorber Plate; 3 year, Window and Selective Ctg. No more than 10% loss of optical efficiency	Daystar 90 Cambridge Street Burlington, MA 01803 617-272-8460
B. Collector Structure	Babylon Iron Works	N.A.	1	1 year	Babylon Iron Works 205 Edison West Babylon, NY 516-645-3311
C. Air Valves	Ventrite	N.A.	2	1 year from Daystar	Daystar 90 Cambridge Street Burlington, MA 01803 617-272-8460
D. Expansion Tank & Air Separator	Bell & Gossett	ATF 12 & AC-2	1	Life time	Bell & Gossett 8200 N. Austin Avenue Morton Grove, IL 60053 312-966-3700
E. Heat Exchanger	Doucette Industries	CSZ-72M9-8	1		Doucette Industries Heat Transfer Div. P.O. Box 1641 York, PA 717-845-8746
F. Antifreeze	Union Carbide/ Foodfreeze	35	Approx. 50% of total fluid in collector loop	None	Dave Turnow Union Carbide 1 University Place Hackensack, NJ 07601 201-676-1111

*Reference schematic

Control Sequence Example

When the differential thermostat senses a positive 20°F difference between sensor t-1, located on the collector absorber plate, and sensor t-3, located at the storage tank, a normally open contact will close, providing a 120 V voltage to the normally open control relay of the pump-motor starter. Relays will close; the primary solar loop pump, P-1, will come on line; and the time-delay relay in the "Auto" circuit of the secondary solar loop pump, P-2, motor starter will start to cycle. After approximately 10 minutes, P-2 will come on line.

When the temperature difference between the two temperature sensors drops to 3°F, contacts in the thermostat will open, and all pumps will stop.

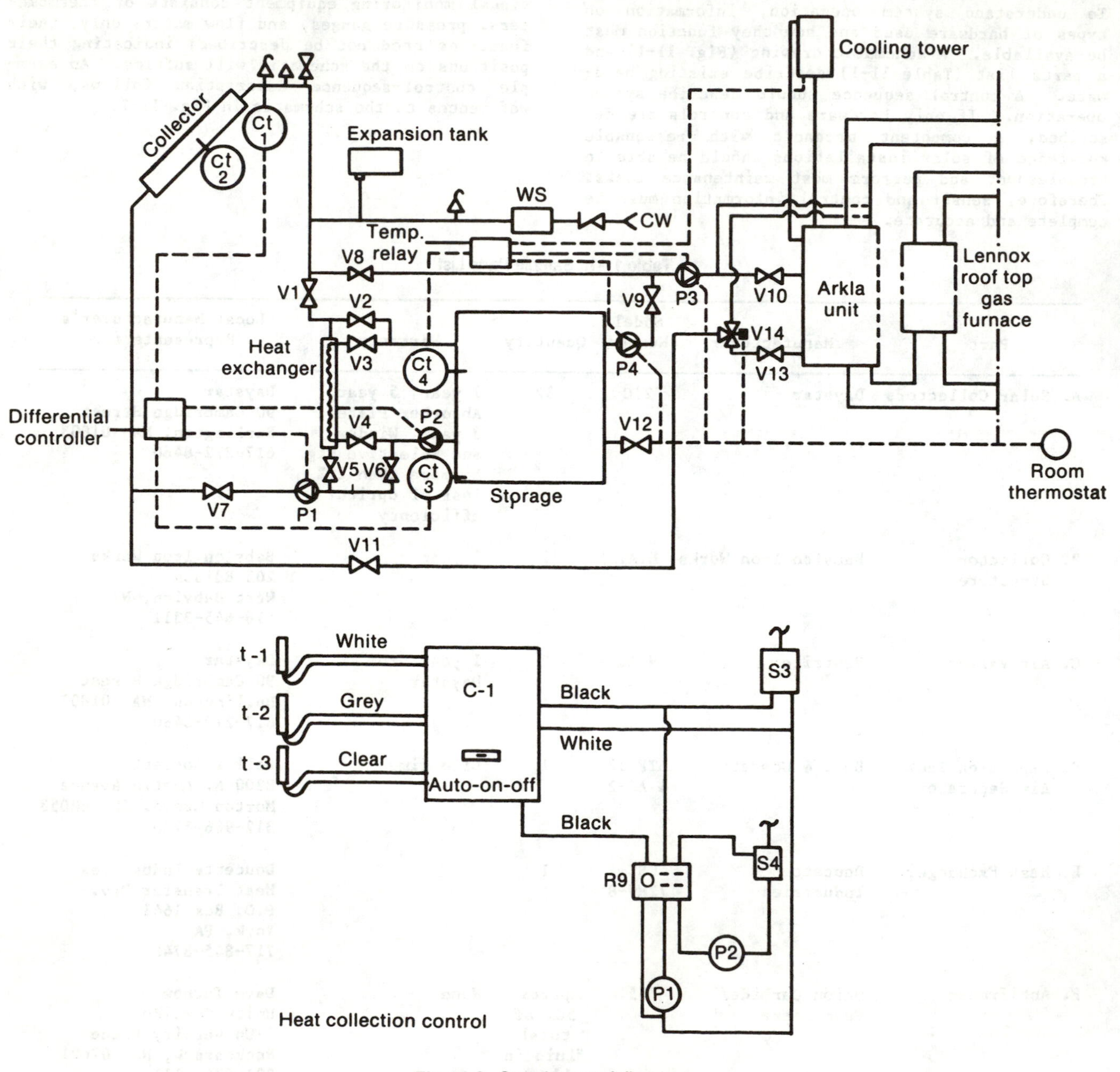

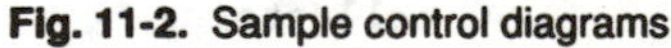
Fig. 11-2. Sample control diagrams

A control description and accompanying schematic should be entered on the blank sheet titled "Overall System Description," located at the end of this chapter.

Monitoring

Pressure gauges on both sides of pumps and thermometers on both sides of the collector array, heat exchangers, and storage tank are the minimum equipment required. Flow meters are also helpful and should be considered, but some meters cause more maintenance problems than they are worth. Some thermometers will function in dual roles; for example, a single thermometer may register temperature both out of collectors and into the heat exchanger.

Monitoring equipment enables instantaneous monitoring of system operation. Some controllers provide visual readouts, but system aspects not controlled with monitors should also be checked. Operating modes and control performance, as described in the "Control Sequence Example," cannot be checked without minimal monitoring devices. However, minimal monitoring will not provide information on total energy collected.

Parts List

A parts list that describes every system component should be included with the schematic. When maintenance is required, a list provides a handy reference to warranty information, component manufacturers, model numbers, quantities of components used, and addresses and phone numbers of local manufacturer's representatives to contact for assistance, repair, or replacement.

Many parts are standard and can be easily obtained or substituted, but numerous parts might be special to solar energy systems. Unique equipment includes collectors, controls, possibly heat exchangers, and some valves. If the schematic has been divided into subsystems (collector, storage, and energy-load), the parts list should be divided accordingly. A blank form, "Parts List," for listing each system component is provided at the end of this chapter.

OPERATION

Purpose

Typically, a relatively untrained person will be involved in system maintenance. Therefore, an operations section should simply and directly describe how the system works and what procedures to follow during routine operation. Procedures are outlined for responsibility (personnel), safety (equipment, personnel), inspection (requirements, equipment available), and operation (initial start-up sequence, collector loop balancing, seasonal changeover, shutdown sequence, component characteristics).

Responsibility for Safety and Damage Avoidance

The person responsible for system operation and maintenance should be named in the manual. Proper safety measures should be described; maintenance personnel should be informed of precautions to take when inspecting equipment and how to prevent accidents. Personnel engaged in operation and maintenance should be thoroughly familiar with the system. Example safety measures to include in the manual are

- To prevent collector thermal shock, do not start a stagnated collector array system during periods of high solar intensity.
- To avoid burns to maintenance personnel from hot pipes or fittings, work when solar radiation is not present, if possible. Avoid touching heated pipes with bare hands.
- To avoid explosion and burns, drain liquid collector arrays immediately after closing isolating valves. (However, any section of piping that can be isolated and is subject to heating should contain a pressure relief valve.)
- To prevent tampering by unauthorized personnel, remove valve handles.
- To ensure safety during collector inspection, where no walkways exist, personnel should not climb over or around collector arrays; ladders, tools, etc. should be used to obtain access to collector arrays.

Inspection Checklist

Periodic inspections of system parts will ensure efficient system functioning. A schedule and a complete checklist should be provided. The checklist should include items to be checked, how to check them, and how to interpret findings. It should be explicit, describing both simple tasks (such as checking for leaks and deterioration of paint and insulation) and more complicated instructions for using monitoring equipment (including details of what to look for if readings are not within reason). The checklist should provide examples of appropriate temperature readings and approximate temperature differentials.

Equipment

Monitoring devices to be described in an O&M manual include thermometers, pressure gauges, flow meters, valve-position indicators, and sight-glass levels. A form, "Visual Monitoring Equipment," for recording temperature ranges and levels is included at the end of this chapter.

Proper system temperature ranges should be developed as a checklist. The following example (using Fig. 11-1) should provide reasonable indications of the type and depth of information that should be included.

Checklist Example

Collector Temperature

In use, t1 = t2. If not, balance rows by adjusting balancing valves.

In use, t1 or t2 > t3. If not, check controls.

In use, t3 = t5. If t3 is much less, check pipe insulation for defects (pipes are losing too much heat).

*In use, t5 < t4 < (t5 + 20°F). If collector entering water (t5) = 120°F @ 16 gpm.

*In use, t7 < (t6 - 9°F). If entering water (t7) = 95°F @ 32 gpm.

*These are design parameters; other entering and leaving temperatures should be proportional. If not, refer to the corrective maintenance procedures section in the O&M manual.

Pressure Meters

"a" and "b" at P1. At rest, a and b ≈25 psig. If not, check relief valves for blow off. Check for leaks (low pressure). Check for blockage in strainers, closed valves, etc. Check expansion tank for waterlogging. Add liquid.

In use: suction b ≈ 15 psig
discharge a ≈ 40 psig

"c" and "d" at P2. At rest, c and d ≈ 60 psig. If not, check for blockage in strainer, close valves.

In use: suction d ≈ 55 psig
discharge c ≈ 65 psig.

Flow Meter

Should indicate 15 to 30 gpm, depending on temperature. The colder the fluid, the lower the flow rate. If not, check for voltage at motor starter (P-1), check motor overloads, breakers, and differential pump control for normal operations (P-1). Check fuse at, and power to, flow meter totalizer; also check flow meter for possible obstruction.

Sight-Glass Level

Under no flow, glass should not be more than 1/2 full. If not, drain tank and recharge to working pressure of system = 25 psi.

System Operation

Requirements for operation of different components are described at this point in the manual. The following procedures serve as an example of a partial checklist for liquid systems. A complete description of procedures for acceptance testing and proper start-up of solar systems is presented in Ch. 10.

1. Filling the Collector Loop
   - Charge the expansion tank
   - Open or close manual valves
   - Pump antifreeze into collector loop
   - Fill the collector loop with water
   - Purge air in collectors and piping
   - Watch pressure gauges, shut off water at specified pressure
   - Purge air in mechanical room equipment
   - Activate pumps
   - Check pressure gauges
   - Check loss of pressure in solar loop.

2. Expansion-Tank Charging Procedure
   - Close valve
   - Drain tank
   - Connect pump or compressor and charge tank
   - Open valve after collector loop is filled.

3. Solar Loop Drain-Out Procedure
   - Turn off differential controls
   - Motor stopping sequence
   - Close supply valves
   - Open air vent at arrays
   - Open drain valves
   - When water is drained from system, close drains.

4. Describe operations that apply:
   - Winter start-up and shut-down procedures

- Summer to winter changeovers
- Winter to summer changeovers
- Summer start-up and shut-down procedures.

5. Valve Schedule, if applicable, for various system applications (e.g., summer or winter).

Seasonal Changeover

If system operation requires a seasonal changeover, changeover requirements should be carefully noted and referenced back to the schematic. Valve tags on equipment should coincide with the schematic. A checklist of all valve changes should be provided in the manual, along with approximate dates when changes should be made. A motor starting and stopping sequence should be included. To record when seasonal changeovers should occur and the position of the valves, a form, "Seasonal Changeover," is provided as the end of this chapter.

MAINTENANCE

Purpose

A maintenance section should simply and directly describe maintenance tasks to be done for each component and list when equipment should be inspected. A permanent record of equipment condition should be kept for scheduling such maintenance tasks as inspection, cleaning, lubrication, and filter replacement.

Responsibility

The manual should specify who is responsible for performing routine maintenance. As inspections are made, maintenance record forms should be filled out for a permanent record. In addition, warranty information should be consulted on questions about component maintenance. Warranties are further discussed in the Individual Maintenance Concerns section, below.

Records and Forms

To document system equipment performance, the maintenance record form should include

- List of items to inspect
- Yearly schedule of when to inspect items
- Date and time
- Work done, or comment, section.

A sample of a completed roof inspection form is provided in Fig. 11-3, and a blank inspection form is located at the end of the chapter.

Suggested instructions to accompany the blank forms include

- Inspection procedures for months indicated by areas left blank. Shaded areas indicate no maintenance is necessary.
- If no maintenance is required, an "OK" in that square will suffice. If any maintenance is made, a footnote description number should be inserted in the square and listed at the bottom of the form.
- Consult individual maintenance sheets for component details concerning specific instruction.

Individual Maintenance Concerns

A sheet with a description of the component and its maintenance schedule should be included in the manual for each system component. All maintenance tasks and periodic schedules should be summarized from individual component sheets onto the record form. These data sheets provide information needed to contact the manufacturer for further information regarding component defects or breakdowns. Information to include on individual component sheets is listed in Fig. 11-4. A blank sheet is provided at the end of the chapter.

The following items should be included:

- Name and model number of component
- Component location and physical relation to subsystem and reference to schematic drawing
- Name of component manufacturer and local manufacturer's representative (address and telephone number) to contact to determine who guarantees warranty and where to buy the component
- Warranty terms
- Shop drawings
- Quantity of parts required
- List of maintenance procedures to be carried out, as noted in periodic table
- Safety precautions, where applicable
- Recommended schedule of inspections for the year.

TROUBLESHOOTING

Problems in solar systems generally result from either design deficiencies or inadequate or improper installation. Typical problems that reduce system efficiency include thermal losses from storage, piping, ducting, and other solar components; air blockage and flow restrictions; control subsystem failures; malfunctions and deviations from designed operating modes; faulty differential thermostats; and improper equipment installation.

To assist in diagnosing and correcting common system failures, the O&M manual should describe corrective maintenance procedures that will restore an item to its specified performance level. Procedures to be described include system testing, fault isolation, unit replacement, component repair, and equipment adjustments or alignments. Table 11-2 provides a sample summary of problems, probable causes, and corrections for common system failures.

A blank corrective maintenance form is provided at the end of this chapter. It should list problems that can occur, with room for notes on probable causes and suggested corrections.

Maintenance Record Form

Roof Inspection

Procedure:	Jan	Feb	Mar	Apr	May	Jun	Jul	Aug	Sep	Oct	Nov	Dec
Date	23	21	19		20	19	21		22	30		18
Time	11	2	10		10	11	2		10	11		2
Check collector cover glass			OK			OK			1			OK
Check collector weep holes			OK			OK			2			OK
Check collector bolts & frame			OK			OK			OK			OK
Check collector for absorber plate leaks or flaking			OK			OK			OK			OK
Check collector for outgassing			OK			OK			OK			OK
Inspect collector structure			OK			OK			3			OK
Inspect hoses between collectors for leaks	OK	OK			OK		OK		OK	OK		
Inspect pipe insulation									OK			
Open and test air vent									OK			
Test pressure relief valves									OK			
Initials	bz	bz	bz		NSR	bz	NSR		bz	NSR		bz

Work Done or Comments

1. Cleaned glass with hose
2. Cleaned
3. Touched up paint

Fig. 11-3. Sample maintenance record form

Maintenance

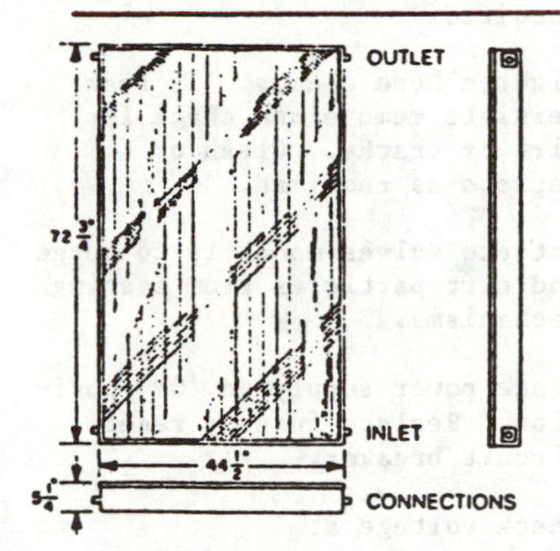

Component Solar Collectors

Location Roof

Manufacturer Daystar 21C

Quantity 32

Warranty
1 yr - 5 yr absorber plate
3 yr - window & selective coating
No more than 10% loss of optical efficiency

Local Manuf. Representative Daystar, 90 Cambridge St., Burlington, MA
617-272-8460

Maintenance Procedures

1. Check cover glass for dirt or debris; clean as necessary with nonabrasive soap; rinse clean
2. Check weep holes for blockage; must be kept open to prevent condensation and freezing
3. Check bolts and collector frame for rust; clean and touch up or replace as necessary

The following items should be checked; if problem exists, see Corrective Maintenance Section.

1. Check cover glass for cracks or breakage
2. Check for internal leaks (condensation or liquid in collector)
3. Check absorber plate for flaking of selective surface
4. Check for outgassing of insulation (fogging of cover glass)

Procedure	Recommended Schedule Jan	Feb	Mar	Apr	May	Jun	Jul	Aug	Sep	Oct	Nov	Dec
Inspect Collectors			X			X			X			X

Fig. 11-4. Sample individual component maintenance sheet

Table 11-2. Example List of Problems, Causes, and Corrections for System Failure

Problem	Probable Cause	Correction
1. System loses pressure/loss of fluid.	1.1 Leaks in piping and joints.	Repair or tighten connections. Replace damaged components as required.
	1.2 Hose connections between collectors leak.	Tighten hose clamps. If leak persists remove and check for dirt or cracks. Clean or replace as required.
	1.3 Leaks in pressure relief valve/air vent. Replace if leak persists.	Actuate valves manually to purge and dirt particles from seating mechanisms.
2. Circulating pumps will not operate.	2.1 Power supply switch in 'OFF' position or blown fuses/ tripped circuit breaker.	Place power supply on 'ON' position. Replace fuse or reset circuit breaker.
	2.2 115 V supply voltage is not available. Check differential thermostat in 'ON' position. If indicator light does not go on, check voltage (if indicator light goes on, then go to item 6).	Check voltage at • Output from electrical panel • Input to differential thermostat (D.T.) • Output of differential thermostat (D.T.) If no voltage at input to differential thermostat (D.T.), correct consistent with local codes. If no voltage at input to differential thermostat, correct cause of interruption between electrical switch and D.T. If no voltage at output of D.T., see manufacturer's literature.
	2.3 Voltage not reaching pump control circuit through auxiliary relay.	Check voltage at input to circulator. If no voltage, check connections for tightness and wiring; correct as required. Check motor overload throw-out switch.
	2.4 Circulator motor failure.	Check manufacturer's literature for proper amperage. If amperage is high, check for motor failure; replace per manufacturer's instructions. Also see item 2.5.
	2.5 Impeller binding or bearing seizure.	Remove motor-impeller cartridge and turn impeller by hand. If impeller shaft does not turn freely, check cartridge or impeller; replace as required per manufacturer's instructions.
	2.6 'AUTO' set-on differential thermostat is malfunctioning.	Check if circulator operates on 'ON' setting. Check voltage of circulator when temperature sensors have been jumped.

Table 11-2. Example List of Problems, Causes, and Corrections for System Failure (continued)

Problem	Probable Cause	Correction
	2.7 Temperature sensors malfunction.	Test sensors by disconnecting sensors, twist leads ends together at sensors' location, check continuity of leads going to both sensors using an ohmmeter. If resistance is zero, replace defective load(s). If continuity is OK, then determine temperatures at collector and bottom of storage tank. Check resistance of leads with ohmmeter. Compare sensor resistance with the resistance temperature chart included with the differential thermostat. If corresponding temperature is too high or too low for resistance, replace defective sensors.
3. Circulators run continually.	3.1 The switch in differential thermostat is in 'ON' position.	Place switch in AUTO position.
	3.2 Differential thermostat malfunction.	Test by disconnecting sensors one at a time. If circulator continues to operate, replace differential thermostat.
	3.3 Sensors malfunction.	Same as item 2.7.
4. Poor performance of solar heat collection system.	4.1 Pressure loss in system.	If pressure is low, turn off circulator and proceed per item 1.
	4.2 Air blockage in system.	Check air vents manually; if manfunctioning, replace.
	4.3 Fluid flow restriction (other than air).	If pipe from the collectors is a lot hotter than pipe to collector, check for devices causing flow thermometer probes, valves; check valves installed improperly.
	4.4 Clogged strainer is blocking fluid flow.	Clean strainers per maintenance section.
	4.5 Poor exterior insulation.	Replace all damaged or deteriorated insulation on all exterior piping.
	4.6 Collector maintenance required.	Follow maintenance instructions for collectors in maintenance section.
	4.7 Balance valves improperly tuned causing too slow or stopped flow rate.	Check temperature rise across collectors per balancing procedure.
5. Roof leaks.	5.1 Pipe penetration through roof improperly flashed.	Repair with roof cement around all pipe penetrations.

Table 11-2. Example List of Problems, Causes, and Corrections for System Failure (concluded)

Problem	Probable Cause	Correction
	5.2 Pitch pocket leaks.	Repair with roof cement all defective pitch pockets.
6. Poor pump performance.	6.1 Worn impeller.	Replace with new impeller or seal.
	6.2 Motor not up to speed - low voltage - worn bearings - overloaded.	Check voltage to motor.
	6.3 Clogged strainer or strainers.	Check for debris and clean per manufacturer's instructions.
7. Noisy operation of pump.	7.1 Worn motor bearings.	Replace.
	7.2 Low discharge head.	Throttle discharge.
	7.3 Debris lodged in impeller.	Remove cover and clean out.
8. Service hot water preheat system not functioning properly.	8.1 Heat exchanger scaled.	Drain and clean heat exchanger.
	8.2 Pump not operating properly.	See items 2 and 3.

SAMPLE FORMS

The following section contains blank forms and subject pages for use in developing an O&M manual:

Title Sheet
Overall System Description
Parts List
Visual Monitoring Equipment
Seasonal Changeover
Maintenance Record Form
Maintenance
Corrective Maintenance

TITLE SHEET

- Project Name ______
  Address ______
  Contact Person ______
  Person Responsible for Maintenance ______
  Telephone Numbers ______
- Designer ______
  Address ______
  Contact Person ______
  Telephone Numbers ______
- Contractor ______
  Address ______
  Contact Person ______
  Telephone Number ______
- Name of Persons Compiling O/M Manual and Affiliations ______

______

OVERALL SYSTEM DESCRIPTION

Collector Subsystem

- general description
- collectors
- collector support
- valve/damper
- venting
- piping/ducts
- heat exchanger (HX)
- expansion tank

Storage Subsystem

- general description
- storage: liquid/ rock/phase-change materials
- tank/enclosure
- pumps/fans
- piping/ducts
- heat exchanger
- sensors and controls

Energy-Load Subsystem

- general description
  - what is the load
  - storage to load
- heating
  - heat exchanger/ liquid/air
- cooling
  - absorption/mechanical
- service hot water
  - whether it involves heat exchanger

Control Sequence Description

- operation
- freeze control
- overheating control

PARTS LIST

Part	Manufacturer	Model Number	Quantity	Warranty	Local Manufacturer's Representative

VISUAL MONITORING EQUIPMENT

Collector Subsystem

	Set 1		Set 2	
• thermometer readings:	___	___	___	___
• pressure gauge readings:	___	___	___	___
• flow meter readings:	___	___	___	___
• valve position:	___	___	___	___
• sight glass level readings:	___	___	___	___

Storage Subsystem

	Set 1		Set 2	
• thermometer readings:	___	___	___	___
• pressure gauge readings:	___	___	___	___
• flow meter readings:	___	___	___	___
• valve position:	___	___	___	___
• sight glass readings:	___	___	___	___

Energy-to-Load Subsystem

	Set 1		Set 2	
• thermometer readings:	___	___	___	___
• pressure gauge readings:	___	___	___	___
• flow meter readings:	___	___	___	___
• valve position:	___	___	___	___
• sight glass readings:	___	___	___	___

SEASONAL CHANGEOVER

Instruction: - Designer will indicate the correct valve position for each season.

- Maintenance person will check box after valve has been placed in the correct position.

Date of Changeover: Winter ____________

Summer ____________

	Winter		Summer	
Valve No.	Correct Valve Position	Position	Correct Valve Position	Position
1	____________	( )	____________	( )
2	____________	( )	____________	( )
3	____________	( )	____________	( )
4	____________	( )	____________	( )
5	____________	( )	____________	( )
6	____________	( )	____________	( )
7	____________	( )	____________	( )
8	____________	( )	____________	( )
9	____________	( )	____________	( )
10	____________	( )	____________	( )
11	____________	( )	____________	( )
12	____________	( )	____________	( )
13	____________	( )	____________	( )
14	____________	( )	____________	( )
15	____________	( )	____________	( )
16	____________	( )	____________	( )
17	____________	( )	____________	( )
18	____________	( )	____________	( )
19	____________	( )	____________	( )
20	____________	( )	____________	( )

Maintenance Record Form

	Jan	Feb	Mar	Apr	May	Jun	Jul	Aug	Sep	Oct	Nov	Dec
Date												
Time												
Procedure:												
Initials												

Work Done or Comments:

Maintenance

Component

Location

Manufacturer

Quantity

Warranty

Local Manuf.
Representative

Maintenance Procedures

Procedure	Recommended Schedule Jan Feb Mar Apr May Jun Jul Aug Sep Oct Nov Dec

Corrective Maintenance

Problem	Probable Cause	Correction

Chapter 12
Instrumentation and Monitoring

INSTRUMENTATION SYSTEMS

Levels of Instrumentation

A wide variety of sensor types and data collection systems are available on the commercial market. Equipment ranges from simple thermometers to sophisticated solid-state devices capable of monitoring energy and controlling system operations. Although many sensors and systems serve both energy monitoring and control functions, only monitoring functions are discussed.

Engineers are often tempted to specify the latest and most sophisticated energy monitoring equipment, though this often results in unnecessary costs. Understanding what information level is appropriate and necessary is vital. The simplest equipment that will provide the type of information required at an acceptable level of accuracy and detail should be specified.

Energy monitoring equipment can be classified according to roughly defined levels of sophistication. Six levels are discussed here: three involve manual data collection over both short and long time spans, and three use automatic data collection over long time periods. These six levels are discussed in order of increasing sophistication and capability. Some level of instrumentation is recommended in all cases.

Visual Monitors

Thermometers, flow meters, and multimeters are used to obtain instantaneous values of system parameters, and sight glasses indicate liquid levels in tanks. Data are collected over time by visually reading and recording values. Depending on their purpose, sensors may or may not be permanently installed. Data readings can be taken nonintrusively (i.e., thermowells or surface measurements) or by inserting probes directly into flowstreams. Appropriate fittings in piping or ductwork are helpful. These data may be sufficiently accurate for determining steady-state conditions.

Energy Meters

These devices monitor and report time-integrated energy quantities that pass through a pair of pipes or wires. By measuring flow rates and temperature differences between fluids in a pair of pipes, Btu meters provide an indication of accumulated energy flow over time. They generally operate without electrical power. Wattmeters provide the same information for electrical energy use. Most energy meters must be read manually, but some provide output to a recorder. Accuracy may be reasonable, but these devices are often expensive.

Energy Meters and Visual Monitors

A combination of these two levels provides both instantaneous and time-integrated data. This level is very versatile and may serve many purposes.

Datalogging

This involves automatically recording data from a number of remote sensors. Although a number of recorders can be used, it is generally most practical to use one centralized recorder. Sensors used at this level typically provide data as electronic signals. Sensor lead wires often must be magnetically shielded and mechanically protected by conduits. Signal processing may be required before recording, and equipment compatibility is a concern. Data are recorded in analog (e.g., strip chart recorder) or in digital form, with processing generally done manually. Data sampling rates must be consistent with typical system behavior and the accuracy level required. With accurate sensors, the accuracy of this level can be quite good. Costs of sensors, sensor leads, and a central recorder make this level more expensive than those above.

Data Recording with Local Processing

An on-site data processor, typically a personal computer, can be added to the datalogger described above. A typical processor will be electronic, will provide some means of integrating data ratings over specified time intervals, and will perform data reduction and manipulation. Processing involves simple algebraic manipulations or more advanced calculations. Output is stored on paper or electronically on magnetic tape or disks. Data accuracy depends primarily on the specific sensors chosen, signal conditioning, and sampling rates. A data processor adds to the total cost, but the relative increase may not be too great.

Data Recording with Remote Processing

All elements of a datalogger system are used, but data are generally stored for a relatively short time. The datalogging device may or may not perform simple preprocessing. Again, accuracy depends primarily on the specific sensors, signal conditioning, and data collection rate. Performance data are often not available locally until after remote processing is completed. Data are typically stored on a cassette tape and are periodically transmitted to a central computer for processing. Considerable computational power is required for large numbers of data. Remote processing is a good approach when many separate sites are to be monitored, as in the National Solar Data Network of the U.S. Department of Energy (Murphy 1978).

Sensors and Equipment

Choosing a specific energy monitoring level is inevitably a compromise between desired accuracy, cost, versatility, reliability, and other constraints. First cost is not a correct indication of total cost. Installation, sensor leads, and signal conditioning equipment may add significantly to cost. Many standard sensors are not suitable for solar application because of the flow rates and temperatures encountered. Within reasonable cost limits, reliability and accuracy are generally most important. Cost considerations should also include the skill level of personnel required for analysis. Detailed descriptions of the more important characteristics of each type of sensor are presented in Tables 12-1 through 12-5.

Data Recording and Processing

Analog electric signals from system probes can be recorded in several ways. Strip chart recorders are probably the oldest. They provide excellent trend indication, but output is difficult to format for numerical analysis. Wide fluctuations in recorded flows and temperatures tend to decrease information accuracy. Chart recorders require extensive maintenance and adjustment. They range in price from about $500 to $1,500, depending on the number of input channels.

A datalogger is another popular data recording device, and manufacturers provide a variety of models. Most either print data on paper or record it on magnetic tape; some do both. Others relay data to a remote computer when called. Some can do limited analysis and can convert inputs into engineering units or compute energy flows. Data sampling rates can be varied with some machines. A datalogger should match user needs and sensor types.

Data sensors should provide the accuracy required by the analysis to be done. Data quality is generally limited by probes and their installation, not by the recording device.

Automatic data analysis provides very useful information on system operation and performance. Numerical analyses are performed on large quantities of data, providing both long-term and transient results to users in easy-to-read formats. Very detailed information can be extracted, but personnel familiar with computer programming are required for system design.

INSTRUMENTATION FOR SPECIFIC PURPOSES

Instrumentation levels and sensors appropriate for each purpose are discussed below. Descriptions are not intended to be definitive but merely to describe a typical cost-effective installation for each purpose. Debugging, acceptance testing, system maintenance, and performance evaluation are the four primary purposes for energy monitoring. Each may be served, to varying degrees, by many of the levels of instrumentation discussed above. Table 12-6 presents a summary of the capability of each instrumentation level for each of the four purposes.

Debugging

Instrumentation helps perform the numerous adjustments and modifications often required to get a new solar system operational. Collected data should indicate instantaneous values and be in a form immediately available to local personnel. Accordingly, a set of portable, visual monitors (thermometers, pressure gauges, flow meters, sight glasses, and multimeters or wattmeters) would be most cost effective. Provisions should be made at appropriate locations in system piping for intrusive measurements of fluid temperatures. Sufficient measurements should be taken to enable proper setting of collector-array balancing valves. With proper care, reasonably accurate temperature readings can be taken from a pipe's surface. Flow measurements are more difficult but can be done with many of the sensors discussed above. Orifice plates or variable-area flow meters are common.

Table 12-1. Solar Radiation Probes

Type of Sensor	Approximate Cost	Accuracy	Rangeability	Maintenance	Convenience	Type of Output	Datalogger Capability	Special Comments
Pyranometer	$150-$300	1%-3% of instantaneous	Adequate	Little	High	Analog electrical	Yes	Mounting point must be unshaded. Some models increase error by tilting.
Integrating pyranometer	$150	5% of integrated value	Adequate	Some	High	Mechanical Totalizer (and analog electrical on some models)	Some models	Some models provide instantaneous reading.

Table 12-2. Thermal Sensors

Type of Sensor	Approximate Cost	Accuracy	Rangeability	Maintenance	Convenience	Type of Output	Datalogger Capability	Special Comments
Bimetallic Dial Thermometer	$25	Fair, 1% of full scale	Adequate	Little	Good, when installed correctly	Visual	No	Not reliable for differential temperature due to low accuracy
Bulb Type Thermometer	$25	High	Adequate	Little	Difficult to read because of small scale	Visual	No	Very fragile
Digital Thermometer	$100	Depends on type of probe(s), typically 1°F	Adequate	Little	Excellent; one indicator can serve several locations (probes)	Visual (digital)	As an option	Probes typically cost $50
Thermocouple	$25-$30	Fair, 2°F	Adequate	Some, periodic recalibration	Excellent, when coupled with indicator	Analog electrical	Must be amplified (increased cost and error)	Not reliable for differential temperatures. Requires special wire for installation.
Resistance Temperature Detectors (RTD)	$60	High, 0.5°F	Adequate	Little	Excellent, when coupled with indicator	Analog electrical	Yes	Matched differential pairs can be accurate to 0.1°F
Thermistors	$10-$15	Moderate, 1°F	Limited	Little	Excellent, when coupled with indicator	Analog electrical	Yes	Not available in proper housing. Can be damaged by high temperature.

Table 12-3. Liquid Flow Sensors and Indicators

Type of Sensor	Approximate Cost	Accuracy	Rangeability	Maintenance	Convenience	Type of Output	Datalogger Capability	Special Comments
Pressure Gauges	$50	Strictly a flow indicator	Adequate	Little, very reliable	Low	Visual	No	
Variable Area	$500	Low, 2% of full scale	Moderate, 12:1	Little, very reliable	Moderate to High	Visual	No	High pressure drop increases pump sizing and energy requirements.
Differential Pressure with Orifice	$800	Fair, 1% of full scale	Poor, 5:1	Little, very reliable	High	Visual or analog electrical		Strictly visual indication available at greatly reduced cost. High pressure drop increases pump sizing and energy requirements.
Turbine	$1,000; $1,500 w/ instantaneous flow indicator	High, 0.25% of instantaneous flow	Excellent, 20:1	Not reliable for continuous operation due to bearing wear	Low	Pulsed	Requires time-base integrator	Portable insertion turbines are very convenient for flow balancing at greatly reduced cost. Prone to damage from high flows and debris.
Target or Impact	$1,000	Moderate, 0.5% of full scale	Moderate, 10:1	No wearing parts, (see comments)	Moderate	Analog electrical	Yes	Transmitters prone to failure. Fluid must be kept free of debris.
Vortex	$800-Analog $1,000-Digital $500-Indicator and Totalizer	High, 1% of instantaneous flow	Excellent, 20:1	Very good, no wearing parts	Low	Analog or digital electrical	Yes	State-of-the-art flowmeter
Magnetic	$3,000	High, 1% of instantaneous flow	Excellent, 30:1	Very good, no wearing parts	Low	Analog electrical	Yes	Very expensive
Positive Displacement	$500	High	Low	Prone to failure	Moderate to high	Mechanical totalizer	As an option; requires time-base integrator	Good for greatly varying flow rates, i.e., domestic water supply

Table 12-4. Air Flow Meters

Type of Sensor	Approximate Cost	Accuracy	Rangeability	Maintenance	Convenience	Type of Output	Datalogger Capability	Special Comments
Hot Wire Anemometer	$600-$1,000	Moderate, 2% of full scale	Moderate, 15:1	Some	Low	Analog electrical	Yes	Some models easily damaged by debris and improper handling. Must be properly located in order to determine mean flow.
Turbine	$300	Good, 1% of flow	Moderate, 15:1	Little	Low	Analog electrical	Yes	Must be properly located in order to determine mean flow.

Table 12-5. Energy Meters

Type of Sensor	Approximate Cost	Accuracy	Rangeability	Maintenance	Convenience	Type of Output	Datalogger Capability	Special Comments
Btu Meter	$400-$700	Moderate, 1% of full scale	Dependent on type of flow sensor	Unproven	High	Mechanical totalizer	No	
Electrical Wattmeter (Hall Probe)	$200	High	Very Good	Excellent	High	Analog electrical	Yes	
Electrical Wattmeter (Induction)	$100	Moderate, 1% of full scale	Good	Excellent	High	Analog electrical	Yes	Requires special installation
kW Demand Meter	$150	Moderate, 1% of full scale	Good	Excellent	High	Mechanical totalizer	An option with time-base integrator	

Table 12-6. Levels of Instrumentation for Each Energy Monitoring Purpose

Data	Debugging	Acceptance Test	System Maintenance	Performance Monitoring
Time Base	Short	Short	Long	Long
Visual Monitors	Adequate	Difficult	Adequate	Difficult
Energy Meters	Difficult	Adequate	Difficult	Adequate
Energy Meters Plus Visual Monitors	Adequate	Adequate	Adequate	Adequate
Data Logging & Local Processing	Good	Good	Good	Adequate
Data Logging & Remote Processing	Difficult	Difficult	Adequate	Very Good

Data accuracy depends primarily on the sensors. Rugged, high-quality equipment is recommended since it is removable and reusable for other applications.

A solar service water heating system can be equipped with permanently installed or removable gauges (Fig. 12-1). In the latter case, temperature measuring stations normally consist of a "thermowell" or "Pete's Plug" into or through which a thermometer is inserted and later removed. Flow measuring stations usually contain a pair of unions between which a flow meter is installed and later replaced by a straight pipe section. Isolation valves are normally installed upstream and downstream of a flow measuring station. If a bypass valve is provided, a replacement pipe section is not needed.

Operation of a new solar water-heating system can be checked and debugged with these gauges. A pair of "thermistor simulating potentiometers" can force the controller into all its possible states (operating modes). An ohmmeter can check for thermistor and leadwire faults.

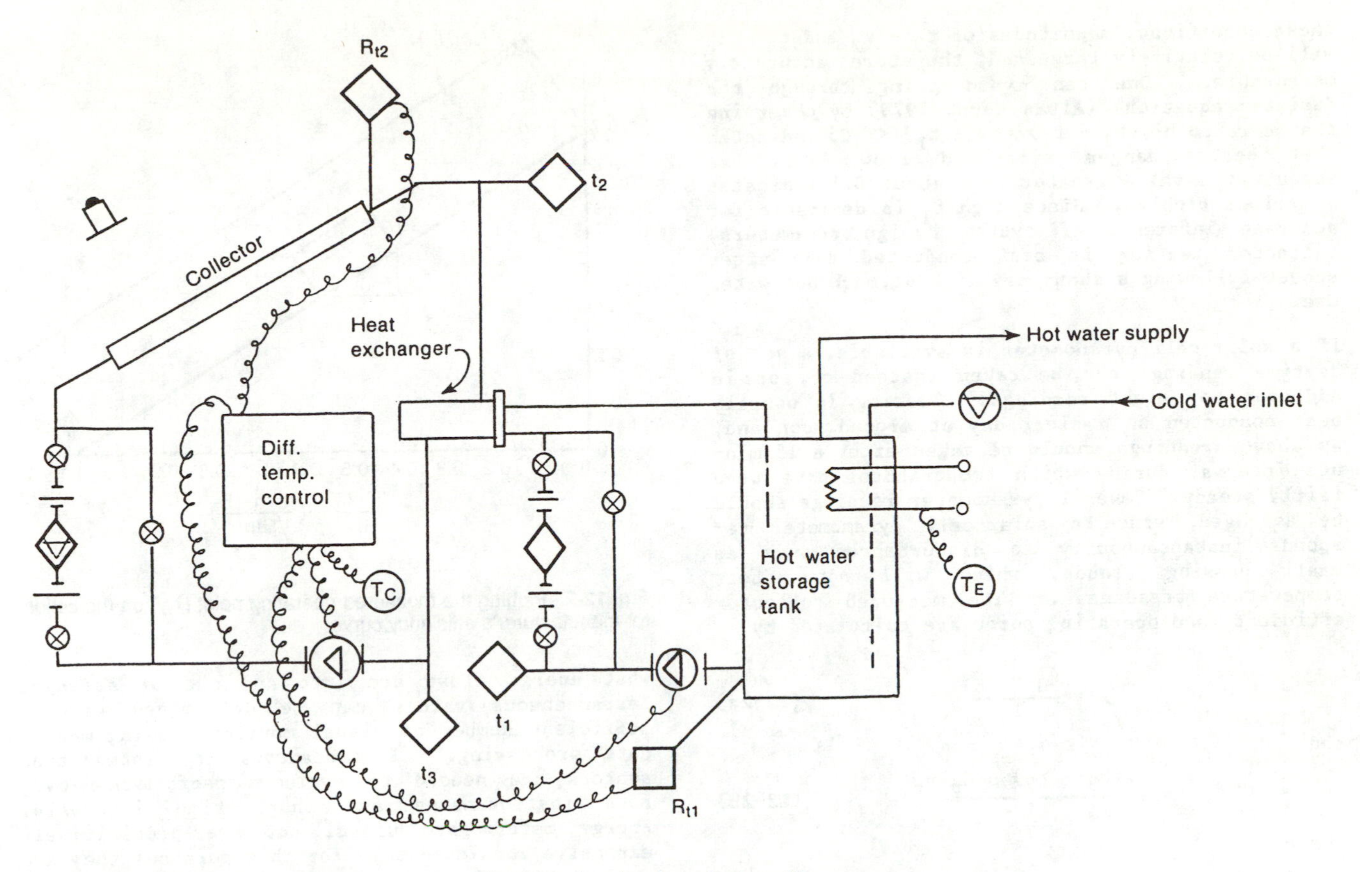

Fig. 12-1. Low level monitoring with gauges

A typical debugging procedure is to short a controller's R_{t2} terminals (simulating a high collector-outlet temperature) or disconnect the leads from the R_{t1} terminals (simulating a low storage-water temperature) to force pumps on. Rotameters are read to verify that flow rates are within design limits. Values of R_{t1} and R_{t2} are then read with an ohmmeter, converted to temperatures with an appropriate thermistor table or equation, and compared to thermometer readings t_1 and t_2. The corresponding temperatures should normally agree within 1.8°F (1°C) after steady-state conditions are indicated by thermistor probes, thermometers, and collector-loop fluid and piping. A final controller test requires substitution of potentiometers for controller thermistors to check Δt_{ON} and Δt_{OFF} parameters of the controller (Winn 1983).

Collector performance is adequately checked with this instrumentation if fairly steady values of outdoor temperature, storage temperature, and flow rate can be maintained for about 15 minutes at night. Readings are taken at the end of a 15-minute interval, and a collector loss coefficient is calculated from

$$A_c F_R U_L = \frac{(t_3 - t_2)\ \rho\ C_p \dot{v}_c}{t_3 - t_o} \qquad (12\text{-}1)$$

where

A_c = collector area
F_R = collector heat removal factor
U_L = collector loss coefficient
t_3 = measured collector inlet temperature
t_2 = measured collector outlet temperature
t_o = measured outdoor temperature
$\dot{v}_c$ = measured collector volumetric flow rate
ρ = density of the collector fluid
C_p = specific heat of the collector fluid

The resulting $A_c F_R U_L$ value should be within about 20% of the manufacturer's quoted value, assuming flow and temperature-difference measurement uncertainties are around 10% of reading. If the flow rate through each collector module differs significantly from the value used in the manufacturer's specification of $A_c F_R U_L$, an F_R correction must be applied.

If storage temperature, t_1, is recorded, it may be substituted for t_3 in the denominator of Eq. (12-1) and a heat-exchanger conductance inferred (Altas Corp. 1979). The temperature measurement accuracy required to infer heat-exchanger properties may be beyond the capabilities of the thermometer probe unless t_1 is quite high and t_o quite low. Under these conditions, magnitudes of $t_3 - t_2$ and $t_3 - t_o$ will be relatively large and, therefore, accurately measurable. One can avoid going through the deWinter equations (Altas Corp. 1979) by observing that a ratio of $(t_3 - t_1)/(t_2 - t_1) < .05$ indicates that heat-exchanger size and flow rates are adequate; a value greater than about 0.1 indicates a serious problem. Since high t_1 is desirable for accurate inference of system design parameters, collector testing is best conducted soon after sunset following a sunny day without much hot water use.

If a solar-cell pyranometer is available, a set of daytime readings can be taken instead of, or in addition to, night readings. Testing is usually best conducted on a clear day at around noon and, as above, readings should be taken after a 15 minute interval during which temperatures have been fairly steady. Several pyranometer readings should be averaged, since a solar-cell pyranometer responds instantaneously to disturbances such as small passing clouds, which will not affect temperature steadiness. The measured collector efficiency and operating point are calculated by

$$x = \frac{t_3 - t_o}{I} \qquad (12\text{-}2a)$$

and

$$y = \frac{(t_3 - t_2)\, \rho\, C_p\, \dot{v}_c}{I} \qquad (12\text{-}2b)$$

where

I = solar radiation (Btu/h-ft^2 or W/m^2), measured on the same tilt as the collector
x = measured collector operating point
y = measured collector efficiency

The resulting (x,y) point is plotted on a manufacturer's collector efficiency curve (Fig. 12-2). The percentage deviation, $100\ \ell/L$, should not exceed about 25%, again assuming that uncertainties in all terms of Eq. (12-2) are less than 10%.

Although a 10% uncertainty has been assumed for individual measurements in the foregoing analyses, better accuracy is possible for this level of monitoring. A 10% figure should be routine for an intelligent, conscientious technician who uses quality gauges and proper installation, test, and analysis procedures. An experienced technician should be able to measure values (except radiation) to accuracies of 2% to 4%, under favorable conditions.

The instrumentation level outlined above is adequate for most debugging, acceptance testing, and maintenance purposes. Typical locations for permanently installed visual monitors, adequate for these purposes, are shown in Fig. 12-3.

Acceptance Test

A new solar energy system should operate as designed. This is verified by monitoring it to determine energy flows through a component or subsystem. For example, it might be desirable to know how much energy the collector loop is delivering or

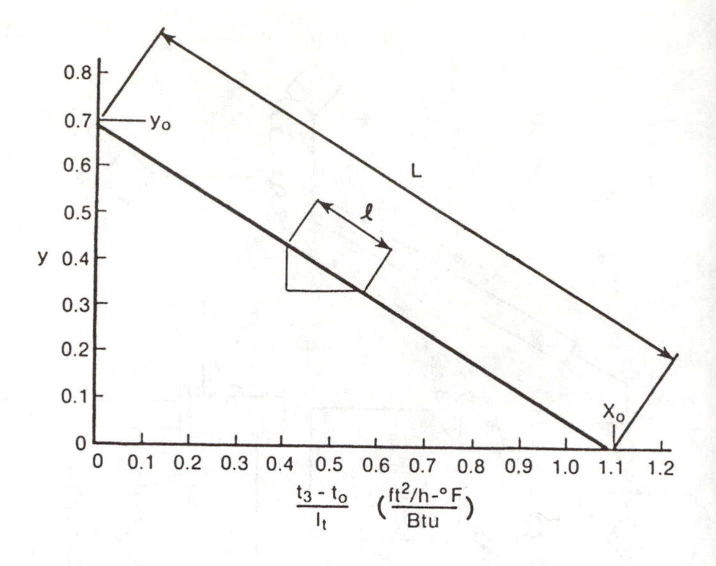

Fig. 12-2. Plotting the measured operating point (x,y) on the collector manufacturer's efficiency curve

what energy flows are into and out of storage. Instantaneous values can be determined with a sufficient number of visual monitors, using manual data processing. Instantaneous and integrating sensors are needed to evaluate performance over both instantaneous and short time intervals. Energy meters are useful but are prohibitively expensive when used only for this purpose; they are generally much less portable than visual monitors. Acceptance testing is described in more detail in Ch. 10.

Maintenance Monitoring

A data monitoring system assists in routine maintenance and repair of a solar energy system, both for isolating problems and for preventive maintenance. Sensors for this purpose should be permanently installed. Data recording, although useful, is generally not necessary; visual monitors, both instantaneous and integrating, are adequate. Sight glasses are used to visually indicate liquid levels. Measurements of intermittent flows, such as service hot water, generally require integrating meters. When sensors are used purely for maintenance, ruggedness and reliability are more important than accuracy.

Liquid System Monitoring with an Energy Meter

Performance monitoring over time requires more elaborate equipment than that discussed above. Although some instantaneous data readings are still useful, the need here is primarily for time-integrated values using either manual or automatic equipment. Manual equipment includes Btu meters and wattmeters; automatic equipment adds electronic sensors and a datalogger. With quality sensors, accurate and reliable data will be provided, either manually or automatically. Manual equipment is generally less expensive but is decentralized, with outputs not amenable to sophisticated processing. Automatic electronic equipment may overcome these

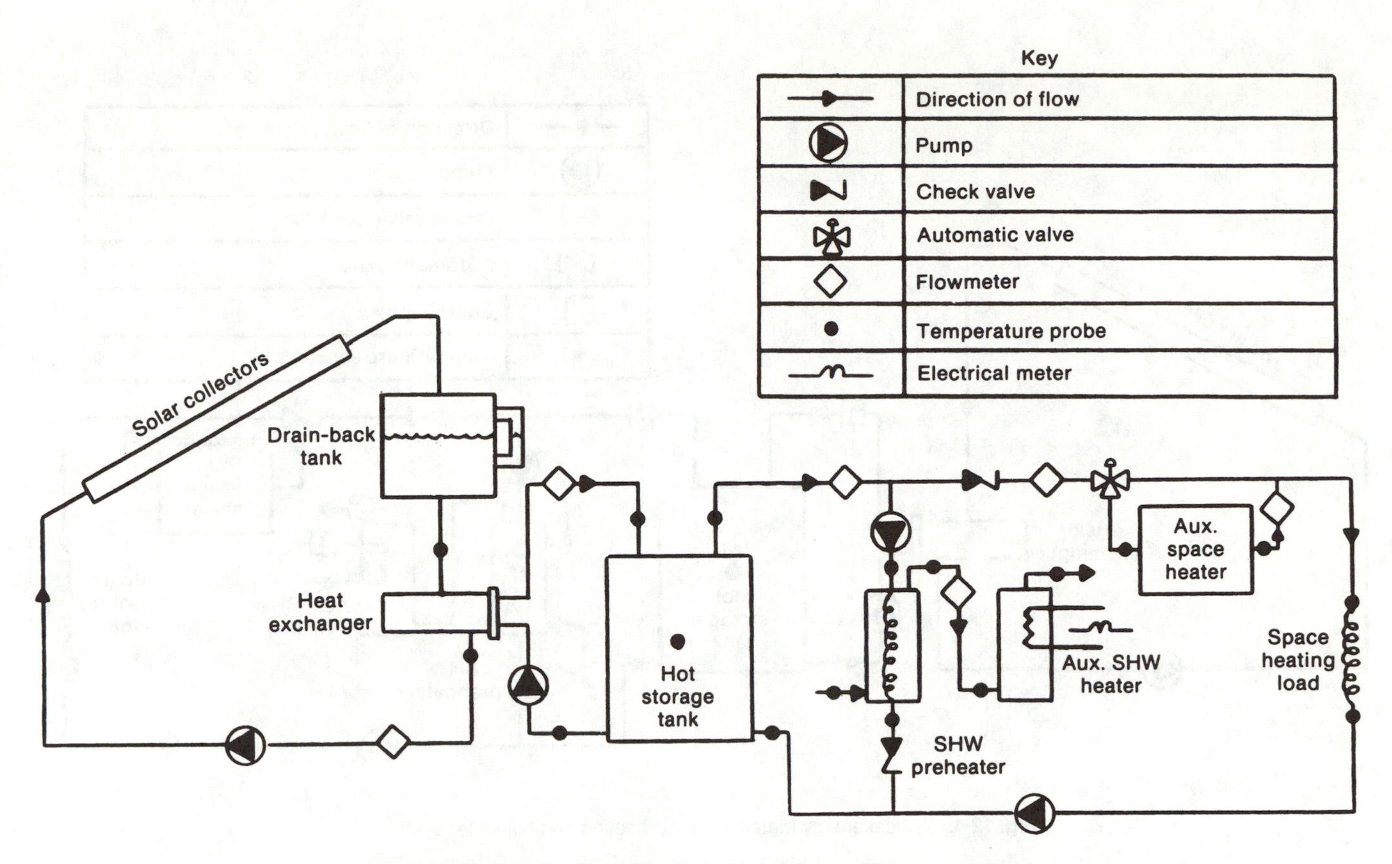

Fig. 12-3. Typical visual monitors for solar heating and hot water systems

problems, but it is more expensive (depending on the specific application). Costs of electronic sensors and dataloggers continue to drop, but costs of running signal wires from sensors to a central recorder remain steady. Btu meters and wattmeters should be completely adequate for many applications. Measurement of integrated solar flux is most reliably done with electronic devices, but this information may not be necessary, or approximate values can be obtained from nearby weather stations.

Energy meters in a typical solar heating and hot water installation should be appropriately located (Fig. 12-4) to determine the net solar contribution to the load(s). For meter locations in Fig. 12-4, the solar contribution to total load is

- Fraction of total load from solar = meter 2 reading ÷ (meter 3 reading + meter 6 reading).

To determine separate solar contributions to space heating and service hot water loads:

- Fraction of total space heating load from solar = (meter 6 reading - meter 5 reading) ÷ meter 6 reading.
- Fraction of total service hot water load from solar = (meter 3 reading - meter 4 reading) ÷ meter 3 reading.

Installation of all six energy meters provides some redundancy, but this allows system performance to be checked by an energy balance. Heat losses through pipes and other sources can be evaluated. For total solar contribution to gross load, at least three energy meters (numbers 2, 3, and 6) are required. To determine separate solar contributions to heating and service hot water loads, four energy meters (numbers 3, 6, and any two of 2, 4, and 5) are required. Storage tank energy loss is calculated from

- Heat lost from tank = meter 1 reading - meter 2 reading - change in internal energy of tank calculated from temperature sensor t-1 readings.

Neglecting pipe losses, the energy balance is

- Meter 6 reading + meter 3 reading = meter 2 reading + meter 4 reading + meter 5 reading.

This equation is used as an approximate check on energy balance and to determine thermal losses from pipes downstream of the storage tank.

Air System Monitoring with a Hard-Copy Output Recorder

Btu meters cannot be used for air-heating systems because commercially available Btu meters use pulse-initiating, positive-displacement water meters, which measure liquid flows. Systems in which major heat flows are not associated with liquid flows must be monitored by other techniques.

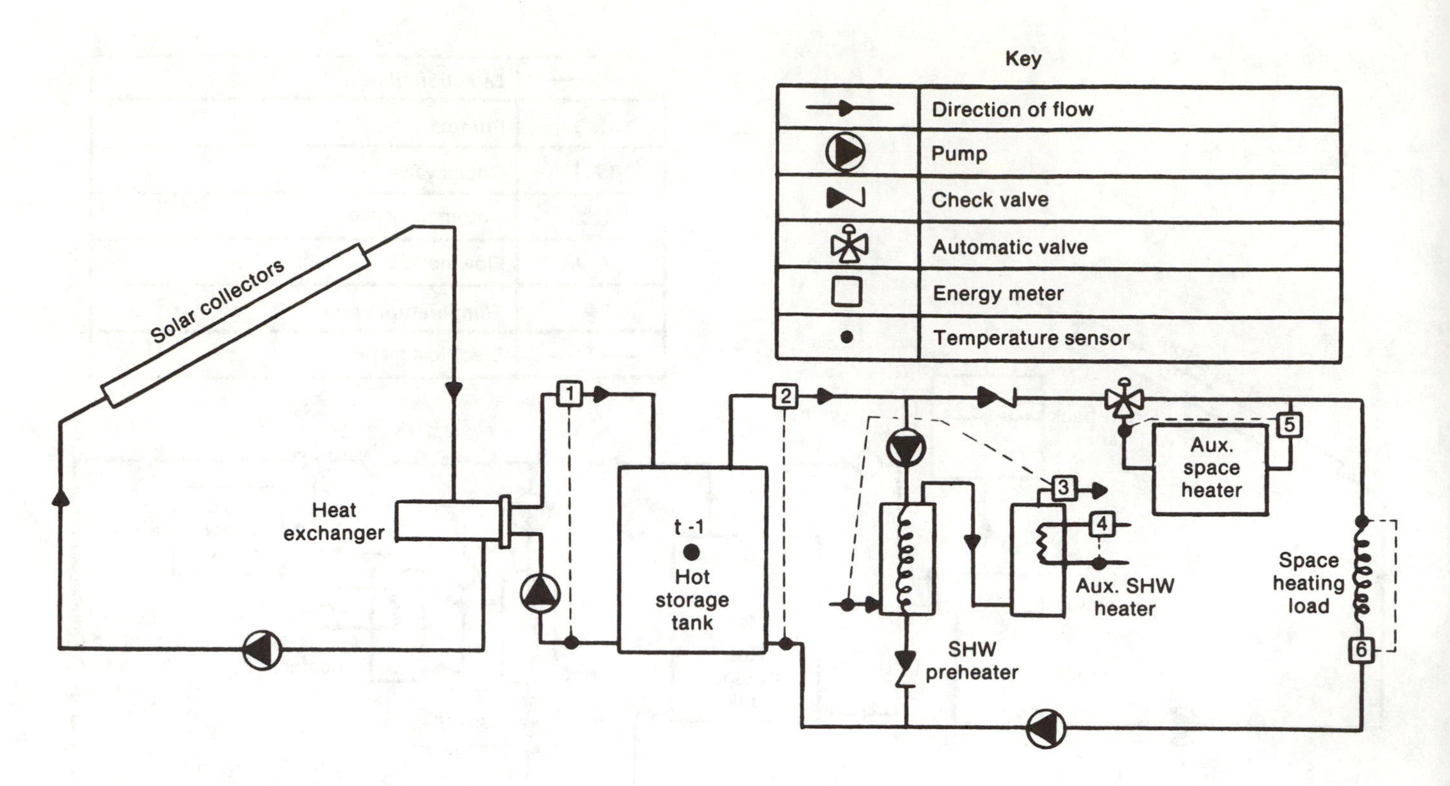

Fig. 12-4. Typical energy meters for solar heating and hot water systems

Air flow rates are more difficult to measure than liquid flow rates, but flow rates for air in active solar energy systems are much less variable than for liquids because air viscosity is much less sensitive to temperature than liquid viscosity is. Air viscosity increases with temperature and compensates for flow increase due to a decrease in air density; thus pressure drop through a forced-air duct network is very insensitive to temperature. If other system parameters (damper adjustments, drive ratios, etc.) remain fixed, performance monitoring can be conducted assuming blower air flow rates will be constant in any operating mode. Air flow rates need be measured only once for each operating mode; operating mode and relevant air temperatures are all that must be continuously monitored (Fig. 12-5). Low-cost strip chart recorders are available to perform this continuous logging function.

An air-heating system has three basic operating modes: (1) heating from collectors, (2) heating from storage, and (3) charging storage. Mode 0 can be used to designate an idle mode. An auxiliary heater can be either on or off in Modes 1 or 2 but is always off in Mode 3. Since the auxiliary heater state does not affect air flow, we will not consider any additional modes.

Flow rates should be measured at the top and bottom of storage in both Modes 2 and 3 to get an indication of leakage rates through the storage container. (The flow rate at the bottom of storage must be corrected to blower temperature.) An average of storage top and bottom flow rates is usually a good value to use in subsequent energy-rate calculations. In Mode 1, flow rates should be measured at a room supply (hot air) duct. Static pressures at inlets and outlets of major system components may be measured in all three modes during the flow-measuring session. Static pressures are often useful in diagnosing system malfunctions and in verifying air leakage observations. Steady-state energy output rates of an auxiliary heater are also measured at this time.

Thermistors, thermocouples, or other transducers can serve as temperature sensors, depending on the datalogger employed. Two channels are used to log the status of blowers B1 and B2. The state of an auxiliary heater is easily determined from the temperature rise across it, $t_H - t_s$.

Energy-transfer rates in the three modes are given by

$$\dot{Q}_{1s} = \dot{v}_1\rho(t_s)C_p(t_s-t_R); \qquad (12\text{-}3)$$

$$\dot{Q}_{1A} = \dot{v}_1\rho(t_s)C_p(t_H-t_s)$$

$$\dot{Q}_{2s} = \dot{v}_2\rho(t_s)C_p(t_s-t_R); \qquad (12\text{-}4)$$

$$\dot{Q}_{2A} = \dot{v}_2\rho(t_s)C_p(t_H-t_s)$$

$$\dot{Q}_{3s} = \dot{v}_3(t_o)C_p(t_o-t_i) \qquad (12\text{-}5)$$

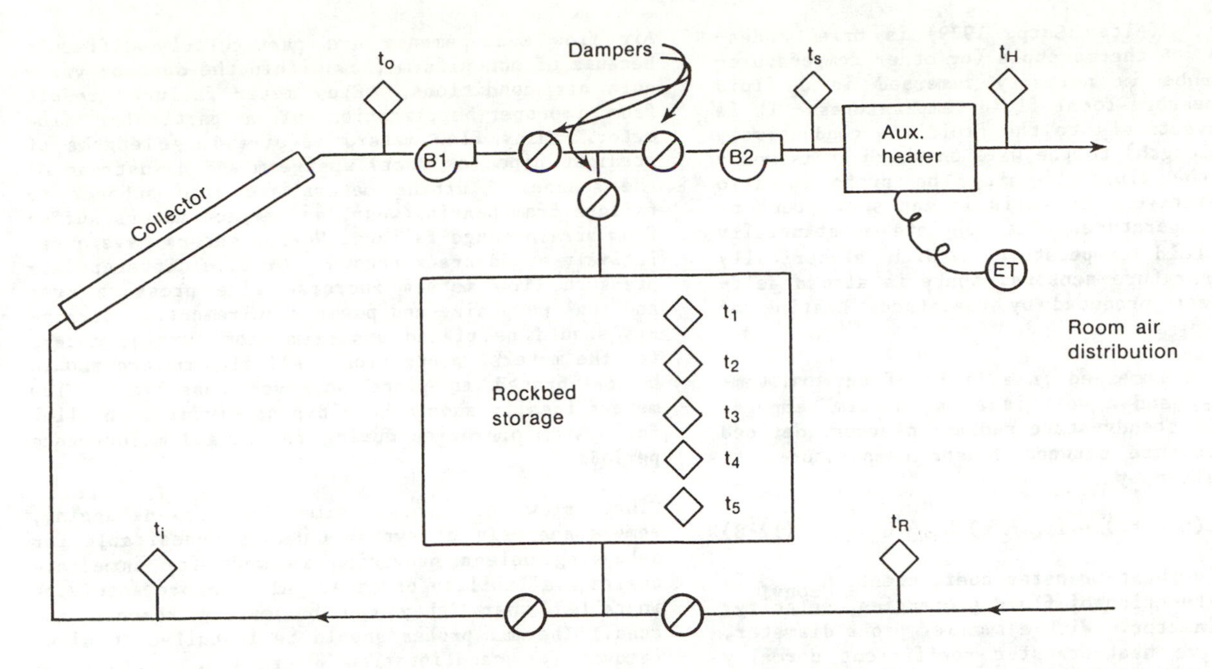

Fig. 12-5. Temperature sensor locations for an air heating system to be monitored by a 10-channel strip chart recorder

where

$\dot{v}_i$ = volumetric flow rate

$\dot{Q}_{is}$ = solar heat rate

$\dot{Q}_{iA}$ = auxiliary heat rate

C_p = air specific heat

$\rho(t_s)$ = air density at blower conditions (temperature and pressure)

i = operating mode

Energy rates are normally integrated by dividing a strip-chart record into time periods between mode changes and subdividing further into periods between which temperature trajectories of interest can be reasonably approximated by straight lines. End points of these straight lines are recorded, corresponding densities are computed, and Simpson's rule is applied. Barometric pressure changes are usually gradual; a daily value obtained from a nearby weather station can be used to calculate air density. A solar load fraction, f, is computed as follows:

$$f = \frac{\sum_{i=1}^{3} \dot{Q}_{is}}{\sum_{i=1}^{3} (\dot{Q}_{is} + \dot{Q}_{iA})} \tag{12-6}$$

Energy lost from storage is

$$Q_{loss} = \int_0^{T_s} \dot{Q}_{3s} - \dot{Q}_{2s} \, dT + \frac{(MC_p)_s}{5} \sum_{j=1}^{5} \left(t_{j,o} - t_{j_s T_s}\right) \tag{12-7}$$

where $(MC_p)_s$ is the total thermal capacitance of the rocks and T_s is the time duration of storage. In the unlikely event that leakage rates are low (less than 5%), the reasonableness of Q_{loss} is an indicator of instrumentation accuracy.

Hard-copy output dataloggers are used for acceptance testing, performance evaluation, and research. For debugging purposes, monitoring should be supplemented with a visual readout that can display any temperature channel. For performance evaluation over periods exceeding a few months, manual data reduction procedures may become prohibitively expensive; a datalogger that has computer-readable output or a real-time data reduction capability, or both, is recommended.

TYPICAL INSTALLATION PROBLEMS

Most problems with temperature probes occur because of poor sensor calibration or improper location. Extraneous heat loss from a probe's extension/weatherhead to the external environment can also be a problem. Insulating the extension may help although the probe should be immersed deep enough in the fluid to prevent conduction error. Thermosiphoning within a pipe under no-flow conditions also produces extraneous heat flows. Better sensor placement or check valves may correct this. Dials and scales of visual probes must be located above pipe insulation. Thermocouples, in particular, suffer from calibration difficulties. Dual thermowells are recommended for side-by-side in-place calibrations with a standard thermometer.

A properly designed temperature probe will allow for biases caused by radiation, conduction, and self-heating effects. A procedure for estimating these effects (Altas Corp. 1979) is briefly described here. A thermocouple (or other temperature-sensing) probe is normally immersed in a fluid stream to measure local fluid temperatures. It is coupled convectively to the fluid and conductively (along its length) to the wall on which it is mounted. If the fluid is air, the probe is also coupled radiatively to walls it can see. Duct or pipe wall temperatures often deviate substantially from bulk-fluid temperature. With electrically excited temperature sensors, there is also a self-heating effect produced by resistance heating of the probe tip.

If a probe is immersed in a fluid of uniform temperature, t_f, and a wall is at a uniform temperature, t_w, the steady-state radiation error, defined as the difference between sensor temperature, t, and t_f, is given by

$$\Delta t_{rad} = (t - t_f) = (t_w - t)\, h_{rad}/h_{conv} \qquad (12\text{-}8)$$

The convective heat-transfer coefficient, h_{conv}, is primarily a function of fluid properties, velocity, and probe diameter. With a smaller probe diameter, the convective heat-transfer coefficient normally increases so the reading error is reduced.

If a resistance temperature difference probe or a semiconductor is used, heat is generated because power is dissipated by the excitation current, and sensor voltage drops. This raises the measured temperature above the temperature that would otherwise be sensed. To estimate this self-heating error, an iterative process can be used to solve for t in terms of t_w and t_f. One should also estimate a probe's time constant, which should be less than 10% of the scan interval.

Air flow measurements are particularly difficult because of nonuniform flow within the duct or variable air conditions. Flow meter failures result from improper application of a particular flow meter. Many flow meters require long lengths of straight pipe (or duct) upstream and downstream of the sensor. Turbine meters are also subject to failure from bearing wear, and impact meters suffer from strain-gauge failure. Vortex meters have a relatively sound track record. Orifice/differential-pressure flow meters increase line pressure drop and thus pump size and power requirements. Strainers should be placed upstream from turbine meters for the meters' protection. All flow meters should be calibrated in place whenever possible. Flow meters ideally should have bypass circuits to allow for system operation during repair and maintenance periods.

Since slow turnaround time hampers debugging, remote analysis of system data is undesirable for debugging unless provision is made for immediate onsite availability of data. All sensors should be installed where they can be easily reached and read. Thermal probes should be installed to allow removal for recalibration or repair. Provisions should be made for electromagnetic interference suppression if high-gain amplifiers are to be used (i.e., with thermocouples).

Component compatibility is very important. A data recorder must accept the signals provided. Care should be taken to ensure that a proper sensor wire is used and that it is adequately shielded from electromagnetic interference and physical damage. Shielded strand wire in rigid conduit may be required.

Table 12-7 presents a partial list of manufacturers of instrumentation and monitoring equipment.

Table 12-7. Partial List of Monitoring Equipment Manufacturers

Orifice and Differential Pressure Flometers

Fisher and Porter Co., Inc.
Warminster, Pennsylvania

Badger Meter Company
Milwaukee, Wisconsin

Foxboro Company
Foxboro, Massachusetts

Vortex Flowmeters

Fisher and Porter Co., Inc.
Warminster, Pennsylvania

Brooks Instrument Division
Emerson Electric Company
Hatfield, Pennsylvania

Foxboro Company
Foxboro, Massachusetts

Magnetic Flowmeters

Brooks Instrument Division
Emerson Electric Company
Hatfield, Pennsylvania

Fisher & Porter Co., Inc.
Warminster, Pennsylvania

Robert Shaw Controls
Long Beach, California

Positive Displacement Flowmeters

Badger Meter Company
Milwaukee, Wisconsin

Brooks Instrument Division
Emerson Electric Company
Hatfield, Pennsylvania

ITT Barton
Monterey Park, California

Bulb Thermometers

Mueller Instrument Co., Inc.
Ivoryton, Connecticut

Kessler Company
Westbury, New York

Palmer Instrument Company
Cincinnati, Ohio

Dial Thermometers

March Instrument Company
Skokie, Illinois

Reotemp Instrument Corporation
San Diego, California

Tel-Tru Manufacturing Division
Germanow-Simon Machine Company
Rochester, New York

Digital Thermometers

Bailey Instruments, Inc.
Saddle Brook, New Jersey

Leeds & Northrup Company
North Wales, Pennsylvania

Analog Devices
Norwood, Massachusetts

RTDs and Thermocouples

Leeds & Northrup Company
North Wales, Pennsylvania

R&D Corporation
Hudson, New York

Minco Products, Inc.
Minneapolis, Minnesota

Dataloggers

Fluke Manufacturing Company
Mountlake Terrace, Washington

Consolidated Controls Corp.
Bethel, Connecticut

Ecosol Ltd.
New York, New York

Chapter 13
Active Solar System Construction Costs

INTRODUCTION

In this chapter, construction cost data for several active solar systems are presented, and a methodology for system cost estimating is provided, based on these data and on experience in active solar system design and analysis. Construction cost data are included for active solar heating and cooling systems from the DOE Commercial Demonstration Program and for solar process heat systems from the DOE Industrial Process Heat Field Engineering Test Program. Costs are divided into construction categories, and factors that affect these costs are identified and discussed.

Guidelines are provided for estimating solar construction costs as input to preliminary feasibility and sizing economic analyses during the conceptual design phase. Additional guidelines are used to assess the effects of design decisions on construction costs during the detailed design phase and to produce a final cost estimate for preparing a bid or for checking the reasonableness of the bids received. Nondesign-related cost factors and useful data sources for cost estimating are identified and discussed.

COST DATA AND ANALYSIS

The commercial space heating and cooling, and industrial process heat federal solar demonstration projects were constructed from 1977 to 1981. Cost data were collected during site visits; data sources included discussions with system designers, construction contractors, and owners plus review of DOE reports and vouchers.

The systems include a variety of system types (see Figs. 13-1 and 13-2). Although not selected to be representative of all solar systems, these systems do provide an adequate basis for some conclusions. In addition, monitored thermal performance data are available for each system.

Two industrial process heat projects that used liquid collectors and a liquid-to-air heat exchanger to provide process air heating are included in the liquid-to-air heating system category.

Construction costs are presented as total construction costs and as disaggregated subsystem costs. Subsystems roughly parallel construction work categories as follows: collector array, support structure, energy transport (piping, ductwork, insulation, and heating or cooling equipment), storage, electrical and controls, and general construction. More detailed construction cost data appear in individual cost reports for the heating and cooling projects and in a report by Mueller Associates (1981) for the process heat systems.

Cost data have been adjusted to account for differences between project contractual arrangements and year of construction. First, data were normalized to account for differences in contractual arrangements by applying a standard overhead and profit rate of 25% to the bare costs. Second, since project construction dates ranged from 1976 to 1980, all but collector array costs were inflated by rates obtained from the Construction Cost Index in the Engineering News Record to express costs in 1981 dollars. To gain insight into how solar system costs changed in years following 1981, the history of the Construction Cost Index can be obtained from the Engineering News Record.

However, during 1977-1981, collector panel costs did not keep pace with inflation. For example, construction bid records for 1977 show a particular brand of flat-plate collector with a bid price of $14.22/ft^2$ ($153.07/m^2$) net aperture area. A nearly identical panel was bid in 1981 at $14.49/ft^2$ ($155.97/m^2$). Therefore, 1981 costs for identical collectors were used where possible. When 1981 costs were not available, 1977 costs were used.

Analyses to identify factors that affected cost were based on multivariate linear regressions, with t-tests for statistical significance. Details on each particular regression are not given, and results are discussed only in general terms.

There are several limitations to the application of these cost data. Caveats include

- In comparing system and subsystem costs, one should not assume that "cheaper is better." A system twice as expensive as another may provide four times as much usable energy. Nonetheless, the measures used here, such as dollars per square foot, can be very useful if their limitations are kept in mind.
- Many projects were designed and built several years ago. Some earlier systems probably are not

representative of the state of the art in solar design. Indeed, a few systems were designed as prototypes and would be much less expensive if built today with off-the-shelf components.

- Cost data do not include design costs or costs of performance instrumentation and monitoring equipment.

The data do provide valuable information about solar system costs. The more costly projects probably would not be reproduced today, but the range of costs of the analyzed projects is reasonable, based on many known private sector projects. Because a relatively large number of systems are represented, statistical analysis can be used to measure the significance of the cost factors studied.

Total-System Cost Analysis

Statistical testing of the relationship between total system unit costs and system application for the 33 sites (Fig. 13-1) showed that system application can be a significant factor affecting total cost. Hot water systems cost the least, followed in order by direct-air heating systems, liquid-to-air heating systems, steam producing systems, and space heating and cooling systems. Unfortunately, statistical tests could not determine whether new systems were more or less expensive than retrofit systems, since most of the more expensive steam and cooling systems were retrofit projects.

Subsystem Cost Analysis

Collector Type

Much attention has been focused on solar collector cost reduction. Site-built collectors are the least expensive (Fig. 13-2), followed by liquid flat-plate, air flat-plate, evacuated-tube, and tracking collectors, in that order.

Costs include collector panel costs, costs of mounting collectors to their support structure, and costs of connecting collectors to a manifold system. For example, air flat-plate panels in general cost less than liquid flat-plate panels, but more labor is required for ductwork manifold connections than for piping manifold connections. As previously noted, "cheaper is not necessarily better," particularly for collector arrays.

Support Structure

Large variations in collector support structure costs have been observed. Support structure costs vary from $0.90/ft^2$ ($9.68/m^2$) of collector area to

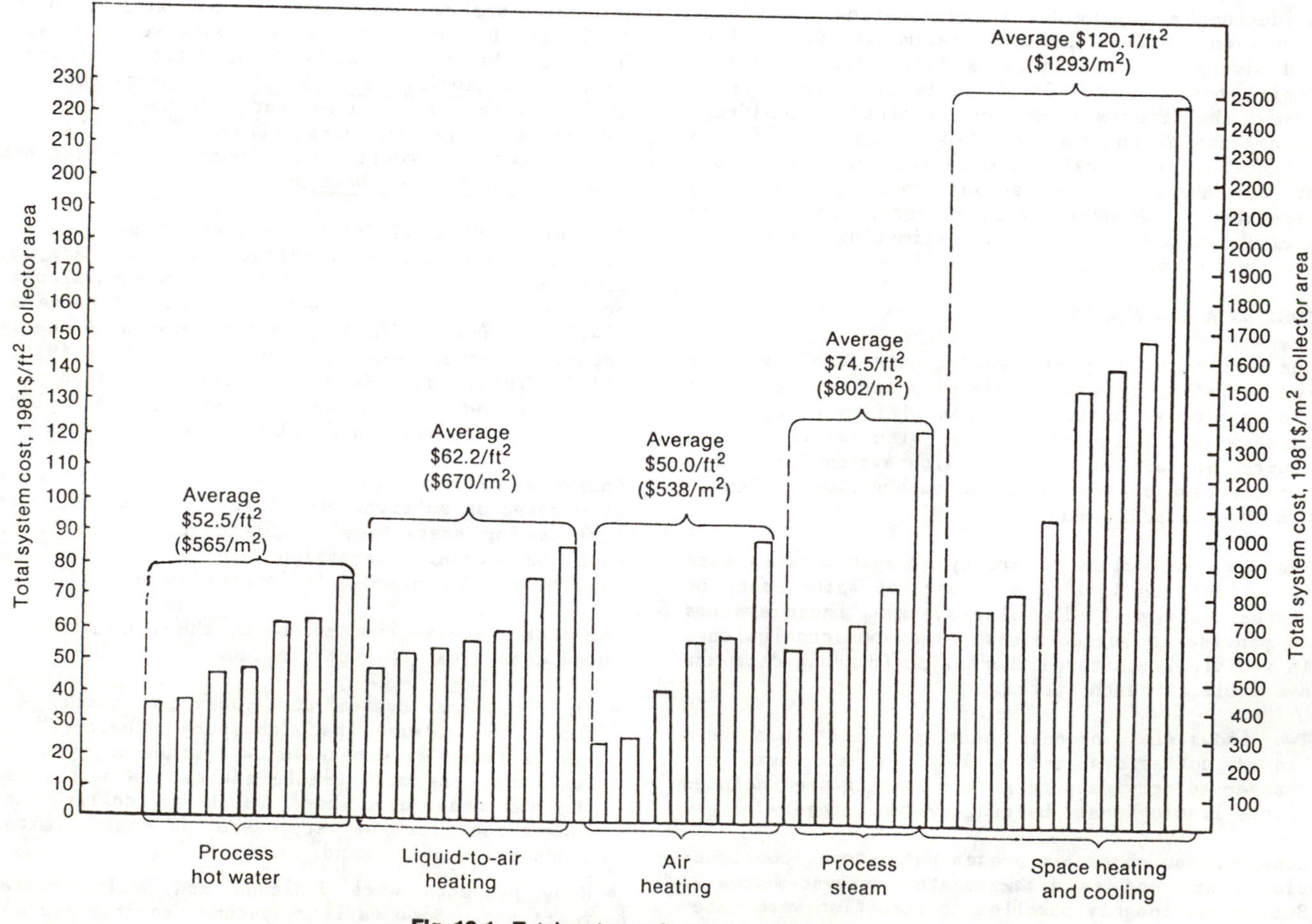

Fig. 13-1. Total system unit costs by application

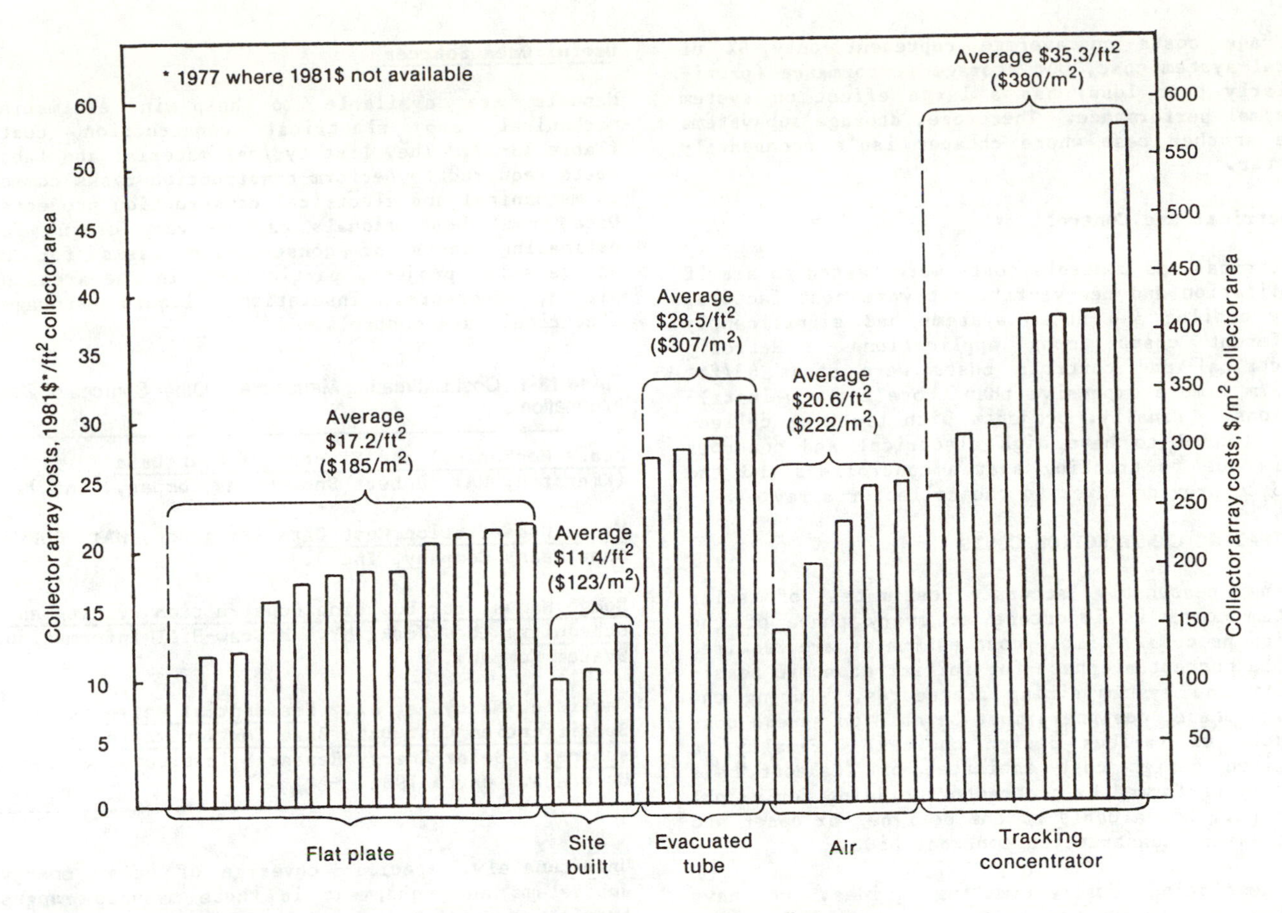

Fig. 13-2. Collector array costs by collector type

$35.70/ft² ($384/m²). Statistical tests show that the single factor affecting support structure costs most is whether the structure is used for additional functions, such as housing other equipment, improving system appearance, accommodating reflectors, forming part of the building roof structure, or forming a building's waterproof membrane.

Performance of multiple functions may serve to reduce costs in other areas or to improve system performance. In most cases, however, the support structure does not contribute to performance, so cheaper may possibly be better for support structure designs. Wide variations in support structure costs certainly suggest that costs of this subsystem can benefit substantially from better design.

Energy Transport (Piping, Ductwork, Insulation, and Heating/Cooling Equipment)

Combined subsystem costs were tested to see whether application and new-vs-retrofit determine costs; both were shown to be significant cost factors.

When combined costs are considered by system application, air systems are least expensive, followed by hot water systems, steam systems, liquid-to-air systems, and cooling systems, in that order. The last two are more expensive partly because of additional heat exchange equipment, although cooling costs in this case do not include absorption chiller costs. Steam systems are more expensive because of high-pressure piping costs.

New systems were about $5/ft² ($54/m²) of collector area cheaper than retrofit systems in this subsystem category because it is difficult to fit an energy transport system into an existing building. In general, support structure costs tend to balance out the energy transport system cost advantage in new construction.

Storage

Because of wide variations in storage capacity per collector area encountered among the projects, storage costs are expressed in terms of storage capacity rather than collector area. Statistical tests showed that storage costs depend on the type and location of storage vessel used. Unpressurized steel tanks were the least expensive, followed by fiberglass tanks, rock bins, and pressurized steel tanks. In terms of location, buried storage was least expensive, followed by exterior and then interior storage. Costs do not include extra piping or ductwork to and from buried and exterior storage. Nevertheless, the results are surprising since most estimation techniques would lead one to believe that interior tanks, which do not require waterproofing, would be least expensive.

Storage costs on average represent only 6% of total-system cost, but storage performance (particularly heat loss) has a large effect on system thermal performance. Therefore, storage subsystems are another case where cheaper isn't necessarily better.

Electrical and Controls

Electrical and controls costs were tested to see if application and new-vs-retrofit were cost factors. Only cooling and steam systems had significantly different costs among applications. Retrofit electrical and controls costs were about $3/ft^2 ($32/m^2) more expensive than those for new installations. Finally, projects with tracking collectors tended to have high electrical and controls costs due to tracking system controllers and the need to provide power to the collector array.

ESTIMATING CONSTRUCTION COSTS

Making reasonably accurate estimates of solar system costs is important at every phase of the design process. Simple cost estimates are required at the conceptual phase for initial economic feasibility and system sizing procedures. During the design phase, designers must be able to assess cost impacts of various design choices. Finally, a detailed design cost estimate, or "takeoff," is usually performed by contractors bidding for a job; this will be valuable to the designer or owner who must assess construction contract bids.

In developing cost-estimating guides, we have relied heavily on the cost data presented above. However, since some projects probably no longer represent state of the art, allowance was made for cost data considered unrepresentative. We have also relied on considerable experience with actual costs, cost estimating, and the design process.

Engineering and designs fees vary widely and are difficult to estimate. Engineering fee data were not collected for the demonstration projects, but some general guidelines for estimating these fees for solar systems are provided here. Engineering fees include the cost of conceptual design, detailed design and specification, bid supervision and evaluation, and construction supervision. Designing a commercial-scale solar system requires at least three man-months if complete detailed specifications and drawings are to be provided. Design costs will increase with construction contract size and system complexity. Design costs will also be higher if the services of an architect or structural engineer are required in addition to those of a mechanical designer.

Cost-estimating manuals commonly indicate a range of 4% to 10% of mechanical construction costs to estimate a mechanical engineer's fee. For solar design, this percentage is somewhat higher, typically from 6% to 20%. Smaller systems (less than 1000 ft^2_c) and very complex systems would normally experience engineering fees greater than 15%. Very large, yet straightforward, system designs would normally experience fees of less than 10%.

Useful Data Sources

Manuals are available to help in estimating mechanical and electrical construction costs (Table 13-1). They list typical material and labor costs required to perform construction tasks common to mechanical and electrical construction projects. Data from these manuals can be very useful for estimating costs of construction tasks for an active solar project, particularly in the areas of piping, ductwork, insulation, liquid storage, electrical, and controls.

Table 13-1. Cost Estimating Manuals and Other Sources of Cost Information

Means Mechanical and Electrical Cost Data (Kingston, MA: Robert Snow Means Company, Inc.).

Means Construction Cost Data (Kingston, MA: Robert Snow Means Company, Inc.).

Dodge Manual for Building Construction Pricing and Scheduling (New York, NY: McGraw-Hill Information System Company.

Carlisle, Nancy, et al. Residential Retrofit Specification Cost Data Base for Renewable Resources. Solar Energy Research Institute, Golden, Colorado. March 1982.

Unfortunately, specific coverage of solar energy subsystems and equipment in these manuals ranges from total costs for domestic or very small commercial sized systems, to costs for panels only, to no coverage at all. In particular, manuals cover collector arrays, support structures, and rock-bed storage inadequately.

Manufacturers' representatives and dealers are useful data sources for material costs of specific components and equipment. Obtaining price lists or simply calling dealers can provide accurate material cost estimates for collector panels, heat exchangers, pumps and fans, control valves and mechanical dampers, and storage tanks.

Cost Estimating in the Schematic Design Phase

During the schematic design phase, a designer needs an estimate of construction costs to assess the feasibility of system-sizing designs. At this point, the system design is rough, and a cost estimate can therefore be only approximate. Guidelines for making initial estimates for use in feasibility and sizing analyses are listed in Table 13-2.

The table lists subsystems that make up a solar system, design categories for that subsystem, cost ranges that might be expected, and a typical cost, based on DOE project results. Unlike cost data presented above (in 1981 dollars), all costs are expressed here in either 1986 dollars/unit net aperture collector area or in 1986 dollars/gallon or ton of rocks for a storage subsystem. No value is given for a fixed-cost component of the total

Table 13-2. Conceptual Phase Cost Estimating Guide*

Subsystem	Category	Range	Typical	Units
Collector Array**	Site Built	7-16	13	$\$/ft^2_c$
	Liquid Flat Plate	15-28	18	
	Air Flat Plate	18-30	21	
	Evacuated Tubes	31-38	36	
	Tracking Concentrator	25-52	38	
Support Structure	Single Function	5-17	11	$\$/ft^2_c$
	Multiple Function	10-28	17	
Energy Transport	Hot Water	11-22	18	$\$/ft^2_c$
	Liquid-to-Air Heat	20-35	27	
	Air Heating	8-22	14	
	Steam	13-40	24	
	Heating and Cooling	26-67	47	
Storage (Liquid)	Unpressurized Steel Tank	1.5-4	2.5	$/gal
	Pressurized Steel Tank	3-8	4	
	Fiberglass Tank	2.5-3.5	3	
Storage (Air)	Rock Bin	120-250	160	$/ton rocks
Electric and Controls	Hot Water	3-8	4	$\$/ft^2_c$
	Liquid-to-Air Heat	4-12	8	
	Air Heating	4-12	6	
	Steam	4-30	17	
	Heating and Cooling	6-29	14	
General Construction	---	0-20	6	$\$/ft^2_c$

*Cost includes materials and labor and contractor's overhead and profit.
**Costs may not correspond with those in Fig. 13-2.

system because statistical tests of demonstration project costs suggest that, for systems having more than 300 ft^2 (27 m^2) of collector, this cost component is negligible. Therefore, system costs can be described accurately solely as a variable cost per collector area.

To use the table to arrive at a total-system cost estimate, add up typical subsystem costs located under the appropriate design category. A designer can use range figures by applying judgment. For example, if an expensive support structure is required because of design constraints imposed by the building's roof structure, or if long piping runs will be required, values should come from the upper end of the ranges for support structure and energy transport subsystems. However, overly optimistic or pessimistic values should be avoided when choosing from the range column.

Some design-related factors that impact construction costs are listed in Table 13-3. These factors may be useful to a designer in two ways. First, they provide guidance for deciding which values to use from the range column in Table 13-2. Initial cost estimates can be refined as the conceptual design is refined. Second, the list can help a designer maintain an attitude of cost consciousness. Though not complete, the list indicates the more important cost-impacting decisions to be faced through all design phases.

Because we have found significant economies of scale among the demonstration projects, we advise multiplying total-system preliminary cost estimates by an economy-of-scale factor (Fig. 13-3).

Cost Estimating During the Design Development Phase

During the design phase, a designer makes many decisions that will affect project construction costs. Many decisions involve cost versus performance tradeoffs; for example, use of high efficiency panels will improve thermal performance but may also increase costs. Analyses of cost versus performance tradeoffs frequently require consideration of cost differentials between more expensive options and incremental increases in energy output. Other decisions may not strongly affect thermal performance, e.g., support structure design. By maintaining a cost-conscious attitude and applying value-engineering concepts, a designer will help keep construction costs down.

Cost Estimating During the Post-Design Phase

After a system has been designed, complete with drawings and specifications, a potential bidder will make a detailed cost estimate, or takeoff. A designer will also prepare a cost estimate, although usually this estimate is less detailed than a contractor's estimate. A designer's estimate can be useful in assessing bids. If all bids are much

Table 13-3. Design Cost Factors

Subsystem	Cost Factor	Comments
Collector Array	Manifold Type	Internally manifolded collectors cost about the same as externally manifolded collectors but require less than half the labor to connect the collectors.
	Panel Size	Larger panels may require less labor per unit of collector area for mounting and connection.
Support Structure	Roof Imposed Structural Constraints	In some retrofit situations, the support structure must span large distances, requiring larger, more expensive structural members.
	Flat/Sloped Roof	In general, support structures for sloped roofs require less material and labor.
	# Penetrations Through Roof Membrane	Roof penetrations are expensive and should be kept to a minimum.
	Reflectors	Reflectors add materials costs.
	"Aesthetic" Constraints	"Cosmetic" details for support structures add significantly to costs.
Energy Transport	Long Pipe/Duct Runs	Long runs to and from an isolated collector array or storage vessel add to materials and labor costs.
	Freeze Protection	Direct systems are probably less expensive. The type and size of heat exchanger used in an indirect system can significantly affect energy transport costs.
	Complexity/# Modes	The simpler the system, the more opportunity for cost savings.
Storage	Size	The most important cost factor is size (storage capacity per collector area).
Electrical and Controls	# Modes/# Actuators	The more complex the control systems, the more cost that will be incurred.
	Monitoring Equipment	Instrumentation and performance monitoring equipment can be very expensive.

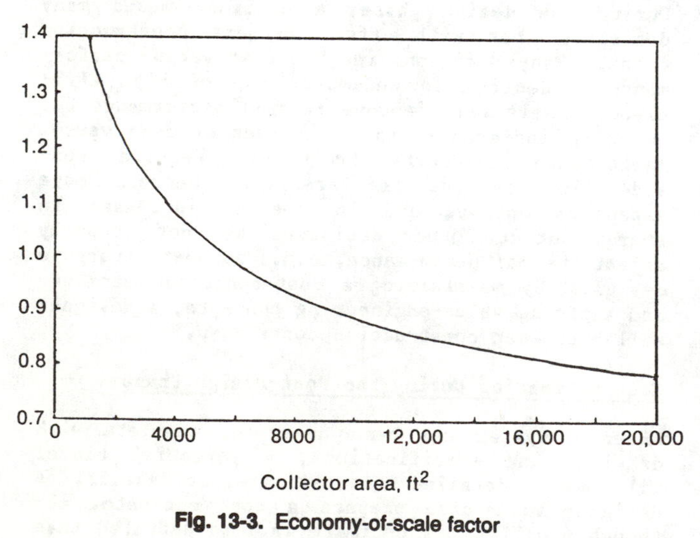

Fig. 13-3. Economy-of-scale factor

higher than this estimate, additional contractors should be invited to bid. If the lowest bid is much lower than the designer's estimate, a contractor's ability to perform might be questioned.

Contractors differ in how they prepare a takeoff, but the following description is representative of what normally happens. A contractor looks at the design drawings, reads the specifications, and formulates a construction plan as a series of tasks. He then prices major components by contacting suppliers and subcontractors and estimates material and labor costs required to complete each task, obtaining estimates or firm bids from any subcontractors he plans to use on the job. He will then add his overhead and profit rate and a contingency to arrive at a bid value. Labor and material cost estimates are often obtained from cost-estimating manuals, although many contractors rely instead on cost data compiled from previous experience.

Nondesign-Related Cost Factors

Cost factors not directly related to system design but that may affect system cost include the following:

- Number of bidders for construction contract - Substantial cost decreases may be obtained by ensuring that a large number of contractors are invited to bid.
- Number of contractors in area and general economic conditions - Contract bid values are sensitive to the laws of supply and demand. Contract bids will be lower if there are many contractors and little current construction in the area. The opposite is also true.
- Transportation requirements - Transportation of materials to an isolated site may increase material costs.
- Time constraints - A tight completion deadline may require overtime labor that will increase labor costs.
- Material delivery delays - Construction schedules may be interrupted if delivery of important materials and equipment must be delayed, leading to increased costs.
- Weather and season - Inclement weather may interrupt construction schedules. Labor productivity may be reduced during cold winters and hot summers.
- Regional variations in labor and material costs - Although this was not found to be a significant factor in the demonstration projects, costs for a particular project may be affected by local cost variations.
- Legal and union restrictions - Local building code restrictions and local union regulations may affect system cost.
- Contingency - A contingency allowance added to a contractor's bid will depend heavily on the extent and nature of his experience with solar and government projects. A contractor's contingency allowance can often be reduced by carefully planned pre-bid conferences and detailed drawings and specifications.
- In-house construction supervision vs. construction management vs. general contractor - Exploring all possible contractual arrangements for the solar system construction can sometimes help reduce total construction cost.

References

"A Baseline for Energy Design," April 1982, Progressive Architecture, pp. 110-116.

Adams, R. C., D. B. Belzer, J. M. Fang, K. L. Imhoff, D. H. Lax, R. J. Moe, J. M. Roop, and A. R. Wusterbarth, 1985, A Retrospective Analysis of Energy Use and Conservation Trends: 1972-1982, PNL-5026, Richland, WA: Pacific Northwest Laboratory.

Ahmed, S. F. and P. W. Scanlon, 1980, "Large Scale Space Heating and DHW System Using Shallow Solar Pond with Heat Pump Assistance," Proceedings of the 1980 Annual Meeting of the American Section of the International Solar Energy Society, Vol. 3.1, pp. 359-363, Newark, DE: ASES.

Altas Corp., 1979, Monitoring the Performance of Solar Heated and Cooled Buildings, EPRI Report ER-1239, Vol. 2, Santa Cruz, CA.

American Council for an Energy-Efficient Economy and the Energy Conservation Coalition, 1986, Federal R&D on Energy Efficiency: A $50 Billion Contribution to the U.S. Economy. A White Paper on the Consequences of Proposed FY87 Budget Cuts, Washington, D.C.

American Institute of Architects, 1981, Energy in Architecture, Washington, D.C.

American Society for Testing Materials, 1986, Standard Specification for Concrete Aggregates, ASTM C33-86, New York.

American Society of Heating, Refrigerating and Air-Conditioning Engineers, 1963, Guide and Data Book, New York.

American Society of Heating, Refrigerating and Air-Conditioning Engineers, 1977, Methods of Testing to Determine the Thermal Performance of Solar Collectors, ASHRAE Standard 93-77, Atlanta, GA.

American Society of Heating, Refrigerating and Air-Conditioning Engineers, 1981a, ASHRAE Handbook. Fundamentals, New York.

American Society of Heating, Refrigerating and Air-Conditioning Engineers, 1981b, Thermal Environmental Conditions for Human Occupancy, ASHRAE Standard 55-81, Atlanta, GA.

American Society of Heating, Refrigerating and Air-Conditioning Engineers, 1982, ASHRAE Handbook. Applications, New York.

American Society of Heating, Refrigerating and Air-Conditioning Engineers, 1984, Handbook. Systems, New York.

American Society of Heating, Refrigerating and Air-Conditioning Engineers, 1985, ASHRAE Handbook. Fundamentals, New York.

Anderson, B., 1977, Solar Energy: Fundamentals in Building Design, New York: McGraw-Hill.

Anderson, J. V., J. W. Mitchell, and W. A. Beckman, 1980, "A Design Method for Parallel Solar-Heat Pump Systems," Solar Energy, Vol. 25, No. 2, pp. 155-163.

Andrews, J. H., 1978, "Development of a Cost-Effective Solar Assisted Heat Pump System," Proceedings of the 1978 Annual Meeting of the American Section of the International Solar Energy Society, Vol. 2.1, pp. 281-287, Newark, DE: ASES.

Arens, E. A., D. H. Nall, and W. L. Carroll, 1979, "The Representativeness of TRY Data in Predicting Mean Annual Heating and Cooling Requirements," ASHRAE Transactions 1979, Vol. 85, Pt. 1, pp. 707-721.

Arens, E., L. Zeren, R. Gonzalez, L. Berglund, and P. E. McNall, 1980, "A New Bioclimatic Chart for Environmental Design," Proceedings of the International Congress on Building Energy Management, Provoa de Varzim, Portugal, May 12-16, 1980, pp. 645-653, Oxford: Pergamon Press. (Also see "A New Bioclimatic Chart for Passive Solar Design," by same authors in Proceedings of the Fifth National Passive Solar Conference, Amherst, Massachusetts, Oct. 1980, Vol. 5.2, pp. 1202-1206, Newark, DE: American Solar Energy Society).

Argonne National Laboratory, 1981, Final Reliability and Materials Design Guidelines for Solar Domestic Hot Water Systems, ANL/SDP-11, Argonne, IL.

Ashby, A., Holtberg, P., and T. Woods, 1985, Comparison of the 1984 DOE/EIA Annual Energy Outlook and the 1984 GRI Baseline Projection, Chicago: Gas Research Institute.

Associated Air Balance Council, 1979, National Standards for Field Measurements and Instrumentation, Washington, D.C.

Atwater, M. A., J. T. Ball, and P. S. Brown, Jr., 1976, "Intraregional Variation of Radiation Data and Applications for Solar Simulation," Paper No. 76-WA/Sol-9, Presented at the American Society of Mechanical Engineers Winter Annual Meeting, Dec. 1976, New York: ASME.

Auh, P. C., 1978, "An Overview on Absorption Cooling Technology in Solar Applications," Proceedings of the Third Workshop on the Use of Solar Energy for the Cooling of Buildings, February 15-17, 1978, San Francisco, California, pp. 14-18, Newark, DE: American Solar Energy Society.

Balcomb, J. D. and J. C. Hedstrom, 1977, "Thermal Performance of the LASL Solar Heated Mobile/Modular Home," Proceedings of the 1977 Annual Meeting of the American Section of the International Solar Energy Society, Vol. 1, pp. 13-22, Newark, DE: ASES.

Balcomb, J. D., J. C. Hedstrom, S. W. Moore, and B. T. Rogers, 1976, Solar Heating Handbook for Los Alamos, LA-5967-MS, Los Alamos, NM: Los Alamos Scientific Laboratory.

Balcomb, J. Douglas, Dennis Barley, Robert McFarland, Joseph Perry, Jr., William Wray, and Scott Noll, 1980, Passive Solar Design Handbook, Volume Two, Passive Solar Design Analysis, DOE/CS-0127/2, Los Alamos, NM: Los Alamos Scientific Laboratory.

Balcomb, J. Douglas, Claudia E. Kosiewicz, Gloria S. Lazarus, Robert D. McFarland, and William O. Wray, 1982, Passive Solar Design Handbook, Volume Three: Passive Solar Design Analysis, Edited by Robert W. Jones, DOE/CS-0127/3, Los Alamos, NM: Los Alamos National Laboratory.

Balcomb, J. Douglas, Robert W. Jones, Robert D. McFarland, and William O. Wray, 1984, Passive Solar Heating Analysis: A Design Manual, Atlanta, GA: American Society of Heating, Refrigerating and Air-Conditioning Engineers.

Banham, R., 1984, The Architecture of the Well-Tempered Environment, 2nd ed., Chicago: University of Chicago Press.

Barley, C. D., 1979, "Load Optimization in Solar Space Heating Systems," Solar Energy, Vol. 23, No. 2, pp. 149-156.

Barley, C. D. and C. B. Winn, 1978, "Optimal Sizing of Solar Collectors by the Method of Relative Areas," Solar Energy, Vol. 21, No. 4, pp. 279-289.

Baumeister, Theodore, ed., 1967, Marks' Standard Handbook for Mechanical Engineers, 7th ed., New York: McGraw Hill.

Beard, J. T., F. L. Huckstep, W. B. May, Jr., F. A. Iachetta, and L. V. Lilleleht, 1976, "Analysis of Thermal Performance of 'Solaris' Water-Trickle Solar Collector," Paper No. 76-WA/Sol-21, Presented at the American Society of Mechanical Engineers Winter Annual Meeting, Dec. 1976, New York: ASME.

Beckman, W. A., S. A. Klein, and J. A. Duffie, 1977, Solar Heating Design by the F-Chart Method, New York: Wiley-Interscience.

Beckman, W. A., J. A. Duffie, S. A. Klein, and J. W. Mitchell, 1982, "F-Load, A Building Heating-Load Calculation Program," ASHRAE Transactions 1982, Vol. 88, Pt. 2, pp. 875-889.

Bedinger, A. F. G., J. J. Tomlinson, R. L. Reid, and D. J. Chaffin, 1981, "Performance of a Solar Augmented Heat Pump," Solar Engineering-1981, Proceedings of the ASME Solar Energy Division Third Annual Conference on Systems Simulation, Economic Analysis/Solar Heating and Cooling Operational Results, Reno, Nevada, April 27-May 1, pp. 58-67, New York: American Society of Mechanical Engineers.

Bennett, I., 1969, "Correlation of Daily Insolation with Daily Total Sky Cover, Opaque Sky Cover, and Percentage of Possible Sunshine," Solar Energy, Vol. 12, No. 3, pp. 391-393.

Bergquam, J. F., M. F. Young, and J. W. Baughn, 1979, "Comparative Performance of Passive and Active Solar Domestic Hot Water Systems," Proceedings of the Fourth National Passive Solar Conference, Kansas City, Missouri, pp. 610-614, Newark, DE: American Section of the International Solar Energy Society.

Besant, R. W., R. S. Dumont, and G. Schoenau, 1979, "Saskatchewan House: 100 Percent Solar in a Severe Climate," Solar Age, Vol. 4, No. 5, pp. 18-23.

Bessler, W. F. and B. C. Hwang, 1980, "Solar Assisted Heat Pump for Residential Use," ASHRAE Journal, Vol. 22, No. 9, pp. 59-63.

Bisset, J. B. and R. F. Monaghan, 1979, "Design Study: Solar Systems for Commercial Buildings," ASHRAE Journal, Vol. 21, No. 5, pp. 37-42.

Booz-Allen and Hamilton, Inc., n.d., "Energy Graphics," Washington, D.C.

Braun, J. E., S. A. Klein, and J. W. Mitchell, "Seasonal Storage of Energy in Solar Heating," Solar Energy, Vol. 26, No. 5, pp. 403-411.

Bridgers, F. H., F. J. Stoltys, and D. R. Broughton, 1979, Passive Solar Energy and the Heat Pump, ASHRAE Paper No. PH-79-2, Atlanta, GA: American Society of Heating, Refrigerating and Air-Conditioning Engineers.

Busch, J. F. et al., 1984, "Measured Heating Performance of New, Low-Energy Homes: Updated Results from the BECA-A Data Base," Doing Better: Setting an Agenda for the Second Decade, Washington, D.C.: American Council for an Energy-Efficient Economy.

Butti, K. and J. Perlin, 1980, A Golden Thread: 2500 Years of Solar Architecture and Technology, Palo Alto, CA: Chershire Books; New York: Van Nostrand Reinhold.

Cash, M., 1978, "Learning from Experience," Solar Age, Vol. 3, No. 11, pp. 14-19.

Chinery, G. T. and F. C. Wessling, Jr., 1981, "Results of ASHRAE 95 Thermal Testing of 32 Residential Solar Water Heating Systems," Solar Engineering-1981, Proceedings of the ASME Solar Energy Division Third Annual Conference on Systems Simulation, Economic Analysis/Solar Heating and Cooling Operational Results, Reno, Nevada, April 27-May 1, pp. 581-588, New York: American Society of Mechanical Engineers.

Choi, M. K., P. J. Hughes, and J. H. Morehouse, 1982, "Status of Several Solar Cooling Systems with Respect to Cost/Performance Goals," Proceedings of the 1982 Annual Meeting of the American Solar Energy Society, Pt. 1, pp. 591-596, Newark, DE: ASES.

Choi, M. K., J. H. Morehouse, and P. J. Hughes, 1980, "Thermal and Economic Assessment of Ground-Coupled Storage for Residential Solar-Heat Pump Systems," Proceedings of the 1980 Annual Meeting of the American Section of the International Solar Energy Society, Vol. 3.1, pp. 349-353, Newark, DE: ASES.

Cinquemani, V., J. R. Owenby, Jr., and R. G. Baldwin, 1978, Input Data for Solar Systems, DOE/TIC-10193, Asheville, NC: National Climatic Center.

Clark, E. F. and F. de Winter, eds., 1978, Proceedings of the Third Workshop on the Use of Solar Energy for the Cooling of Buildings, February 15-17, 1978, San Francisco, California, Newark, DE: American Solar Energy Society.

Clarke, R. S., 1980, "Shadows at the Site: Solar Access Measurement," Solar Age, Vol. 5, No. 10, pp. 12-15.

Cole, Roger L., Kenneth J. Nield, Raymond R. Rohde, and Ronald M. Wolosewicz, eds., 1979, Design and Installation Manual for Thermal Energy Storage, ANL-79-15, Argonne, IL: Argonne National Laboratory.

Costello, F. A., T. Kusuda, and S. Aso, 1982, TI-59 Program for Calculating the Annual Energy Requirements for Residential Heating and Cooling, Vol. I—Users Manual, Vol. II—Program Reference Manual, DOE/NBB-0011, Washington, D.C.: National Bureau of Standards.

Crall, G. Christopher P., Bibliography on Available Computer Programs in the General Area of Heating, Refrigerating, Air Conditioning and Ventilating, Research Project Report GRP-153, New York: American Society of Heating, Refrigerating and Air-Conditioning Engineers.

Curran, H. M., 1975, Assessment of Solar-Powered Cooling of Buildings, NSF-RA-N-75-012. Washington, D.C.: Energy Research and Development Administration.

Curran, H. M., 1977, "Coefficient of Performance for Solar-Powered Cooling Systems," Solar Energy, Vol. 19, No. 5, pp. 601-603.

Curran, H. M., 1978, "Overview of Solar Rankine Cooling," Proceedings of the Third Workshop on the Use of Solar Energy for the Cooling of Buildings, February 15-17, 1978, San Francisco, California, pp. 177-185, Newark, DE: American Solar Energy Society.

"Deskbook Directory of Solar Product Manufacturers," Solar Engineering, Dec. 1977, p. 68.

de Winter, F. and J. W. de Winter, 1976, Use of Solar Energy for the Cooling of Buildings," SAN/1122-76-2, Washington, D.C.: Energy Research and Development Administration.

Diamond, S. C., B. D. Hunn, and C. C. Cappiello, 1981, DOE-2 Verification Project, Phase 1, Interim Report, LA-8295-MS, Los Alamos, NM: Los Alamos National Laboratory.

Diamond, S. C., C. C. Cappiello, and B. D. Hunn, 1986, DOE-2 Verification Project, Phase I, Final Report, LA-10649-MS, Los Alamos, NM: Los Alamos National Laboratory.

Dickinson, W. C. and P. N. Cheremisinoff, eds., 1980a, Solar Energy Technology Handbook, Part A, Engineering Fundamentals, New York: Marcel Dekker, Inc.

Dickinson, W. C. and P. N. Cheremisinoff, eds., 1980b, Solar Energy Technology Handbook, Part B, Applications, Systems Design, and Economics, New York: Marcel Dekker, Inc.

DOE-2 Reference Manual, Parts 1 and 2, Version 2.1A, 1981a, LA-8520-M Ver. 2.1a, Los Alamos, NM: Los Alamos National Laboratory; LBL-8706 Rev. 2, Berkeley, CA: Lawrence Berkeley Laboratory.

DOE-2 Engineers Manual (Version 2.1A), 1981b, LA-8520-M, Los Alamos, NM: Los Alamos National Laboratory; LBL-11353, Berkeley, CA: Lawrence Berkeley Laboratory.

Drew, M. S. and R. B. G. Selvage, 1980, "Sizing Procedure and Economic Optimization Methodology for Seasonal Storage Solar Systems," Solar Energy, Vol. 25, No. 1, pp. 79-83.

Duff, W. S., T. M. Conway, G. O. G. Löf, D. B. Meredith, and R. B. Pratt, 1978, "Performance of Residential Solar Heating and Cooling System with Flat-Plate and Evacuated Tubular Collectors: CSU Solar House I," Proceedings of the 1978 Annual Meeting of the American Section of the International Solar Energy Society, Vol. 2.1, pp. 385-390, Newark, DE: ASES.

Duffie, J. A., 1978, "Simulation and Design Methods," Proceedings of the 1978 Annual Meeting of the American Section of the International Solar Energy Society, Inc, Vol. 2.1, pp. 142-144, Newark, DE: ASES.

Duffie, John A. and William A. Beckman, 1980, Solar Engineering of Thermal Processes, New York: Wiley.

Eads, W. G., 1982, Testing, Balancing and Adjusting of Environmental Systems, Vienna, VA: Sheet Metal and Air Conditioning Contractors' National Association.

Eden, A. and M. Morgan, 1980, A Comparison of DOE-2 and TRNSYS Solar Heating System Simulation, SERI/TR-721-822, Golden, CO: Solar Energy Research Institute.

Energy Research and Development Administration, 1976, Pacific Regional Solar Heating Handbook, 2nd ed., TID-27630, Washington, D.C.

Erbs, D. G., S. A. Klein, and J. A. Duffie, 1982, "Estimation of the Diffuse Radiation Fraction for Hourly, Daily and Monthly-Average Global Radiation," Solar Energy, Vol. 28, No. 4, pp. 293-302.

ESG, Inc., 1984, Survey of Failure Modes from 122 Residential Solar Water Heaters, SERI/STR-253-2531, Golden, CO: Solar Energy Research Institute.

Evans, B. H., 1981, Daylight in Architecture, Hightstown, NJ: Architectural Record Books.

Fanger, P. O., 1972, Thermal Comfort: Analysis and Applications in Environmental Engineering, New York: McGraw-Hill.

Feldman, Scott J. and Richard L. Merriam, 1979, Building Energy Analysis Computer Programs with Solar Heating and Cooling System Capabilities, EPRI-ER-1146, Cambridge, MA: Arthur D. Little, Inc.

Florida Solar Energy Center, 1982, Thermal Performance Ratings, FSEC-GP-14-82, Cape Canaveral, FL.

Freeman, T. L., M. W. Maybaum, and S. Chandra, 1978, "A Comparison of Four Solar Simulation Programs in Solving a Solar Heating Problem," Proceedings of the Conference on Systems Simulation and Economic Analysis for Solar Heating and Cooling, San Diego, California, June 27, SAND 78-1927, pp. 4-7, Washington, D.C.: DOE.

Freeman, T. L., J. W. Mitchell, and T. E. Audit, 1979, "Performance of Combined Solar-Heat Pump Systems," Solar Energy, Vol. 22, No. 2, pp. 125-135.

Fried, J. R., May 28, 1973, "Heat Transfer Agents for High Temperature Systems," Chemical Engineering, Vol. 80, pp. 89-98.

Gardiner, B. L. and M. A. Piette, 1985, "Measured Results of Energy-Conservation Retrofits in Non-residential Buildings: Interpreting Measured Data," ASHRAE Transactions 1985, Vol. 91, Pt. 2B, pp. 1488-1498.

Gary, W. M., 1978, "Performance Comparison Between F-CHART Simulations and Several Operating Solar Systems," Proceedings of the Conference on Systems Simulation and Economic Analysis for Solar Heating and Cooling, San Diego, California, June 27, SAND-78-1927, pp. 202-204, Washington, D.C.: DOE.

Geiger, R., 1965, The Climate Near the Ground, Cambridge, MA: Harvard University Press.

Gilman, S. F., ed., 1975, Solar Energy Heat Pump Systems for Heating and Cooling Buildings, Proceedings of a workshop conducted by the Pennsylvania State University, College of Engineering, June 12-14, 1975, University Park, Pennsylvania, ERDA COO-2560-1.

Gilman, S. F., E. R. McLaughlin, and M. W. Wildin, 1976, "Field Study of a Solar Energy Assisted Heat Pump Heating System," Solar Cooling and Heating: Architectural Engineering and Legal Aspects, Proceedings of the Solar Cooling and Heating Forum, Miami Beach, Florida, Dec. 13-15, 1976, Vol. 2, pp. 577-614.

Givoni, B., 1977, "Underground Longterm Storage of Solar Energy--An Overview," Solar Energy, Vol. 19, No. 6, pp. 617-623.

Gordon, H. T., J. Estoque, G. K. Hart, and M. Kantrowitz, 1985, Performance Overview: Passive Solar for Non-Residential Buildings. Berkeley, CA: Solar Energy Group, Lawrence Berkeley Laboratory.

Grassie, S. L. and N. R. Sheridan, 1977, "Modelling of Solar-Operated Absorption Air Conditioner System with Refrigerant Storage," Solar Energy, Vol. 19, No. 6, pp. 691-700.

Griffith, J. W., 1978, "Benefit of Daylighting--Cost and Energy Savings," ASHRAE Journal, Vol. 20, No. 1, pp. 53-56.

Guinn, G. R., B. J. Novell, and L. Hummer, 1981, "Commercial Solar Water Heating Systems Operational Test," Solar Engineering-1981, Proceedings of the ASME Solar Energy Division Third Annual Conference on Systems Simulation, Economic Analysis/Solar Heating and Cooling Operational Results, Reno, Nevada, April 27-May 1, pp. 81-89, New York: American Society of Mechanical Engineers.

Hall, I. J., R. R. Prairie, H. E. Anderson, and E. C. Boes, 1979, "Generation of Typical Meteorological Years for 26 SOLMET Stations," ASHRAE Transactions 1979, Vol. 85, Pt. 2, pp. 507-518.

Henniger, R. H., 1975, NECAP: NASA's Energy-Cost Analysis Program, Part 1. User's Manual; Part 2. Engineering Manual, NASA-CR-2590, Parts 1 and 2, Niles, IL: General American Transportation Corp.

Hewitt, H. C. and E. I. Griggs, 1976, "Optimal Flow Rates Through Flat Plate Solar Collector Panels," Paper No. 76-WA/Sol-19, Presented at the American Society of Mechanical Engineers Winter Annual Meeting, Dec. 1976, New York: ASME.

Hill, J. E. and E. R. Streed, 1977, "Testing and Rating of Solar Collectors," Applications of Solar Energy for Heating and Cooling of Buildings, Chapter X, ASHRAE GRP 170, New York: American Society of Heating, Refrigerating and Air-Conditioning Engineers.

Hirst, E., J. Clinton, H. Geller, and W. Kroner, 1986, Energy Efficiency in Buildings: Progress and Promise, Edited by F. M. O'Mara. Washington, D.C.: American Council for an Energy-Efficient Economy.

Hittle, D. C., 1979, BLAST, the Building Loads Analysis and System Thermodynamics Program, Vol. I, User's Manual, Report No. E-153, Champaign, IL: U.S. Army Construction Engineering Research Laboratory.

Hittle, D. C., 1981, BLAST, the Building Loads Analysis and System Thermodynamics Program, Vol. I Supplement, Version 3.0, Report No. E-171, Champaign, IL: U.S. Army Construction Engineering Research Laboratory.

Hittle, D., D. Holshouser, and G. Walton, 1976, Interim Feasibility Assessment Method for Solar Heating and Cooling of Army Buildings, CERL-TR-E-91, Champaign, IL: U.S. Army Construction Engineering Research Laboratory.

Hittle, D. C., G. N. Walton, D. F. Holshouser, and D. J. Leverenz, 1977, Predicting the Performance of Solar Energy Systems, CERL-IR-E-98, Champaign, IL: U.S. Army Construction Engineering Research Laboratory.

Hollands, K. G. T., 1976, "Solar Collectors," Sharing the Sun: Solar Technology in the Seventies, Proceedings of the Joint Meeting of the American Section of the International Solar Energy Society, and the Solar Energy Society of Canada, Inc., Vol. 2, pp. R1-R7, Newark, DE: ASES.

Hubbell, R. H., W. A. Vachon, R. F. Patel, and G. Purcell, 1982, "A Summary of Recent Test Results for Solar and Load-Managed Cooling Experiments in Commercial Buildings," Proceedings for the 1982 Annual Meeting of the American Solar Energy Society, Pt. 1, pp. 621-626, Newark, DE: ASES.

Hulstrom, R. L., 1981, Solar Radiation Energy Resource Atlas of the United States, SERI/SP-642-1307, Golden, CO: Solar Energy Research Institute.

Hunn, B. D., 1980, "A Simplified Method for Sizing Active Solar Space Heating Systems," W. C. Dickinson and P. N. Cheremisinoff, eds., Solar Energy Technology Handbook, Part B, Applications, System Design, and Economics, Ch. 44, pp. 639-666, New York: Marcel Dekker.

Hunn, B. D. and D. O. Calafell, 1977, "Determination of Average Ground Reflectivity for Solar Collectors," Solar Energy, Vol. 19, No. 1, pp. 87-89.

Illuminating Engineering Society, 1979, Recommended Practice of Daylighting, RP-S, New York.

Illuminating Engineering Society, 1981, IES Lighting Handbook. Vol. 1, 1981 Reference Volume, New York.

International Telephone and Telegraph Corporation, 1977, Solar Heating Systems Design Manual, Fluid Handling Division, Bulletin TESE-576, Morton Grove, IL.

Jones, R. E. Jr. and M. D. Worley, 1982, "Investigations of Vertical Fin Shading of Windows," Solar Engineering-1982, New York: American Society of Mechanical Engineers.

Jordan, R. C. and B. Y. H. Liu, eds., 1977, Applications of Solar Energy for Heating and Cooling of Buildings, ASHRAE GRP 170, New York: American Society of Heating, Refrigerating and Air-Conditioning Engineers.

Jorgensen, G. J., 1984, A Summary and Assessment of Historical Reliability and Maintainability Data for Active Solar Hot Water and Space Conditioning Systems, SERI/TR-253-2120, Golden, CO: Solar Energy Research Institute.

Jurinak, J. J., J. W. Mitchell, and W. A. Beckman, 1984, "Open-Cycle Desiccant Air Conditioning as an Alternative to Vapor Compression Cooling in Residential Applications," J. Solar Energy Eng., Vol. 106, No. 3, pp. 252-260.

Kaehn, H. D., M. Geyer, D. Fong, F. Vignola, and D. K. McDaniels, 1978, "Experimental Evaluation of the Reflector-Collector System," Proceedings of the 1978 Annual Meeting of the American Section of the International Solar Energy Society, Vol. 2.1, pp. 654-659, Newark, DE: ASES.

Karaki, S., P. R. Armstrong, and T. N. Bechtel, 1977, Evaluation of a Residential Solar Air Heating and Nocturnal Cooling System, COO-2868-3, Ft. Collins, CO: Solar Energy Applications Laboratory, Colorado State University.

Kauffman, K. W., 1977, "Non-corrosive, Non-freezing and Non-toxic Heat Transfer Fluids for Solar Heating and Cooling," Proceedings of the 1977 Annual Meeting of the American Section of the International Solar Energy Society, Vol. 1, pp. 5-8 to 5-12, Newark, DE: ASES.

Kent, T. A., C. B. Winn, and W. G. Huston, 1981, Controllers for Solar Domestic Hot Water Systems, AH-9-8189-1, draft prepared for Solar Energy Research Institute by Solar Environmental Engineering Co., Ft. Collins, CO.

Kern, D. Q., 1950, Process Heat Transfer, New York: McGraw-Hill.

Kettleborough, C. F., Dec. 1983, "Solar-Assisted Comfort Conditioning," Mechanical Engineering, pp. 48-55.

Kimura, K. and D. G. Stephenson, 1969, "Solar Radiation on Cloudy Days," ASHRAE Transactions, Vol. 75, Pt. 1., pp. 227-234.

Kinloch, D., N. Hinsey, and J. Tichy, 1981, "Computer Simulation of a Passive Solar Assisted Heat Pump," Solar Engineering-1981. Proceedings of the ASME Solar Division Third Annual Conference on Systems Simulation, Economic Analysis/Solar Heating and Cooling Operational Results, Reno, Nevada, April 27-May 1, pp. 738-745, New York: American Society of Mechanical Engineers.

Kirkpatrick, Donald L., 1983, Flat-Plate Solar Collector Performance Data Base and User's Manual, SERI/STR-254-1515, Golden, CO: Solar Energy Research Institute.

Kissner, F., 1980, "The Suntime™ Passive, Phase-Change Solar Water Heating System: One Year of Operating Data," Proceedings of the Fifth National Passive Solar Conference, Amherst, Massachusetts, pp. 1061-1065, Newark, DE: American Section of the International Solar Energy Society.

Klein, S. A. and W. A. Beckman, 1979, "A General Design Method for Closed-Loop Solar Energy Systems," Solar Energy, Vol. 22, No. 3, pp. 269-282.

Klein, S. A., W. A. Beckman, and J. A. Duffie, 1976, "TRNSYS - A Transient Simulation Program," ASHRAE Transactions, Vol. 82, Pt. 1, pp. 623-633.

Klein, S., W. Beckman, and J. Duffie, 1977, Solar Heating and Cooling of Residential Buildings - Design of Systems, Fort Collins, CO: Solar Energy Applications Laboratory, Colorado State University.

Klucher, T. M., 1979, "Evaluation of Models to Predict Insolation on Tilted Surfaces," Solar Energy, Vol. 23, No. 2, pp. 111-114.

Knapp, C. L., T. L. Stoffel, and S. D. Whitaker, 1980, Insolation Data Manual, SERI/SP-755-789, Golden, CO: Solar Energy Research Institute.

Knebel, David E., 1983, Simplified Energy Analysis Using the Modified Bin Method, ASHRAE RP-363, Atlanta, GA: American Society of Heating, Refrigerating and Air-Conditioning Engineers.

Kreith, Frank and Jan F. Kreider, 1978, Principles of Solar Engineering, New York: McGraw-Hill.

Kusuda, T., 1974, NBSLD, Computer Program for Heating and Cooling Loads in Buildings, NBSIR 74-574, Washington, D.C.: National Bureau of Standards.

Kusuda, Tamami, 1980, Review of Current Calculation Procedures for Energy Analysis, NBSIR-80-2068, Washington, D.C.: National Bureau of Standards.

Kutscher, C. F., R. L. Davenport, D. A. Dougherty, R. C. Gee, P. M. Masterson, and E. K. May, 1982, Design Approaches for Solar Industrial Process Heat Systems: Nontracking and Line-Focus Collector Technologies, SERI/TR-253-1356, Golden, CO: Solar Energy Research Institute.

Lantz, L. J. and C. B. Winn, 1978, Model Validation Studies of Solar Systems, Phase III, COO-4229-T1, Ft. Collins, CO: Solar Environmental Engineering Company.

Leckie, J., G. Masters, H. Whitehouse, and L. Young, 1975, Other Homes and Garbage: Designs for Self-Sufficient Living, San Francisco, CA: Sierra Club Books.

Leflar, J. A., D. E. Hull, III, and C. B. Winn, 1978, "Design of Liquid Shell and Tube Heat Exchangers for Use in Solar Systems," Conference on Systems Simulation and Economic Analysis for Solar Heating and Cooling, San Diego, California, June 27, SAND 78-1927, Washington, D.C.: DOE.

Libbey-Owens-Ford Co., 1976, How to Predict Interior Daylight Illumination: Conserve Energy and Increase Visual Performance, Toledo, OH.

Liu, B. Y. H. and R. C. Jordan, 1963, "A Rational Procedure for Predicting the Long-Term Average Performance of Flat-Plate Solar-Energy Collectors," Solar Energy, Vol. 7, No. 2, pp. 53-74.

Liu, S. T. and A. H. Fanney, 1980, "Comparing Experimental and Computer-Predicted Performance for Solar Hot Water Systems," ASHRAE Journal, Vol. 22, No. 5, pp. 34-38.

Lunde, P. J., 1978, "Prediction of Monthly and Annual Performance of Solar Heating Systems," Solar Energy, Vol. 20, No. 3, pp. 283-287.

Lunde, P. J., 1979, "Prediction of the Performance of Solar Heating Systems Over a Range of Storage Capacities," Solar Energy, Vol. 23, No. 2, pp. 115-121.

Lunde, P. J., 1980, Solar Thermal Engineering: Space Heating and Hot Water Systems, New York: John Wiley and Sons.

Lunde, P. J., 1982, "More On-Time for Solar Collectors," Solar Age, Vol. 7, No. 3, pp. 63-64; "Control of Active Systems Using the Stagnation Temperature," Solar Age, Vol. 7, No. 4, pp. 69-70.

MacDonald, J. M., H. P. Misuriello, D. Goldenberg, M. P. Ternes, and J. O. Kolb, 1986, Commercial Retrofit Research Multi-year Plan, FY1986-FY1991; Building Energy Retrofit Research, ORNL/CON-218, Oak Ridge, TN: Oak Ridge National Laboratory.

Manton, B. E. and J. W. Mitchell, 1981, "A Regional Comparison of Solar-Heat Pump and Solar-Heat Pump Systems," Solar Engineering-1981, Proceedings of the ASME Solar Energy Division Third Annual Conference on Systems Simulation, Economic Analysis/Solar Heating and Cooling Operational Results, Reno, Nevada, April 27-May 1, pp. 196-203, New York: American Society of Mechanical Engineers.

Mazria, E., 1979, The Passive Solar Energy Book: A Complete Guide to Passive Solar Home, Greenhouse, and Building Design, Emmaus, PA: Rodale Press.

McGraw, B. A., A. F. G. Bedinger, and R. L. Reid, 1981, "Experimental Evaluation of a Series Solar Assisted Heat Pump System," Proceedings of the 1981 Annual Meeting of the American Section of the International Solar Energy Society, Vol. 4.1, pp. 562-566, Newark, DE: ASES.

Meckler, M., 1978, "Simulations of Solar Assisted Multiple Zone Water Source Heat Pump System with Diversity Cooling Loop," Proceedings of the 1978 Annual Meeting of the American Section of the International Solar Energy Society, Vol. 2.1, pp. 319-330, Newark, DE: ASES.

Mehta, J. and Z. Lavan, 1982, "Rankine Cycle Cooling with Shaped Tube Collectors," Proceedings of the 1982 Annual Meeting of the American Section of the International Solar Energy Society, Pt. 1, pp. 615-620, Newark, DE: ASES.

Meinel, A. B. and M. P. Meinel, 1976, Applied Solar Energy: An Introduction, Reading, MA: Addison-Wesley Publishing Company.

Merriam, R. L., 1981, Solar Heating and Cooling (SHAC) Simulation Programs: Assessment and Evaluation. Vol. 1: Summary Report; Vol. 2: Detailed Findings, EPRI-EM-1866, Cambridge, MA: Arthur D. Little, Inc.

Merriam, R. L. and R. J. Rancatore, 1982, Evaluation of Existing Programs for Simulation of Residential Building Energy Use, EPRI EA-2575, Cambridge, MA: Arthur D. Little, Inc.

Mertol, A., W. Place, T. Webster, and R. Greif, 1981, "Detailed Loop Model (DLM) Analysis of Liquid Solar Thermosiphons with Heat Exchangers," Solar Energy, Vol. 27, No. 5, pp. 367-386.

Metz, P. D., 1978, "The Potential for Ground Coupled Storage within the Series Solar Assisted Heat Pump System," Proceedings of the 1978 Annual Meeting of the American Section of the International Solar Energy Society, Vol. 2.1, pp. 331-334, Newark, DE: ASES.

Milne, M. and B. Givoni, 1979, "Architectural Design Based on Climate," Energy Conservation Through Building Design, edited by D. Watson, New York: McGraw-Hill.

Mitchell, J. W., W. A. Beckman, and M. J. Pawelski, 1978, "Comparisons of Measured and Simulated Performance for CSU House I," Proceedings of the Conference on Systems Simulation and Economic Analysis for Solar Heating and Cooling, San Diego, California, June 27, SAND 78-1927, pp. 198-201, Washington, D.C.: DOE.

Mitchell, J. W., T. L. Freeman, and W. A. Beckman, 1978, "Heat Pumps: Do They Make Economic and Performance Sense with Solar?" Solar Age, Vol. 3, No. 7, pp. 24-28.

Morehouse, J. H. and P. J. Hughes, 1979, "Residential Solar Heat Pump Systems: Thermal and Economic Performance," Paper No. 79-WA/Sol-25, Presented at the American Society of Mechanical Engineers Winter Annual Meeting, Dec. 1979, New York: ASME.

Morrison, G. L. and C. M. Sapsford, 1983, "Long Term Performance of Thermosyphon Solar Water Heaters," Solar Energy, Vol. 30, No. 4, pp. 341-350.

Mueller Associates, Inc., 1981, The Analysis of Construction Costs of Ten Solar Industrial Process Heat Systems, SERI/TR-09144-1, Golden, CO: Solar Energy Research Institute.

Mueller Associates, Inc., 1982, "Westover Job Corps Center Solar System Acceptance Test Plan," Baltimore, MD.

Murphy, L. J., 1978, The National Solar Data Network, SOLAR/0009-78/10, Washington, D.C.: DOE.

Murray, H. S., J. C. Hedstrom, and J. D. Balcomb, 1978, Solar Heating and Cooling Performance of the Los Alamos National Security and Resources Study Center, LA-UR-78-1206, Los Alamos, NM: Los Alamos Scientific Laboratory.

Mutch, J. J., 1974, Residential Water Heating: Fuel Conservation, Economics, and Public Policy, Report No. R 1498-NSF, Santa Monica, CA: Rand Corporation.

O'Dell, M. P., J. W. Mitchell, and W. A. Beckman, 1983, "Solar Heat Pump Systems with Refrigerant-Filled Collectors," ASHRAE Transactions, Vol. 89, Pt. 1B, pp. 519-525.

Olgyay, V. G., 1963, Design with Climate: Bioclimatic Approach to Architectural Regionalism, Princeton, NJ: Princeton University Press.

Olson, T. J., D. M. Beekman, W. A. Beckman, and J. W. Mitchell, 1977, "Performance of Solar Source Rankine Cycle Engine Cooling Systems," Proceedings of the 1977 Annual Meeting of the American Section of the International Solar Energy Society, Vol. 1, pp. 7-15 to 7-19, Newark, DE: ASES.

Oonk, R. L., L. E. Shaw, and H. H. Hopkinson, 1978, "Modelling of Combined Air Base Solar/Heat Pump Heating and Cooling Systems," Proceedings of the 1978 Annual Meeting of the American Section of the International Solar Energy Society, Vol. 2.1, pp. 308-312, Newark, DE: ASES.

Oonk, R. L., L. E. Shaw, B. E. Cole-Appel, and G. O. G. Löf, 1976, "A Method of Comparing Flat-Plate Air and Liquid Solar Collectors for Use in Space Heating Applications," Sharing the Sun: Solar Technology in the Seventies, Proceedings of the Joint Conference of the American Section, International Solar Energy Society and the Solar Energy Society of Canada, Inc., Vol. 2, pp. 83-93, Newark, DE: ASES.

Orgill, J. F. and K. G. T. Hollands, 1977, "Correlation Equation for Hourly Diffuse Radiation on a Horizontal Surface," Solar Energy, Vol. 19, No. 4, pp. 357-359.

Phillips, W. F. and R. N. Dave, 1982, "Effects of Stratification on the Performance of Liquid-Based Solar Heating Systems," Solar Energy, Vol. 29, No. 2, pp. 111-120.

Place, W., M. Daneshyar, and R. Kammerud, 1979, "Mean Monthly Performance of Passive Solar Water Heaters," Proceedings of the Fourth National Passive Solar Conference, Kansas City, Missouri, pp. 601-604, Newark, DE: American Section of the International Solar Energy Society.

Popplewell, J. M., 1975, "Corrosion Considerations in the Use of Aluminum Copper, and Steel Flat-Plate Collectors," paper presented at the 1975 International Solar Energy Congress and Exposition, UCLA.

Prigmore, D. and R. Barber, 1976, "Cooling with the Sun's Heat. Design Considerations and Test Data for a Rankine Cycle Prototype," Solar Energy, Vol. 17, No. 3, pp. 185-192.

Robbins, C. L. and K. C. Hunter, 1983, A Model for Illuminance on Horizontal and Vertical Surfaces, SERI/TR-254-1703, Golden, CO: Solar Energy Research Institute.

Rohles, F. M., B. W. Jones, and S. A. Konz, 1983, "Ceiling Fans as Extenders of the Summer Comfort Envelope," ASHRAE Transactions, Vol. 89, No. 1A, pp. 245-263.

Roschke, M. A., B. D. Hunn, and S. C. Diamond, 1978, "A Component-Based Simulator for Solar Systems," Proceedings of the Conference on Systems Simulation and Economic Analysis for Solar Heating and Cooling, San Diego, California, June 27, SAND 78-1927, pp. 8-18, Washington, D.C.: DOE.

Rubin, A. I., 1982, Thermal Comfort in Passive Solar Buildings—An Annotated Bibliography, NBSIR 82-2585, Washington, D.C.: National Bureau of Standards.

Saha, H., 1978, "Heat Transfer Characteristics of Water Filled Cans as Solar Thermal Storage Medium: A Comparative Test Data Analysis," Proceedings of the 1978 Annual Meeting of the American Section of the International Solar Energy Society, Vol. 2.1, pp. 664-670, Newark, DE: ASES.

Schreyer, J. M., 1981, "Residential Application of Refrigerant-charged Solar Collectors," Solar Energy, Vol. 26, No. 4, pp. 307-312.

Schurr, N. M., B. D. Hunn, and K. D. Williamson, 1981, "The Solar Load Ratio Method Applied to Commercial Building Active Solar System Sizing," Solar Engineering-1981, Proceedings of the ASME Solar Energy Division Third Annual Conference on Systems Simulation, Economic Analysis/Solar Heating and Cooling Operational Results, Reno, Nevada, April 27-May 1, pp. 204-215, New York: American Society of Mechanical Engineers.

Sherman, M. H. and D. T. Grimsrud, 1980, Measurement of Infiltration Using Fan Pressurization and Weather Data, LBL-10852, Berkeley, CA: Lawrence Berkeley Laboratory.

Shiran, Y., A. Shitzer, and D. Degani, 1982, "Computerized Design and Economic Evaluation of an Aqua-Ammonia Solar Operated Absorption System," Solar Energy, Vol. 29, No. 1, pp. 43-54.

Sillman, S., 1981, "Performance and Economics of Annual Storage Solar Heating Systems," Solar Energy, Vol. 27, No. 6, pp. 513-528.

Solar Energy Research Institute, 1980, Analysis Methods for Solar Heating and Cooling Applications: Passive and Active Systems, SP-35-232R, Golden, CO.

Solar Energy Research Institute, 1981, Microcomputer Methods for Solar Design and Analysis: Passive and Active Systems, SERI/SP-722-1127, Golden, CO.

Solar Energy Research Institute, 1984, Passive Solar Performance: Summary of 1982-1983 Class B Results, SERI/SP-271-2362, Golden, CO.

Solar Energy Research Institute, 1985, Solar Energy Computer Models Directory, SERI/SP-271-2589, Golden, CO.

Solar Environmental Engineering Company, 1981, Investigation of Residential Solar Systems in Three Regions of the United States, Fort Collins, CO.

Solar Rating and Certification Corporation, 1982, Directory of SRCC Certified Solar Collector Ratings, Presentation of Thermal Performance Ratings of Solar Collectors Certified by the Solar Rating and Certification Corporation, 2nd ed., Washington, D.C.

Stamper, E., 1979, "Air-Conditioning Usage Study: Equivalent Rated Load Hours," ASHRAE Journal, Vol. 21, No. 6, pp. 43-45.

Stoever, Herman J., 1941, Applied Heat Transmission, New York: McGraw-Hill.

Strickford, G. H., 1976, "An Averaging Technique for Predicting the Performance of a Solar Energy Collector System," Sharing the Sun: Solar Technologies in the Seventies, Proceedings of the Joint Conference of the American Section of the International Solar Energy Society and the Solar Energy Society of Canada, Inc., Vol. 4, pp. 295-315, Newark, DE: ASES.

Strock, C., 1959, Handbook of Air Conditioning, Heating, and Ventilating, New York: Industrial Press.

Sud, I., R. W. Wiggins, Jr., J. B. Chaddock, and T. D. Butler, 1979, Development of the Duke University Building Energy Analysis Method (DUBEAM) and Generation of Plots for North Carolina, NCEI-0009, Durham, NC: Duke University Center for the Study of Energy Conservation.

Swanson, S. R. and R. F. Boehm, 1977, "Calculation of Long-Term Solar Collector Heating System Performance," Solar Energy, Vol. 19, No. 2, pp. 129-138.

Swisher, J., 1981, Active Charge/Passive Discharge Solar Heating Systems: Thermal Analysis and Performance Comparisons, SERI/TR-721-1104, Golden, CO: Solar Energy Research Institute.

Tennessee Valley Authority, 1980, Climatic Data Base, Great Valley/Zone II, Chattanooga, Tennessee, Knoxville, TN.

Ternoey, Steven, Larry Bickle, Claude Robbins, Robert Busch, and Kitt McCord (Solar Energy Research Institute), 1984, The Design of Energy-Responsive Commercial Buildings, New York: John Wiley and Sons.

Terrell, R. E., 1979, "Performance and Analysis of a "Series" Heat Pump-Assisted Solar Heated Residence in Madison, Wisconsin," Solar Energy, Vol. 23, No. 5, pp. 451-453.

Trenschel, D. and P. Goetze, 1981, Test Results from TIPSE (Testing and Inspection Program for Solar Equipment), P-500-80-056, Sacramento, CA: California Energy Resources Conservation and Development Commission.

U.S. Departments of the Air Force, Army, and Navy, 1978, Engineering Weather Data, AFM 88-29, Washington, D.C.

U.S. Department of Commerce, U.S. Coast and Geodetic Survey, 1965, Isogonic Chart of the U.S., Chart No. 3077, Washington, D.C.

U.S. Department of Commerce, Environmental Sciences Services Administration, Environmental Data Services, 1968, Climatic Atlas of the United States. Washington, D.C.

U.S. Department of Commerce, National Climatic Center, 1973, Annual Degree Days to Selected Bases for First-Order Type Stations, Asheville, NC.

U.S. Department of Commerce, National Bureau of Standards, 1976, Interim Performance Criteria for Solar Heating and Cooling Systems in Commercial Buildings, NBSIR-76-1187, Washington, D.C.

U.S. Department of Commerce, National Bureau of Standards, 1978, Interim Performance Criteria for Solar Heating and Cooling Systems in Residential Buildings, 2nd ed., NBSIR 78-1562, Washington, D.C.

U.S. Department of Commerce, 1980, Solar Heating and Cooling of Residential Buildings: Design of Systems. Washington, D.C.

U.S. Department of Energy, Office of Conservation and Solar Applications, Solar Heating and Cooling R&D Branch, 1977, Unpublished minutes of the Solar Heating and Cooling System Simulation and Economic Analysis Working Group Meeting, Golden, CO, November 14-15, 1977.

U.S. Department of Energy, 1978a, DOE Facilities Solar Design Handbook, DOE/AD-0006/1, Washington, D.C.

U.S. Department of Energy, 1978b, Preliminary Issue: Solar Heating and Cooling Project Experiences Handbook, TID-28722, Washington, D.C.

U.S. Department of Energy, 1978c, "Summary of the Problems Encountered in the Solar Heating and Cooling Commercial Demonstration Program," Proceedings of the Department of Energy's Solar Update, Conf. 780701, Washington, D.C.

U.S. Department of Energy, 1979a, Active Solar Energy System Design Practice Manual, SOLAR/0802-79/01, Washington, D.C.

U.S. Department of Energy, Energy Information Administration, 1979b, Annual Report to Congress, Vol. 2, DOE/EIA-0173(79)/2, Washington, D.C.

U.S. Department of Energy, Energy Information Administration, 1983a, Nonresidential Buildings Energy Consumption Survey: 1979 Consumption and Expenditures, Part 1, Natural Gas and Electricity, and Part 2, Steam, Fuel Oil, LPG, and All Fuels, Washington, D.C.

U.S. Department of Energy, Energy Information Administration, 1983b, Monthly Energy Review, Dec., DOE/EIA-0035(83/12), Washington, D.C.

U.S. Department of Energy, Assistant Secretary for Conservation and Renewable Energy, 1984, Renewable Technologies Program Summaries, DOE/CE-0105, Washington, D.C.

U.S. Department of Energy, Office of Conservation, 1985a, Energy Conservation Multi-year Plan, FY 1987, DOE/CE/29999-T1, Washington, D.C.

U.S. Department of Energy, Energy Information Administration, 1985b, Nonresidential Buildings Energy Consumption Survey: Characteristics of Commerical Buildings, 1983. Washington, D.C.

U.S. Department of Energy, Energy Information Administration, 1986, Monthly Energy Review, Sept., DOE/EIA 0035(86/09), Washington, D.C.

U.S. Department of Housing and Urban Development, 1977, Intermediate Minimum Property Standards, Solar Heating and Domestic Hot Water Systems, Vol. 5, PB-82-144692, Washington, D.C.

U.S. Department of Housing and Urban Development, 1980, Installation Guidelines for Solar DHW Systems in One- and Two-Family Dwellings, HUD-PDR-407(2), Washington, D.C.

University of Wisconsin-Madison, 1983, TRNSYS Version 12.1: A Transient Simulation Program, Report No. 38-12, Madison, WI: Solar Energy Laboratory, Engineering Experiment Station.

University of Wisconsin-Madison, 1985, F-Chart 4.2, A Design Program for Solar Heating Systems, Madison, WI: Solar Energy Laboratory, Engineering Experiment Station.

Utzinger, D. M. and S. A. Klein, 1979, "A Method of Estimating Monthly Average Solar Radiation on Shaded Receivers," Solar Energy, Vol. 23, No. 5, pp. 369-378.

Vliet, G. C. and J. L. Askey, 1984, "The Influence of Residential Solar Water Heaters on Electric Peak Demand," Proceedings of the First Annual Symposium--Efficient Utilization of Energy in Residential and Commercial Buildings, College Station, TX: Texas A&M University.

Waksman, David, Elmer R. Streed, Thomas W. Reichard, and Louis E. Cattaneo, 1978, Provisional Flat-Plate Solar Collector Testing Procedures, First Revision, NBSIR 78-1305A, Washington, DC: National Bureau of Standards.

Ward, D. S., 1979, "Solar Absorption Cooling Feasibility," Solar Energy, Vol. 22, No. 3, pp. 259-268.

Ward, D. S. and H. S. Oberoi, 1981, "Experiences in Solar Cooling Systems," Solar Engineering-1981. Proceedings of the ASME Solar Division Third Annual Conference on Systems Simulation, Economic Analysis/Solar Heating and Cooling Operational Results, Reno, Nevada, pp. 15-27, New York: American Society of Mechanical Engineers.

Ward, D. S. and J. C. Ward, 1979, "Design Considerations for Residential Solar Heating and Cooling Systems Utilizing Evacuated Tube Solar Collectors," Solar Energy, Vol. 22, No. 2, pp. 113-118.

Ward, D. S., G. O. G. Löf, and T. Uesaki, 1978, "Cooling Subsystem Design in CSU Solar House III," Solar Energy, Vol. 20, No. 2, pp. 119-126.

Ward, J. C., 1976, "Minimum Cost Sizing of Solar Heating Systems," Sharing the Sun: Solar Technologies in the Seventies, Proceedings of the Joint Conference of the American Section of the International Solar Energy Society and the Solar Energy Society of Canada, Inc., Vol. 4, pp. 336-346, Newark, DE: ASES.

Wessling, F. C., Jr., 1981, "Correlations of Laboratory Test Results with Field Performance of Solar Domestic Hot Water Systems Under Controlled Draw Rates," ASHRAE Transactions 1981, Vol. 87, Pt. 2, pp. 835-841.

White, N. M., J. H. Morehouse, and T. D. Swanson, 1980, "Sensitivity Analysis of Solar Assisted Heat Pump Systems," Proceedings of the 1980 Annual Meeting of the American Section of the International Solar Energy Society, Vol. 3.1, pp. 324-328.

Wilbur, P. J. and T. R. Mancini, 1976, "A Comparison of Solar Absorption Air Conditioning Systems," Solar Energy, Vol. 18, No. 6, pp. 569-576.

Winn, C. B., 1980, "Solar Simulation Computer Programs," W. C. Dickinson and P. N. Cheremisinoff, eds., Solar Energy Technology Handbook, Part B, Applications, System Design, and Economics, Ch. 42, pp. 481-515, New York: Marcel Dekker.

Winn, C. B., 1983, "Controls in Solar Energy Systems," Advances in Solar Energy, Boulder, CO: American Solar Energy Society.

Winn, C. B. and N. Duong, 1976, The Validation of Solar House Design Programs, Phases I and II, COO-2928-2, Ft. Collins, CO: Solar Environmental Engineering Company.

York, D. A., E. F. Tucker, and C. C. Cappiello, 1982, DOE-2 Reference Manual (Version 2.1A), LA-7689-M Ver. 2.1A (LBL-8706 Rev. 2), Los Alamos, NM: Los Alamos National Laboratory; DOE-2 Supplement, Version 2.1B, LBL-8706, Rev. 3, Suppl.; DOE-2 Supplement, Version 2.1C, LBL-8706, Rev. 4 Suppl., Berkeley, CA: Lawrence Berkeley Laboratory.